Electronic Thin Film Science

For Electrical Engineers and Materials Scientists

Electronic Thin Film Science

For Electrical Engineers and Materials Scientists

King-Ning Tu
IBM T. J. Watson Research Center

James W. Mayer
Cornell University

Leonard C. Feldman
AT&T Bell Laboratories

Macmillan Publishing Company
NEW YORK

Maxwell Macmillan Canada
TORONTO

Maxwell Macmillan International
NEW YORK OXFORD SINGAPORE SYDNEY

Editor: David Johnstone
Production Supervisor: John Travis
Production Manager: Paul Smolenski
Text and cover designed by Sheree Goodman
Cover photograph by King-Ning Tu, D. A. Smith, and B. Z. Weiss, from *Physical Review B*, 36, 8948 (1987).
Illustrations provided by the authors.

This book was set in Times Roman by Waldman Graphics, Inc.,
and was printed and bound by Book Press, Inc.
The cover was printed by Phoenix Color Corp.

Macmillan Publishing Company
866 Third Avenue, New York, New York 10022

Macmillan Publishing Company is part of the Maxwell Communication Group of Companies.

Collier Macmillan Canada, Inc.
1200 Eglinton Avenue East
Suite 200
Don Mills, Ontario M3C 3N1

Library of Congress Cataloging in Publication Data

Tu, K. N. (King-Ning)
Electronic thin film science : for electrical engineers and materials scientists / King-Ning Tu, James W. Mayer, Leonard C. Feldman.
p. cm.
Includes bibliographical references and index.
ISBN 0-02-421575-9
1. Thin, films, Multilayered. 2. Electric engineering—Materials.
I. Mayer, James W., (date). II. Feldman, Leonard C. III. Title.
QC176.9.M84T83 1992
621.381'52—dc20 91-3507
CIP

Printing: 1 2 3 4 5 6 7 8 Year: 2 3 4 5 6 7 8 9 0 1

Dedicated
to
our families

Preface

The rapid advance in data and information processing is the single greatest achievement in modern technology. It has been accomplished through the development of devices to unprecedented limits and the growth of new materials with fundamentally new properties. The foundation underlying these innovations is thin film science and technology. Understanding the materials aspect of thin film processing, characterization, and stability at the atomic level is essential to the success of this enterprise. Thin film science is a multi-disciplinary subject; it brings together chemistry, electrical engineering, materials science and physics. While its development has required a sound understanding of the science of surfaces and bulk single crystals, it nevertheless has its own unique characteristics. Further advances in solid state electronic technology will undoubtedly depend on the ability to manipulate thin film materials with even greater control. Thus, there is an urgent need to teach the subject to future electrical engineers and materials scientists.

When we planned to offer an "electronic thin films" course at Cornell four years ago, we were unable to find a suitable textbook. Instead we developed our own lecture notes, which form the basis of this book. Although intended as a textbook, we believe, on the basis of our own working experience in the field, that it will be a useful reference to the professionals who have a keen interest in the applications of thin film technology to microelectronics.

This text is aimed at undergraduate seniors and graduate students interested in electronic thin films. The original course was offered by the Department of Materials Science and Engineering. Approximately half of the students came from Electrical Engineering or disciplines other than Materials Science. Their background and training varied. Hence, a firm grasp of the underlying concepts of materials science needed in the course could not be assumed. For this reason, the first four chapters of the book are devoted to review of topics on deposition, surface energy, diffusion, and stress, which are important in understanding the formation and behavior of thin films. The purpose of these chapters is to provide a sufficient coverage of the fundamentals for the subsequent chapters.

The remainder of the book is divided into two parts. The first part (Chapters 5 to 9) deals with the growth of homoepitaxial and heteroepitaxial structures and superlattices, and their electrical and optical properties. The second part (Chapters 10 to

15) is about kinetics, phase changes and reliability behavior of single and multilayer thin film structures.

Throughout the text, we have emphasized a combination of descriptive and quantitative presentations. To parallel the discussions of a physical concept a simple mathematical derivation and equation is given. The mathematical level is no more than that of college calculus. Whenever desirable, we have also attempted to carry out an illustrative calculation in order to reduce the equation to a number with proper units.

To enhance students involvement and interaction in the class, we have selected a set of "case studies" to accompany each chapter. The case studies are reprints of journal articles along with notes to the instructor. These are available from Macmillan Publishing company in a separate book "Case Studies in Electronic Thin Film Science." Our experience has been that students appreciate the opportunity to critique and present case studies.

In writing the book, we have benefited immensely from the help of students in our classes. Their inquiries and responses to our lectures have strengthened the contents and organization of the book. We would like to thank those who have allowed us to use their figures, tables and micrographs. We are grateful to David Turnbull for constructive comments on Chapter 3, Che-Yu Li and L. T. Shi for comments on Chapter 4, and H. B. Huntington and A. B. Pippard on Chapter 14. At our request the original draft was reviewed by several outside experts. Their critical comments have greatly improved the text, and we are grateful to them. The warm and expert editing support by David Johnstone and John Travis at Macmillan is appreciated. The superb typing of the manuscript, with many revisions which we wish not to recall, by Mrs. Teddy Oxton (IBM Yorktown) and Mrs. Noreen Ocello (Cornell University) is sincerely acknowledged. Jane Jorgesen and Ali Avcisoy (Cornell University) and Lou Kristiansen (IBM Yorktown) supplied many of the drawings and art work.

King-Ning Tu
James W. Mayer
Leonard C. Feldman

Contents

8: Electrical and Optical Properties of Heterostructures, Quantum Wells, and Superlattices 194

9: Schottky Barriers and Interface Potentials 222

Standard Symbols

a	lattice constant (nm)
A	area (cm^2), (m^2)
b	Burgers vector (nm)
C	concentration of atoms (cm^{-3})
D	diffusion coefficient (cm^2/sec), (m^2/sec)
D_0	pre-exponential factor of D (cm^2/sec), (m^2/sec)
D_s	surface diffusion coefficient (cm^2/sec), (m^2/sec)
e	charge on electron (coulomb, C)
E	energy (general expression) (eV), (Joule, J), (erg)
$\mathcal{E}$	electric field (V/cm), (V/m)
f	misfit (epitaxy)
F	force (Newtons, N)
F	Helmholtz free energy
G	Gibbs function or Gibbs free energy
ΔG_m, ΔG_f	motion, formation energy of diffusion (eV), (J)
g	gravity constant (dyne/gm)
h	height (height of monolayer) (nm)
h	Planck's constant (eV·sec), (Joule·sec)
H	enthalpy (eV), (J)
h_c	critical thickness (epitaxy) (nm)
J	flux (cm^2-sec)$^{-1}$
J_S	surface flux (cm-sec)$^{-1}$
k	Boltzmann's constant (eV/K)
K	Kelvin temperature unit
L	length (m), (cm), (micron, μm), (nm)
L_0	step spacing (nm)
m	mass (gram), (kg)
M	atomic mobility (D/kT), (cm^2/eV-sec)
n	atomic or carrier concentration (cm^{-3}), (m^{-3})
N_A	Avogadro's number (atoms/mole)
N	surface concentration (cm^{-2})
N_S	number of sites (or atoms) per unit area

p	pressure (N/m^2), (Torr), ($dyne/cm^2$)
p_0	equilibrium pressure on plane surface (N/m^2), ($dyne/cm^2$)
R	gas constant (J/mole·K), (cal/mole·K)
S	entropy (eV/K)
t	time (sec)
T	absolute temperature (K)
T_θ	Debye temperature
U	internal energy (eV), (J)
v	velocity (cm/sec)
V	volume (cm^3), (m^3)
W	work (eV), (N-m), (dyne-cm)
W	total energy to remove an atom from surface (eV)
Y	Young's modulus (N/m^2), ($dyne/cm^2$)

Greek Symbols

alpha α	thermal expansion coefficient
gamma γ	surface energy (eV/cm^2) (J/cm^2) (eV/atom)
epsilon ε	strain
ϵ	permittivity
theta θ	contact angle
θ	shear strain
lambda λ	characteristic length, $(Dt)^{1/2}$, (cm)
mu μ	chemical potential (eV), (J)
μ	Shear modulus (N/m^2), ($dyne/cm^2$)
μ	carrier mobility (cm^2/volt-sec)
nu ν	frequency (sec^{-1})
ν	Poisson's ratio
ν_s	surface vibration frequency
rho ρ	resistivity (ohm-cm)
ρ	mass density (gm/cm^3)
sigma σ	stress (N/m^2) ($dyne/cm^2$)
tau τ	characteristic time (sec)
phi ϕ	potential (volts)
omega Ω	atomic volume (volume/atom), (cm^3)

CHAPTER 1

Thin Film Deposition and Layered Structures

1.0 Introduction

Thin films and layered structures are critical to electronic and opto-electronic technologies. Semiconductor devices, whether the transistor or the solid-state laser, are produced by growth of thin layers on semiconductor substrates. Electrical connections between circuit elements are made by thin metal films patterned into micron-wide lines. Lasers are made by sandwiching thin layers of light-emitting semiconductors between layers of a different semiconductor. In electronic/optical systems, the active elements lie within the top few microns of the surface—this is the province of thin film technology.

Thin films bridge the gap between monolayer and bulk structures. They span the thickness range from 1 nanometer to 1 micron. This book deals with the science and technology of thin films as they apply to electronic materials. Chapters are devoted to surface energy, atomic diffusion, stress, and energy bands. These are subjects of importance in understanding the processing and properties of thin films. The remaining chapters deal with cluster formation, epitaxy, superlattices, and kinetic processes in thin films such as crystallization of amorphous layers, interfacial reactions, electromigration and morphology changes.

1.1 Applications in Electronic Devices

Advances in thin film and layered structure technology have been pivotal in the evolution of integrated circuits and opto-electronics. For example, a deposited metal on silicon forms a compound, known as a silicide, when heated to temperatures of 300–400°C. A silicide contact used in a transistor is shown in Fig. 1.1. The contact is formed by depositing a metal such as Pt on the silicon substrate and heating to form the metal–silicon compound—a silicide such as PtSi. It is critical that only a small amount of silicon be consumed in the formation process, so that the contact does not penetrate deeply into the underlying silicon. The silicide/silicon interface should be planar without irregular penetration spikes which could short out the underlying electrical junctions.

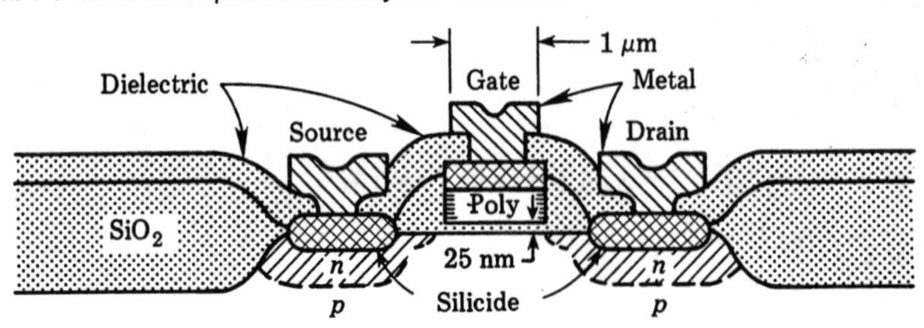

FIGURE 1.1 Field-effect transistor with silicide contacts to *n*-type source and drain and to polycrystalline Si gate.

Another example is that of electrical connections between individual transistors. These connections are made by means of a metal film, often aluminum (Al), chosen because of its low resistivity and stability. When Al is deposited on the silicide, stability is lost during thermal processing. Aluminum reacts with the silicide and degrades its electrical properties. To preserve the advantages of the low resistivity of Al and of the stability of silicides, Al is prevented from making direct contact to the silicide by an interposed diffusion barrier such as TiN. This barrier must prevent interdiffusion of Al into the silicide (Al penetrates silicides readily) through all the subsequent high temperature processing steps required to make a finished integrated circuit. Interdiffusion, silicide formation, the influence of grain boundary diffusion in barriers, and deformation due to stress all form the subjects of later chapters in this book. It is the understanding of these phenomena that allows thin film technology to be broadly applied in device fabrication.

1.2 Deposition and Growth

Epitaxial growth of semiconductor layers occurs when the deposited atoms are aligned with (are commensurate with) the atoms in the underlying single-crystal semiconductor substrate. Epitaxy is central to the fabrication of almost all semiconductor devices. An outstanding example is the solid-state laser, the basis of optical communications. Lasers are made from compound semiconductors (GaAs is an example), composed almost always of elements from Columns III and V of the periodic table and known as III–V semiconductors. The distinguishing characteristic of many III–V semiconductors is their ability to emit light efficiently as compared to the group IV semiconductors Si and Ge. Most lasers make use of a layered epitaxial structure consisting of two different III–V materials (Fig. 1.2). Because different materials are in contact, the device is called a heterostructure. The growth process is known as *heteroepitaxy*. Successful heteroepitaxy occurs in materials with nearly the same lattice constant, a condition well met in the AlGaAs/GaAs system in Fig. 1.2. *Homoepitaxy* refers to the epitaxial growth of a material on itself. It is used extensively in silicon-based integrated circuits to form layers with controlled amounts of electrically active impurities which are called dopants. In both hetero- and homoepitaxial layers, the concentration of dopant atoms is low (in the range of

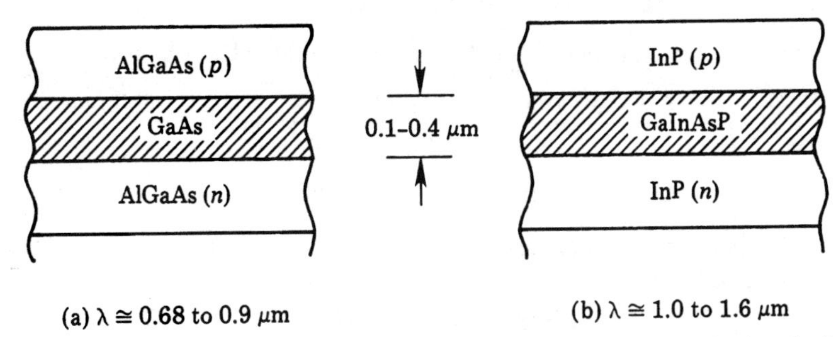

(a) $\lambda \cong 0.68$ to $0.9\ \mu m$ (b) $\lambda \cong 1.0$ to $1.6\ \mu m$

FIGURE 1.2 Layered compound semiconductors for efficient optical emission.

10^{15} to 10^{20} atoms/cm^3). Silicon has an atomic concentration of 5×10^{22} atoms/cm^3; dopant concentrations are below 0.2 atomic percent and range down to 2×10^{-6} atomic percent. The growth of epitaxial layers requires that the growth process itself be sufficiently pure and uncontaminated so that the concentration of unwanted impurities is well below dopant concentrations.

The considerations of purity set the requirements for growth conditions. Thin films used in electronics are almost invariably deposited in vacuum systems. All vacuum systems have a finite background pressure which establishes the purity of the growth. In this section we calculate the impinging flux of atoms from the background gas pressure. This impurity flux will then be compared to the flux of deliberately deposited atoms.

The basic relation describing residual gas in an imperfect vacuum is the ideal gas law,

$$pV = RT = N_A kT \tag{1.1}$$

where

- p is pressure (N/m^2)
- V is molar volume or the volume of one mole of gas
- R is the gas constant = 8.31 Joule/mole-K
- N_A is the number of molecules in one mole of gas (i.e., Avogadro's number = 6.02×10^{23} molecules/mole)
- k is Boltzmann's constant = 1.38×10^{-23} Joule/K
- T is absolute temperature K (K = °C + 273.16)

Commonly used units of pressure are:

$$1 \text{ Torr} = 1 \text{ mm Hg} = 1333 \text{ dyne/cm}^2 = 133.3 \text{ N/m}^2$$
$$1 \text{ atmosphere} = 760 \text{ Torr} = 1.013 \times 10^6 \text{ dyne/cm}^2 = 1.013 \times 10^5 \text{ N/m}^2$$
$$1 \text{ Pascal (Pa)} = 1 \text{ Newton/m}^2 = 7.5 \times 10^{-3} \text{ Torr}$$

Using the ideal gas law, we derive the relation between gas density and pressure. This relation is then used to estimate the number of gas atoms impinging on the surface. At 1 atmosphere and 0°C, the molar volume of an ideal gas is 22.4×10^3

cm^3 or 22.4 liters. This quantity is obtained by using the ideal gas law,

$$V = \frac{N_A kT}{p} = \frac{6.02 \times 10^{23}\ \text{(molecule/mole)} \times 1.38 \times 10^{-23}\ \text{(J/K)} \times 273\ \text{K}}{1.013 \times 10^5\ \text{N/m}^2}$$

$$= 2.24 \times 10^{-4}\ \text{m}^3 = 22.4 \times 10^3\ \text{cm}^3$$

The reader should note that the centimeter-gram-sec (ergs and dynes) and meter-kilogram-sec (Joules and Newtons) systems are freely mixed throughout this book. Many defining terms are expressed in M.K.S. units while current-day technology tends to favor c.g.s. units.

The gas density at 25°C (298 K) is

$$n = \frac{N_A}{V} = \frac{6.02 \times 10^{23}}{2.44 \times 10^4} = 2.46 \times 10^{19}\ \text{molecules/cm}^3$$

where n is the number of molecules (or atoms) per unit volume. Since

$$n = \frac{N_A}{V} = \frac{p}{kT} \tag{1.2}$$

we see that n is directly proportional to p at a given temperature. This is an important relation in vacuum technology. At 1 Torr and 25°C, we have $n = n_1$,

$$n_1 = \frac{2.46 \times 10^{19}}{760} = 3.24 \times 10^{16}\ \text{molecules/cm}^3$$

In industrial processes, a vacuum of 10^{-7} Torr is common; in this vacuum the particle density is

$$n_1 \times 10^{-7} = 3.24 \times 10^9\ \text{molecules/cm}^3$$

In an ultrahigh vacuum system, a vacuum of 10^{-11} Torr is achieved and we have

$$n_1 \times 10^{-11} = 3.24 \times 10^5\ \text{molecules/cm}^3$$

The flux of atoms impinging on a solid surface is given by the product of particle concentration and velocity. The mean kinetic energy of a molecule in an ideal gas is

$$\overline{E}_k = \frac{3}{2}\,\text{k}T = \frac{1}{2}\,mv_a^2 \tag{1.3}$$

$$v_a = \left(\frac{3kT}{m}\right)^{1/2} = \left(\frac{3RT}{M}\right)^{1/2}$$

where M is molar weight and v_a is the root mean square velocity (Appendix A). We take $M = 28$ g/mole for nitrogen gas and obtain

$$v_a = \left(\frac{3 \times 8.31\ \text{J/K-mole} \times 10^7\ \text{erg/J} \times 298\ \text{K}}{28\ \text{g/mole}}\right)^{1/2}$$

$$= (26.5 \times 10^8\ \text{erg/g})^{1/2} = 5.2 \times 10^4\ \text{cm/sec}$$

The magnitude of v_a is of the order of the speed of sound in the gas.

The rate of molecules impinging on a surface per unit area per unit time is derived by considering that only those molecules within a distance $v_a t$ from the surface can hit the surface in the time t. For a surface of area A, the molecules in the volume of $v_a tA$ can hit the surface in time t, and the number of molecules in the volume is $nv_a tA$, so the rate of impingement per unit area per unit time is

$$\frac{nv_a tA}{tA} = nv_a \text{ molecules/cm}^2\text{-sec} \tag{1.4}$$

In the derivation of Eq. (1.4), we did not consider the velocity distribution of the particles. If we take the velocity distribution of an ideal gas (i.e., the Maxwell distribution of velocities), we can show (Appendix A) that the flux is

$$J_c = \frac{1}{4} nv_a \tag{1.5}$$

Equation 1.5 corresponds to a flux of atoms impinging on the surface. The flux is equal to the product of concentration (atoms/cm^3) times velocity (cm/sec) and has units of atoms/cm^2-sec. In Appendix A, a more complete derivation shows that the flux J is given by:

$$J = n(kT/2\pi m)^{1/2} \tag{1.6}$$

Using Eq. 1.5, we can estimate the flux of molecules impinging on a substrate surface when exposed to one atmosphere of air,

$$J_c = \frac{1}{4}(2.46 \times 10^{19} \times 5.2 \times 10^4) = 3.2 \times 10^{23} \text{ molecules/cm}^2\text{-sec}$$

Most solids have atomic densities n of 5 to 9 $\times$ 10^{22} atoms/cm^3. There are then about 10^{15} atoms/cm^2 in a monolayer (ML) using the estimate that one ML $= (n)^{2/3}$. It will take about

$$t = \frac{10^{15}}{3.2 \times 10^{23}} = 3.1 \times 10^{-9} \text{ sec}$$

in air for a monolayer of gas atoms to impinge on a surface. Then, in an ultra-high vacuum of 10^{-11} Torr, we have

$$J_c = \frac{1}{4}(3.24 \times 10^5 \times 5.2 \times 10^4) = 4.2 \times 10^9 \text{ molecules/cm}^2\text{-sec}$$

and it will now take

$$t = \frac{10^{15}}{4.2 \times 10^9} = 2.4 \times 10^5 \text{ sec } (\sim \text{three days})$$

for the residual gas to form a monolayer of deposit on the surface assuming that the gas sticks to the surface.

For epitaxial growth rates of one monolayer/sec (about 10^{15} atoms/cm^2-sec) we require the rate of impingement of gas atoms to be at least 10^{-4} times the deposition rate to maintain high purity films. This rate is equivalent to a vacuum of 10^{-10} Torr.

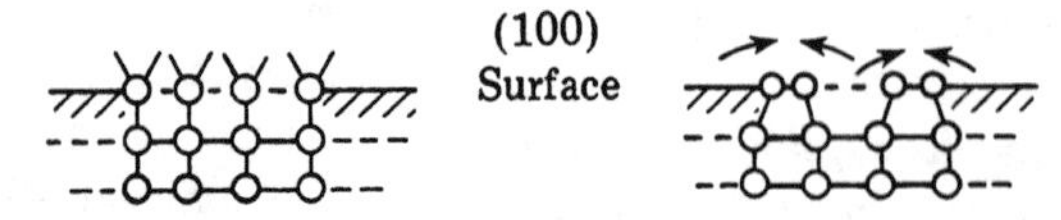

FIGURE 1.3 Surface reconstruction on the Si (100) surface.

1.3 Surfaces

The surface of a semiconductor is the starting place for epitaxial growth. The surface has a microscopic structure associated with atomic displacements and a macroscopic structure associated with surface steps and other surface defects.

A ball-and-stick crystal model would depict a (100) surface (the face of a diamond lattice) as a portion of a plane of infinite extent consisting of a square array of atoms. For the diamond lattice, each of these atoms have two unpaired electron bonds. In silicon, the atoms displace laterally (as shown schematically in Fig. 1.3) to satisfy the bonding requirement. Such surfaces are called *reconstructed.*

On a larger scale, surfaces can contain terraces, steps, and other defects, as shown in Fig. 1.4. Steps are far from perfect. The atomic structure of steps has been revealed by scanning tunneling microscopy. Figure 1.5 shows an array of steps with highly irregular step edges. Steps and kinks, visible in Fig. 1.5, form high-energy binding sites for deposited materials.

Steps and terraces form the boundary between regions of low-index planes, such as the (100) plane. It is the displacement laterally of the edges of the sheets of low-index planes that defines the surface. The sheets of planes appear in sharp contrast in Fig. 1.5.

The surface layers of compound semiconductors must be further characterized by their atomic composition. For example, the (100) surface of GaAs contains either all Ga atoms (the A face) or all As atoms (the B face) of this A–B compound. Other surfaces can contain a mixture of atoms. The monoatomicity of the (100) surface has made it the preferred growth face for GaAs epitaxy; the growth proceeds monolayer by monolayer in the sequence A, B, A, B.

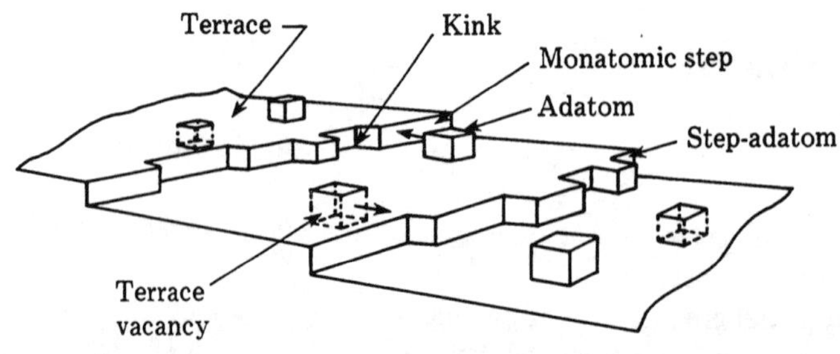

FIGURE 1.4 Schematic diagram of surface steps and kinks (from Somorjai).

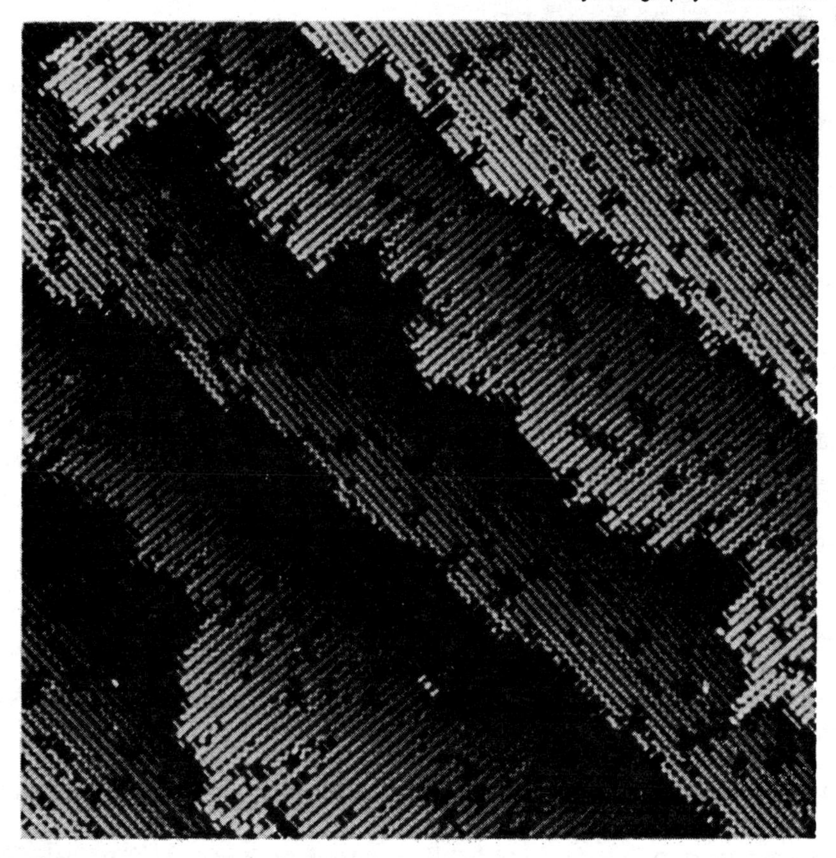

FIGURE 1.5 A scanning tunneling microscopic picture of the Si (100) surface. The average distance between steps is 17 nm (Courtesy of R. J. Hamers, U. K. Kohler, and J. E. Demuth, *J. of Vac. Sci. Tech.* A8 (1990): 195).

1.4 Crystallography and Notation

The previous sections have described the deposition of thin films and edges and steps on the surface. At this point we summarize the crystal structure of solids so that the number per cm^2 N_s of atoms on a surface and the height h of a monolayer can be determined.

The atomic volume can be calculated without use of crystallography. The atomic density n of atoms/cm^3 is given by

$$n = N_A \rho / A \tag{1.7}$$

where N_A is Avogadro's number, ρ is the mass density in gram/cm^3 and A is the atomic mass. Taking Si as an example, $n = (6.02 \times 10^{23} \times 2.33)/28 = 5 \times 10^{22}$ atoms/cm^3. The semiconductors Ge and GaAs have atomic densities of about

4.4×10^{22} atoms/cm^3; the metals Al and Au, about 6×10^{22} atoms/cm^3, and metals such as Co, Ni, and Cu, about 9×10^{22} atoms/cm^3. The volume Ω occupied by an atom is given by

$$\Omega = (1/n) \tag{1.8}$$

with a typical value of 20×10^{-24} cm^3.

1.4.1 Crystal Structure

A crystal is composed of atoms arranged in a periodic pattern in space and is defined by a set of lattice points. The space containing this set of points can be divided by three sets of planes into a set of cells each identical in size, shape, and orientation; such a cell is called a *unit cell.* This cell can be described by three unit vectors **a**, **b**, and **c**, called *crystallographic axes,* which are related to each other in terms of their lengths a, b, and c, and the angles α, β, and γ (see Fig. 1.6). Any direction in the cell can be described as a linear combination of the three axes:

$$\mathbf{r} = n_1\mathbf{a} + n_2\mathbf{b} + n_3\mathbf{c}. \tag{1.9}$$

where n_1, n_2 and n_3 are integers.

Seven different cells are necessary to describe all possible point lattices. These define the seven crystal systems, shown in Fig. 1.6. Each corner of the unit cell of these seven systems has a lattice point, but not the interior of the cells nor the cell faces. It is possible to place more points either in the center of the unit cell or on the cell faces without violating the general definition of a lattice point. A lattice point has the same surrounding in the lattice as every other lattice point. Based on this arrangement of points, a total of fourteen Bravais lattices can be produced for the seven crystal systems. Figure 1.7 shows the face-centered cubic (fcc) and the associated diamond structure of silicon and germanium.

The number of lattice points in a unit cell, n_u, is given by

$$n_u = n_i + \frac{n_f}{2} + \frac{n_c}{8} \tag{1.10}$$

where n_i is the number of points in the interior, n_f is the number of points on faces (each n_f is shared by two cells), and n_c is the number of points on corners (each n_c point is shared by eight cells).

In considering the Bravais lattice of the face-centered cubic lattice shown in Fig. 1.7, it is customary to use the conventional unit cell of the fcc cell rather than the primitive cell. Since it is a cubic system $a = b = c$, $\alpha = \beta = \gamma = 90°$, and the length a is called the lattice parameter. The number of atoms per unit cell with $n_i = 0$, $n_f = 6$, and $n_c = 8$ is

$$n_u = \frac{6}{2} + \frac{8}{8} = 4 \text{ atoms/unit cell} \tag{1.11}$$

The height of a monolayer on the top surface of the fcc structure in Fig. 1.7 is $a/2$.

A number of commonly used metals in thin film technology, such as Al (a = 0.405 nm) and Au (a = 0.408 nm), possess fcc crystal structure. Many semicon-

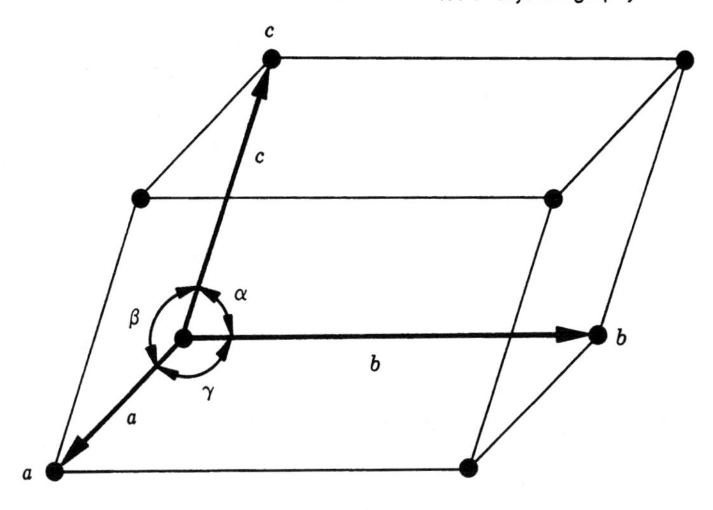

System	Axial Lengths and Angles	Bravais Lattice
Cubic	Three equal axes at right angles $a = b = c$, $\alpha = \beta = \gamma = 90°$	Simple Body-centered Face-centered
Tetragonal	Three axes at right angles, two equal $a = b \neq c$, $\alpha = \beta = \gamma = 90°$	Simple Body-centered
Orthorhombic	Three unequal axes at right angles $a \neq b \neq c$, $\alpha = \beta = \gamma = 90°$	Simple Body-centered Base-centered Face-centered
Rhombohedral*	Three equal axes, equally inclined $a = b = c$, $\alpha = \beta = \gamma \neq 90°$	Simple
Hexagonal	Two equal coplanar axes at 120°, third axis at right angles $a = b \neq c$, $\alpha = \beta = 90°$, $\gamma = 120°$	Simple
Monoclinic	Three unequal axes, one pair not at right angles $a \neq b \neq c$, $\alpha = \gamma = 90° \neq \beta$	Simple Base-centered
Triclinic	Three unequal axes, unequally inclined and none at right angles $a \neq b \neq c$, $\alpha \neq \beta \neq \gamma \neq 90°$	Simple

*Also called trigonal.

FIGURE 1.6 Unit cell, seven crystal systems, and fourteen Bravais lattices.

ductors have a diamond cubic structure, which is not one of the Bravais lattices. The diamond structure can be considered as two interpenetrating fcc lattices, having two atoms associated with one lattice point. The crystal structure of Si is diamond cubic with a lattice parameter a of 0.543 nm at room temperature. The number of atoms/unit cell for the diamond lattice is found from $n_i = 4$, $n_f = 6$, and $n_c = 8$, so that

$$n_u = 4 + \frac{6}{2} + \frac{8}{8} = 8 \text{ atoms/unit cell} \tag{1.12}$$

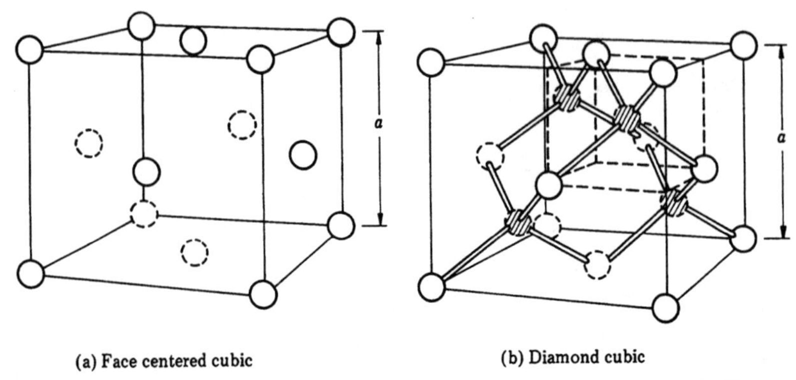

FIGURE 1.7 The face-centered cubic lattice and the diamond lattice of silicon with lattice parameters a. The shading on the atoms denotes Ga atoms in the zinc blende structure of GaAs.

The height h of a monolayer of atoms on the Si surface of Fig. 1.7 is $a/4$.

The atomic density or the number of atoms/cm^3, n, is given by

$$n = \frac{n_u}{a^3} \text{ atoms/unit cell} \tag{1.13}$$

For Si, $n = 8/(0.543 \times 10^{-7})^3 = 5 \times 10^{22}$ atoms/cm^3; Eq. (1.13) and Eq. (1.7) give the same result. To determine the number N_s of atoms/cm^2 on a crystal surface, the number of surface atoms/unit cell for a given orientation is divided by the surface area,

$$N_s = \frac{\text{surface atoms/unit cell}}{\text{surface cell area}} \tag{1.14}$$

For example, the surface of the Si lattice, Fig. 1.7b, has an area a^2 and 2 atoms/unit cell (one center atom and four corner atoms shared by four adjacent cells so that each contributes one-quarter atom to this calculation). Thus $N_s = 2/a^2$. The values of N_s and h depend on the crystal structure and the orientation of the surface plane.

The III–V compounds such as GaAs and AlSb and the II–VI compounds such as ZnS have a crystal structure, the zinc blende structure, very similar to the diamond lattice. In zinc blende structures, the interior atoms of the cubic unit cell are different from those on the faces and at the corners; otherwise, the zinc blende structure is identical in atom position and stacking sequence to the diamond lattice. In Fig. 1.7b, one of the fcc sublattices of zinc blende structure consists of shaded atoms of one element (e.g., Ga), and the other fcc sublattice consists of atoms of a different element (e.g., As).

1.4.2 Directions and Planes

The direction of any line in a lattice may be described by drawing a line through the origin parallel to the given line and then assigning the coordinates of any point

TABLE 1.1 Conventions Used to Indicate Directions and Planes in Crystallographic Systems

A. Directions: Line from origin to point at u, v, w

1. Specific directions are given in brackets, $[uvw]$.
2. Indices uvw are the set of smallest integers. $[\frac{1}{2}\,\frac{1}{2}\,1]$ goes to $[112]$.
3. Negative indices are written with a bar, $[\bar{u}vw]$.
4. Directions related by symmetry are given by $\langle uvw \rangle$. $[111]$, $[1\bar{1}1]$, and $[\bar{1}\bar{1}1]$ are all represented by $\langle 111 \rangle$.

B. Planes: Plane that intercepts axes at $1/h$, $1/k$, $1/l$

1. Orientation is given by parentheses, (hkl).
2. hkl are Miller indices.
3. Negative indices are written with a bar, $(\bar{h}kl)$.
4. Planes related by symmetry are given by $\{hkl\}$. (100), (010), and $(\bar{1}00)$ are planes of the form $\{100\}$.

C. In cubic systems: bcc, fcc, diamond

1. Direction $[hkl]$ is perpendicular to plane (hkl).
2. Interplanar spacing: $d_{hkl} = a/\sqrt{h^2 + k^2 + l^2}$.

on the line. If the line goes through the origin and the point with coordinates u, v, w, where these numbers are not necessarily integers, the directions $[uvw]$, written in brackets, are the indices of the direction of the line. Since this line also goes through $2u$, $2v$, $2w$, and $3u$, $3v$, $3w$, and so on, it is customary to convert u, v, w to a set of smallest integers.

The orientation of planes in a lattice can also be defined by a set of numbers called the Miller indices. We can define the Miller indices of a plane as the reciprocals of the fractional intercepts that the plane makes with the crystallographic axes. Table 1.1 gives conventions used to indicate directions and planes in crystallographic systems. In these conventions, if the plane is parallel to the axis, the intercept is taken to be at infinity; the reciprocal of ∞ is 0 and is the number used to designate the plane.

Figure 1.8 shows the Miller indices of the three lattice planes most referred to in electronic materials technology. Planes are conventionally designated by (hkl). The plane (hkl) is parallel to the $(\bar{h}\bar{k}\bar{l})$ plane, which is on the opposite side of the origin.

1.4.3 Surface Structure

The structure of the surface layer of atoms generally differs from that of the bulk crystal. The surface may reconstruct (Fig. 1.3) so that surface atoms can share bonds. This reconstruction results in a two-dimensional symmetry with a periodicity differing from that of the underlying atoms of the crystal. The surface atoms may also seek new equilibrium positions and thus change the distance between the first and second layers of atoms. This is called *relaxation*; it changes the bond angles but not the number of nearest neighbors.

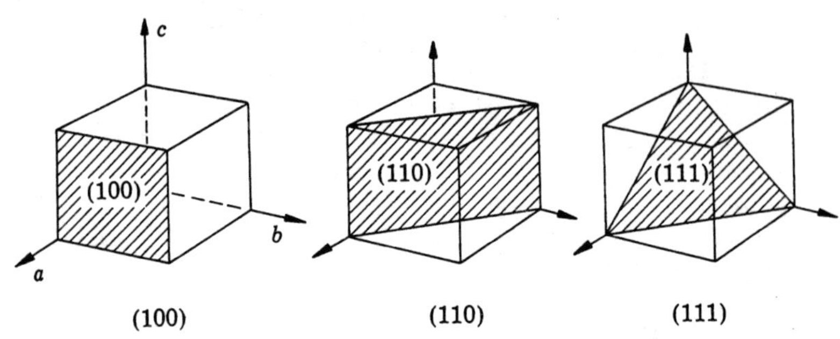

FIGURE 1.8 Miller indices of the three principal planes in the cubic structure.

Surface structure can be revealed directly by scanning tunneling microscopy (STM). However, the notation used to describe surface structure is based on low-energy electron diffraction (LEED). In LEED, electrons of well-defined energy (in the range of 10 to 500 eV) and well-defined direction of propagation diffract from the crystal surface. The low-energy electrons are scattered mainly by individual atoms on the surface and produce a pattern of spots on the fluorescent observation screen because of wave interference. The spots in the pattern correspond to the points of the two-dimensional reciprocal lattice of the repetitive surface structure. As in any diffraction experiment, the LEED pattern is a reciprocal map of the surface periodicity as determined by the size and orientation of the surface unit cell.

Although the assignment of atom positions on the basis of LEED patterns is not unique, it is possible to predict the symmetry of a LEED pattern from the real space configuration of atoms. Figure 1.9 gives examples of overlayers on the (100) surface of a cubic crystal. The letter "p" in the figure indicates that the unit cell is primitive; the LEED pattern for p(2×2) has extra, half-order spots. The letter "c" indicates that the unit has an additional scatterer in the center which gives rise to $\frac{1}{2}, \frac{1}{2}$ spots in the diffraction pattern.

In general, changes in the periodicity of the surface will result in changes in the diffraction pattern that are easily observable and interpretable in terms of the new two-dimensional symmetry. Such changes are often observed, for example, when gases are absorbed on crystal surfaces. Electron diffraction is routinely used to evaluate surface cleanliness in ultrahigh vacuum chambers where contamination-free substrates are a prerequisite for subsequent crystal growth.

1.4.4 Polycrystalline and Amorphous Layers

Silicon dioxide and other oxides are often amorphous; that is, they do not have the periodic structure of a crystal.

In deposition of metals on most oxides, there is no crystal template to orient the atoms. These become arranged in crystallite grains of (usually) 10- to 100-nm size. The deposited film is polycrystalline with each crystallite having the structure of the parent metal but misaligned with respect to other crystallite grains. The crystallites in some cases have one crystal axis aligned perpendicular to the surface plane of the substrate but other crystal axes rotated with respect to each other. The boundaries

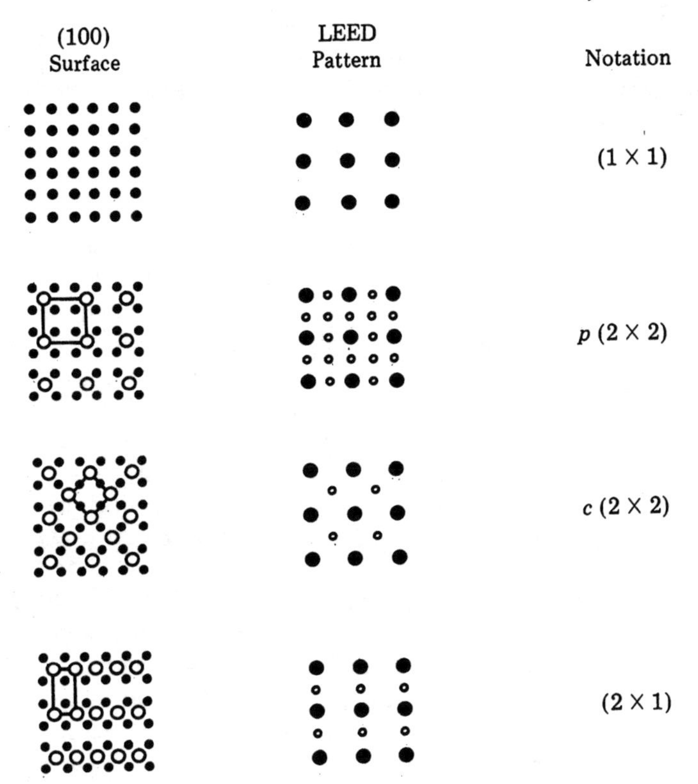

FIGURE 1.9 The (100) surface of a cubic crystal with different atom configurations and the associated LEED patterns in reciprocal space. The notation for the patterns is indicated.

between the crystallites are typically 0.5 to 1 nm wide and provide easy diffusion paths for impurities.

Semiconductors such as silicon, when deposited at low temperatures to suppress surface migration, form amorphous layers. A crystalline solid has long-range atomic order whereas an amorphous solid has only short-range order (of the distance between nearest neighbors). In amorphous silicon, the covalent bond lengths are preserved and hence nearest-neighbor distances are maintained. The bond angles are, however, distorted so that second-nearest-neighbor distances are not sharply defined.

1.5 Layered Structures

Electronic devices of the future will be multilayered and fabricated using the technology and control inherent in ultrahigh vacuum systems. High-density integrated circuits will have lateral feature sizes in the range down to 0.1 micron (100 nm). Growth processes must then be controlled to at least one-tenth and preferably one-hundredth the feature size. Control dimensions will be equivalent to several lattice parameters.

The chapters in this book deal with the materials science underlying the control of thin film structures. Three topics form a thread linking the book: epitaxy, interdiffusion, and stability.

1.5.1 Epitaxy

Homoepitaxial growth of silicon layers of controlled electrical properties on silicon substrates, or of gallium–arsenide on gallium–arsenide substrates, is an accomplished art in semiconductor technology. The aim of this technology is to take full advantage of the properties of each material, or of a combination of two materials in heteroepitaxy, by producing a fault-free interface between two layers.

The objective, shown in Fig. 1.10a, is a uniform layer of one semiconductor whose atoms line up across the interface with atoms of another semiconductor. The lattice parameter of the film does not, however, match that of the substrate except in very specialized cases. The film is strained by the requirement for lattice-matched epitaxy. If the difference in lattice parameters is 1 percent, for every 100 lattice parameters laterally along the film one plane must be removed if the film is to be unstrained (Fig. 1.10b). The missing plane represents a defect at the interface which degrades electronic performance. During high-temperature process steps a strained film may relax to an equilibrium—but undesirable—unstrained state.

If the surface energies of the film and substrate differ, the atoms in the film may prefer to bond to each other rather than to the substrate. This leads to formation of clusters of film atoms on the substrate. Clustering is, of course, not an acceptable thin film growth mode for electronic devices. Clustering requires lateral transport of atoms and can be suppressed by reduction of film deposition temperatures.

1.5.2 Silicides

Thin film structures of metal atoms, such as nickel on a silicon substrate, tend to lower the thermodynamic free energy by interdiffusing and forming a metal–silicon compound, a silicide, such as Ni_2Si, PtSi or Co_2Si (Fig. 1.11). The reaction between

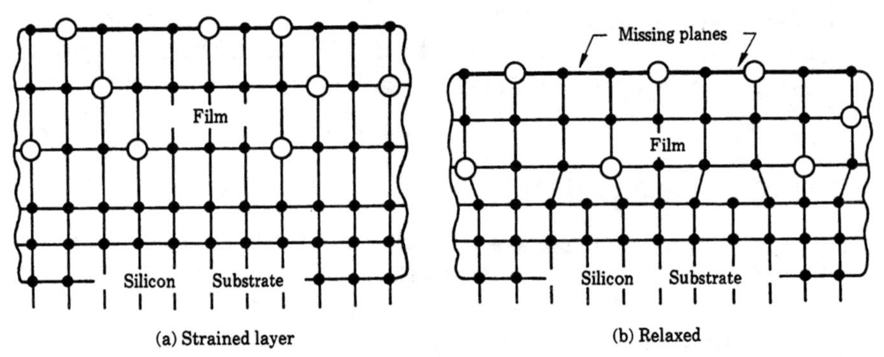

FIGURE 1.10 Epitaxial structures for Ge_xSi_{1-x} (Ge open circles) on a (100) silicon substrate as (a) strained-layer with distorted unit cell and (b) relaxed with missing planes to accommodate the lattice mismatch.

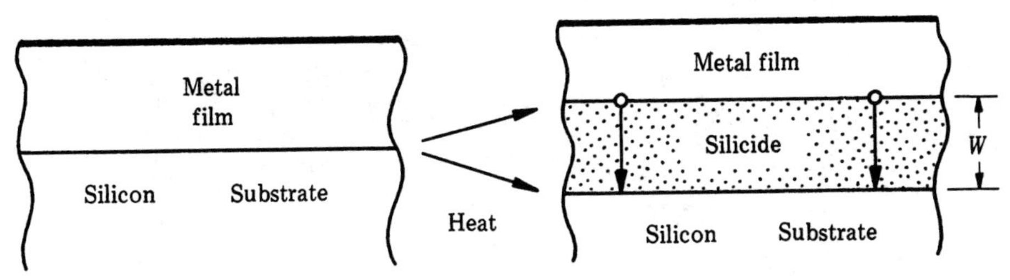

FIGURE 1.11 Deposited metal film on a silicon substrate reacts with the silicon during heat treatment to form a metal–silicon compound, a silicide. In this case, the metal atoms diffuse across the growing silicide layer to react with silicon atoms at the silicide/silicon interface.

metal and silicon can occur during deposition and thus provides an alternative mechanism (to surface diffusion to a step) for incorporating the atom in the substrate.

Uniform and flat interfaces are required in small-volume electronic devices. These planar interfaces arise naturally during the formation of silicides. After the metal layer is deposited on the silicon substrate, the sample is heated and the metal–silicon reaction continues. As shown schematically in Fig. 1.11, the silicide layer grows by diffusion of metal or Si atoms across it. The width W of the growing silicide phase is often given by

$$W = 2(\tilde{D}t)^{1/2} \tag{1.15}$$

where $\tilde{D}$ is the interdiffusion coefficient and t the time. Here again good vacuum and contamination-free interfaces are required during deposition. Impurities, hydrocarbon layers and oxides can all impede the metal–silicon reaction and lead to rough interfaces.

1.5.3 Stability and Metastability

All electronic devices are metastable with respect to equilibrium; semiconductor devices rely on abrupt changes in concentrations of dopant atoms. Thin film structures have abrupt changes in composition, namely, metal layers in contact with semiconductors or with other reactive metal layers. Nonequilibrium structures are used to make integrated circuits. Consequently, heat treatment or even long periods at room ambients can lead to degradation of these structures. Metastable states tend toward stable states which have the lowest free energy and which are defined by equilibrium phase diagrams.

The deposition of amorphous silicon on a silicon substrate and the subsequent epitaxial growth, Fig. 1.12, is an example of the transformation of a metastable (in this case amorphous) state to a stable (single-crystal) state by thermal treatment. Amorphous silicon retains its covalent bonds and nearest-neighbor spacing, but because of the distortion in bond angles, long-range order is lost. In contact with a single-crystal silicon substrate, amorphous silicon rearranges its bonds and grows epitaxially in the solid phase at temperatures about one-half its melting temperature.

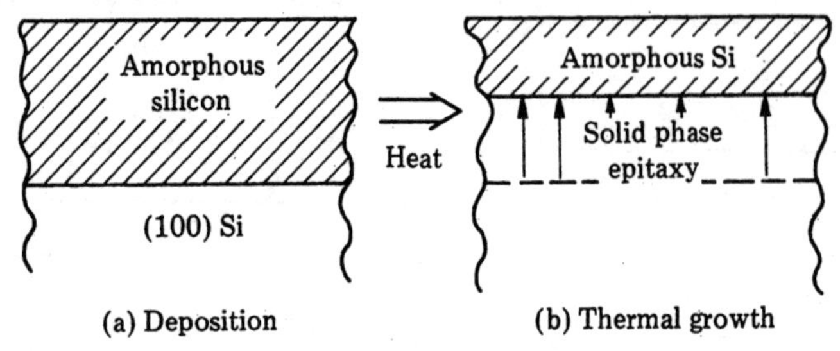

FIGURE 1.12 Deposited amorphous film on a (100) silicon substrate crystallizes by epitaxial growth during heat treatment.

As indicated in Fig. 1.12, the epitaxial growth proceeds in a layer-by-layer fashion, maintaining a planar interface, with a velocity v given by

$$v = v_0 \exp(-E_A/kT) \tag{1.16}$$

where the activation energy E_A is 2.7 eV for silicon. Ion implantation of dopants, a process used to introduce electrically active species into semiconductors, leads to the formation of amorphous layers in silicon. Epitaxial regrowth of the amorphous layer is a required process step in the manufacture of integrated circuits. The amorphous-to-crystalline transformation and solid-phase epitaxy are treated in detail in a later chapter.

The metal lines, Fig. 1.13, used to interconnect circuit elements in integrated circuits are often under compressive or tensile stress during actual device operation. This stress is due, in part, to the difference in thermal expansion of metal and substrate arising in deposition of the metal or in operation of the circuit at temperatures around 100°C. Metal lines conduct large current densities of 10^5 to 10^6 amps/cm^2 which lead to electromigration within these lines.

Stress can cause yield and reliability problems in microelectronic devices. Stress may be considered a metastable condition; during relaxation, the response of an interconnect line to an applied or thermal stress can lead to structural and morphological changes. The atomic arrangement in the metal line relaxes to accommodate stress through transport of atoms and formation of voids. The transport of atoms leads to formation of hillocks in the lines, which can break through insulating layers

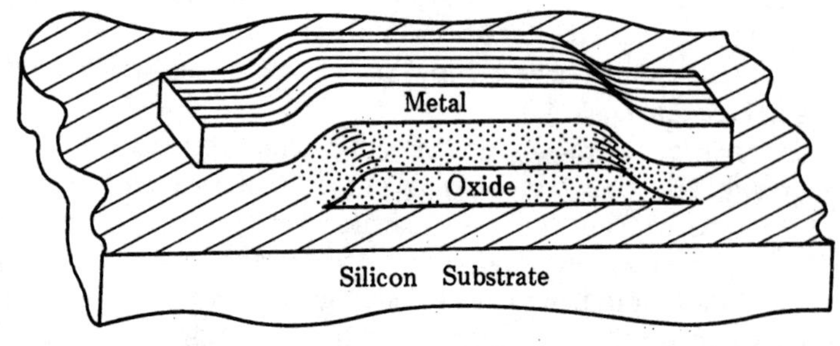

FIGURE 1.13 Metal line deposited over an oxide layer acts as an interconnection between two regions on a silicon substrate.

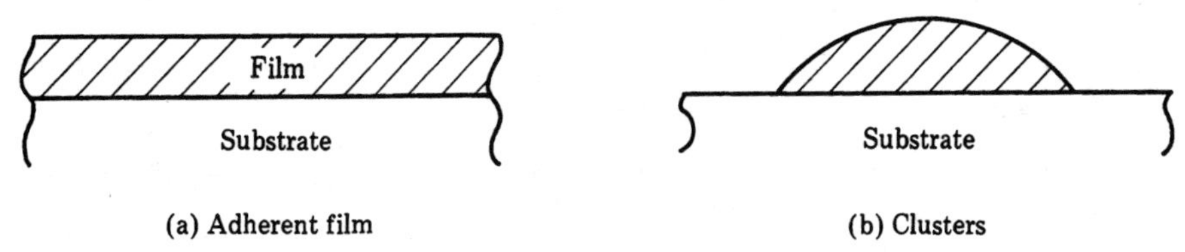

FIGURE 1.14 Deposited film on a substrate forms either (a) an adherent film or (b) a series of clusters, depending on the relative magnitude of film and substrate surface energies.

to form shorts, whereas voids lead to high electrical resistance and may even open portions in the interconnect line itself. The last two chapters of this book are concerned with morphological changes that occur in thin films due to electromigration and stress.

1.6 Conclusions

The application of thin film technology in the electronics industry is based on knowledge of the physical processes occurring in the deposition, operation and utilization of thin film structures. In this chapter we discussed deposition, nature and structure of the surface, and examples of topics to be treated in subsequent chapters.

The critical aspects of modern thin film technology are the growth of epitaxial layers and the interdiffusion and reactions that occur in thermal processing of thin film structures. We establish a firm scientific basis for the understanding of these thin film phenomena in later chapters. First, we provide an overview of the basic concepts of materials science which underlie thin film processes. These chapters concern surface energies, diffusion in solids, and stress. Surface energies, for example, determine whether a film deposited on a substrate forms a uniform adherent layer or forms clusters (Fig. 1.14). Thin film technology requires uniform, planar films of controlled thickness; consequently, we discuss surface energies in Chapter 2.

References

1. L. Eckertova, *Physics of Thin Films* (Plenum Press, New York, 1986).

2. L. C. Feldman and J. W. Mayer, *Fundamentals of Surface and Thin Film Analysis* (Elsevier Science Publishing, New York, 1986).

3. B. H. Flowers and E. Mendoza, *Properties of Matter* (Wiley, New York, 1970).

4. C. Kittel and H. Kroemer, *Thermal Physics*, 2nd ed. (W. H. Freeman, New York, 1980).

5. L. Maissel and R. Glang, eds., *Handbook of Thin Film Technology* (McGraw-Hill, New York, 1970).

6. J. M. Poate, K. N. Tu and J. W. Mayer, eds., *Thin Films: Interdiffusion and Reactions* (Wiley-Interscience, New York, 1978).

7. D. P. Seraphim, R. C. Laskey and C.-Y. Li, eds., *Principles of Electronic Packaging* (McGraw-Hill, New York, 1989).

8. J. C. Slater, *Introduction to Chemical Physics* (McGraw-Hill, New York, 1939).

9. G. A. Somorjai, *Chemistry in Two Dimensions: Surfaces* (Cornell University Press, Ithaca, 1981).

10. K. N. Tu and R. Rosenberg, eds., *Analytical Techniques for Thin Films*, Vol. 27 of *Treatises on Materials Science and Technology* (Academic Press, Boston, 1988).

11. K. N. Tu and R. Rosenberg, eds., *Preparation and Properties of Thin Films*, Vol. 24 of *Treatises on Materials Science and Technology* (Academic Press, New York, 1981).

12. D. Turnbull, *Solid State Physics* **3**, 225 (1956).

13. A. Zangwill, *Physics at Surfaces* (Cambridge University Press, Cambridge, 1988).

Problems

1.1 Germanium is a diamond lattice with atomic density of 4.42×10^{22} atoms/cm^3, while aluminum is a face-centered cubic with atomic density of 6.02×10^{22} atoms/cm^3.
(a) Calculate the lattice parameter a for both.
(b) Determine the number of atoms/cm^2 on the (100) surface for both.
(c) What is the height of a monolayer in terms of a for both?
(d) What is the atomic volume Ω for both in terms of the lattice parameter a?

1.2 The equilibrium vapor pressure above a silicon surface at 1123K corresponds to 6.9×10^{-9} Pa. At equilibrium the fluxes of atoms leaving and returning to the surface are equal.
(a) What is the flux of atoms/m^2-sec?
(b) What is the number of Si atoms/m^3 in the vapor?
(c) What is the velocity of the Si atoms in m/sec?

1.3 What is the surface density of atoms, N_S, in terms of the lattice parameter a for (111) planes in the diamond lattice?

1.4 On (100) Si (a = 0.543 nm), what is the surface density of atoms and what is the miscut angle $\Delta\theta$ to produce a monolayer-high step h every 50 nm? What would the angle be in units of a if the structure were a face-centered cubic or simple cubic? Use the relationship, step length $L_0 = h/\tan\theta$.

1.5 Nickel is a face-centered cubic metal with atomic density of 9.14×10^{22} atoms/cm^3, atomic weight = 58.73, and density ρ of 8.91 gm/cm^3.

(a) What is the lattice parameter a, height h of a monolayer, and the atomic volume Ω?

(b) What is the number of atoms/cm^2 on the (110) plane in terms of the lattice parameter a.

1.6 **(a)** What are the interplanar spacings d for the (100), (110), and (111) planes of Al ($a = 0.405$ nm)?

(b) What are the Miller indices of a plane that intercepts the x-axis at a, the y-axis at $2a$, and the z-axis at $2a$?

1.7 **(a)** Sketch the LEED spot pattern for a crystal where the rectangle of surface atoms have dimensions of a and $2a$.

(b) Sketch the LEED patterns from a normal (not reconstructed) (100) surface and a reconstructed (100) surface where the reconstruction consists of surface dimers (atom pairs) formed by adjacent atoms undergoing displacements toward one another along the $\langle 100 \rangle$ direction.

1.8 The figure for this problem shows the notation, number of electrons, and binding energies for inner-shell electrons in silicon and copper. Three analytical techniques in thin film analysis are x-ray photo-electron spectroscopy (XPS), Auger electron spectroscopy (AES), and electron microprobe analysis (EMA).

(a) In XPS an incident x-ray ejects an electron. For incident Al K_α x-rays ($E = 1.49$ keV), what are the kinetic energies of the ejected Si and Cu 2s electrons?

(b) For KL_1L_2 Auger electrons, an L_1 electron makes a transition to a vacancy in the K-shell and an L_2 electron is ejected. What are the energies of the Si and Cu L_2 electrons?

(c) For K_{α_1} x-rays, a $2p_{3/2}$ electron makes a transition to a vacancy in the K-shell. What are the energies of the Si and Cu K_{α_1} x-rays?

Shell	Level	E_B(eV) Si	Cu
L_3	$2p_{3/2}$	99	931
L_2	$2p_{1/2}$	100	951
L_1	$2s$	149	1096
K	1s	1839	8979

Binding Energies

PROBLEM 1.8

1.9 The figure for this problem shows the escape depth or mean free path of electrons from solids without loss of energy. These escaping electrons have energies characteristic of the Auger and photoelectric processes and hence identify the target atoms.

(a) What are the escape depths for Si and Cu KL_1L_2 electrons?

(b) What are the escape depths of Ge (LMM) (1147 eV), Ge (MVV) (52 eV), and Si (LVV) (92 eV) Auger electrons? In these processes, LVV means a valence band; V-electron makes a transition to a vacancy in the L-shell and a valence band V-electron is ejected.

(c) For 10 keV incident x-rays, what are the escape depths of Si and Cu 1s photoelectrons?

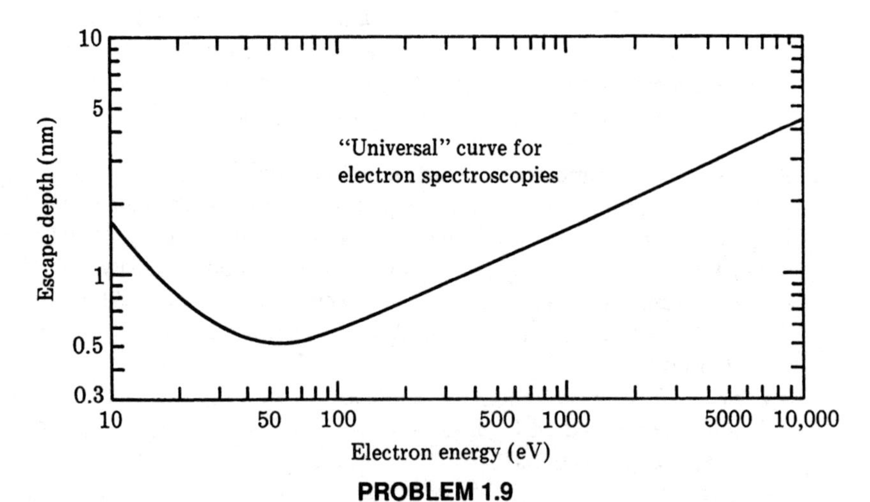

PROBLEM 1.9

CHAPTER 2

Surface Energies

2.0 General Concepts

Surface energy is an underlying concept in thin film science. In general terms, it is the extra energy expended to create a surface. It is important to know that metals have high surface energies and oxides have low surface energies. The surface energy determines whether one material wets another and forms a uniform adherent layer. A material with a very low surface energy will tend to wet a material with a higher surface energy. On the other hand, if the deposited material has a higher surface energy it tends to form clusters (''ball up'') on the low-surface-energy substrate.

Waterproofing is a good example of manipulating surface energies. Organic materials tend to have low surface energies, so a car is waxed with an organic substance and water droplets form on the wet surface.

Most treatises on surface energy start with the famous story of Benjamin Franklin, who was looking at a large pond and noticed the surface roughness of the water. When he put a spoonful of oil on the surface of the pond, the pond became very smooth. One cubic centimeter of such material (about one spoonful) over a large pond forms a film of about 1 nanometer thickness. A film only a few monolayers thick wets a surface and dramatically changes its surface properties.

Surface energy is defined as the energy spent to create a surface; it is a positive quantity because energy is added. In nature, a liquid tends to ball up to reduce its surface area, and crystals tend to facet in order to expose those surfaces of lowest energy. When we break a solid, two new surfaces are created and bonds are broken. It is clear that surface energy is related to bond energy and to the number of bonds broken in creating the surface. This in turn is related to the binding energy of the material.

2.1 Binding Energy and Interatomic Potential Energy

The binding energy is defined as the energy needed to transform one mole of solid or liquid into gas at a low pressure. It is nearly the same in magnitude as the energy of sublimation (transforming solid to gas) or the energy of evaporation (transforming liquid to gas), except that the latter two are generally measured at one atmosphere

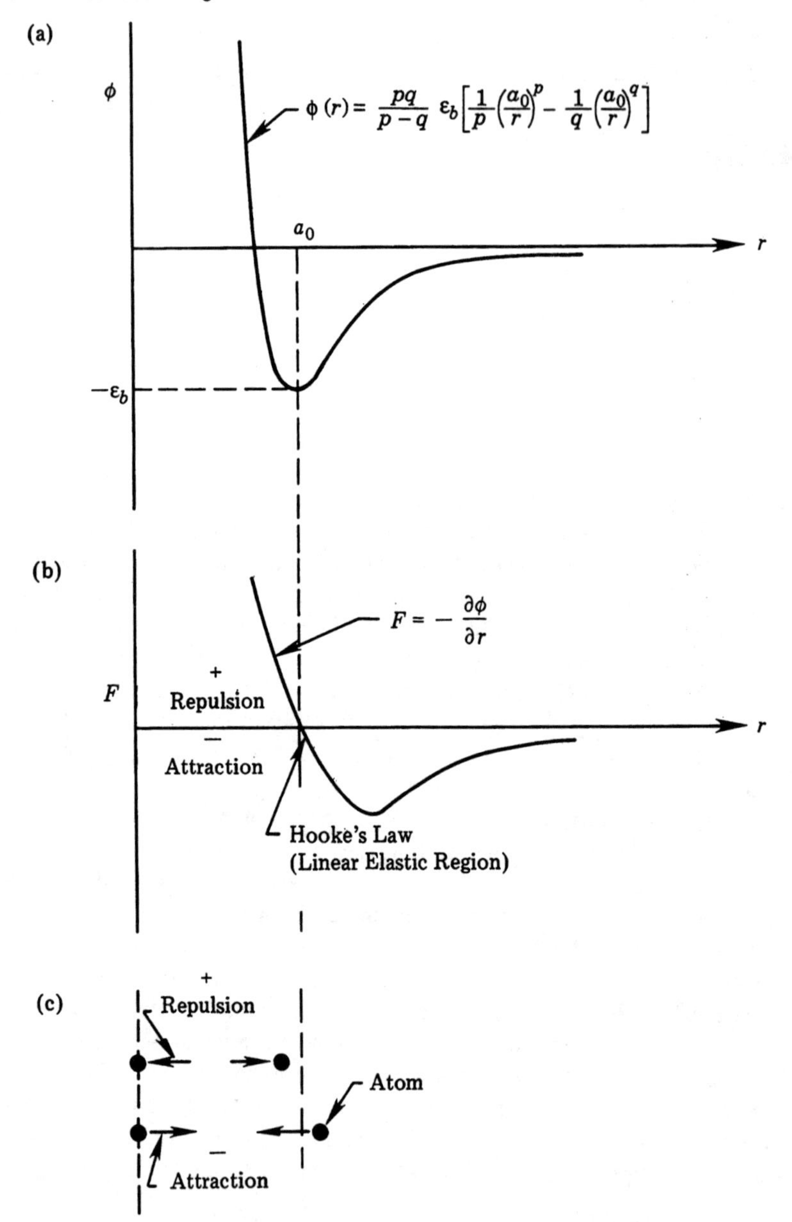

FIGURE 2.1 (a) Interatomic potential function ϕ plotted against interatomic distance r. The attractive potential is of short range. (b) Interatomic force plotted against r. (c) The direction and the sign of force by convention.

pressure rather than at a low pressure. These energies are related to a fundamental energy in materials, that is, the interatomic potential energy between atoms.

A schematic curve of interatomic potential energy ϕ as a function of interatomic distance r is shown in Fig. 2.1a. The minimum potential energy $-\varepsilon_b$ corresponds to the equilibrium interatomic separation a_0. For crystalline solids, the equilibrium interatomic distance can be measured by x-ray diffraction. We estimate a_0 from the lattice parameter a. Aluminum forms a face-centered cubic (fcc) lattice with lattice parameter $a = 0.405$ nm (4.05 Å). Since the densest packing direction is along $\langle 110 \rangle$, the equilibrium interatomic distance in Al is $a/\sqrt{2} = 0.29$ nm. We can also calculate the distance approximately from the atomic volume Ω where Ω is the reciprocal of the atomic density n in atoms/cm^3 ($\Omega = 1/n$; $\Omega = a^3/4$ for fcc lattices). The equilibrium distance $a_0 \simeq \Omega^{1/3} = 0.26$ nm in this approximation. The density calculation can be used for liquids as well.

An atom in a solid, displaced from its equilibrium position, experiences a restoring force given by

$$F = -\frac{d\phi}{dr} \tag{2.1}$$

A schematic curve of the force as a function of distance is shown in Fig. 2.1b. For displacements on the order of 0.1% of the interatomic spacing, the displacement is proportional to the force. In solids, this linear displacement is the origin of Hooke's Law where strain is directly proportional to stress.

The direction of the force is indicated in Fig. 2.1c. In the figure, an atom is placed at the origin and another at the equilibrium separation. If the atoms are displaced toward each other, a repulsive force acts to increase the interatomic distance. The force is defined to be positive or negative according to whether it increases or decreases the interatomic distance. The repulsive force is positive.

While the shape of the interatomic potential energy curve (or the force) controls many of the physical properties of an aggregate of atoms, such as bulk modulus and thermal expansion, we shall consider only whether the potential is short range or long range and how the surface energy is affected by it. In simple cases, the potential can in general be represented by

$$\phi(r) = \frac{pq}{p-q}\,\varepsilon_b\left[\frac{1}{p}\left(\frac{a_0}{r}\right)^p - \frac{1}{q}\left(\frac{a_0}{r}\right)^q\right] \tag{2.2}$$

where ε_b is the minimum potential energy and p and q are numbers whose value depends on the shape of the potential. For short-range interactions in some simple solids, such as a frozen inert gas, solid Ar, the Lennard–Jones potential ($p = 12$ and $q = 6$) applies, and Eq. (2.2) reduces to

$$\phi(r) = \varepsilon_b\left[\left(\frac{a_0}{r}\right)^{12} - 2\left(\frac{a_0}{r}\right)^6\right] \tag{2.3}$$

At the equilibrium position, $r = a_0$, the potential is at its minimum, and $\phi(a_0) = -\varepsilon_b$. The attractive interaction is due to the van der Waals force because each atom is electrically neutral. The interaction is of short range because of the power-law dependence of the potential. When the interatomic separation is twice the equilibrium distance, that is, $r = 2a_0$, the potential energy changes by a factor of about 32. For this reason, we can ignore the interaction energy beyond nearest neighbors and approximate the binding energy by taking only the nearest-neighbor bonds. This simple estimate shows why we are interested in the short-range interactions here. We note that the repulsive interaction is of even shorter range.

We define n_c to be the coordination number, that is, the number of nearest neighbors of an atom in a liquid or solid, and the binding energy for one mole is given by

$$E_b = \frac{1}{2} n_c N_A \varepsilon_b \tag{2.4}$$

where N_A is Avogadro's number and the factor of $\frac{1}{2}$ arises because we have counted each bond twice in the product $n_c N_A$. The maximum number of rigid spheres that can be brought into contact with another sphere of the same radius is twelve in a crystalline solid. The coordination number n_c equals 12 for hexagonal close-packed (hcp) and face-centered cubic (fcc) structures. For the more open, covalently bonded structure of silicon, $n_c = 4$, i.e., there are four nearest neighbors. For body-centered cubic (bcc) structures, $n_c = 8$.

For a long-range interaction, we consider ionic crystals where the attractive interaction between a positive ion and a negative ion is Coulombic. The Coulomb potential is proportional to $1/r$ and is long range since it decays very slowly with increasing r. For an ionic crystal, we can take the exponential parameters $p = 12$ (repulsive interaction remains the same) and $q = 1$ in Eq. (2.2). The profile of the potential is shown in Fig. 2.2.

For metals, the cohesion is due to the interaction of the regularly arranged positive ions with the "electron sea" wherein the electrons move freely. In the free-electron model, the positive ions are shielded by the free electrons from interacting with each other; electrical neutrality is achieved locally and the attractive interaction between atoms is again of short range. The parameter q in the pair-interaction potential of metals can be taken to be close to 6 rather than 1. For this reason, we can use the

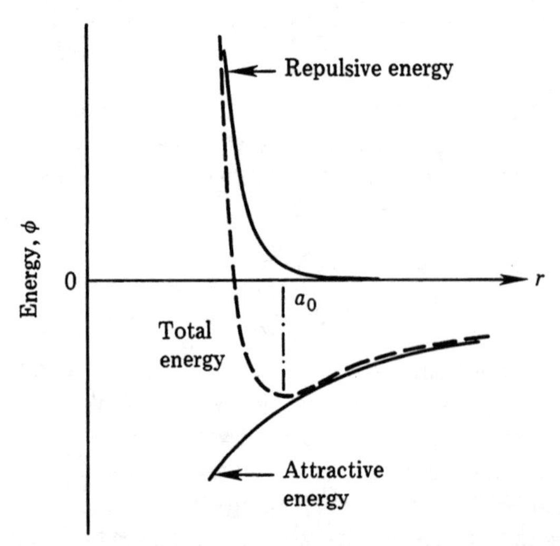

FIGURE 2.2 The broken curve shows the interatomic potential energy where the attractive component is of long range.

nearest-neighbor interaction approximation to estimate the binding energy and the surface energy for metals.

In covalent solids such as semiconductors and inorganic materials where electrons are shared between neighboring positive ions, the screening effect again leads to short-range attractive interaction and the chemical bond picture prevails. Nevertheless, the bonds are more directional.

Using the nearest-neighbor interaction approximation, we can compare magnitudes of the heats (energies) of sublimation, evaporation, melting, and crystallization, with surface energies. The heat of sublimation ΔE_s at low pressures (i.e., the binding energy),

$$\Delta E_s = \frac{1}{2} n_c N_A \varepsilon_b \tag{2.5}$$

depends on the coordination number n_c. In the molten state, metal atoms may have 10 or 11 nearest neighbors instead of 12, so the heat of evaporation will be about 10% less than the heat of sublimation because there are about 10% fewer bonds to be broken. Consequently, the heat of melting or heat of crystallization is only about 10% of the heat of sublimation.

2.2 Surface Energy and Latent Heat

To evaluate surface energy from the point of view of nearest-neighbor bonds, we first define N_S to be the number of atoms per unit area and E_S/A to be the surface energy per unit area. Across an arbitrary atomic plane each atom has on the average $n_c/2$ nearest neighbors on each side, so that the number of bonds to be broken per unit area when we cleave along the plane is $\frac{1}{2}n_c N_S$. Since we create two surfaces in cleaving, the surface energy per unit area is $n_c N_S \varepsilon_b/4$. The argument of $n_c N_S/2$ broken bonds is oversimplified for crystalline solids since it has ignored the packing configuration of atoms in a crystal. Crystals tend to show faceted surfaces, indicating that they have different surface energies on different atomic planes.

The ratio of surface energy/atom, E_S/AN_S, to latent heat of sublimation/atom, $\Delta E_S/N_A$, of a solid is

$$\frac{E_S/AN_S}{\Delta E_S/N_A} = \frac{\frac{1}{4} n_c \varepsilon_b}{\frac{1}{2} n_c \varepsilon_b} = \frac{1}{2} \tag{2.6}$$

From measured values, Table 2.1, we can calculate the ratio of surface energy per atom to latent heat per atom, and also the interatomic potential energy. The latent heat of Au is 60 kcal/mole, or 2.6 eV/atom. To convert the surface energy per cm^2 to units of eV/atom, we use the lattice parameter of Au, 0.4078 nm, to calculate a value of 1.39×10^{15} atoms/cm^2 in the (111) plane. If we assume that the measured surface energy per unit area of 1400 erg/cm^2 is for the (111) surface of Au, we obtain

TABLE 2.1 Relationship Between Solid–Vapor Surface Energy and Latent Heat of Evaporation

Metal	*Solid–vapor surface energy (erg/cm^2)	*Latent heat of evaporation (Kcal/mole)	Ratio of surface energy per atom to latent heat per atom	Interatomic potential energy (eV/atom)	$^\Delta$Cohesive energy (Kcal/mole)
Copper	1700	73.3	0.22	0.58	80.4
Silver	1200	82	0.15	0.65	68
Gold	1400	60	0.24	0.47	87.96

*Data from B. Chalmers, *Physical Metallurgy* (Wiley, New York, 1959). Similar surface energies are given in Table 2.2.

$^\Delta$Data from C. Kittel, *Introduction to Solid State Physics,* 6th ed. (Wiley, New York, 1986), 55.

$$1400 \text{ erg/cm}^2 = \frac{1400 \times \dfrac{1}{1.6} \times 10^{12} \text{ eV/cm}^2}{1.39 \times 10^{15} \text{ atom/cm}^2} = 0.63 \text{ eV/atom}.$$

Thus, the ratio of $0.63/2.6 = 0.24$, as given in Table 2.1. We use the (111) plane which has only 3 broken bonds out of 12 nearest neighbors. Hence, we expect the ratio to be 0.25, which agrees well with the calculated value of 0.24.

To calculate the interatomic potential energy ε_b from the measured latent heat, we rearrange Eq. (2.5),

$$\varepsilon_b = \frac{2\Delta E_s}{n_c N_A} \tag{2.7}$$

Since the heat of evaporation is between liquid and gas, we shall take the coordination number n_c to be 11 (following the same argument used for the molten state). We have for Au,

$$\varepsilon_b = \frac{2 \times 60}{11 \times 23} = 0.47 \text{ eV/atom}$$

In the last column of Table 2.1, we list the calculated theoretical cohesive energies at 0 K and low pressure. This is not the same as the value obtained from latent heat, since the latter is measured at the melting point and at 1 atm.

2.3 Surface Tension

Surface properties can be described by a thermodynamic variable, the surface tension γ. The basic definition is that the reversible work dW on the material upon increasing its surface area dA is

$$dW = \gamma dA \tag{2.8}$$

The surface tension has dimensions of work/area or erg/cm^2. Three quantities enter the scientific literature in this connection: surface energy, surface tension and surface stress. All have the same units, energy/area or force/length. The interrelationship of these quantities arises because solids can change their "surface energy" in two ways; either by increasing the physical area, as may occur from a cleave, or by changing the arrangement of atoms on a surface as in a surface reconstruction. The former case simply involves creating (forming) more surface area, the latter case involves the detailed arrangements of atoms within a solid surface area and may be thought of as the work involved in stretching a surface. The different processes are related through the surface stress tensor. A diagonal element of the stress tensor may be written in the form $S = \gamma + d\gamma/d\varepsilon$ where the first quantity is the surface energy and the second quantity is the change in the surface energy with a variation in strain. In liquids no elastic deformation is possible and $d\gamma/d\varepsilon = 0$. Then the surface stress is identically equal to the surface tension.

In this book we use the following notation and concepts. The phrases surface tension and surface energy shall be used interchangeably. (This convention was originally introduced by Gibbs in his historic work). This quantity will be denoted by "γ". We shall occasionally refer to the total surface energy defined below. The concept of "surface tension" also implies a force. For example the shape of a droplet on a planar surface is an equilibrium configuration set-up by the balance of surface forces acting on the drop. (see Prob. 2.11.) The force is a vector quantity, acting in the surface plane, with the magnitude of the surface tension, γ. The total surface energy E_S is an energy while surface tension is an energy/area. By definition, for area A

$$E_S = \gamma A \tag{2.9}$$

The surface tension can also be thought of as a force/length since erg/cm^2 is equivalent to dynes/cm. In this sense the force $|\mathbf{F}|$ along a line of length l is

$$|\mathbf{F}| = \gamma l \tag{2.10}$$

In the following discussions we speak of surface tension as the scalar quantity γ, the surface energy/area; it is a vector quantity acting in the surface plane when considering phenomena such as the balance of forces acting on a liquid under capillary action.

The relationship between surface energy and tension can also be shown directly by the example of stretching a soap film. In Fig. 2.3, we show a soap film covering the rectangular area formed by a U-shaped wire and a straight wire. If we pull the straight wire by a distance d, the work done is

$$W = 2F_S ld \tag{2.11}$$

where F_S is the force per unit length and l is the width of the soap film. The factor 2 arises because the film has two surfaces. Since the increase in area is $2ld$, the

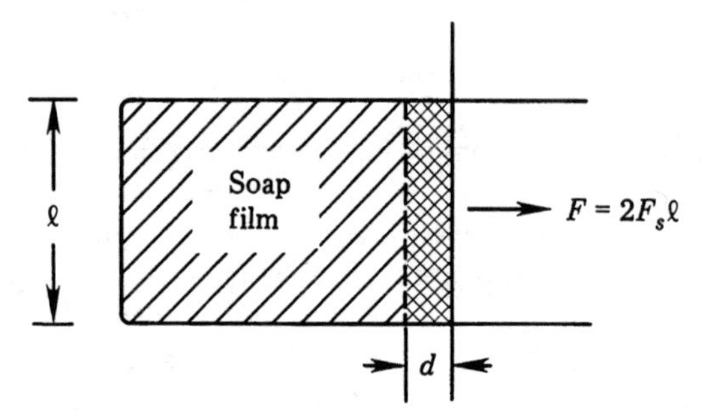

FIGURE 2.3 A soap film stretched by a force.

increase in surface energy is $2ld\gamma$, which should equal the work done,

$$2ld\gamma = 2F_S ld \tag{2.12}$$

So the surface energy per unit area and the surface tension per unit length have the same magnitude.

Review of surface energy per area units shows:

$$\frac{\text{erg}}{\text{cm}^2} = \frac{\text{dyne}}{\text{cm}} = \frac{10^{-3}\ \text{Joule}}{\text{m}^2} = \frac{10^{-3}\ \text{Newton}}{\text{m}}$$

A more physical unit is in terms of eV/atom. Since 1 eV = 1.6×10^{-12} ergs and there are approximately 10^{15} atoms/cm^2 on a typical surface and surface tension is on the average 1000 ergs/cm^2,

$$\gamma = \frac{1000\ \text{ergs}}{\text{cm}^2} \times \frac{\text{cm}^2}{10^{15}\ \text{atoms}} \times \frac{1\ \text{eV}}{1.6 \times 10^{-12}\ \text{ergs}} = 0.6\ \text{eV/atom}$$

The value of 0.6 eV/atom is the order of the bonding energy of an atom in a solid: it takes roughly that much energy to take an atom out of the surface.

2.4 Liquid Surface Energy Measurement by Capillary Effect

Surface energies are often measured in the liquid state by taking the material up to its melting point, and watching either how droplets form or how a meniscus forms in interaction with a solid wall. In Fig. 2.4a, we consider the rise of a liquid column in a capillary tube of diameter $2r$ to reach the equilibrium height h. The driving force of the rise is the reduction of surface energy of the inside wall of the tube. The rise, however, increases the potential energy of the liquid column. The change in total energy of the process is

$$E = \rho V g \frac{h}{2} - 2\pi r h(\gamma_{SV} - \gamma_{SL}) \tag{2.13}$$

where ρ and V ($= \pi r^2 h$) are the density and volume of the liquid column, g is the gravitational constant, and γ_{SV} and γ_{SL} are surface energy per unit area of the wall unwetted (surface–vapor) and wetted (surface–liquid) by the liquid. The first term on the right-hand side of Eq. (2.13) is the potential energy of the liquid column. The mass of the column is $\rho V g$ and the center of gravity of the mass is at $h/2$. The second term is due to the change in surface energy of the inside wall of the tube by the wetting of the liquid. At equilibrium, we have

$$\frac{dE}{dh} = 0$$

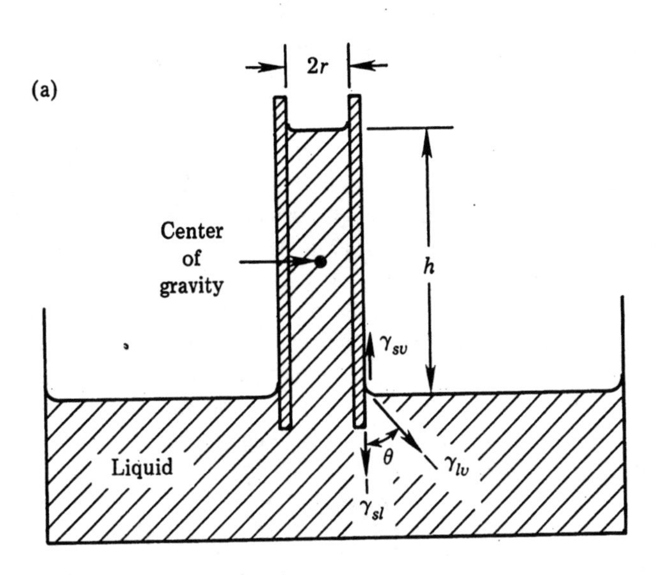

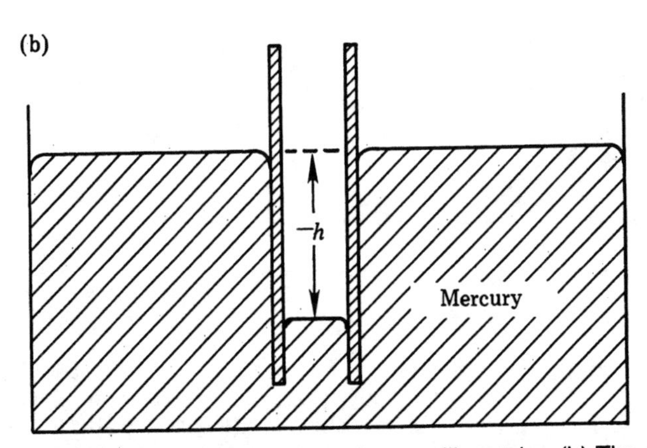

FIGURE 2.4 (a) The rise of a liquid column in a capillary tube. (b) The sink of a liquid column in a capillary tube.

or

$$\rho g \pi r^2 h - 2\pi r(\gamma_{SV} - \gamma_{SL}) = 0$$
$$\gamma_{SV} - \gamma_{SL} = \frac{\rho g h r}{2} \tag{2.14}$$

At the edge of the liquid column as shown in Fig. 2.4a, the surface tensions (energies) are balanced,

$$\gamma_{SV} - \gamma_{SL} = \gamma_{LV} \cos \theta \tag{2.15}$$

where γ_{LV} is the liquid-to-vapor surface energy, and θ is the contact angle. Combining the last two equations, we have

$$h = \frac{2\gamma_{LV} \cos \theta}{\rho r g} \tag{2.16}$$

This relates the liquid–surface energy to measurable quantities of the liquid column (r, h, θ, and ρ). The measurement of θ deserves further discussion, but we shall first illustrate an application of capillary effect in electronic packaging technology.

The capillary effect has been used to fill Cu-plated holes in thick multilayer printed circuit boards with molten solder for mechanical and electrical connection. The board has multilayers of embedded wires, and holes are drilled through the board and plated with Cu for interconnection of the wires. Metallic pins are inserted into the top side of the holes for contacts to external circuits. Then the bottom side of the board is dipped into low-melting-point and hence low-surface-energy molten solder to fill the holes and solder the pins. The success of the process depends on

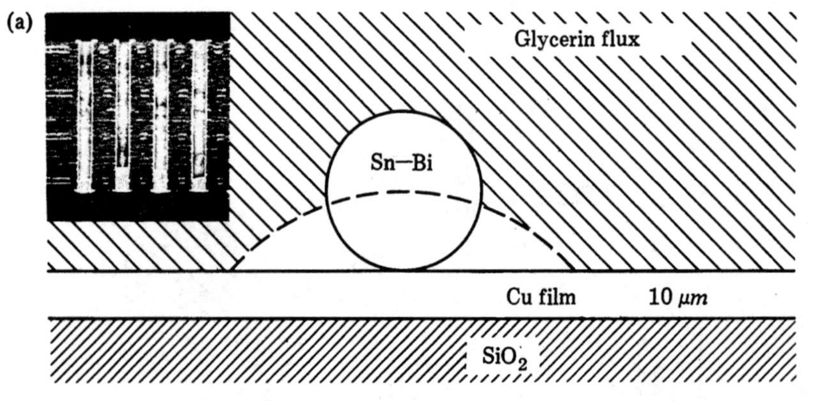

FIGURE 2.5 (a) A schematic diagram of a bead of eutectic Sn–Bi solder placed on a Cu film and immersed in glycerin flux. The broken curve shows the spreading of the bead when it melts. The insert is a cross-section of Cu-plated through-holes in a circuit board of 0.5 cm in thickness. (b) Scanning electron microscopic image of a molten Sn–Bi bead on a Cu surface. (c) Side view of the same bead. A cross-section of a printed circuit board showing the Cu-plated holes is displayed in the upper left of the figure. (Courtesy of K. N. Tu.)

(b)

(c)

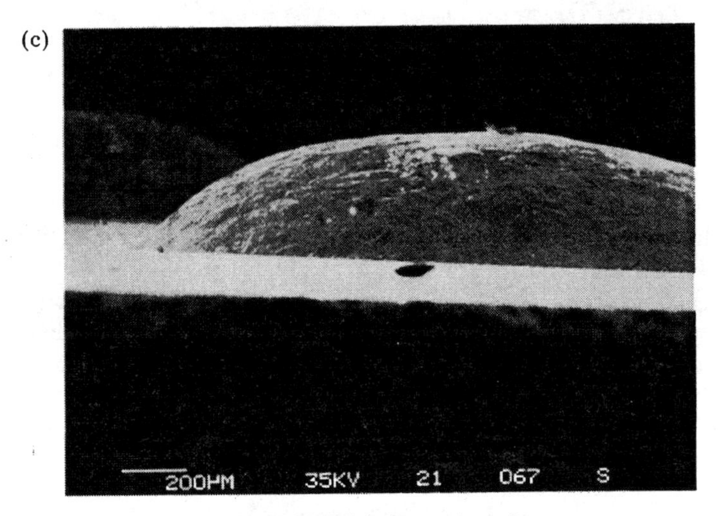

FIGURE 2.5 continued

the capillary effect, which in turn depends on the surface energy of the molten solder, its contact angle to the Cu, and the aspect ratio (height/diameter) of the hole. To define the process, we need to measure the contact angle.

Fig. 2.5a shows a schematic diagram of a bead of Sn–Bi solder placed on a Cu surface and immersed in glycerin flux. The flux removes surface oxides. Upon heating to 137°C, the melting point of the solder, the Sn–Bi solder spreads out to wet the Cu (broken curve in Fig. 2.5a). After equilibrium is reached the temperature is lowered to solidify the solder, and the contact angle can be measured. Scanning electron micrographs of the top view and side view of the wetting of the solder on a Cu surface are shown in Fig. 2.5b and 2.5c, respectively; the contact angle θ = 40°. To calculate h in Eq. (2.16), we assume λ_{LV} = 250 erg/cm^2, $\rho \cong 10$ gm/cm^3, the gravitational constant g = 980 dyne/gm, and the hole diameter = 0.5 mm. We obtain h = 1.6 cm, which should be greater than the thickness of the board. The capillary effect can pull the molten solder all the way through the hole. In practice, to protect the Cu surface from oxidation and to enhance the capillary action, the Cu surface is coated with a thin layer of immersion Sn. The contact angle of the Sn–Bi solder to the Sn surface is zero. This can be shown experimentally by replacing the Cu layer in Fig. 2.5a with a Sn layer. On the other hand we see in Eq. (2.15) that if $\gamma_{SL} > \gamma_{SV}$, the contact angle θ will be greater than 90°, which means that the liquid will ball up and will not wet the solid surface. In this case, cos θ is negative and it gives a negative value to the height h in Eq. (2.16). For example, when we insert a glass tube into mercury, we see the negative capillary effect as the column of mercury goes below the surrounding mercury level (see Fig. 2.4b).

2.5 Solid Surface Energy Measurement by Zero Creep

We can extend the capillary technique of measuring liquid surface energies to measuring solid surface energies by turning the arrangement in Fig. 2.4a upside down (i.e., by hanging a wire from the ceiling and measuring its rate of elongation or *creep rate* under its own weight). Zero creep means the strain rate is zero; that is, at zero creep the weight of the wire is balanced by the surface tension of the wire surface, and thus the surface tension of the wire can be determined by knowing its weight.

If we take a glass fiber or a wire made of metallic glass (amorphous alloy), there are no grain boundaries in the wire, and the analysis is quite similar to that in the last section. In Fig. 2.6a, we consider a glassy wire of diameter $2r$ and length l hanging down. In order to reduce surface energy the wire shortens in length, but the weight of the wire balances this tendency. We assume that the wire shortens a small length $d\ell$ and that the diameter is increased by dr in order to reach equilibrium. Since the volume of the wire must remain the same, we have

$$\pi r^2 \ell = \pi (r + dr)^2 (l - dl) \tag{2.17}$$

By ignoring the higher order terms, we obtain

$$dr = -\frac{r}{2l} dl \tag{2.18}$$

where dl is negative (decrease in length) and dr is positive (increase in radius). To calculate the energy change of the shrinkage, we start with the total energy E of the wire,

$$E = \rho V g (l/2) - 2\pi r l \gamma_{SV} \tag{2.19}$$

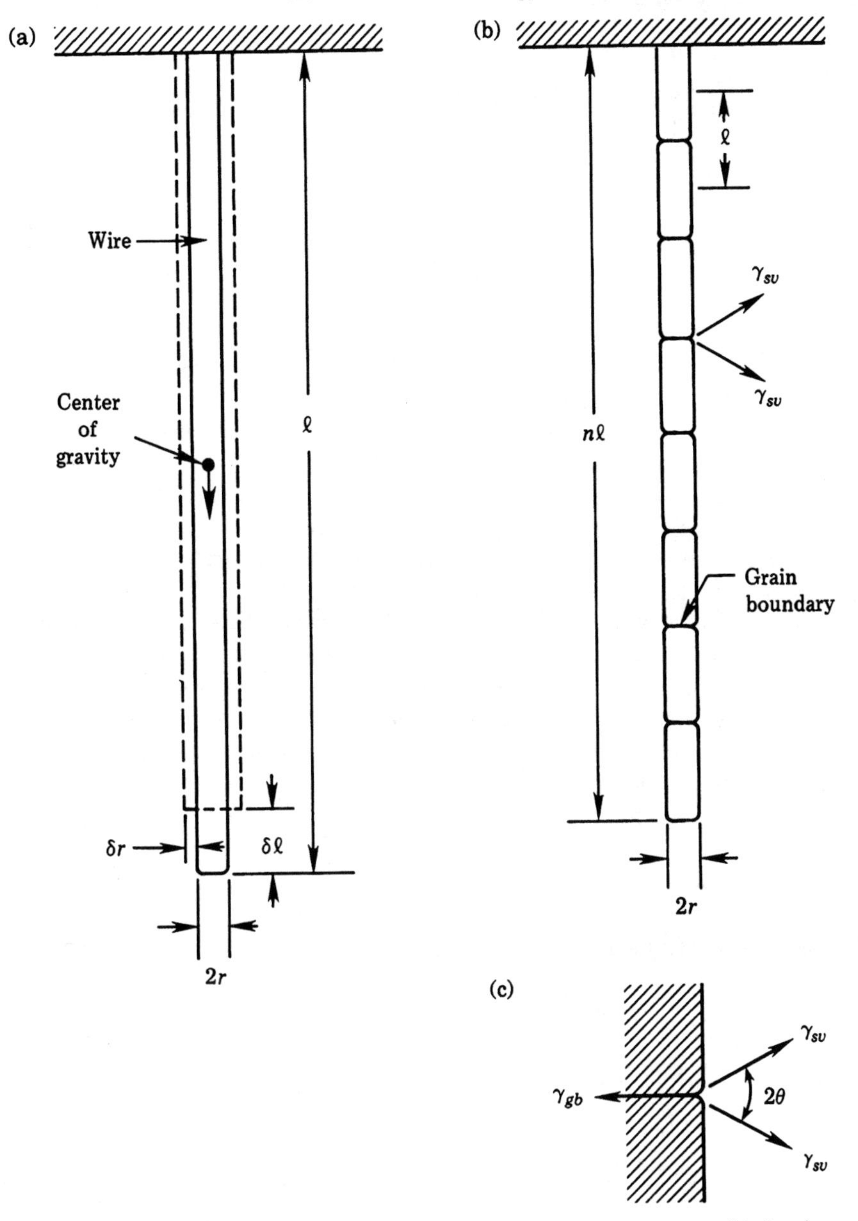

FIGURE 2.6 (a) A schematic diagram of a glassy wire in zero creep. (b) A schematic diagram of a crystalline wire having a bamboo-type grain structure in zero creep. (c) Equilibrium at a triple point where a grain boundary meets the surface.

where ρ and V ($= \pi r^2 l$) are the density and volume of the wire (the factor of $\frac{1}{2}$ comes in because the center of gravity of the wire is at $l/2$) and γ_{SV} is the surface energy per unit area of the wire. Then,

$$dE = \tfrac{1}{2}\rho g \pi (2l^2 r\, dr + 2r^2 l\, dl) - 2\pi\gamma_{SV}(r\, dl + l\, dr) \tag{2.20}$$

By substituting Eq. (2.18) into Eq. (2.20), we have

$$dE = (\tfrac{1}{2}\pi r^2 l \rho g - \pi r \gamma_{SV})\, dl \tag{2.21}$$

At equilibrium,

$$\frac{dE}{dl} = 0$$

or

$$l = \frac{2\gamma_{SV}}{\rho r g} \tag{2.22}$$

which has the same form as Eq. (2.16). It shows that at zero creep we can determine γ_{SV} by measuring l and r, which gives the weight of the wire when its density is known. From the density, 19.3 g/cm^3, and surface energy, 1400 ergs/cm^2, the length of an Au wire of diameter of 0.02 cm at zero creep can be calculated to be about 14.8 cm with weight of about 0.1 g. If we perform the creep experiment in an ultrahigh vacuum environment and if we can increase the temperature without introducing crystallization of the glassy wire, we may be able to compare the surface energies of a glass in the liquid state and in the solid state at similar temperatures.

When the wire crystallizes, it develops grain boundaries and we must take the grain-boundary energy into account. Actually, this is common for crystalline wires. In Fig. 2.6b, we consider a wire having a bamboo-type grain structure and we assume that there are η number of grains in the wire. If we assume that the grains have an average length of l and diameter of $2r$, the changes in dimensions (r and l) upon shortening are again given by

$$dr = -\frac{r}{2l}\, dl$$

The total energy of the wire is

$$E = \eta \rho \vartheta g \frac{\eta l}{2} - \eta 2\pi r l \gamma_{SV} - \eta \pi r^2 \gamma_{gb} \tag{2.23}$$

where ϑ ($= \pi r^2 l$) is the volume of a single grain and γ_{gb} is the grain boundary energy per unit area. The energy change upon a small change of wire length is

$$dE = \frac{\eta^2 \pi \rho g}{2} r^2 l\, dl - \eta \pi r \gamma_{SV}\, dl + \eta \pi \frac{r^2}{l} \gamma_{gb}\, dl \tag{2.24}$$

The last term in Eq. (2.24) is the increase of the cross-sectional area of the wire upon shortening of its length. The increase can also be obtained by

$$\begin{aligned} dA &= \pi(r + dr)^2 - \pi r^2 \\ &\cong 2\pi r\, dr = -\pi \frac{r^2}{l}\, dl \end{aligned} \tag{2.25}$$

where dA is positive when dl is negative. Since there are η grain boundaries in the wire, the corresponding energy change is given by the last term in Eq. (2.24). At equilibrium, we have

$$\frac{dE}{dl} = 0$$

$$\frac{\eta\pi\rho g}{2} r^2 l - \pi r \gamma_{SV} + \frac{\pi r^2}{l} \gamma_{gb} = 0 \qquad (2.26)$$

If we substitute for $\eta\pi r^2 l = \eta\vartheta$ the total volume of the wire V, we have

$$\rho V g = 2\pi r(\gamma_{SV} - \frac{r}{l} \gamma_{gb}) \qquad (2.27)$$

To relate γ_{SV} and γ_{gb}, we consider a joint where a grain boundary meets the wire surface (see Fig. 2.6c). If we assume the two surface tension vectors are equal, we have

$$\frac{\gamma_{gb}}{\sin 2\theta} = \frac{\gamma_{SV}}{\sin (180° - \theta)}$$

or

$$\gamma_{gb} = 2\gamma_{SV} \cos \theta \qquad (2.28)$$

By substituting this relationship into Eq. (2.27), we obtain

$$\gamma_{SV} = \frac{\rho V g}{2\pi r \left(1 - \dfrac{2r \cos \theta}{l}\right)} \qquad (2.29)$$

We note that if the grains are long (i.e., $l >> r$) we can drop the second term in the denominator, and Eq. (2.29) becomes the same as Eq. (2.22).

In zero creep experiments, the major portion of the wire often is replaced by a weight so that a short wire can be used. In this case, Eq. (2.29) is still applicable except that the term $2\pi r$ in the denominator is replaced by πr. The factor of 2 comes from the fact that the center of gravity is located at half of the length of the wire when no external weight is used. When we use a weight, potential energy is measured at the full length of the wire from the ceiling.

2.6 Surface Energy Systematics

We now consider the systematics of surface energies. Figure 2.7 is organized with respect to atomic number and gives the surface tension values of liquid materials at their melting point. Table 2.2 gives surface tensions of some liquid halides, oxides, and sulfides, as well as polymers. The oxides, halides, and sulfides have lower surface tensions than most of the metals. Under normal evaporation conditions many metals will ball up upon deposition on a halide.

It is also interesting to consider semiconductor/insulator problems. Silicon tends to form a thin oxide. The formation and control of oxide layers on Si is one of the key features in fabrication of integrated circuits. On the other hand, silicon deposited

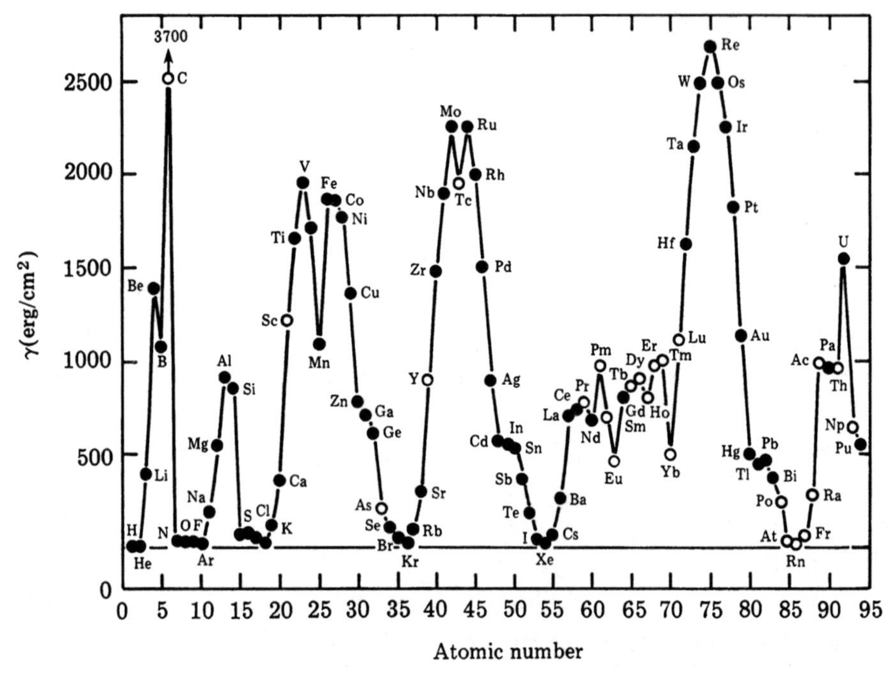

FIGURE 2.7 Surface tension of elements in the liquid phase (from Zangwill, 1988).

on an insulator would tend to ball up because it has a higher surface energy than that of oxides. Formation of heterostructures is always difficult, particularly if the heterostructure involves a superlattice of different types of materials. A superlattice requires forming the structure twice (i.e., first material B on material A, and then A on B). For one interface the surface energy balance will be unfavorable. It is easier to grow superlattices if both materials are semiconductors, because many semiconductors have similar surface energies. Silicon and germanium, for example, are not very different in this regard; therefore superlattices of thin layers of silicon and germanium can be formed.

2.7 Magnitudes of Surface Energies

The magnitude of the surface tension (or surface energy per unit area) for many materials used in device technology is about 1,000 ergs/cm^2. In the following sections, we consider different kinds of formal approaches—thermodynamic, mechanical, and atomic—to describe surface energy and the magnitude of surface tension.

2.7.1 Thermodynamic Approach

From the point of view of thermodynamics, there are two important relations:

$$\frac{d\gamma}{dT} = \frac{-S_S}{A} \tag{2.30}$$

TABLE 2.2 Surface Tension of Selected Solids and Liquids*

Material	γ (ergs/cm^2)	T (°C)
W (solid)	2900	1727
Nb (solid)	2100	2250
Au (solid)	1410	1027
Ag (solid)	1140	907
Ag (liquid)	879	1100
Fe (solid)	2150	1400
Fe (liquid)	1880	1535
Pt (solid)	2340	1311
Cu (solid)	1670	1047
Cu (liquid)	1300	1535
Ni (solid)	1850	1250
Hg (liquid)	487	16.5
LiF (solid)	340	−195
NaCl (solid)	227	25
KCl (solid)	110	25
MgO (solid)	1200	25
CaF_2 (solid)	450	−195
BaF_2 (solid)	280	−195
He (liquid)	0.308	−270.5
Na (liquid)	9.71	−195
Xenon (liquid)	18.6	−110
Ethanol (liquid)	22.75	20
Water (liquid)	72.75	20
Benzene (liquid)	28.88	20
n-Octane (liquid)	21.80	20
Carbon tetrachloride (liquid)	26.95	20
Bromine (liquid)	41.5	20
Acetic acid (liquid)	27.8	20
Benzaldehyde (liquid)	15.5	20
Nitrobenzene (liquid)	25.2	20
Perfluoropentane (liquid)	18.6	−110

*From G. A. Somorjai, *Chemistry in Two Dimensions* (Cornell, Ithaca, NY, 1981).

and

$$E_S = \gamma A - T\frac{d\gamma}{dT}A \tag{2.31}$$

where S_S is the entropy of the surface, A is the area, γ is the surface tension, and T is the temperature. These relations allow estimation of the surface tension at room temperature from tabulated values of $d\gamma/dT$ and measured values of γ taken at the melting temperature. The values of $d\gamma/dT$ are assumed to be independent of temperature.

TABLE 2.3 Surface Tension of Liquid Metals*

Metal	γ_{LV} (ergs/cm²)	$d\gamma_{LV}/dT$ ergs/cm²/°C
Al	866	−0.50
Cu	1300	−0.45
Au	1140	−0.52
Fe	1880	−0.43
Ni	1780	−1.20
Si	730	−0.10
Ag	895	−0.30
Ta	2150	−0.25
Ti	1650	−0.26

*From L. E. Murr, *Interfacial Phenomena in Metals and Alloys* (Addison–Wesley, Reading, MA, 1975).

To estimate the surface energy for silicon, use Table 2.3 to find:

$$\gamma = 730 \text{ ergs/cm}^2 \text{ at the melting point}$$

$$\frac{d\gamma}{dT} = -0.1 \text{ ergs/cm}^2/^\circ\text{C}$$

The melting point for silicon $T_m(\text{Si}) = 1410°\text{C}$. Then at room temperature, the surface energy of silicon is given by

$$\gamma_{RT}(\text{Si}) = 730 + (1410 - 25) \times 0.1$$
$$= 869 \text{ ergs/cm}^2$$

The surface energy at the melting point is not very different from the surface energy at temperatures of interest. The small change of the surface energy is associated with the entropy contribution, which is small.

2.7.2 Mechanical Approach

The mechanical approach (Murr, 1975) uses the mechanical properties of solids to make an estimate of the surface tension. We separate a solid into two pieces which are far apart. Two new surfaces are created and twice the surface energy is added to the system. The surface tension is:

$$\gamma = \frac{E(\infty)}{2A} = \frac{1}{2}\int_0^{R_F} \frac{F_y}{A}\, dy \tag{2.32}$$

where $E(\infty)$ = energy required to bring surfaces to infinity

F_y = applied force normal to the surface created

R_F = range over which force operates

A = area

$$\gamma = \frac{Y}{2} \int_0^{R_F} \frac{y}{a}\, dy = \frac{YR_F^2}{4a} \tag{2.33}$$

where we have expressed the stress F_y/A in terms of Young's Modulus Y and the strain as y/a. Young's Modulus is the material constant that relates the applied force/area, F_y/A, to the fractional change in length in the y direction, $\Delta y/y$, of the solid; $F_y/A = Y\Delta y/y$. If we assume that the force is short range and that $R_F \simeq 10^{-8}$ cm,

$$\gamma = \frac{10^{-16}\, Y}{4a}$$

As shown in Table 2.4, Y is about the same value for many materials. Using a = 0.25 nm and $Y = 10^{12}$ dynes/cm^2, we obtain the canonical number of 1,000 ergs/cm^2 for γ.

This approach connects γ to bulk properties of materials. The strain associated with Young's modulus is a valid concept for small displacements, but we have used it for an enormous displacement. Hooke's law for small displacements, which is the basis for elasticity theory of solids, is violated here. The force between two atoms is essentially zero when we get to distances of the order of one angstrom. The interesting dependence is the surface energy γ as a function of Y values (Table 2.4). In general surface energies scale approximately with Young's modulus. It is not exact, by factors of six or seven, but it is a way of connecting the bulk properties to the surface properties of the material. To estimate a surface energy we might first look at Young's modulus, which is tabulated for almost every material known.

Another way to look at the connection between Young's modulus and surface energy is to see that both of them are linearly related to the interatomic potential energy discussed in Section 2.1. Briefly, let us consider the bulk modulus K which is similar to Young's modulus and is defined as

$$\Delta p = -K \frac{\Delta V}{V} \tag{2.34}$$

where ΔV is the volume change upon a pressure change of Δp, say in compression.

TABLE 2.4 Values of Young's Modulus and γ_{LV}

Material	Y (dynes/cm^2)	γ_{LV} (ergs/cm^2)
Aluminum	6.0×10^{11}	866
Gold	7.8×10^{11}	1410
Iron (Cast)	9.1×10^{11}	1880
Tantalum	18.6×10^{11}	2150

If the compression is carried out adiabatically (i.e., $dQ = 0$), we can use the first law of thermodynamics expressed by Eq. (B.4) in Appendix B to obtain

$$K = -V\left(\frac{dp}{dV}\right) = V\left(\frac{d^2E}{dV^2}\right)_S \tag{2.35}$$

To evaluate the second derivative of E,

$$\begin{aligned}\frac{dE}{dV} &= \frac{dE}{dr}\frac{dr}{dV}\\ \frac{d^2E}{dV^2} &= \frac{d^2E}{dr^2}\left(\frac{dr}{dV}\right)^2 + \frac{dE}{dr}\frac{d^2r}{dV^2}\end{aligned} \tag{2.36}$$

If we take $V = N_A r^3$ and $E(r) = \frac{1}{2}n_c N_A \phi(r)$ where $\phi(r)$ is the interatomic potential function, and if we also assume that $\phi(r)$ obeys the Lennard–Jones potential as given by Eq. (2.3), we obtain

$$\begin{aligned}\left(\frac{dr}{dV}\right)^2 &= \left(\frac{1}{3N_A r^2}\right)^2 = \frac{1}{9N_A a_0 V}\bigg|_{r=a_0}\\ \frac{dE}{dr} &= \frac{1}{2}n_c N_A \frac{d\phi(r)}{dr}\bigg|_{r=a_0} = 0\\ \frac{d^2E}{dr^2} &= \frac{1}{2}n_c N_A \frac{d^2\phi(r)}{dr^2}\bigg|_{r=a_0} = \frac{36n_c N_A \varepsilon_b}{a_0^2}\end{aligned} \tag{2.37}$$

Hence,

$$K = \frac{4n_c N_A \varepsilon_b}{V} = \frac{8\Delta E_s}{V} \tag{2.38}$$

where ΔE_s is the latent heat of sublimation as given by Eq. (2.6). Thus there is a linear relation between K and ε_b as shown by Eq. (2.38).

2.7.3 Atomic Approach

Neither the thermodynamic nor the mechanical approach shows specifically the crystallographic orientation dependence of the surface energy. A third way to think about a calculation of a surface energy which reveals this orientation dependence focuses upon the interaction between atoms. Consider a bulk array of atoms with a pairwise potential energy ϕ. The potential energy represents the binding energy of an atom to all the other atoms in the solid. Then define

$$\gamma_0 = \sum_{k\neq l}\frac{\phi_{kl}}{2A} \tag{2.39}$$

where A is the surface area. For a simple cubic crystal, each atom has six nearest neighbors, twelve second-nearest neighbors, eight third-nearest neighbors, and so

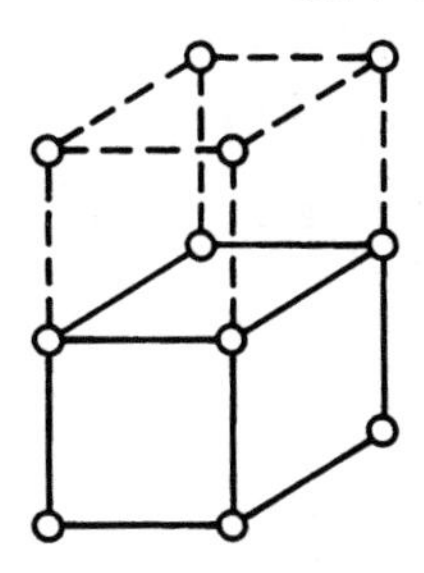

FIGURE 2.8 Atomic bonds across a (100) plane in a simple cubic structure.

on. The energy required to remove an atom from the bulk is then

$$\phi_0 = 6\phi_1 + 12\phi_2 + 8\phi_3 + \ldots. \tag{2.40}$$

The quantity γ_0 is the total binding energy/area of a bulk-like surface atom. If bonds are broken there is an *increase* in the potential energy (a decrease in the binding energy) of the atom. The surface energy is the excess in potential energy over a bulk-like atom. The energy ϕ_0 is related to the sublimation energy, ΔE_s which is equal to $\phi_0/2$.

If we take a simple cubic and cleave it along the (100) plane (Fig. 2.8), the bond is broken for only one nearest neighbor, 4 second-nearest neighbors, 4 third-nearest neighbors and so forth. This modifies Eq. (2.40) when applied to a surface and it shows that surface energies are not equal to bulk binding energy ϕ_0. Nearest neighbors are a lattice constant away, second-nearest neighbors are $\sqrt{2}$ lattice constants away. Thus

$$\gamma_{100} = (\phi_1 + 4\phi_2 + 4\phi_3 + \ldots)/a_0^2 \tag{2.41}$$

is the difference between bulk bindings and surface bindings and a_0^2 is the area/surface atom.

The ratio R is

$$R = \frac{\gamma_{001}}{\gamma_0} = \frac{\phi_1 + 4\phi_2 + 4\phi_3 + \ldots}{6\phi_1 + 12\phi_2 + 8\phi_3} \tag{2.42}$$

It is clear that different crystallographic surfaces have different surface energies. R cannot be evaluated without a specific surface potential. For illustrative purposes we take the Lennard–Jones potential, Eq. (2.3). In this case

$$R_{001} = 0.224$$

$$R_{011} = 0.235$$

$$R_{111} = 0.212$$

These values represent the ratio of the surface energy to the bulk binding energy for the different crystal faces, assuming a Lennard–Jones potential and a face-centered cubic crystal. For semiconductors, there is no simple potential as given by Eq. (2.3). It is assumed that the surface energies are proportional to the number of unpaired bonds/cm^2. For a semiconductor like Ge this corresponds to

$$\text{Ge}(100) = 1.25 \times 10^{15} \text{ bonds/cm}^2$$

where there are two dangling bonds per surface atom and to

$$\text{Ge}(111) = 0.72 \times 10^{15} \text{ bonds/cm}^2$$

where there is one dangling bond per surface atom. The (111) surface is the lowest-energy surface of the principal surfaces, and in general, is the lowest surface energy in germanium. The first-nearest-neighbor bonds are by far the dominant contribution.

References

1. A. W. Adamson, *Physical Chemistry of Surfaces,* 4th ed. (Wiley, Newark, NJ, 1982).
2. H. Brooks, "Theory of Internal Boundaries" in *Metal Interfaces* (American Society for Metals, Cleveland, OH, 1952), 20–65.
3. B. Chalmers, *Physical Metallurgy* (Academic Press, New York, 1959).
4. L. Eckertova, *Physics of Thin Films* (Plenum Press, New York, 1986).
5. B. H. Flowers and E. Mendoza, *Properties of Matter* (Wiley, New York, 1970).
6. D. L. Goodstein, *States of Matter* (Prentice-Hall, Englewood Cliffs, NJ, 1975).
7. M. H. Grabow and G. H. Gilmer, in *Semiconductor Based Heterostructures,* M. L. Green, J. E. E. Baglin, G. Y. Chin, H. W. Deckman, W. Mayo, and D. Narasinham, eds. (The Metallurgical Society, Warrendale, PA, 1986), 3–20.
8. A. Guinier and R. Jullien, *The Solid State* (Oxford University Press, Oxford, 1989).
9. P. Haasen, Physical Metallurgy (Cambridge University Press, Cambridge, 1978).
10. D. Haneman, "Atomic Structures of Surfaces," Chapter 1 in *Surface Physics of Phosphors and Semiconductors,* C. G. Scott and C. E. Reed, eds. (Academic Press, New York, 1975).
11. C. Kittel, *Introduction to Solid State Physics,* 6th ed. (Wiley, New York, 1986).
12. C. Kittel and H. Kroemer, *Thermal Physics,* 2nd ed. (W. H. Freeman, New York, 1980).
13. J. W. Mayer and S. S. Lau, *Electronic Materials Science* (Macmillan, New York, 1989).
14. L. E. Murr, *Interfacial Phenomena in Metals and Alloys* (Addison-Wesley, Reading, MA, 1975).

15. A. B. Pippard, *The Elements of Classical Thermodynamics* (Cambridge University Press, Cambridge, 1966).

16. W. T. Read, *Dislocations in Crystals* (McGraw-Hill, New York, 1953).

17. G. A. Somorjai, *Chemistry in Two Dimensions: Surfaces* (Cornell University Press, Ithaca, NY, 1981).

18. R. A. Swalin, *Thermodynamics of Solids,* 2nd ed. (Wiley, New York, 1972).

19. D. Turnball, in *Impurities and Imperfections* (American Society for Metals, Cleveland, OH, 1955).

20. H. Udin, "Measurement of Solid/Gas and Solid/Liquid Interfacial Energies," in *Metal Interfaces* (American Society for Metals, Cleveland, OH, 1952).

21. J. R. Waldman, *The Theory of Thermodynamics* (Cambridge University Press, Cambridge, 1989).

22. A. Zangwill, *Physics at Surfaces* (Cambridge University Press, Cambridge, 1988).

Problems

2.1 **(a)** Calculate the surface tension γ of Ni ($T_m = 1453°C$) at 300K.
(b) Would Ni deposited on Si or Si deposited on Ni form clusters? In both cases a uniform adherent layer is formed. Suggest an explanation.
(c) Ag ($T_m = 962°C$) wets Cu. Compared to the Sn–Bi solder discussed in the text, how far up a Cu hole ($d = 0.5$ mm) would Ag solder be pulled (contact angle $\theta = 40°$ and density of Ag $= 10.5$ gm/cm^3)?

2.2 Plot the Lennard–Jones potential (Eq 2.3) from $r = 0.8a_0$ to $r = 4a_0$ for Au where $\varepsilon_b = 0.47$ eV/atom.

2.3 For Au fcc metal with 6.0×10^{22} atoms/cm^3 and a (100) surface energy of 0.5 eV/atom, calculate
(a) latent heat of sublimation ΔE_s.
(b) Interatomic potential energy ε_b.

2.4 What is the surface tension of benzene in the liquid state if in a liquid column measurement, the column height $h = 1.2 \times 10^{-2}$ meter, contact angle $\theta =$ 0 degrees, $r = 5 \times 10^{-4}$ m and the benzene density $= 800$ kg/m^3 (neglect the infuence of air and its density). Compare your answer with that of Table 2.2.

2.5 For an fcc metal, determine the ratio $\gamma_{111}/\gamma_{100}$ of the surface energies on the (111) and (100) surfaces by considering the first- (ϕ_1) and second- (ϕ_2) nearest-neighbor bond energies and assuming $\phi_2 = (\frac{1}{4})\phi_1$.

2.6 Two elastically isotropic fcc materials A and B have the properties shown in the table below.

	a (nm)	Y (10^{12} dynes/cm^2)
A	0.566	1.03
B	0.543	1.30

(a) Calculate the surface energies γ assuming $R_F = a$.
(b) Which material wets the other?

2.7 Using the values for γ in Table 2.2, find the following for a drop of liquid mercury (Hg) on solid sodium chloride (NaCl):
(a) The surface tension γ given a contact angle θ of 80°.
(b) Maximum allowable surface tension and minimum allowable contact angle.
(c) Compare answers in (b) with those in (a).

2.8 If you have a liquid droplet on a substrate surface, what will be the range of contact angles in the following:
(a) $\gamma_{LV} > \gamma_{SV} > \gamma_{SL}$?
(b) $\gamma_{LV} > \gamma_{SV} = \gamma_{SL}$?
(c) $\gamma_{LV} = \gamma_{SV} < \gamma_{SL}$?

2.9 For Cu, an fcc structure with density of 8.93 g/cm^3 and atomic mass of 63.55, determine the atomic density n and lattice parameter a. Using the values in Table 2.1, determine the solid–vapor surface tension and heat of sublimation in terms of eV/atom for the (100) surface.

2.10 For zero creep measurements, what is the length of a 0.01 cm radius Au wire, density 19.3 g/cm^3. Use the value of surface energy in Table 2.1 and ignore the influence of grain boundaries.

2.11 **(a)** Consider the "spherical cap" droplet shown below with surface tensions as indicated. Show that the total surface energy of the system can be expressed as:

$$E_T = (\gamma_{SL} - \gamma_{SV})\left[\frac{6V - \pi h^3}{3h}\right] + \gamma_{LV}\left[\frac{2V}{h} + \frac{2\pi h^2}{3}\right] + \gamma_{SV}A$$

In this expression the volume of the droplet, $V = \frac{\pi}{6}(h^3 + 3ha^2)$ or $V = \frac{\pi h^2}{3}(3R - h)$ and A is the total area of the slab. The surface area of such a spherical cap is $S = 2\pi Rh$ and $R = (a^2 + h^2)/2h$.

(b) Show that the surface energy minimization condition, $dE_T/dh = 0$, (holding V and A constant), results in

$$h^3 = \frac{3V}{\pi}\frac{(\gamma_{SL} + \gamma_{LV} - \gamma_{SV})}{(-\gamma_{SL} + 2\gamma_{LV} + \gamma_{SV})}$$

(c) Show that this relationship for h, the height of the droplet, is consistent with the "balance of surface tensions" equation given in the figure: i.e. for $\theta = 0°$, complete wetting, ($h = 0$) and for $\theta = 90°$ hemispherical cluster ($h = R$).

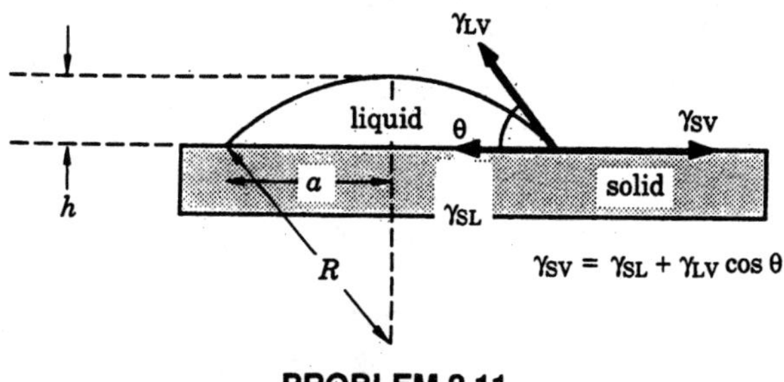

PROBLEM 2.11

CHAPTER 3

Diffusion in Solids

3.0 General Concepts

We introduce solid-state diffusion first with a few examples of macroscopic phenomena, then we give the microscopic picture, and at the end of the chapter we link microscopic and macroscopic aspects.

It is very easy to see the effect of diffusion on a macroscopic scale. If we take a glass of water and put a drop of ink into it, we will see the spread of ink due to diffusion plus a certain amount of convective fluid flow. If a piece of iron is kept in the air over a period of time, it rusts because oxygen diffuses into the iron and forms iron oxide. The oxide has a brown color and has a different density than iron, so it stresses and breaks into pieces. Another example of macroscopic diffusion phenomena in solids is the introduction of electrically active impurities, called dopants, into semiconductors. Pure silicon is not very useful until we diffuse dopants into it. In fact the fundamental behavior of a transistor (i.e., *p-n* junctions in silicon) is due to the non-uniform distribution of two kinds of dopants. Diffusion has been a very important subject in the microelectronics industry. Therefore, we have a keen interest in understanding diffusion in solids.

To relate the macroscopic picture of diffusion to a microscopic description, we first have to understand atomic motion and rearrangement in a crystalline solid where atoms are arranged in an ordered pattern with translational symmetry. In Fig. 3.1 we sketch several atomic diffusion mechanisms in a two-dimensional square lattice or simple cubic solid. If we look at the lower left corner, we see that the atoms are arranged in an ordered structure. At equilibrium, each atom is located in a minimum energy position. To move an atom away from its equilibrium position, we have to do work.

The first concept necessary for understanding atomic diffusion is that a solid always contains defects. The reason is due to entropy. A solid at a finite temperature must have some disorder. In Fig. 3.1, we show the defects known as vacancies and interstitials at (a) and (b), respectively.

A second important concept is that diffusion occurs in a solid by exchange of an atom with a neighboring defect. One of the surrounding atoms can exchange positions with a vacancy for net motion (Fig. 3.1a). This is comparable to driving in a parking lot. When we want to move a car, we must have an empty space next to it.

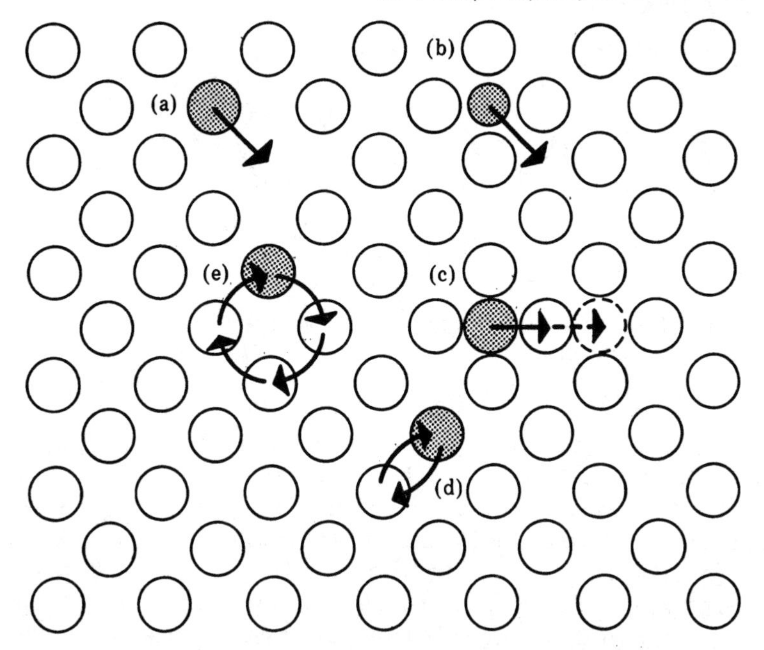

FIGURE 3.1 Sketch of atomic diffusion mechanisms in a two-dimensional square lattice. (a) An atom diffuses by jumping into a neighboring vacant lattice site. (b) An interstitial goes to a neighboring interstitial site. (c) An interstitial pushes an atom from its lattice site to an interstitial site. (d) Two neighboring atoms swap position directly. (e) Ring rotation of four atoms.

If all of the space is occupied, it is difficult to move. However, if we ride a motorcycle we can get through a crowded lot. The motorcycle corresponds to an interstitial (see Fig. 3.1b). Its small size is important: we can't drive a truck through a full parking lot. Interstitial diffusion can occur without exchange with a vacancy, but only for very small atoms in the solid. There are several other mechanisms of atomic diffusion; for example, an impurity-interstitial pair can displace an atom (Fig. 3.1c).

Atoms can also exchange positions with their neighbors (Fig. 3.1e) without involving defects. The atoms make a rotation leading to rearrangement of their positions. Energy calculations show that this requires high energy, so the probability for it to occur is very small, particularly in a close-packed structure. As we show in Appendix C, the concentration of vacancies near the melting point of metals is relatively high (for Al it is 10^{-4}). Therefore we limit our discussion on diffusion in a close-packed crystal structure to defects, particularly to vacancies. In vacancy-mediated diffusion in solids, the jump distance is constant; this is unlike molecular diffusion in a gas phase, where we use the concept of mean free path between collisions.

3.1 Jump Frequency and Diffusional Flux

To quantify the picture of diffusion, we shall take a very simple one-dimensional case, and consider a typical atom in a minimum-energy position in a lattice. We cut

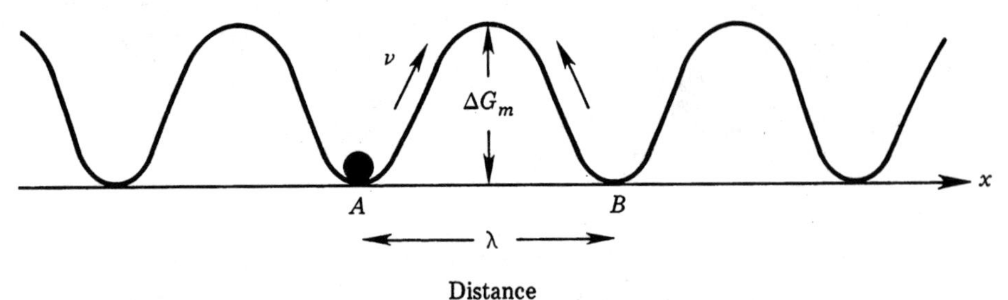

FIGURE 3.2 Potential energy diagram of an atom along a lattice row with barrier energy ΔG_m and jump distance λ.

a line along a row of atoms and draw an ideal potential energy diagram (Fig. 3.2). This represents the potential energy of an atom at any position along the row. We consider a vacancy at B and next to the vacancy an atom A which will exchange positions with the vacancy. We first define an exchange frequency ν at which an atom goes over the energy barrier that we call the activation energy, ΔG_m. In the process, the atom has to do work because, first, the atom is in a minimum-energy position and, second, for it to jump, the surrounding structure has to open up a little bit. Now, let us write the exchange frequency as the product of an attempt frequency and a probability of success:

$$\nu = \nu_0 \exp\left(\frac{-\Delta G_m}{kT}\right) \tag{3.1}$$

where ν_0 is the vibrational frequency of the solid, or the Debye frequency (discussed in Section 3.9). The exponential follows the Boltzmann law of distribution, which indicates that the probability of finding an atom in a given position varies exponentially with the negative of the potential energy of that position divided by kT. We should emphasize that we are going to use Boltzmann's relationship very often. This is the distribution function of classical particles (identical and distinguishable). Here, it is the probability of a successful jump (i.e., the probability of finding an atom in the "activated" position, at the top of the barrier.)

In equilibrium, the base line is level, which means that each atom has the same probability of jumping left or right. As a function of time there is an equal number of atoms going either way, but this condition does not lead to a net flux of diffusion. To transport atoms, we need a net diffusional flux under a driving force (see Fig. 3.3). If we have a driving force F to move an atom the interatomic distance λ, the jump frequency in the forward direction is given by

$$\nu^{+} = \nu \exp\left(+\frac{\lambda F}{2kT}\right)$$

Since the jump distance is half of λ because when the atom has jumped further than $\lambda/2$ it is in the next site, the work done or energy gain (force times distance) is $\lambda F/2$. The jump frequency in the reverse direction is

$$\nu^{-} = \nu \exp\left(-\frac{\lambda F}{2kT}\right)$$

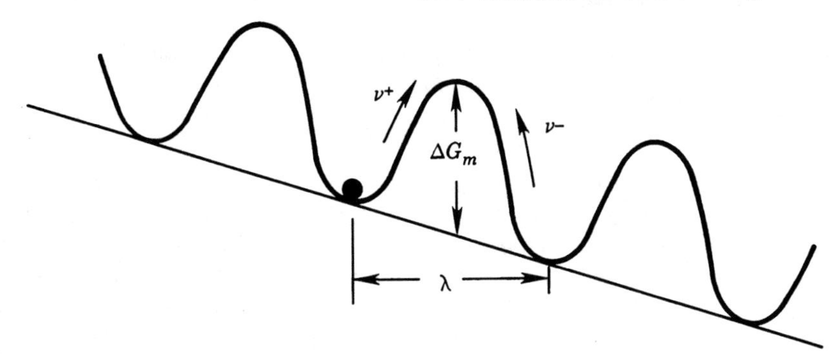

FIGURE 3.3 Potential energy diagram under the influence of a driving force given by the gradient in the chemical potential.

The net frequency ν_n, is equal to the difference of those two,

$$\nu_n = \nu^+ - \nu^- = 2\nu \sinh\left(\frac{\lambda F}{2kT}\right) \tag{3.2}$$

Now, if we assume the term $\lambda F/kT$ is small compared to unity, we can approximate sinh $(\lambda F/2kT)$ by $\lambda F/2kT$, then

$$\nu_n = \nu \frac{\lambda F}{kT} \tag{3.3}$$

The net jump frequency is linearly proportional to the driving force. If we define a velocity v (= frequency times jump distance) and a flux J (the amount of material through a unit area per second), the flux is equal to the velocity times the local concentration C at the unit area of consideration, so that we have

$$v = \lambda\nu_n = \frac{\nu\lambda^2}{kT} F \tag{3.4}$$

and

$$J = Cv = \frac{C\nu\lambda^2}{kT} F \tag{3.5}$$

In chemical diffusion, the flux is in the direction of a high-concentration region to a low-concentration region, that is, the driving force acts to homogenize the concentration gradient.

3.2 Chemical Potential and Driving Force

We now describe the driving force F used in Eq. (3.5). Clearly, we are talking about a chemical force; nevertheless it should be similar in nature to the gravitational force or electrical force. A force can be defined by a field (which is a vector quantity) or a potential (which is a scalar quantity). For example, when a charge q is placed in an electrical potential ϕ, it is acted upon by a force F which is equal to the charge times the negative gradient of the potential:

$$F = -q\nabla\phi$$

In Fig. 3.3 we represent the force acting on the diffusing atom by a gradient in the potential. If the electrical potential is a function of state, independent of the paths going from one potential to another, we can define $-\nabla\phi = \mathscr{E}$ and $\mathscr{E}$ is called the electrical field. So we can also define $\mathbf{F} = q\mathscr{E}$. Another way to look at the potential ϕ is that if a charge q is placed in it, the charge has a potential energy $U = q\phi$. The potential ϕ gives numerically the potential energy of a charge in that potential. The same concept applies to a mass placed in a gravitational field or in a gravitational potential.

We define a chemical potential μ so that we have the force acting on a single atom in the one-dimensional case

$$F = -\frac{\partial\mu}{\partial x} \tag{3.6}$$

In the same sense as when a charge q is placed in an electrical potential ϕ and is acted upon by a force equal to $-q(\partial\phi/\partial x)$, an atom can be placed in a chemical potential μ and acted upon by a force $-\partial\mu/\partial x$. We can obtain the driving force of diffusion from the gradient of μ, and in turn find the velocity and the flux of diffusing atoms. Since the chemical force is typically short range in nature (as discussed in the last chapter), the net atomic flux is the result of a combination of the short-range force and random walk.

For thermodynamic functions and variables of a closed system which has a fixed number of particles (reviewed in Appendix B), the concept of chemical potential is not a factor because the number of particles is not a variable of the energy functions. In diffusion, when we diffuse some solute atoms into a piece of solid, the number of solute atoms inside the solid changes with time. Thus the energy of an open system must include a new independent variable N, the number of particles. We write the change in internal energy dE in terms of change in entropy S, volume V, and number of particles N as

$$dE = T\,dS - p\,dV + \mu\,dN \tag{3.7}$$

where μ is called the chemical potential and its unit is energy per particle. We see that μ and N are a new pair of variables, just like T and S, p and V. Also μ, like T and p, is an intensive property which does not depend on the size of the system. We use $\mu\,dN$ for simplicity. It should be $\Sigma_i\,\mu_i dN_i$ for systems having binary or multiple components.

For the energy function E, we can treat S, V, and N as the three independent variables, so we write

$$dE = \left.\frac{\partial E}{\partial S}\right|_{V,N} dS + \left.\frac{\partial E}{\partial V}\right|_{S,N} dV + \left.\frac{\partial E}{\partial N}\right|_{S,V} dN \tag{3.8}$$

By comparing Eq. (3.8) to Eq. (3.7), we have

$$\mu = \left(\frac{\partial E}{\partial N}\right)_{S,V} \tag{3.9}$$

Similarly, the differential of the other three energy functions described in Appendix B can be obtained:

(a) Enthalpy

$$dH = T\,dS + V\,dp + \mu\,dN$$

(b) Helmholtz free energy

$$dF = -S\,dT - p\,dV + \mu\,dN \tag{3.10}$$

(c) Gibbs function

$$dG = -S\,dT + V\,dp + \mu\,dN$$

(There should be no confusion between the use of F as a driving force in the early part of this chapter and dF as Helmholtz free energy here.) From Eq. (3.10), the chemical potential can be defined as

$$\mu = \left(\frac{\partial E}{\partial N}\right)_{S,V} = \left(\frac{\partial H}{\partial N}\right)_{S,p} = \left(\frac{\partial F}{\partial N}\right)_{T,V} = \left(\frac{\partial G}{\partial N}\right)_{T,p} \tag{3.11}$$

This shows that chemical potential is the partial derivative of energy functions with respect to composition (in other words, it is the slope of the energy function plotted against composition). Physically, it represents the change of energy when changing the composition by one particle or one atom.

If we now consider the process of a binary system at constant temperature and at constant pressure (e.g., to interdiffuse Ag and Au at 400°C and at atmospheric pressure), we have from the last equation in Eq. (3.10) the Gibbs free energy change

$$dG = \mu_1 dN_1 + \mu_2 dN_2 \tag{3.12}$$

where the subscripts 1 and 2 stand for the two components of the system (Ag and Au). Since $N_1 + N_2 =$ constant, we have

$$dG = (\mu_1 - \mu_2)dN_1 \tag{3.13}$$

At $dG = 0$, this implies $\mu_1 = \mu_2$. Hence the equilibrium condition defined by a minimum free energy where $dG = 0$ means that temperature, pressure, and chemical potential are uniform in the system. For alloying Ag and Au, the equilibrium condition is a homogeneous alloy. If the alloy is not homogeneous, μ_1 does not equal μ_2, and there exists a driving force to homogenize it due to the chemical potential difference. The question now is how to calculate the chemical potential in the alloy.

In an alloy of A and B, we replace an A atom by a B atom or vice versa. It is the difference in their potential energies which is of interest, and μ can be defined as the change of potential energy in replacing atoms. This becomes more clear if we compare a homogeneous alloy to an inhomogeneous alloy. A homogeneous alloy is

defined as a system where there is no concentration gradient or no chemical potential gradient. At a finite temperature, atoms diffuse randomly in the alloy, executing a random walk. In the inhomogeneous alloy, there is a concentration or potential gradient, and atoms are acted upon by a force to diffuse in a direction to reach equilibrium or to become homogeneous. The net effect of the diffusion is to replace A atoms by B atoms or vice versa. The same applies to substitutional solute atoms in a solvent.

Let us now consider diffusion in an alloy or in a solid solution. To proceed, we shall use the expression for μ of a dilute ideal solution. We note that to take an ideal solution is physically inconsistent to the consideration of diffusion under a chemical driving force; nevertheless let us proceed for the reason of simplicity and we shall come back to correct it later. For a dilute ideal solution, the chemical potential equals

$$\mu = \mathrm{k}T \ln C \tag{3.14}$$

What is an ideal solution? An ideal solution is defined such that the internal energy is independent of composition. We mix A and B, and whether there is 1% or 2% of B, the internal energy remains the same. In other words, the interatomic potential energies in the pairs of atoms A–A, B–B and A–B are the same. A very simple example of an ideal solution is an element mixed with a small amount of its isotope (for example, cobalt metal and a little bit of cobalt isotope). It is an ideal solution, because there is no chemical interaction between isotope and element and thus no chemical driving force. On the other hand entropy, or disorder, still exists in the system. In Eq. (3.14), $\mathrm{k} \ln C$ is the entropy. Diffusion or mixing of the isotope is an effect of random walk. We will not discuss that subject here. For those who are interested, a good discussion of random walk and Brownian movement has been given in Chapters 6 and 41 of the *Feynman Lectures on Physics,* (Feynman et al., 1963).

3.3 Fick's First Law

We can derive the relationship between the flux J and the driving force F using the chemical potential of a dilute ideal solution,

$$F = -\frac{\partial \mu}{\partial C}\frac{\partial C}{\partial x}$$

$$= -\frac{\mathrm{k}T}{C}\frac{\partial C}{\partial x}$$

Then we have

$$J = \frac{C\nu\lambda^2}{kT} F = \frac{C\nu\lambda^2}{\mathrm{k}T}\left(-\frac{\mathrm{k}T}{C}\frac{\partial C}{\partial x}\right)$$

$$= -\nu\lambda^2 \left(\frac{\partial C}{\partial x}\right)$$

$$= -D\left(\frac{\partial C}{\partial x}\right)$$

We have obtained Fick's First Law of Diffusion,

$$\frac{J}{-\left(\frac{\partial C}{\partial x}\right)} = D = \nu\lambda^2 \text{ cm}^2/\text{sec} \tag{3.15}$$

where D is the diffusion coefficient (or diffusivity) in units of cm^2/sec. Also the velocity can be expressed as (see Eq. 3.4),

$$v = \frac{D}{kT} F = MF \tag{3.16}$$

where M is defined as mobility of the diffusing atom, or $1/M$ is the friction coefficient.

Equation (3.15) is in a macroscopic form, yet it has been derived from a microscopic picture. This is the link between atomic diffusion and macroscopic quantities. We measure flux, the amount of flowing material per unit area per second, and if we know the concentration gradient, we can determine the diffusion coefficient D of an isotope in its element.

Let us now go back to consider chemical diffusion; for a non-ideal solution the chemical potential is not equal to $kT \ln C$, but

$$\mu = kT \ln (C\gamma) \tag{3.17}$$

where γ is the activity coefficient. The activity coefficient represents a chemical interaction in the alloy, or the departure from an ideal solution. A detailed discussion of the subject can be found in the book by Swalin (1972). In an ideal solution we have no chemical effect, but in the real solution there must be some chemical effect between A and B. Atoms can interact to form an alloy or compound or they can repel each other; therefore we must introduce a factor of γ. Repeating the procedure for the "isotope problem" we have

$$\frac{J}{-\left(\frac{\partial C}{\partial x}\right)} = \nu\lambda^2\left(1 + \frac{\partial \ln \gamma}{\partial \ln C}\right) \tag{3.18}$$

We have considered in the above a chemical driving force coming from the gradient of chemical potential, as defined in Eq. (3.6). In general, we may have other driving forces acting simultaneously on atomic diffusion, such as temperature gradient or centrifugal force. If the effects of these forces on atomic jump frequency are linear so that we can apply the principle of superposition, we write

$$J = -D\frac{\partial C}{\partial x} + \sum_i C\, M_i F_i \tag{3.19}$$

where $M_i F_i = \langle v_i \rangle$ and $\langle v_i \rangle$, M_i and F_i are the drift velocities, mobilities and corresponding driving forces, respectively.

3.4 Non-Linear Diffusion

So far we have discussed linear diffusion when the driving force is very small. We have assumed that

$$\frac{\lambda F}{kT} << 1$$

This assumption is true for diffusion in bulk materials, yet it may not be true for diffusion in ultrathin films. There is a unique thin film structure, a man-made superlattice, in which the concentration gradient can be extremely large. In this kind of superlattice, for example, gallium arsenide/aluminum gallium arsenide, each layer can be of the order of 10 angstroms. If we heat the superlattice, the layers will interdiffuse. The above derivation may not apply to diffusion in such a short distance because the gradient is too large, or the driving force is too large. In other words, we shall have a non-linear effect.

We shall discuss the non-linear effect of the activity coefficient based on the diagram in Fig. 3.4. If we consider the diffusion between the two points, x_1 and x_2,

$$\mu_1 = kT \ln C_1\gamma_1$$

$$\mu_2 = kT \ln C_2\gamma_2$$

$$\Delta\mu = \mu_2 - \mu_1 = kT \ln (C_2\gamma_2/C_1\gamma_1).$$

The driving force is

$$F = -\frac{\Delta\mu}{\Delta x} = \frac{kT \ln (C_2\gamma_2/C_1\gamma_1)}{N\lambda} \tag{3.20}$$

noting that

$$\Delta x = x_2 - x_1 = -N\lambda$$

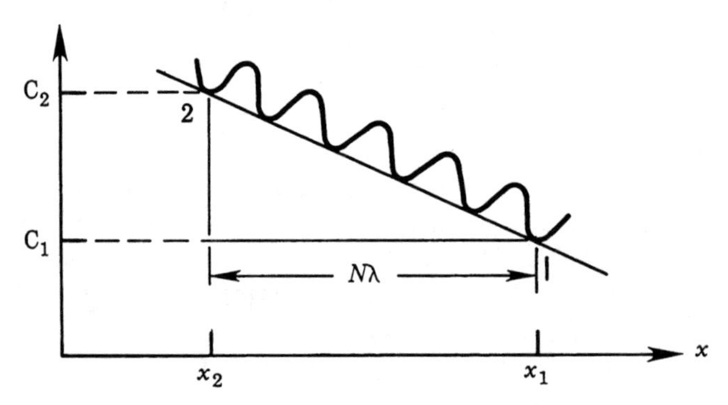

FIGURE 3.4 Potential energy diagram in which there are N lattice spacings λ between x_2 and x_1.

We recall that if $\lambda F/kT$ is very small, we obtain a linear relation. But here we can evaluate how small it is in terms of

$$\frac{\lambda F}{kT} = \frac{\ln (C_2\gamma_2/C_1\gamma_1)}{N} \tag{3.21}$$

No matter how large $C_2\gamma_2/C_1\gamma_1$ is, whether it be a factor of 100 or 1,000, the ln terms all become very small (i.e., $2 \times 2.3 = 4.6$ or $3 \times 2.3 = 6.9$). The ln term is always of the order of 10. The ratio $\lambda F/kT$ is indeed small if N is of the order of 100 or larger. On the other hand, in a superlattice structure, if the diffusion distance is only 10 spacings or 10 atomic layers, the number N equals 10, so that the ratio almost equals one. These ratios are not small and do not satisfy the linear condition that we imposed on our derivation of the classical diffusion equation. So if we consider interdiffusion in a superlattice structure, we cannot use Fick's First Law. In the expansion of the hyperbolic sine expression we have to include higher order terms—at least the second term. This will lead to a third-order term for the flux and in turn a fourth-order diffusion equation. Since

$$\sinh x = x + \frac{x^3}{3!} + \ldots\ldots$$

it is easy to show that if the second term in the expansion of sinh $(\lambda F/2kT)$ is included, we can derive the expression of the net frequency to be

$$\nu_n = \nu \frac{\lambda F}{kT} + \frac{\nu}{24}\left(\frac{\lambda F}{kT}\right)^3 \tag{3.22}$$

This will lead to a fourth-order non-linear diffusion equation.

3.5 Continuity Equation (Fick's Second Law)

We derived a flux equation, Eq. (3.15), under a constant driving force. That equation can describe diffusion phenomena under a constant chemical potential, or constant concentration gradient, or with a constant flux. However, it cannot be used to describe diffusion where the flux varies with position, or where the flux has a changing driving force. A simple example is a drop of ink in water. It spreads out and eventually reaches homogenization; the concentration gradient changes with time and position. To handle such a non-steady-state problem we derive the continuity equation from the principle of conservation of mass.

We start with Cartesian coordinates. Consider the flux going in and out of a cubic element, as shown in Fig. 3.5. If we have the flux represented by a vector, flux vector $\mathbf{J}$, we can decompose the vector into three components, $\mathbf{J}_x$, $\mathbf{J}_y$, and $\mathbf{J}_z$. The amount of material flowing into the cubic box through the surface x_1 per unit time is equal to

$$\mathbf{J}_{x_1} \cdot \mathbf{x}_1 = J_{x_1} x_1 \cos 180° = -J_{x_1}\, dydz$$

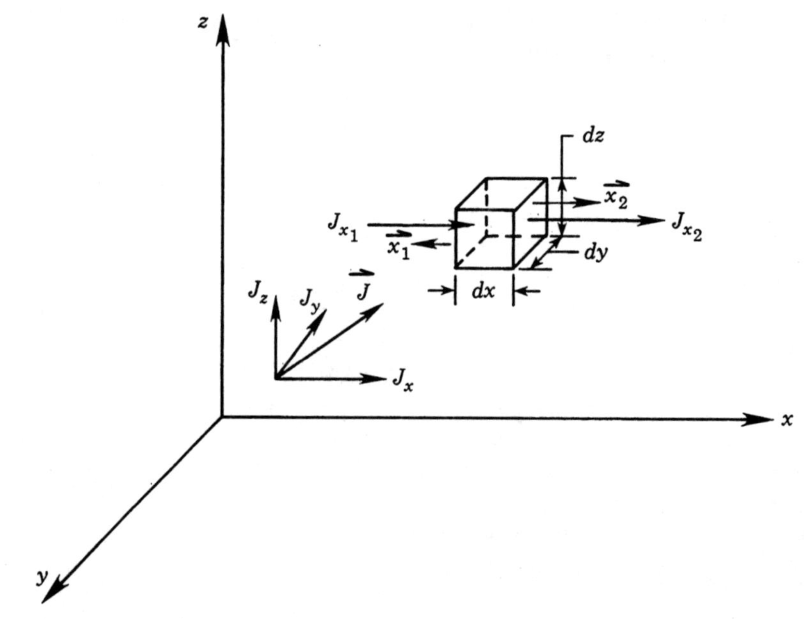

FIGURE 3.5 The flux **J** through a cubic element *dx dy dz* in a Cartesian coordinate system.

Similarly, the amount of material flowing out of the cubic box through the surface x_2 (opposite to x_1) is equal to

$$\mathbf{J}_{x_2} \cdot \mathbf{x}_2 = (J_{x_2} + \frac{\partial J_x}{\partial x} dx) x_2 \cos 0^\circ$$

$$= J_{x_2}\, dy\, dz + \frac{\partial J_x}{\partial x} dx\, dy\, dz$$

If we add these together the net flux out of the box in the x-direction becomes

$$(J_{x_2} - J_{x_1})\, dy\, dz = \left(\frac{\partial J_x}{\partial x}\right) dx\, dy\, dz$$

If we follow this approach, in the y and z directions, we have

$$(J_{y_2} - J_{y_1})\, dx\, dz = \left(\frac{\partial J_y}{\partial y}\right) dx\, dy\, dz$$

$$(J_{z_2} - J_{z_1})\, dx\, dy = \left(\frac{\partial J_z}{\partial z}\right) dx\, dy\, dz$$

If we sum all of these together, we can write the following

$$\sum_{i=1}^{6} J_i A_i = \left(\frac{\partial J_x}{\partial x} + \frac{\partial J_y}{\partial y} + \frac{\partial J_z}{\partial z}\right) dV \tag{3.23}$$

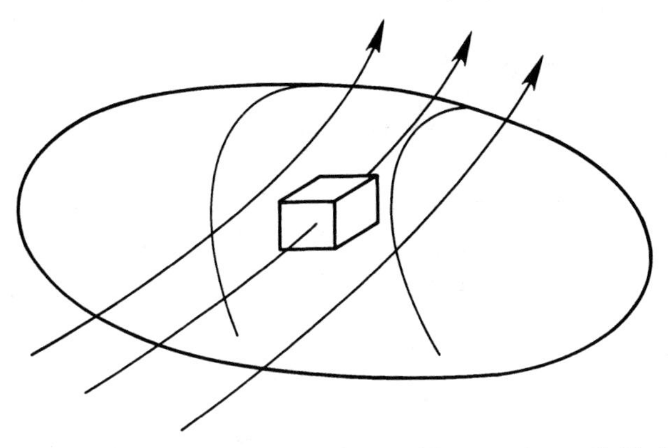

FIGURE 3.6 Schematic of the flux through an arbitrary volume divided into small cubes.

Now, instead of a cube, consider an arbitrary volume. An arbitrary volume can always be cut up into small cubes as shown in Fig. 3.6, and we see that across all the internal surfaces the flux going out is equal to the flux going in, so that they all cancel. The flux we have to consider is the outer surface flux. For an arbitrary volume bounded by area A, the summation can be expressed as an integral,

$$\int_A \mathbf{J} \cdot \mathbf{n} \, dA = (\nabla \cdot \mathbf{J})V \tag{3.24}$$

This is the well-known Gauss theorem, and the right hand side is known as the divergence of the flux, where

$$\nabla \cdot \mathbf{J} = \frac{\partial J_x}{\partial x} + \frac{\partial J_y}{\partial y} + \frac{\partial J_z}{\partial z}$$

Now this quantity, divergence of $\mathbf{J}$ times V, must by mass conservation equal the change of composition inside the volume. It has a negative sign since we derived this quantity under the conditions that the outflux is greater than the influx and that there is no source within the volume (we lose material from this volume). Therefore it must equal to the time rate of change of matter. This is the continuity equation in differential form,

$$-(\nabla \cdot \mathbf{J}) = \frac{\partial C}{\partial t} \tag{3.25}$$

This describes a non-steady-state flux flow. It is a very well-known equation in fluid mechanics and in heat conduction, as well as here in diffusion. Since we have already derived the flux equation of

$$J = -D \frac{\partial C}{\partial x}$$

in the one-dimensional case, then

$$\frac{\partial C}{\partial t} = \frac{\partial}{\partial x} D \left(\frac{\partial C}{\partial x} \right)$$

and if D is independent of position

$$\frac{\partial C}{\partial t} = D \frac{\partial^2 C}{\partial x^2} \tag{3.26}$$

This is Fick's second law of diffusion.

3.6 A Solution of the Diffusion Equation

As mentioned in Section 3.5, an example of three-dimensional diffusion is the case of a drop of ink spreading out in water. A two-dimensional example would be a drop of gasoline that spreads out on the surface of water. Another example of two-dimensional diffusion is grain-boundary diffusion, where atoms diffuse on grain boundaries rather than in the bulk of a material.

Here we consider a one-dimensional problem using Eq. (3.26). We take a long rod of a pure metal and place a small amount of its isotope, a tracer, on the end

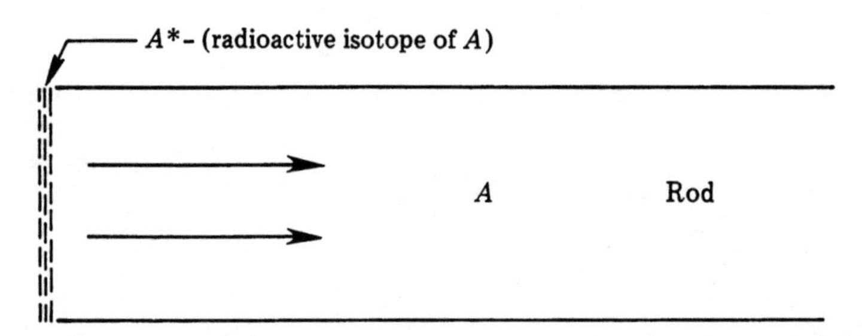

FIGURE 3.7a Diffusion of an isotopic tracer into a rod of a pure metal.

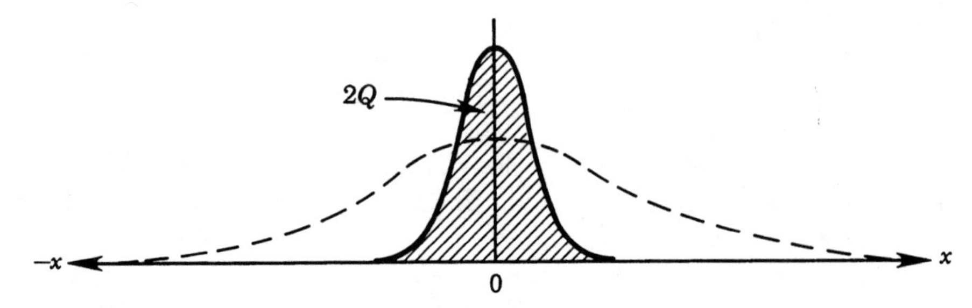

FIGURE 3.7b The composition curve for the one-dimensional symmetrical diffusion case with a total amount of material $2Q$.

surface, as shown in Fig. 3.7a. We wish to determine how the isotope diffuses into the rod as a function of time and temperature. The standard method of setting up the problem is to put a reflecting barrier at the end surface, and thus change the problem into a symmetrical one as shown in Fig. 3.7b. The solution is then a standard one,

$$C(x, t) = \frac{Q}{(\pi Dt)^{1/2}} \exp\left(\frac{-x^2}{4Dt}\right) \tag{3.27}$$

The constant Q satisfies the boundary condition that

$$\int_0^\infty C(x)\, dx = Q$$

for a fixed amount of material. The initial conditions of the problem are that

$$\text{at } x = 0,\ C \rightarrow Q \text{ as } t \rightarrow 0$$
$$\text{for } |x| > 0,\ C \rightarrow 0 \text{ as } t \rightarrow 0$$

Two important values are

$$C(0, t) = C_0 = \frac{Q}{(\pi Dt)^{1/2}} \quad \text{at } x = 0 \tag{3.28}$$

$$C(\lambda_D, t) = C_\lambda = C_0/\mathrm{e} \quad \text{at } x = \lambda_D = (4Dt)^{1/2} \tag{3.29}$$

That is, the position where the ratio of local concentration to concentration at the source point is $1/e$ always occurs where $x^2 = 4Dt$. Here we present only one solution, but other solutions using different boundary conditions always involve x^2 proportional to Dt. This proportionality is one of the most important relationships of diffusion. If a kinetic process is controlled by diffusion it must obey this relationship, $x^2 \sim Dt$. Based on this equation we can measure diffusivity, for example, from the concentration profile shown in Fig. 3.8. Since we know the time of diffusion, we plot ln C versus x^2, and the slope equals $1/(4Dt)$ according to Eq. (3.27), as shown in Fig. 3.9.

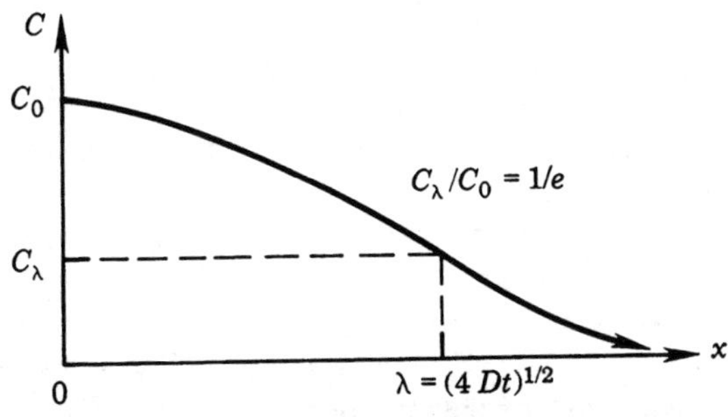

FIGURE 3.8 Composition as a function of distance where $C_\lambda/C_0 = 1/e$ at $\lambda^2 = 4Dt$.

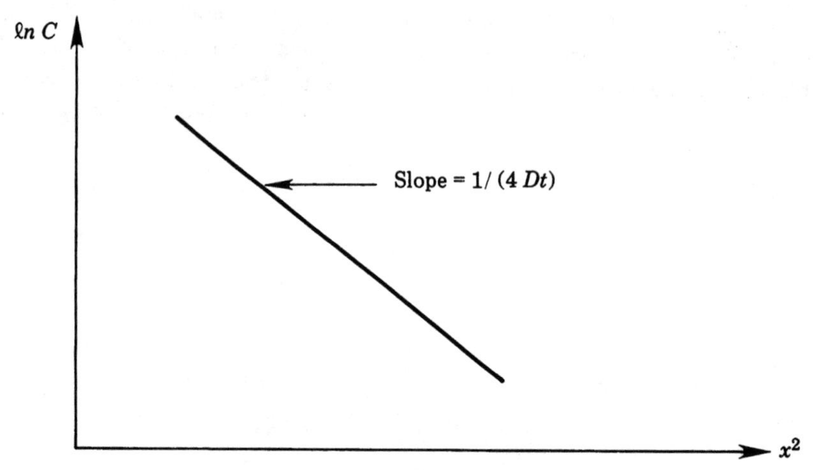

FIGURE 3.9 The ln C versus x^2 plot for diffusion relationships following equation (3.27). The slope equals $1/(4Dt)$.

3.7 Diffusion Coefficient

In general, it is found that the measured diffusion coefficient can be expressed as

$$D = D_0 \exp\left(-\frac{\Delta H}{kT}\right) \tag{3.30}$$

Again, this is a Boltzmann distribution function. We can separate the coefficient into two parts: the pre-exponential factor D_0 and the exponential term $\Delta H/kT$ where

1. D_0 depends on temperature only slightly
2. ΔH does not depend on temperature

Previously we derived a simple flux equation (3.15), where we saw that

$$D = \lambda^2 \nu \tag{3.31}$$

which gives us the dimension of cm^2/sec, using the relationship of

$$x^2 = 4Dt \tag{3.32}$$

We derived Eq. (3.31), by assuming that an atom exchanges position with a neighboring vacancy, and that the exchange is characterized by a jump frequency and an interatomic jump distance. We assumed that there was a vacancy next to the atom, but this is just not true in most cases. There are a very small number of vacancies in a solid. Usually there are no vacancies around an atom such that it could diffuse. The probability that a vacancy is available to any particular atom is based on the Boltzmann statistical distribution. The probability of the existence of a vacancy in a solid is equal to the number n of atoms per cm^3 times an exponential,

$$n_v = n \exp\left(-\frac{\Delta G_f}{kT}\right) \tag{3.33}$$

The energy ΔG_f is the formation energy of the vacancy. This energy is the energy change when we take an atom from inside the solid and put it on the surface. The number of atoms in the solid is conserved while we create a vacancy. The energy state of this aggregate as compared to the one without the vacancy is the change in potential energy, ΔG_f. From the Boltzmann distribution function, the ratio

$$\frac{n_v}{n} = \exp\left(-\frac{\Delta G_f}{kT}\right)$$

is the probability of finding a vacancy somewhere in the solid. A statistical evaluation of the defect concentration is given in Appendix C.

In a face-centered cubic (fcc) solid each atom has twelve nearest neighbors, so that there are twelve positions where a vacancy can exist next to an atom. Therefore any arbitrary atom has a coordination factor n_c (with $n_c = 12$ for an fcc solid and $n_c = 8$ for a body-centered cubic (bcc) solid), and we have

$$n_c \frac{n_v}{n} = n_c \exp\left(-\frac{\Delta G_f}{kT}\right) \tag{3.34}$$

We modify the diffusion coefficient D, which was derived assuming a vacancy already exists, by the above probability function.

$$D = n_c \lambda^2 \nu \exp\left(-\frac{\Delta G_f}{kT}\right) \tag{3.35}$$

However, to make a jump to a neighboring vacancy requires an energy ΔG_m, the motion energy, as shown in Fig. 3.2. If we substitute the relation in Eq. (3.1) for ν in Eq. (3.35), we have

$$D = n_c \lambda^2 \nu_0 \exp\left(-\frac{\Delta G_f + \Delta G_m}{kT}\right) \tag{3.36}$$

Finally, we still have to multiply D by one more factor, the correlation factor f.

$$D = f n_c \lambda^2 \nu_0 \exp\left(-\frac{\Delta G_f + \Delta G_m}{kT}\right) \tag{3.37a}$$

The correlation factor means that if a particular atom exchanges positions with a neighboring vacancy, its most likely next jump is to exchange back (see Fig. 3.10). In a solid the atom and the vacancy often jump back and forth before separating further. If correlation is very strong, they will always exchange back, so that no net diffusion occurs. In fcc solids it is found that the correlation factor is about 0.78 (that is, about 20% exchange back, while 80% keep jumping randomly). We define the correlation factor f to be the fraction of jumps which lead to random walk, or (1-f) to be the fraction of correlated walk. When f is close to zero, there is a very strong correlation effect and no net flux of diffusion, and when f is about unity, the correlation effect is weak and permits random walk.

Since the Gibbs free energy ΔG can always be separated into the enthalpy part ΔH and the entropy part ΔS, we observe that at constant temperature,

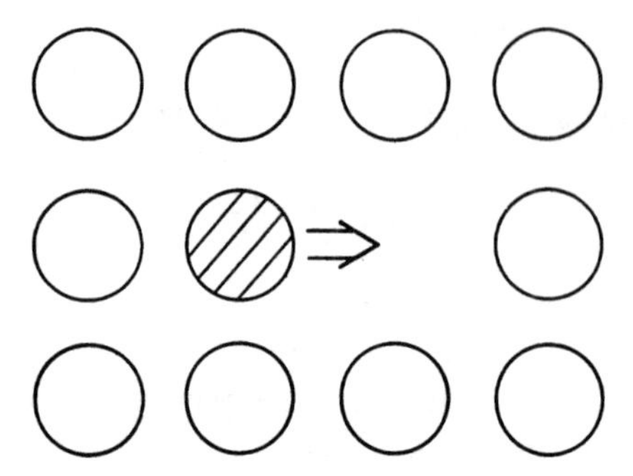

FIGURE 3.10 Two-dimensional diagram showing an atom exchanging positions with a neighboring vacancy.

$$\Delta G = \Delta H - T\,\Delta S$$

$$\Delta G_m + \Delta G_f = (\Delta H_m + \Delta H_f) - T(\Delta S_m + \Delta S_f)$$

Therefore we can express the diffusion coefficient as

$$D = fn_c\lambda^2\nu_0 \exp\left(\frac{\Delta S_m + \Delta S_f}{\mathrm{k}}\right) \exp\left(-\frac{\Delta H_m + \Delta H_f}{kT}\right) \tag{3.37b}$$

or

$$D = D_0 \exp\,(-\Delta H/\mathrm{k}T)$$

which is the form of D presented at the beginning of this section.

We have connected the continuum mechanics approach to the atomistic picture of diffusion. We started with an atomic jump, derived the flux and diffusion equations, and then came to diffusivity. From diffusivity we returned to the atomic picture.

3.8 Calculation of Diffusion Coefficient

Table 3.1 gives the self-diffusion coefficient D_0 (in $\mathrm{cm^2/sec}$) for some elements. The activation energies ΔH, ΔH_m, and ΔH_f are given in units of eV/atom.

We calculate the self-diffusivity of *Al* ($D_0 = 0.047\ \mathrm{cm^2/sec}$, $\Delta H = 1.28$ eV/atom) at 140°C. When *Al* is used as an electrical conductor on Si devices, Joule heating during device operation raises the temperature, and 140°C has often been chosen as the upper limit for reliability reasons.

$$D = D_0 e^{-\Delta H/\mathrm{k}T} \cong 0.047 \times 10^{-\left[\frac{1.28 \times 5000}{(140 + 273)}\right]}$$

$$\cong 0.15 \times 10^{-16}\ \mathrm{cm^2/sec}$$

Knowing D at a given temperature, we can easily estimate the diffusion distance for a given time by using the relation of $x^2 = 4Dt$. In the above case of A*l*, if we take the time to be 1 day (= 86,400 sec $\cong 10^5$ sec), we have

$$x = 2(0.15 \times 10^{-16} \times 10^5)^{1/2} \simeq 2.4 \times 10^{-6}\ \mathrm{cm}$$

TABLE 3.1 Lattice Self Diffusion in Some Important Elements*

Element	D_0(cm²/sec.)	ΔH(eV)	ΔH_f(eV)	ΔH_m(eV)	$\Delta S/k$
FCC					
Al	0.047	1.28	0.67	0.62	2.2
Ag	0.04	1.76	1.13	0.66	—
Au	0.04	1.76	0.95	0.83	1.0
Cu	0.16	2.07	1.28	0.71	1.5
Ni	0.92	2.88	1.58	1.27	—
Pb	1.37	1.13	0.54	0.54	1.6
Pd	0.21	2.76	—	—	—
Pt	0.33	2.96	—	1.45	—
BCC					
Cr	970	4.51	—	—	—
α-Fe	0.49	2.95	—	0.68	—
Na	0.004	0.365	0.39/0.42	—	—
β-Ti	0.0036	1.35	—	—	—
V	0.014	2.93	—	—	—
W	1.88	6.08	3.6	1.8	—
β-Zr	0.000085	1.2	—	—	—
HCP					
Co	0.83	2.94	—	—	—
α-Hf	0.86/0.28	3.84/3.62	—	—	—
Mg	1.0/1.5	1.4/1.41	0.79/0.89	—	—
α-Ti	0.000066	1.75	—	—	—
Diamond Lattice					
Ge	32	3.1	2.4	0.2	10
Si	1460	5.02	~3.9	~0.4	

*From Chapter I of *Diffusion Phenomena in Thin Films and Microelectronic Materials* (Gupta and Ho, 1988) (Courtesy of D. Gupta).

which means that by lattice diffusion, A*l* atoms can diffuse a distance of about 24 nm at 140°C in a day.

We shall make a rough estimate of the upper bound and lower bound of diffusivity in solids. From Table 3.1, we estimate the diffusivity of the fcc metals at their melting points to be around 10^{-8} cm²/sec. This value is smaller than the diffusivity found in liquid or molten metals which is about 10^{-5} cm²/sec. As a lower bound if we take x to be an atomic distance of ~0.1 nm and the time to be 10 days, we have $x^2/t = 10^{-22}$ cm²/sec. Any diffusivity of this order of magnitude is not of practical interest and is difficult to measure. Using superlattice structures or layer removal techniques for concentration profiling, we can measure diffusivity around 10^{-19} to 10^{-21} cm²/sec. In an intermediate range of diffusivities, we find interdiffusion distances of 10^{-6} to 10^{-5} cm in thin film reactions in times of 1000 sec, giving values of $x^2/t = 10^{-15}$ to 10^{-13} cm²/sec.

At room temperature, most metals and semiconductors have a diffusivity smaller than 10^{-22} cm²/sec, except the low-melting-point metals such as Pb, Sn and Bi.

Therefore most metals and semiconductors are quite stable in regard to lattice diffusion.

Substitutional dopants in Si have diffusivities close to that of self-diffusion in Si. In Fig. 3.11, we show a plot of diffusivity versus temperature for dopants in Si. To produce a p-n junction in Si by dopant diffusion, the diffusion temperature must be close to 1000°C. There is a question about the mechanism of such diffusion. The measured diffusivity could be a combination of two mechanisms (e.g., two kinds of defects such as vacancies and divacancies may coexist in the sample). Whether self-diffusion in Si is by vacancies exclusively or may occur by interstitials or direct exchanges is also unclear. However, Fig. 3.11 shows that metals such as Cu and Li have diffusivities several orders of magnitude higher than that of Si self-diffusion. They diffuse interstitially in Si. For bcc metals, the activation energy of diffusion

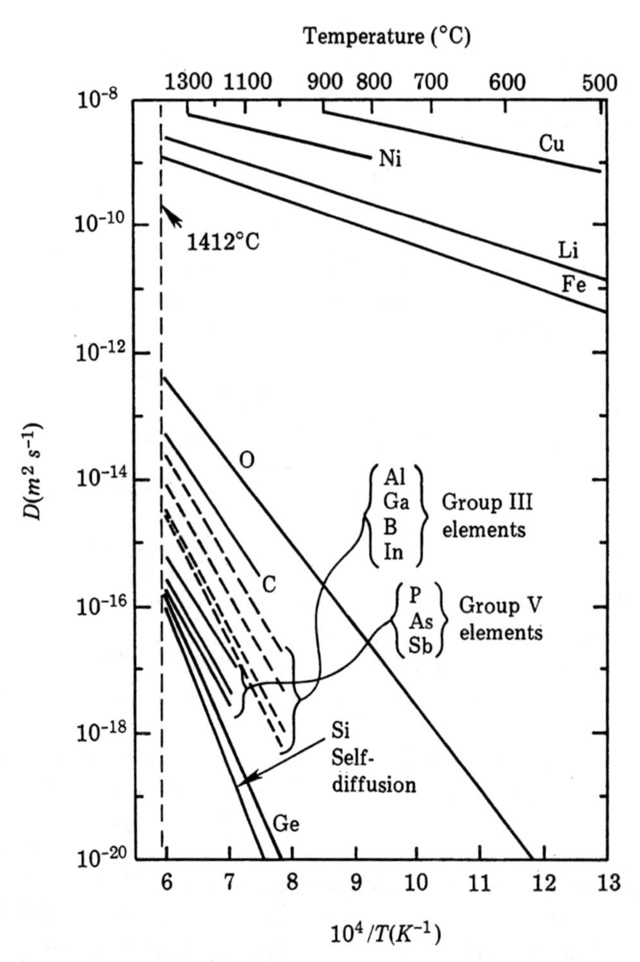

FIGURE 3.11 Survey of the diffusivities of foreign atoms in silicon (see W. Frank, U. Gösele, H. Mehrer, and A. Seeger in "Diffusion in Silicon and Germanium" in *Diffusion in Crystalline Solids,* edited by G. Murch and A. S. Nowick, Academic Press, Orlando, FL, 1984).

shows a small dependence on temperature. It is a subject to be resolved and will not be discussed here.

In the Simmons-Balluffi experiment (see Section 3.10) the lattice parameter expansion and the change of sample dimension due to the defect generation at high temperatures were determined simultaneously. It was concluded that in fcc metals such as Al, vacancies are the dominant point defects which mediate diffusion, and that the concentration of vacancies near the melting point is about 10^{-4}. Next, we shall consider the parameters in the diffusion coefficient.

3.9 Atomic Vibrational Frequency

We have derived an expression for the diffusion coefficient in Eq. (3.30),

$$D = D_0 \exp\left(-\frac{\Delta H}{kT}\right)$$

with the pre-exponential factor,

$$D_0 = f n_c \lambda^2 \nu_0 \exp\left(\frac{\Delta S_m + \Delta S_f}{k}\right)$$

and the activation energy

$$\Delta H = \Delta H_m + \Delta H_f$$

Among all the parameters in the pre-exponential factor and the activation energy, the atomic vibrational frequency ν_0 is a fundamental parameter of the solid. We present here an order of magnitude calculation of ν_0. It serves to provide a simple physical picture of atomic vibration in the solid and an estimate of the magnitude of the diffusion coefficient.

In Fig. 3.12a, we sketch a single surface atom which is bonded to a solid by a potential ϕ. The potential function is given by, for example, the one shown in Fig. 2.1a. We assume that the atom undergoes harmonic motion. It is a simplified model, yet it gives a vibrational frequency of the same order of magnitude as that of atoms in the solid. We approximate the bottom portion of ϕ (i.e., around $\phi(a_0)$), by a parabolic potential and move the point $(-\varepsilon_b, a_0)$ to the origin of the coordinates as shown in Fig. 3.12b. For small amplitude of vibration, this approximation is reasonble, and we write

$$\phi = \frac{1}{2} kr^2 \tag{3.38}$$

Thus

$$F = -\frac{\partial \phi}{\partial r} = -kr \tag{3.39}$$

where k is the force (or spring) constant. It describes a simple harmonic motion; the equation of motion is

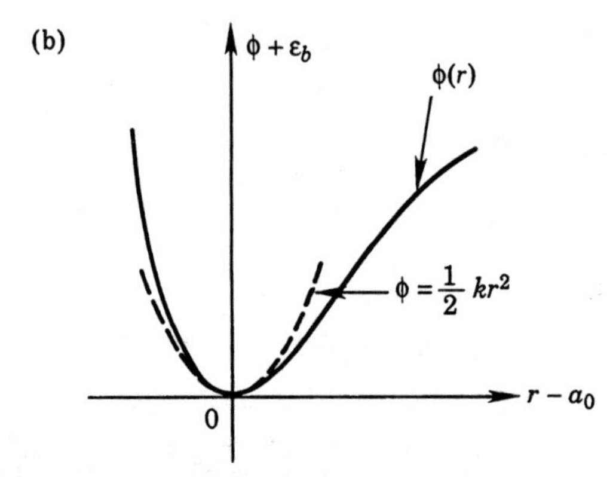

FIGURE 3.12 (a) A surface atom undergoes a simple harmonic vibration. (b) The interatomic potential $\phi(r)$ of the surface atoms is plotted on the coordinates of $r - a_0$ and $\phi(r) + \varepsilon_b$. Near the origin, $\phi(r)$ can be approximated by $\phi = kr^2/2$.

$$m \frac{d^2r}{dt^2} = -kr \tag{3.40}$$

where m is the mass of the atom. Eq. (3.40) has a solution in the simplest form,

$$r = \cos \omega t$$

where $\omega = \sqrt{k/m}$ is the "phase" of the motion. Since we know that the function cosine has a period of 2π, the time needed to complete a period (or a cycle) of motion is

$$\omega t = 2\pi \qquad \text{or} \qquad t = \frac{2\pi}{\omega}$$

Hence, the frequency of vibration

$$\nu_0 = \frac{1}{t} = \frac{\omega}{2\pi} = \frac{1}{2\pi}\sqrt{\frac{k}{m}} \tag{3.41}$$

We can calculate ν_0 if we know k and m. The latter is given by knowing the atomic weight of the solid and Avogadro's number. To determine k, we take the second derivative of ϕ in Eq. (3.38) and we have

$$\frac{\partial^2 \phi}{\partial r^2} = +k \tag{3.42}$$

In Chapter 2, we have given Eq. (2.3) for a solid which obeys the Lennard–Jones potential, from which we obtain

$$\frac{\partial^2 \phi}{\partial r^2} = \frac{12\varepsilon_b}{a_0^2}\left(\frac{a_0}{r}\right)^8\left[13\left(\frac{a_0}{r}\right)^6 - 7\right]$$

Then

$$k = \frac{72\varepsilon_b}{a_0^2} \text{ at } r = a_0 \tag{3.43}$$

Substituting k into Eq. (3.41), we have

$$\nu_0 = \frac{3}{\pi}\sqrt{\frac{2\varepsilon_b}{ma_0^2}} \tag{3.44}$$

Now, to calculate ν_0, we consider the atoms shown in Fig. 3.12a to be gold atoms, then

$$\begin{aligned} m &= \frac{197 \text{ g/mole}}{6.02 \times 10^{23} \text{ atoms/mole}} \\ &= 32.8 \times 10^{-23} \text{ g/atom} \end{aligned}$$

The interatomic distance a_0 in face-centered cubic Au is given by $a_0/\sqrt{2} = 0.288$ nm, so that $a_0^2 = 8.3 \times 10^{-16}$ cm^2 and

$$\begin{aligned} ma_0^2 &= 2.72 \times 10^{-37} \text{ gm-cm}^2\text{/atom} \\ &= 2.72 \times 10^{-37} \text{ dyne-cm-sec}^2\text{/atom} \\ &= 2.72 \times 10^{-37} \text{ erg-sec}^2\text{/atom} \end{aligned}$$

where we have used the conversion 1 dyne $=$ 1 gm-cm/sec^2. From Table 2.1, the interatomic potential energy of Au is

$$\begin{aligned} \varepsilon_b &= 0.47 \text{ eV/atom} \\ &= 0.47 \times 1.6 \times 10^{-12} \text{ erg/atom} \end{aligned}$$

We obtain

$$\nu_0 = \frac{3}{\pi}\sqrt{\frac{2\varepsilon_b}{ma_0^2}} = 2.24 \times 10^{12} \text{ sec}^{-1} \tag{3.45}$$

If we extend the above simple calculation to an atom within the fcc lattice where it has twelve nearest neighbors, the force constant has to be multiplied by a factor of 6 for vibration along a close-packed direction. The factor of 6 comes in because we use the principle of superposition and sum the projections ($\cos\theta$) of the interatomic force of all twelve atoms. In turn, we have to multiply ν_0 by $\sqrt{6}$, and we have

$$\nu_0 = \frac{6}{\pi}\sqrt{\frac{3\varepsilon_b}{ma_0^2}} \tag{3.46}$$

This is called the Einstein frequency. For a Au atom within its lattice, we have $\nu_0 = 5.5 \times 10^{12}$ cycles/sec.

The formal treatment of elastic vibrations (phonons) in a finite piece of solid has been given by Debye, and the subject is covered in textbooks of solid state physics. The Debye frequency ν_D is defined by

$$h\nu_D = kT_\theta \tag{3.47}$$

where h is Planck's constant ($h = 6.626 \times 10^{-27}$ erg-sec) and T_θ is the Debye temperature at which all the $3N$ modes of elastic waves are operative. For metal Au, $T_\theta = 165$ K which is given, for example, in Table 4.1 of *Thermal Physics* by Kittel and Kroemer. Hence

$$\nu_D = \frac{kT_\theta}{h} = \frac{1.38 \times 10^{-16}\ \text{erg-K}^{-1} \times 165\ \text{K}}{6.626 \times 10^{-27}\ \text{erg-sec}}$$
$$= 3.42 \times 10^{12}\ \text{sec}^{-1}$$

which is not far from the frequencies we have calculated. Since the Debye temperatures of common metals vary only by a factor of 2 to 3 (e.g., $T_\theta = 428$ K for Al), the atomic vibrational frequency of metals from the viewpoint of diffusion is typically taken to be 10^{13} cycles/sec.

3.10 Activation Enthalpy

As we have shown in Eq. (3.37), the activation enthalpy of vacancy diffusion consists of two components,

$$\Delta H = \Delta H_f + \Delta H_m$$

Since knowing the activation energies is important in understanding the mechanism of diffusion and in identifying the type of defects which mediates the diffusion, their measurement has been a key activity in studying diffusion. The values ΔH can be determined by measuring D at several temperatures and by plotting $\ln D$ versus $1/kT$; from the slope of the straight-line plot we obtain ΔH. Experimental techniques of thermal expansion, quenching plus resistivity measurement, and positron annihilation have been used to determine ΔH_f. A quenching technique has also been used to measure ΔH_m. These techniques are well covered in textbooks and reference books on diffusion. Here we shall only discuss briefly the measurement of ΔH_f by thermal expansion.

The concentration of vacancies in a solid is an equilibrium quantity. The concentration n_v/n increases with temperature as given by Eq. (3.33),

$$\frac{n_v}{n} = \exp\left(-\frac{\Delta G_f}{kT}\right) = \exp\left(\frac{\Delta S_f}{k}\right)\exp\left(-\frac{\Delta H_f}{kT}\right)$$

where n_v and n are respectively the number of vacancies and atoms in the solid. As temperature increases, more vacancies are formed by removing atoms from the interior to the surface of the solid. Consequently, the volume of the solid increases. Provided that we can decouple this volume increase from that due to thermal expansion, we can measure the vacancy concentration. Thermal expansion can be determined by measuring the lattice parameter change Δa as a function of temperature using x-ray diffraction, and we can express the fractional change by $\Delta a/a$. Similarly, for the volume change, we can use a wire of length L and measure the length change ΔL as a function of temperature. We then have

$$\frac{\Delta n_v}{n} = 3\left(\frac{\Delta L}{L} - \frac{\Delta a}{a}\right) \tag{3.48}$$

The factor of 3 comes in because both ΔL and Δa are linear changes. Fig. 3.13 shows $\Delta L/L$ and $\Delta a/a$ of Al wire as a function of temperature up to the melting

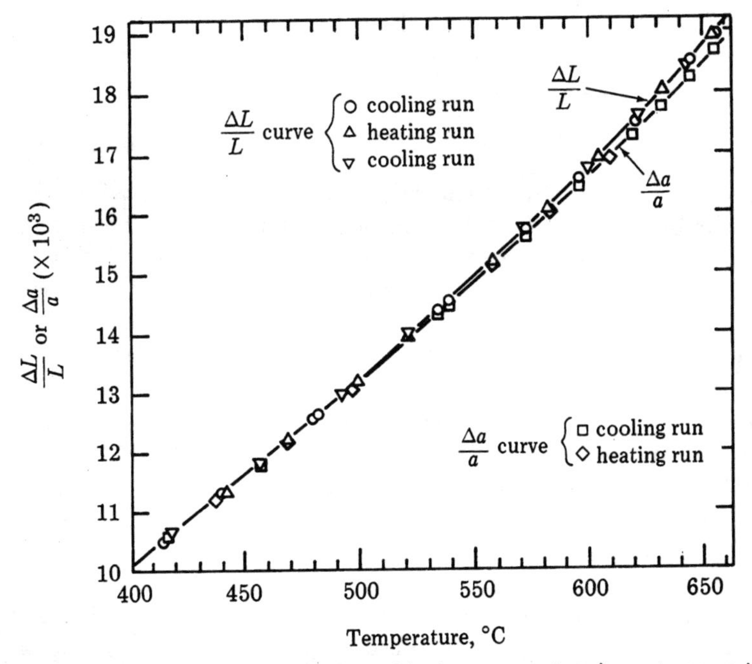

FIGURE 3.13 Values of length change and lattice parameter change versus temperature for aluminum, taking $\Delta L/L$ and $\Delta a/a$ equal to zero at 20°C. The difference between the two lines is directly proportional to the concentration of vacant atomic sites (From R. Simmons and R. Balluffi, *Physical Review,* 117 (1960): 52).

point, as measured by Simmons and Balluffi. The results show that near the melting point, the vacancy concentration $n_v/n = 10^{-4}$. Since $(\Delta L/L - \Delta a/a)$ is positive, the defect is predominantly vacancies. For interstitials, the difference is expected to be negative. In the derivation, we have ignored the effect of compensation between a vacancy and an interstitial, and also the effect of divacancies. From the two curves shown in Fig. 3.13,

$$\frac{\Delta n_v}{n} = \exp(2.4) \exp(-0.76 \text{ eV}/kT) \tag{3.49}$$

so we have $\Delta H_f = 0.76$ eV and $\Delta S_f/k = 2.4$ for Al. These values are in good agreement with those listed in Table 3.1.

3.11 The Pre-exponential Factor

Using $\nu_0 \cong 10^{13}$ Hz, we can estimate the entropy factor by measuring the pre-exponential factor of diffusion D_0, provided that we accept the theoretical value of the correlation factor f (= 0.78 for face-centered cubic metals). This is shown by the form of Eq. (3.37b) presented near the end of Section 3.7:

$$\exp\left(\frac{\Delta S}{\text{k}}\right) = \exp\left(\frac{\Delta S_f + \Delta S_m}{\text{k}}\right) = \frac{D_0}{f\nu_0\lambda^2 n_c}$$

Since ΔS has to be positive, we have $D_0 > f\nu_0\lambda^2 n_c$. For Au,

$$D_0 > 0.78 \times 3.42 \times 10^{12} \times (2.88 \times 10^{-8})^2 \times 12 = 0.027 \text{ cm}^2/\text{sec}.$$

In general, D_0 for self-diffusion in metals is of the order of 0.1 to 1 cm^2/sec, so the entropy change per atom is of the order of unity times k. In textbooks on diffusion, ΔS is sometimes given in terms of R, the gas constant (instead of Boltzmann's constant k), when the activation enthalpy is given in Kcal/mole or KJ/mole rather than eV/atom.

We consider the theoretical calculation of ΔS (given by Zener) for interstitial diffusion such as carbon in iron. In such a case, diffusion requires only the activation energy of motion because in Fe the interstitial sites neighboring an interstitial C solute atom are always available. To calculate the entropy change in this case, for a constant pressure process, we have

$$\Delta S_m = -\frac{\partial \Delta G_m}{\partial T} \tag{3.50}$$

Zener reasoned that the Gibbs free energy change during the diffusion of a carbon atom is essentially the strain energy needed to push out the Fe atoms in order to open up a passage wide enough for the interstitial carbon atom to pass to a neighboring interstitial site. We shall consider the diffusion process shown schematically in Fig. 3.14a, where two unit cells of body-centered cubic Fe contain an interstitial carbon atom at the center position. The carbon atom is surrounded by six Fe atoms

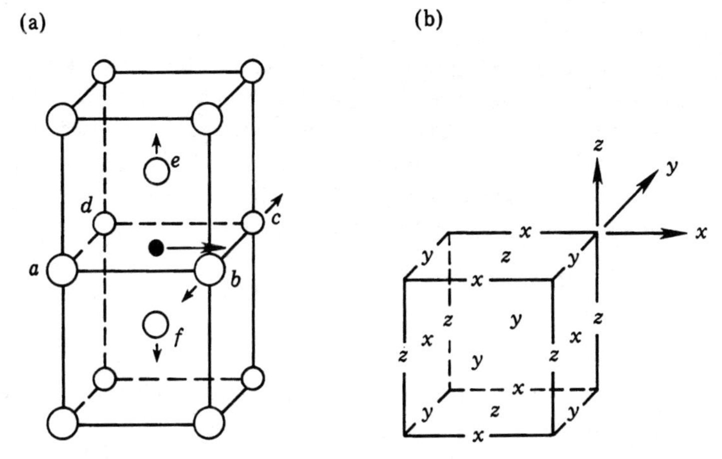

FIGURE 3.14 (a) Two body-centered cubic unit cells of Fe containing an interstitial carbon atom in the center of four Fe atoms labelled a, b, c, and d. The carbon atom has caused a displacement of the other two neighboring Fe atoms labelled e and f. When the carbon atom jumps to the interstitial site between b and c, they move apart as shown by the pair of short arrows. (b) The equivalent interstitial sites on the bcc lattice are shown by the labels of *x, y,* and *z*.

labeled a, b, c, d, e, and f, where the pair e-f is along the vertical direction, and there is a strain in this direction because of the carbon atom.

The lattice parameter of the body-centered cubic (bcc) unit cell of Fe is $a = 0.2866$ nm. The distance of closest approach between two Fe atoms is 0.2481 nm; we regard this distance as the diameter of an iron atom. The distance of closest approach between two carbon atoms in the basal plane of graphite is 0.142 nm, and we shall take this value to be the diameter of a carbon atom. On the bcc structure shown in Fig. 3.14a, the interstitial carbon atom fits comfortably with the four neighboring Fe atoms labelled a, b, c and d. This is because the distance between a and c, or between b and d, is $0.2866 \times \sqrt{2} = 0.4052$ nm, which is slightly greater than the sum of the diameters of an iron atom and a carbon atom (which is 0.3901 nm). On the other hand, the other two neighboring Fe atoms e and f must be stretched out (as indicated by the pair of short arrows) by an amount of about $0.3901 - 0.2866 = 0.1035$ nm (i.e., by about $0.1035/0.2866 = 36\%$). It is reasonable to expect that the distance between the carbon atom and the two iron atoms can be compressed a bit closer because of hybridization, but we note that in iron carbide, Fe_3C, the closest interatomic distance between Fe and C is 0.39 nm. Therefore the strain in the e-f pair is of the order of 36%.

To consider the diffusion of the interstitial carbon atom to another interstitial site, we first note that the equivalent interstitial sites in the bcc structure are located in the face-centered and edge-centered positions in the lattice; they are indicated in Fig. 3.14b. They are equivalent because, except for a rotation of 90°, they have similar surrounding Fe atoms. That is, the interstitial site between b and c is the same as the center one, except that when the centered carbon atom jumps to the site between b and c, the stretch (as indicated by the pair of short arrows) is now along the b-c direction (y-direction) rather than the e-f direction (z-direction). For such a diffusion jump, we see that before and after the jump, the e-f and b-c pairs are strained respectively. Yet, during the diffusion, both must be strained; this is because without stretching the b-c pair, the carbon atom cannot jump into the interstitial positions between b and c. So an activation energy of diffusion is needed.

If we assume the process is elastic, we can regard the activation energy as the strain energy involved,

$$\Delta G_m = -\frac{1}{2} K\varepsilon^2 \tag{3.51}$$

where K is the bulk modulus and ε is the volume strain. Since the strain can be regarded as independent of temperature, we have

$$\Delta S_m = \frac{1}{2} \varepsilon^2 \frac{\partial K}{\partial T} \tag{3.52}$$

In Chapter 2, we have shown that surface energy and Young's modulus are closely related as illustrated by Table 2.2 Both are proportional to the interatomic potential energy. This relation is also true for the binding energy and the bulk modulus K. Specifically, if a solid obeys a Lennard–Jones potential, we have shown that

$$K = \frac{8\Delta E_s}{V}$$

where ΔE_s is the latent heat of sublimation and V ($=N_A\Omega$, where Ω is the atomic volume and N_A is Avogadro's number) is the molar volume. Then,

$$\begin{aligned} \Delta S_m &= \frac{4\varepsilon^2}{V} \frac{\partial \Delta E_s}{\partial T} \\ &= \frac{4\varepsilon^2}{V} c_v = \frac{4\varepsilon^2(3N_A k)}{N_A \Omega} \\ &= 12\varepsilon^2 \mathrm{k}/\Omega \end{aligned} \tag{3.53}$$

where c_v is the heat capacity at constant volume and we take $c_v = 3N_A\mathrm{k}$. For interstitial carbon in iron, the strain occurs along one direction, so the volume strain can be approximated by the linear strain without a factor of 3. Then if we take ε = 0.36, we have

$$\Delta S_m = 1.6\mathrm{k/atom}$$

which is of the right order of magnitude even though $\varepsilon = 0.36$ is unreasonably large.

Empirically, we can approximate Eq. (3.50) by

$$\Delta S_m = \beta \frac{\Delta H}{T_m} \tag{3.54}$$

where ΔH is the activation energy of diffusion, T_m is the melting point, and β is a proportionality constant. This relationship holds well for interstitial diffusion of carbon, nitrogen, and oxygen in bcc transition metals.

3.12 Surface Diffusion

The process of surface diffusion refers to the migration of atoms or molecules across a surface at a finite temperature. Surface diffusion is an important mechanism in thin film growth and epitaxy and is discussed in detail in Chapter 5. At this point however it is useful to compare surface diffusion to bulk diffusion, which has been in the main subject of this chapter. Surface diffusion not only applies to thin film growth but has relevance to grain boundary diffusion, a significant source of material intermixing in thin films.

The picture of surface diffusion is nicely represented by Fig. 3.2 where the x-axis is a surface coordinate containing a periodic array of "potential wells" for atoms to reside. The description of surface diffusion has two important characteristics that distinguish it from bulk diffusion: (1) the mathematics is mostly two-dimensional and (2) surface diffusion does not require the existence of a nearby vacancy as in the case of substitutional diffusion through a solid.

Two-dimensionality is relatively easily incorporated via the two-dimensional diffusion equation, although mathematical difficulties can arise. Artificial boundary conditions may be necessary since some solutions may not converge as the distance parameter approaches infinity. This issue arises directly in the solution of the surface cluster growth problem discussed in Chapter 5. Dimensionality also affects the migration distance in a fundamental way. In n-dimensions the root mean square (rms) diffusion depends on the square root of the dimensionality. That is if $\langle x^2 \rangle = 4Dt$ for the one-dimensional case, then $\langle R^2 \rangle = \langle x^2 \rangle + \langle y^2 \rangle = 8Dt$ for a two-dimensional isotropic solid. The rms distance for two dimensions is then $\sqrt{2/3}$ smaller than the rms distance for three dimensions, for the same time and assuming an isotropic diffusion coefficient.

By far the most significant quantitative difference between surface and bulk diffusion is the aspect of vacancy formation, required for the bulk case and not required in the surface case. An inspection of Table 3.1 shows that the vacancy formation energy is at least comparable to the migration energy for many solids. For self-diffusion in Ge and Si the vacancy formation energy is the dominant term. Since both the formation energy, ΔH_f, and migration energy, ΔH_m, enter the exponential, the difference in diffusion coefficient (with and without ΔH_f) can be enormous. For example the ratio of the exponential factors in Si (Table 3.1), with and without the vacancy formation term is: $(e^{0.4\text{eV}/kT}/e^{4.3\text{eV}/kT}) \simeq 10^{24}$ at T = 550°C.

The pre-exponential factor (Eq. 3.31) for surface diffusion is relatively simple since the statistics of vacancy formation are not involved. Thus an estimate of the pre-exponential factor, $D = \lambda^2\nu$ with $\lambda \simeq 10^{-8}$ cm and $\nu = 10^{13}$/sec. yields $\lambda^2\nu = 10^{-3}$ cm^2/sec. This value is close (within a factor of 10) to the pre-exponential factor of many measured surface diffusion coefficients. An order-of-magnitude estimate for the ratio of the surface diffusion coefficient to the bulk diffusion coefficient for the Si/Si system is $\sim 10^{18}$ at T = 550°C. More graphically $\sqrt{4D_{surf}t} = 1.1 \times 10^{-2}$ cm and $\sqrt{4D_{bulk}t} = 1.1 \times 10^{-11}$ cm for T = 550°C, using Table 3.1 for the bulk diffusion coefficient and $D_{surf} = 10^{-3} e^{-0.4/kT}$ cm^2/sec. Thus there is extensive surface diffusion at 550°C while the bulk diffusion is essentially turned-off at this temperature, i.e. $\sqrt{4D_{bulk}t}$ is less than an atom spacing.

The large value of the surface diffusion coefficient relative to the bulk diffusion coefficient permits the formation of structures with sharp interfaces. There is sufficient surface mobility for good film formation yet bulk diffusion (intermixing) is negligible. On the other hand thin metallic films are often deposited in the form of small crystallites (microcrystalline) with "grain boundaries" between the crystallites. This relatively fast interdiffusion between two polycrystalline metal layers can be a genuine difficulty in preserving multilayer metal structures used in Si device formation.

References

1. R. J. Borg and G. J. Dienes, *An Introduction to Solid State Diffusion* (Academic Press, Boston, MA, 1988).
2. H. S. Carslaw and J. C. Jaeger, *Conduction of Heat in Solids*, 2nd ed. (Clarendon Press, Oxford, 1980).
3. J. Crank, *Mathematics of Diffusion* (Oxford University Press, Fair Lawn, NJ, 1956).
4. R. P. Feynman, R. B. Leighton, and M. Sands, *The Feynman Lectures on Physics,* Vol. I (Addison-Wesley, Reading, MA, 1963).
5. S. Glasstone, K. J. Laidler, and H. Eyring, *The Theory of Rate Processes* (McGraw-Hill, New York, 1941).
6. D. Gupta and P. S. Ho, eds., *Diffusion Phenomena in Thin Films and Microelectronic Materials* (Noyes Publications, Park Ridge, NJ, 1988).
7. C. Kittel and H. Kroemer, *Thermal Physics* (Wiley, New York, 1970).
8. J. R. Manning, *Diffusion Kinetics for Atoms in Crystals* (Van Nostrand, Princeton, NJ, 1968).
9. J. W. Mayer and S. S. Lau, *Electronic Materials Science* (Macmillan, New York, 1990).
10. P. G. Shewmon, *Diffusion in Solids,* 2nd ed. (The Minerals, Metals, and Materials Society, Warrendale, PA, 1989).
11. J. C. Slater, *Introduction to Chemical Physics* (McGraw-Hill, New York, 1939).
12. R. A. Swalin, *Thermodynamics of Solids,* 2nd ed. (Wiley, New York, 1972).

Problems

3.1 Using the data in Table 3.1 for copper (Cu) and Aluminum (Al),
(a) Calculate D_{Cu} and D_{Al} at 600K.
(b) At the melting temperature, T_m, calculate D_{Cu} (T_m = 1083°C) and D_{Al} (T_m = 660°C).
(c) In comparing diffusion coefficients, which is the more appropriate scaling factor, T or T_m?

3.2 A sample is diffused at 1100°C for 20 min. with a total amount of radiotracer of 2×10^{15} atoms/cm^2. The diffusion length λ_D is 10^{-4} cm.
(a) What is the diffusion coefficient?
(b) What is the surface concentration C_0?
(c) Calculate the concentration at $0.3\lambda_D$ and $0.4\lambda_D$ and estimate the flux J.

3.3 Using the data in Table 3.1, calculate the fraction of vacancies at 0.75 T_m for Al (T_m = 660°C) and for Ge (T_m = 937°C),
(a) Assuming the entropy term is negligible.
(b) Including the entropy term.
(c) Compare the ratio of vacancy fractions and the ratio of self-diffusion coefficients for the two materials.

3.4 Calculate D_0 for an fcc lattice, a = 0.4 nm, and compare with values in Table 3.1.

3.5 Based on the values in the table below, calculate the Einstein frequency and the Debye frequency for copper (Cu) and silver (Ag) and compare these values with those given in the text for gold (Au).

	$n(10^{22}/\text{cm}^3)$	Atomic Weight	T_θ(°K)	ε_b(eV/atom)
Cu	8.45	63.5	343	0.58
Ag	5.85	107.9	225	0.65
Au	5.9	197	165	0.47

3.6 For a diffusion of 10^{15} radiotracer Cu atoms into Cu at 800°C to a diffusion length λ_D of 10^{-5} cm, calculate using data in Table 3.1
(a) The diffusion coefficient D and time t.
(b) At $0.5\lambda_D$, the force F, the mobility M, and the velocity v.

3.7 It has been proposed that ancient civilizations may be dated by the diffusion of water vapor into their carvings made of obsidian, a volcanic glass. An obsidian arrowhead is discovered atop a mountain in central Africa where they find the average temperature to be 40°F. The obsidian is tested and it is found that the activation energy for diffusion of water vapor into it is 1.25 eV, and

$D_0 = 50\ cm^2/sec$. Upon measuring the arrowhead's hydrated layer, the archaeologists decide that the civilization is 2,000,000 years old. However, the archaeologists are unaware that recent deforestation in the area has produced a smog layer that blocks out the sun's warmth. Prior to the existence of the smog layer the average temperature on the mountain was 53.5°F. How old is the civilization really?

3.8 A film of gold (198 isotope) is applied to one side of a gold disc. The assembly is raised to 900°C and the isotope begins to diffuse into the disc. After 1 hour the assembly is quenched. The concentration of the isotope at a depth of 10 microns into the disc is found to be 4.0×10^{-5} atom fraction. At an 80-micron depth the isotope concentration is 2.3×10^{-6} atom fraction. **(a)** What is the diffusivity? **(b)** Given an activation energy of diffusion of 1.84 eV, what is the pre-exponential factor D_0?

3.9 With the solution given by Eq. (3.27) and its boundary condition, evaluate

$$\bar{x} = \frac{\int_0^\infty x\, C(x)\, dx}{\int_0^\infty C(x)\, dx} = \frac{1}{\sqrt{\pi}} \sqrt{4Dt}$$

and explain the meaning of $\bar{x}$.

3.10 ^{57}Fe is a γ-ray absorbing nucleus used in Mössbauer spectroscopy. It has a natural abundance of $\simeq 2.3$ atomic percent. Backscatter Mössbauer can be used to examine the ^{57}Fe environment of a surface to a depth of $\simeq 100$ nm. To give an enhanced signal, a monolayer of ^{57}Fe was sputter-deposited on the (100) surface of an Fe sample. Fe has a bcc unit cell parameter of $\simeq 0.38$ nm. Calculate ^{57}Fe concentration at the surface and at 5 nm after 100 hours of heating at 450°C in an inert atmosphere. (Fe density $= 7.86\ g/cm^3$)

CHAPTER 4

Stress in Thin Films

4.0 Introduction: The Theoretical Strength

Thin films are seldom used in electronic devices as structural parts to carry mechanical loads. Nevertheless, stress or strain does commonly exist in thin films as a result of constraints imposed by their substrates. A thin film and its substrate generally have different thermal expansion coefficients, so stress is produced during temperature changes occurring in deposition and annealing. Stress in thin films is known to cause yield and reliability problems in microelectronic devices. In some optoelectronic devices the stress can affect the actual device properties. In this chapter, we shall discuss the nature of stress in thin films, the chemical potential in a stressed solid, and the time-dependent response of a solid to applied stresses.

A piece of solid is under stress when its atoms are displaced from their equilibrium positions by a force. The displacement is governed by the interatomic potential. It is well known that the potential ϕ and the internal force F ($F = -\partial\phi/\partial r$) between two atoms as a function of interatomic distance generally obey the schematic relations shown in Fig. 2.1a and Fig. 2.1b. When we consider instead an applied external force, we define

$$F_{ex} = +\frac{\partial\phi}{\partial r} \tag{4.1}$$

where we have changed the sign to be positive compared to the internal force between two atoms. An external tensile force tends to lengthen the solid and in turn to increase the interatomic distance. On the basis of the sign convention given in Chapter 2, Section 2.2, a force which increases the interatomic distance is positive, and hence the tensile force (or stress) is positive. An external compressive force (or stress) which tends to shorten the solid is negative. The interatomic potential, the external force, and the sign of the force are shown schematically in Fig. 4.1a to 4.1c, respectively.

Clearly, Fig. 4.1b is an inverted diagram of Fig. 2.1b. We define the point F_{max} to be the maximum force which corresponds to the dissociative distance r_D. F_{max} is the maximum tensile force needed to pull the solid apart, because the force needed to increase the interatomic distance beyond r_D is less than F_{max}. We can regard F_{max} as the theoretical strength of the solid. To calculate F_{max}, we require that

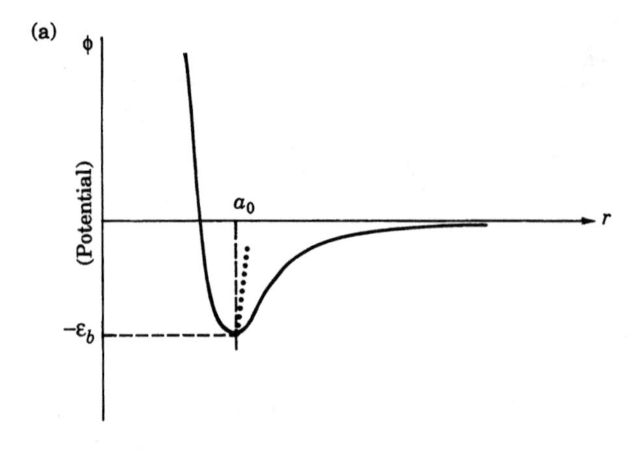

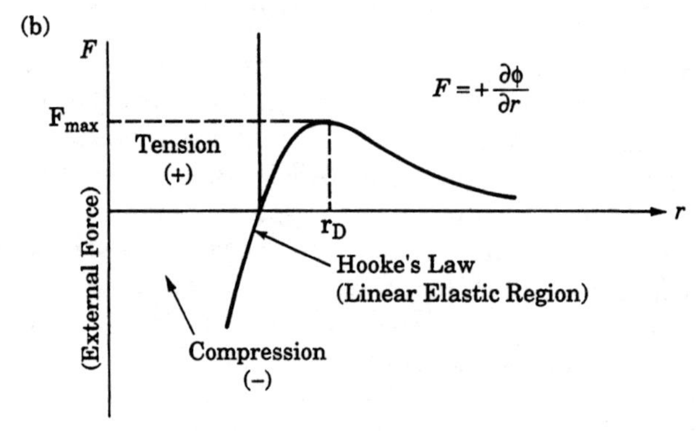

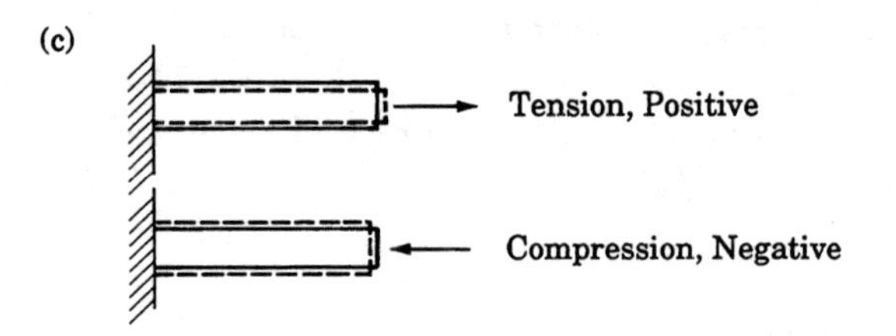

FIGURE 4.1 (a) Interatomic potential function plotted against interatomic distance. The dotted curve shows the anharmonicity of atomic vibration. (b) Applied force plotted against atomic displacement. (c) The direction and sign of applied force by convention.

$\partial^2\phi/\partial r^2 = 0$ at $r = r_D$. If we assume that the solid obeys the Lennard–Jones potential and that the function ϕ is given by Eq. (2.3), we obtain its second derivative with respect to r,

$$\frac{\partial^2\phi}{\partial r^2} = \varepsilon_b \frac{12}{a_0^2}\left(\frac{a_0}{r}\right)^8\left[13\left(\frac{a_0}{r}\right)^6 - 7\right] = 0 \qquad \text{at } r = r_D \tag{4.2}$$

where a_0 is the equilibrium interatomic distance. The solution of Eq. (4.2) shows that:

$$r_D = 1.11a_0$$

Theoretically the solid can be stretched (strained) by about 11% before it breaks! Furthermore, if stretched just below that strain, it would return to the original condition when the external force is removed. Experimentally, these observations are not true at all. Most polycrystalline metals, whether they obey the Lennard–Jones potential or not, have an elastic limit of only 0.2%; beyond that, plastic deformation sets in. We consider elastic behavior in 4.1.

At the equilibrium position a_0, the external force is zero, and the potential corresponds to the minimum potential energy ε_b between the atoms. At a small displacement in either direction from a_0, the force is linearly proportional to the displacement. This is the origin of the elastic behavior in a solid aggregate of atoms under stress. The elastic behavior observed is described by Hooke's law. Within the elastic region, the displacement disappears when the force is removed. Beyond the elastic limit, permanent deformation occurs. In permanent damage, a structural ductile solid such as steel deforms by dislocation motion, but a brittle solid such as glass will deform by fracture via crack propagation. The major difference is due to the nature of chemical bonds and crystal structure in these solids.

4.1 Elastic Stress–Strain Relationship

Consider a piece of thin solid film of dimensions $l \times W \times t$ as shown in Fig. 4.2. If we apply a force F to the area $A = Wt$ to stretch the film length l by Δl, we have

$$\frac{F}{A} = Y\frac{\Delta l}{l} \quad \text{or} \quad \sigma = Y\varepsilon \tag{4.3}$$

where $\sigma = F/A$ and $\varepsilon = \Delta l/l$ are stress and strain, respectively, and Y is Young's modulus. This is Hooke's law. In addition

$$\frac{\Delta t}{t} = \frac{\Delta W}{W} = -\nu\frac{\Delta l}{l} \tag{4.4}$$

where ν is Poisson's ratio. This ratio is a positive number for almost all materials and is less than one-half. Notice that there is a negative sign before ν which means

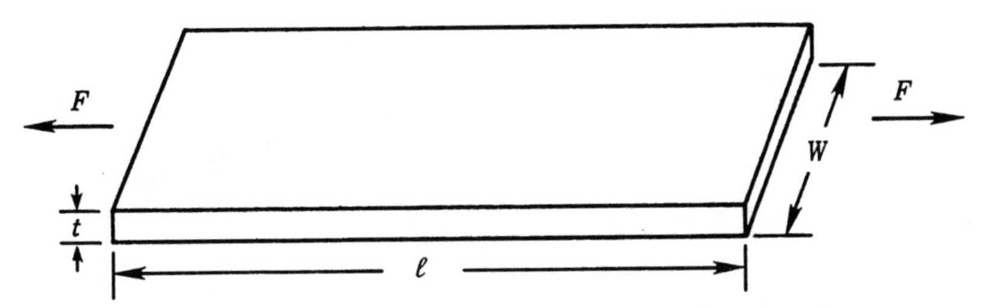

FIGURE 4.2 A piece of thin film of dimensions $l \times W \times t$ under tension.

that while we stretch l, both W and t shrink. An easy way to measure Poisson's ratio is to observe the change in lattice parameter by x-ray diffraction in the direction normal to the tensile stress. For example, take a single-crystal film of cubic crystal structure and stretch it by bending as shown in Fig. 4.3. The cubic unit cell deforms into a tetragonal cell (dashed rectangle). The interplanar spacing normal to the substrate surface decreases. This decrease is measured by the shift in the x-ray diffraction angle.

Different materials have different values of Y and ν. The elastic behavior of a polycrystalline material is characterized by just these two parameters. Sometimes other parameters such as shear modulus and bulk modulus are given, but they are interrelated. An example is given in the following discussion.

We consider shear strain and illustrate it with the example shown in Fig. 4.4a. It is a schematic diagram of the cross-sectional view of a Si chip joined by two solder joints to a ceramic pad in the flip-chip packaging scheme. During operation, the device will experience a temperature rise of ~100°C. Since Si expands more than the ceramic, the solder joints experience a shear strain. The strain is actually cyclic because the device is being turned on and off frequently. The cyclic strain has been found to cause reliability failure of the solder joints. Furthermore, we can imagine that if we increase the chip size, the solder joints at the edges of the chip will feel a greater shear strain. Therefore, we cannot increase the chip size arbitrarily. Shear strain is a critical factor limiting the yield of the device and also the size of the Si chip.

By shear strain, we mean (Fig. 4.4b) that the square block is deformed by a pair of forces in such a way that the bottom side is held down to prevent rotation. At equilibrium the net force and torque are zero. The shear strain θ is defined by

$$\theta = \frac{\delta}{l} \tag{4.5}$$

where δ is the shear displacement of material of length l (Fig. 4.4b). To relate the strain to the shear stress, we translate the shear stress to a combination of tensile and compressive stresses on the rectangular block as shown in Fig. 4.4c. The tensile force is equal to $\sqrt{2}\,S$, yet the length and hence the area is also increased by a factor of $\sqrt{2}$. So, the tensile stress is S/A, which is the same magnitude as the shear stress. The compressive stress is the same except for the sign. Consider the stress–strain

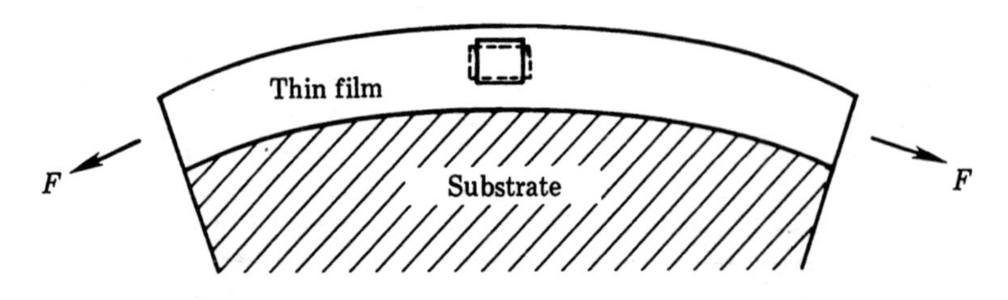

FIGURE 4.3 A thin film is stretched by bending the substrate.

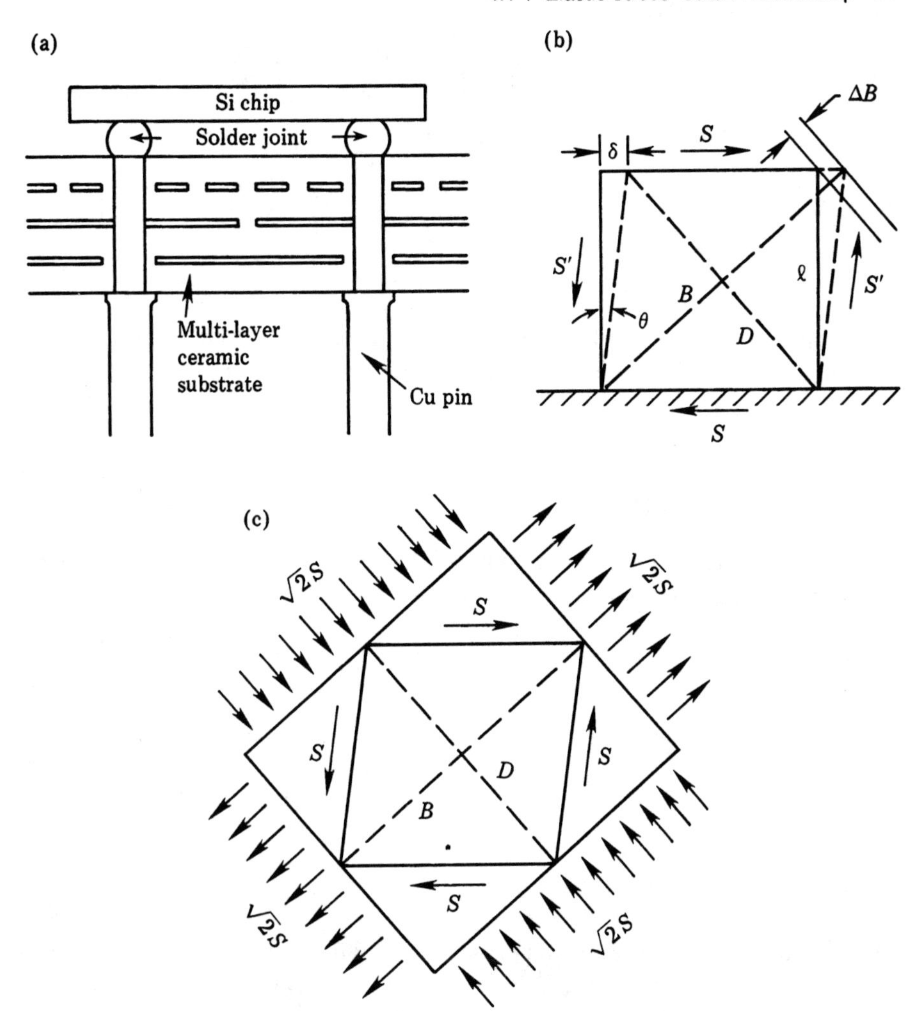

FIGURE 4.4 (a) A schematic diagram of a Si chip solder-joined to a ceramic substrate. (b) A square is sheared by a pair of shear forces S. (c) The shear stress is converted to a combination of tensile and compressive stresses.

relation along the diagonal of the rectangular block in Fig. 4.4c; we have

$$\frac{\Delta B}{B} = \frac{1}{Y}\frac{S}{A} + \frac{-\nu}{Y}\left(\frac{-S}{A}\right) = \frac{1+\nu}{Y}\frac{S}{A} \tag{4.6}$$

On the other hand, we have defined

$$\theta = \frac{\delta}{l} = \frac{\sqrt{2}\,\Delta B}{l} = \frac{2\Delta B}{B}$$

Then

$$\theta = 2\left(\frac{1+\nu}{Y}\right)s \tag{4.7}$$

where $s = S/A$ is the shear stress. If we define the shear modulus as

$$\mu = \frac{s}{\theta}$$

then

$$\mu = \frac{Y}{2(1 + \nu)} \tag{4.8}$$

Shear stress can be regarded as a pair of normal stresses, and vice versa.

The relationship between the elastic constants for single crystals is given in Appendix E.

4.2 Strain Energy

It is interesting to estimate the magnitude of the energy involved in elastic strain. Consider the case at the elastic limit. The elastic energy is given by

$$E_{\text{elastic}} = \int \sigma \, d\varepsilon = \frac{1}{2} Y \varepsilon^2 \tag{4.9}$$

To estimate the elastic energy, we take values of Young's Modulus from Table 2.4 in Chapter 2, or choose one of the stiffest materials, steel, with $Y = 2.0 \times 10^{12}$ dyne/cm^2 and 8.4×10^{22} atoms in 1 cm^3. We take $\varepsilon = 0.2\%$, then

$$E_{\text{elastic}} = \frac{1}{2} Y \varepsilon^2 = 4 \times 10^6 \frac{\text{dyne}}{\text{cm}^2} \cong 3 \times 10^{-5} \frac{\text{eV}}{\text{atom}}$$

The value of elastic energy obtained is three to four orders of magnitude smaller than the typical chemical energy, say the formation energy of silicide, which is about 0.5 eV/atom. Therefore, in chemical reactions such as compound formation, the effect of elastic strain energy or stress effect is negligible. This is the reason that in measuring interdiffusion coefficients during silicide formation, the part of the driving force due to stress is ignored. We can also conclude that elastic energies are small from Fig. 4.1a. The potential energy corresponding to a strain of 0.2% is still very close to the binding energy.

Strain energy, although small, is important in cases where solids are near equilibrium. At equilibrium the forces are balanced, so any small additional force will be able to tilt the balance; it affects the equilibrium. Strain energy due to epitaxial misfit can stabilize metastable phases. Furthermore, in an epitaxial structure where the dislocation slip system is nonoperative or the nucleation of dislocations is difficult, the elastic limit can be greatly extended (for instance, up to a few percent), so that the strain energy can be two orders of magnitude greater. But when a solid is far away from equilibrium, the strain energy tends to be unimportant in most kinetic processes.

4.3 The Origin of Stress in Thin Films

There are intrinsic and extrinsic stresses in a thin film. Intrinsic stress comes from defects such as dislocations in the film. The origin of extrinsic stress in a thin film comes mainly from adhesion to its substrate. Stress can be introduced in a thin film due to differential thermal expansion between the film and its substrate, due to lattice misfit with its substrate, or due to chemical reaction with its substrate when the intermetallic compound formed is coherent to the film but has a slight lattice misfit. It has also been suggested that in thin film grain growth, the removal of grain boundaries—and hence the reduction of the excess volume in the grain boundaries—will induce stress in the film when it is constrained by the substrate.

To illustrate the stress induced by differential thermal expansion, we shall consider a Pb thin film on a Si substrate. The coefficients of thermal expansion of Pb and Si are $29.5 \times 10^{-6}/°C$ and $2.6 \times 10^{-6}/°C$, respectively. We deposit Pb on Si at room temperature and then cool them to 4.2 K where the Pb becomes superconducting. The sample experiences a temperature drop of about 300 K, and the net change in linear dimension is 0.86% for Pb. While the Pb tries to shrink, the Si substrate restricts it from doing so; hence in cooling, the Pb is under tension. The tensile stress will be relaxed to some extent because of the yielding of the Pb film. Upon heating the sample back to room temperature, the Pb tends to expand and again it is restricted by the Si substrate. The Pb is under compression upon heating. Since room temperature is about half the melting point of Pb, atomic diffusion is substantial. The Pb film will release its compressive stress partly by atomic diffusion, hence hillock formation occurs. A picture of a Pb hillock is shown in Fig. 15.4 in Chapter 15. This is a well-known phenomenon in Josephson junction devices where Pb has been used as electrodes and has experienced the temperature cycling between room temperature and 4.2 K. Hillock formation causes rupture of the ultrathin oxide layer used for junction tunneling and the device fails.

The difference in thermal expansion coefficient between the film and the substrate causes thermal stress. The thermal expansion coefficient is an intrinsic property of a pure element. Indeed large differences in thermal expansion coefficients between different kinds of materials (metals, semiconductors, and insulators) may be one of the limiting factors in growing high-quality epitaxial structures which combine these kinds of materials.

If the interatomic potential well of an atom is parabolic, the atom's displacement will be linearly proportional to the driving force. The atom undergoes a harmonic oscillation, so that its average position does not change and there is no thermal expansion upon heating. This is the same as a simple pendulum whose mean position remains unchanged in oscillation. But in reality, the interatomic potential well is not parabolic; the Lennard–Jones potential shows that resistance to compression is stronger than tension. Thus, thermal vibration tends to drive atoms apart. The stronger the vibration, the greater the separation. This leads to thermal expansion by the anharmonicity of the atomic potential. The dotted line in Fig. 4.1a depicts the increase of a_0 with energy or temperature. Since we can also change a_0 by stress and the change is described by Young's modulus, the thermal expansion coefficient and this modulus are related. We recall that in Section 2.7 the relation between Young's modulus and interatomic potential was discussed. The equation of state of solids which relates changes of pressure, temperature, and volume is given by Grüneisen's equation (see Mott and Jones).

4.4 Biaxial Stress in Thin Films

On a planar substrate, the stress due to differential thermal expansion experienced by the thin film is biaxial. As shown in Fig. 4.5a, the stresses act along the two principal axes in the plane of the film, but there is no stress in the direction normal to the film free-surface, and yet there is strain in the normal direction. Another example of biaxial stress is the stresses on the surface of a balloon. The in-plane stress results in strain normal to the balloon surface and it becomes thinner as it expands. To express the biaxial stress, we start with a three-dimensional isotropic system in which the stress σ and strain ε are related by the following basic equations

$$\begin{aligned} \varepsilon_x &= \frac{1}{Y}[\sigma_x - \nu(\sigma_y + \sigma_z)] \\ \varepsilon_y &= \frac{1}{Y}[\sigma_y - \nu(\sigma_x + \sigma_z)] \\ \varepsilon_z &= \frac{1}{Y}[\sigma_z - \nu(\sigma_x + \sigma_y)] \end{aligned} \tag{4.10}$$

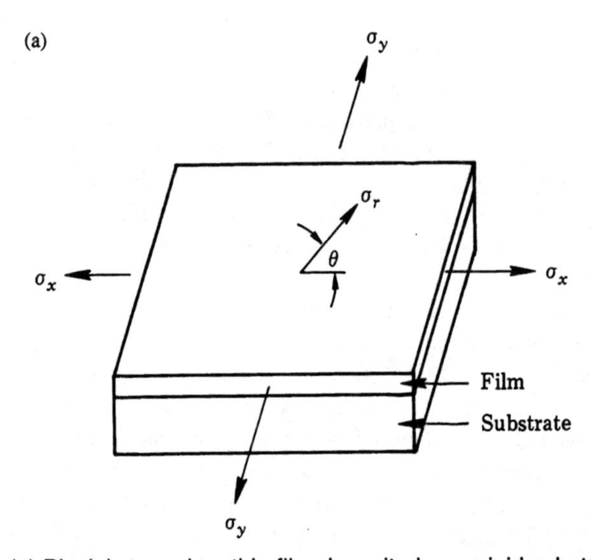

FIGURE 4.5 (a) Biaxial stress in a thin film deposited on a rigid substrate. (b) Cross-sectional view of a thin film under compression on a bent substrate. (c) A schematic diagram showing the stress distribution in film and substrate and the corresponding forces and bending moments.

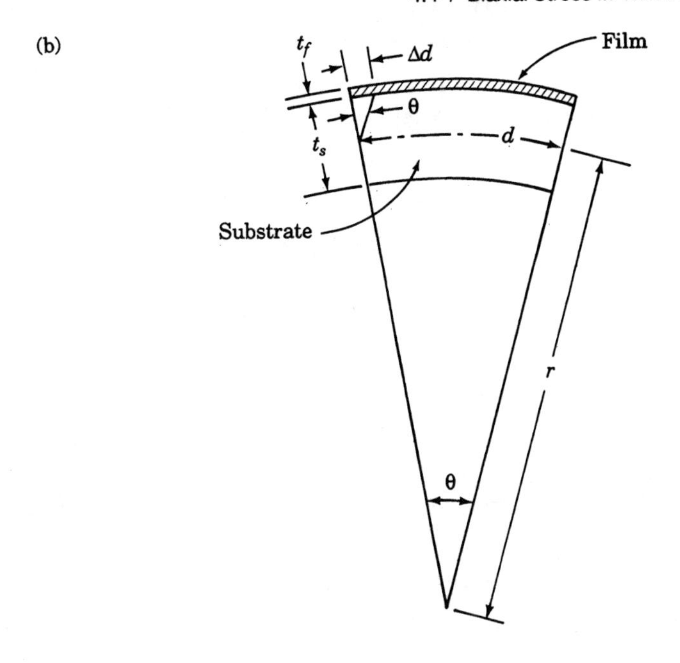

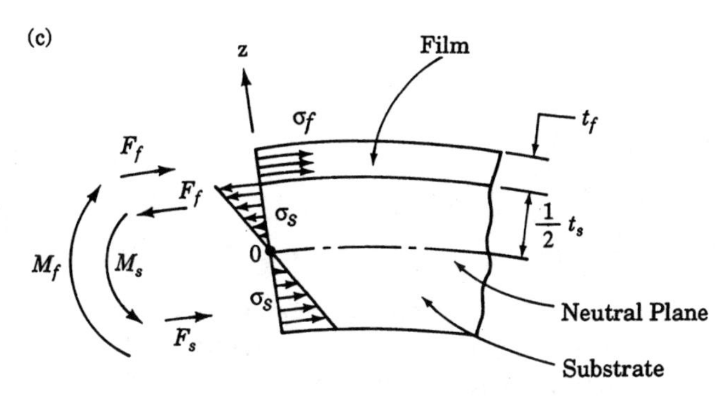

FIGURE 4.5 *(continued)*

In thin films, there is stress within the plane of the film (x and y) but no stress in the z direction ($\sigma_z = 0$). Therefore

$$\varepsilon_x = \frac{1}{Y}(\sigma_x - \nu\sigma_y)$$

$$\varepsilon_y = \frac{1}{Y}(\sigma_y - \nu\sigma_x)$$

$$\varepsilon_z = \frac{-\nu}{Y}(\sigma_x + \sigma_y)$$

From these equations, we have

$$\varepsilon_x + \varepsilon_y = \frac{1 - \nu}{Y} (\sigma_x + \sigma_y) \tag{4.11}$$

and

$$\varepsilon_z = \frac{-\nu}{1 - \nu} (\varepsilon_x + \varepsilon_y)$$

In two-dimensional isotropic systems where $\varepsilon_x = \varepsilon_y$,

$$\varepsilon_z = -\frac{2\nu}{1 - \nu} \varepsilon_x$$

We shall apply these relations in Chapter 7 to calculate the tetragonal distortion (lattice constant change) in heteroepitaxial growth of thin films.

With thin films on a circular substrate, it is convenient to use cylindrical rather than Cartesian coordinates, and we have

$$\begin{aligned} \sigma_r &= \sigma_x \cos^2\theta + \sigma_y \sin^2\theta + 2\tau_{xy} \sin\theta \cos\theta \\ \sigma_\theta &= \sigma_x \sin^2\theta + \sigma_y \cos^2\theta - 2\tau_{xy} \sin\theta \cos\theta \end{aligned} \tag{4.12}$$

where τ_{xy} is the shear stress. If there is no curl in the stress field ($\sigma_\theta = 0$), we obtain

$$\sigma_r = \sigma_x + \sigma_y$$

Similarly

$$\varepsilon_r = \varepsilon_x + \varepsilon_y$$

Then

$$\varepsilon_r = \frac{1 - \nu}{Y} \sigma_r \qquad \text{and} \qquad \varepsilon_z = -\frac{\nu}{Y} \sigma_r \tag{4.13}$$

We shall use these relationships, for example, when we consider the growth of a circular hillock in thin films under a compressive stress in Chapter 15.

The biaxial stress discussed above assumes that the substrate is rigid. We shall now consider the bending of the substrate under the biaxial stress. For example, if we deposit an Al film on a Si wafer at liquid nitrogen temperature and bring them up to room temperature, the Si wafer bends because Al has a larger thermal expansion coefficient than Si. Their cross section with a concave curvature is sketched in Fig. 4.5b. The Al film is constrained by the substrate, assuming a very good adhesion, and is under compression. The compressive stress in the Al film can be determined by measuring the curvature of the Si wafer and using the following analysis.

We begin the analysis by assuming that the film thickness t_f is much less than that of the substrate t_s, so the neutral plane where there is no stress can be taken to be at the middle of the substrate. In Fig. 4.5c, we enlarge one end of the substrate to show the neutral plane, the stress distribution in the film and in the substrate, and the corresponding forces and moments. At equilibrium, the moment produced by the stress in the film must equal to that produced by the stress in the substrate. Since we have assumed that the film thickness is thin, the stress σ_f is uniform across the film thickness. The moment M_f (force times perpendicular distance) due to the force in the film with respect to the neutral plane is

$$M_f = \sigma_f W t_f \frac{t_s}{2} \tag{4.14}$$

where W is the width of film normal to t_f. To calculate the moment of the substrate, we first obtain the geometrical relation,

$$\frac{d}{r} = \frac{\Delta d}{\frac{t_s}{2}} \tag{4.15}$$

and so

$$\frac{1}{r} = \frac{\Delta d}{\frac{t_s d}{2}} = \frac{\varepsilon_{\max}}{\frac{t_s}{2}} \tag{4.16}$$

where r is the radius of curvature of the substrate measured from the neutral plane, d is an arbitrary length of the substrate measured at the neutral plane, and $\Delta d/d = \varepsilon_{\max}$ is the strain measured at the outer surfaces of the substrate. In the substrate, the elastic strain is zero at the neutral plane, yet it increases linearly with distance z measured from the neutral plane (i.e., it obeys Hooke's law and increases linearly with stress), so that

$$\frac{\varepsilon_s(z)}{z} = \frac{\varepsilon_{\max}}{\frac{t_s}{2}} = \frac{1}{r} \tag{4.17}$$

where $\varepsilon_s(z)$ is the strain in a plane which is parallel to and at a distance of z from the neutral plane. Then, by assuming a state of biaxial stress in the substrate, we have from Eq. (4.11) that

$$\sigma_s(z) = \left(\frac{Y}{1-\nu}\right)_s \varepsilon_s(z) = \left(\frac{Y}{1-\nu}\right)_s \frac{z}{r} \tag{4.18}$$

Therefore the moment produced by the stress in the substrate is

$$\begin{aligned} M_s &= \mathrm{W} \int_{-t_s/2}^{t_s/2} z\sigma(z)\, dz = \mathrm{W} \int_{-t_s/2}^{t_s/2} \left(\frac{Y}{1-\nu}\right)_s \frac{z^2}{r}\, dz \\ &= \left(\frac{Y}{1-\nu}\right)_s \frac{Wt_s^3}{12r} \end{aligned} \tag{4.19}$$

By equating M_s to M_f, we have Stoney's equation,

$$\sigma_f = \left(\frac{Y}{1-\nu}\right)_s \frac{t_s^2}{6rt_f} \tag{4.20}$$

where the subscripts f and s refer to film and substrate, respectively. Eq. (4.20) shows that by measuring the curvature and the thicknesses of the film and the substrate, and by knowing Young's modulus and Poisson's ratio of the substrate, we can determine the biaxial stress in the film. The curvature can be measured by laser interference or by stylus profiling.

Equation (4.20) has been applied to measure surface stress during epitaxial growth of a film on a substrate. The pseudomorphic (commensurate) growth as shown in Fig. 1.4b induces a stress between the film and the substrate. When the substrate is sufficiently thin, the misfit stress can bend the substrate as discussed. In the extreme case of one monolayer pseudomorphic growth of Ge on a 0.1 mm thick (001) Si strip, the bending is large enough to be detected by laser reflection (Schell-Sorokin and Tromp, 1990). In essence, the force on the cross section Wt_f of a film is $F_f = \sigma_f W t_f$, or $F_f/W = \sigma_f t_f$. Rearranging Eq. (4.20), we have

$$r = \left(\frac{Y}{1-\nu}\right)\frac{t_s^2}{6\left(\dfrac{F_f}{W}\right)} \tag{4.20a}$$

This shows that by measuring r (or determining r as a function of t_f), we determine F_f/W.

The dimension of F_f/W is force per unit width of the film (i.e., it is a measure of surface stress). Recall the discussion of surface energy in Section 2.3 where we have shown that surface energy and surface tension of a liquid have the same magnitude. This is not so for solids. Liquids cannot take shear stresses and the surface of a liquid cannot sustain a compressive stress along the surface, yet solids can. On the other hand, neither solid nor liquid surfaces can have a normal stress.

4.5 Chemical Potential in a Stressed Solid

If the stress is maintained (either under constant stress or constant load) and even if it is within the elastic limit, the film or the material will respond by a slow deformation (i.e., relaxation by diffusional creep). A typical example is the sagging of lead pipes by their own weight in some very old houses. Room temperature is a relatively high temperature for lead, which melts at 327°C, therefore atomic diffusion is sufficient for creep to occur. A modern application of creep is the use of pure and well-annealed copper rings as pressure seals in ultrahigh vacuum systems. In general, creep is a high-temperature phenomenon except where grain-boundary diffusion becomes dominant, and then only a moderate temperature is required. We have discussed in Section 4.2 that elastic strain energy is small as compared to chemical energy, so we shall deal with creep in pure elements only, in absence of chemical reactions.

We consider the chemical potential in a stressed solid. In Appendix B, we have the change of Helmholtz free energy F as

$$dF = -S\,dT - p\,dV$$

If the change occurs at constant temperature as in room temperature creep, we eliminate the first term on the right-hand side and rewrite

$$p = -\frac{\partial F}{\partial V} \tag{4.21}$$

The last equation can be interpreted to mean that stress (pressure) is an energy density (i.e., energy per unit volume). For a given volume, the energy change equals the energy density times the given volume. So for an atomic volume Ω we have

$$p\Omega = -\frac{\partial F}{\partial V}\Omega = -\frac{\partial F}{\partial\left(\dfrac{V}{\Omega}\right)} = -\frac{\partial F}{\partial N} \tag{4.22}$$

where N is the number of atoms in volume V. The last term is the chemical potential by definition, where the negative sign is used to indicate that a decrease in volume by pressure results in an increase in energy. Pressure is a compressive stress which is negative. Since chemical potential is defined as Helmholtz (or Gibbs) free energy per atom, the chemical potential change in a stressed solid is

$$\mu = \pm\sigma\Omega \tag{4.23}$$

where the positive and negative sign refer respectively to tensile and compressive stress, following the sign convention given in Section 4.0. In other words, we can express the Helmholtz free energy of a stressed solid as

$$dF = -S\,dT - (p \pm \sigma)\,dN\Omega \tag{4.24}$$

where p is the ambient pressure and σ is the external applied stress.

To gain a quantitative feeling of $\sigma\Omega$, we shall consider a piece of Al stressed at the elastic limit (i.e., strain is 0.2%). Young's modulus for Al is $Y = 6 \times 10^{11}$ dyne/cm^2, so that the stress is

$$\begin{aligned}\sigma = Y\varepsilon &= 1.2 \times 10^9 \text{ dyne/cm}^2 \\ &= 1.2 \times 10^9 \text{ erg/cm}^3\end{aligned}$$

Since Al has a face-centered cubic lattice with a lattice parameter of 0.405 nm, there are four atoms in a unit cell of (0.405 nm)^3, or 0.602×10^{23} atoms/cm^3. Then

$$\sigma\Omega = \frac{1.2 \times 10^9 \text{ erg}}{0.602 \times 10^{23} \text{ atom}}$$

$$= 2 \times 10^{-14} \text{ erg/atom} = 0.0125 \text{ eV/atom}$$

It is interesting to compare this value to the value of strain energy per atom calculated by Eq. (4.9). The latter is much smaller with values around 10^{-5} eV/atom. The strain energy is the increase in energy per atom due to the strain. The chemical potential energy is the energy change in removing one atom from the stressed solid.

For thermally activated processes such as diffusion the chemical potential, $\sigma\Omega$, enters as an exponential factor. For Al stressed to the elastic limit at 400°C, we have

$kT = 0.058$ eV and

$$\exp(\sigma\Omega/kT) = \exp\left(\frac{0.0125}{0.058}\right) = 1.23$$

Usually, creep occurs at a much lower stress ($\sigma\Omega << kT$), so we can linearize the exponential term by

$$\exp\left(-\frac{\sigma\Omega}{kT}\right) \cong 1 - \frac{\sigma\Omega}{kT} \tag{4.25}$$

However, we have to be careful in using this. For example, let us consider a different case where we deposit an Al thin film on a thick fused quartz substrate kept at 400°C. Then we lower the temperature to 100°C and observe the relaxation of the Al film under a tensile stress. The tensile stress in the Al film is due to the much smaller thermal expansion of the quartz substrate. The linear thermal expansion coefficients of Al and quartz in the temperature range of 100°C are $\alpha = 25 \times 10^{-6}$/°C and 0.5×10^{-6}/°C, respectively. The thermal strain is

$$\varepsilon = \Delta\alpha\Delta T = 25 \times 10^{-6} \times 300 = 0.75\%$$

which is greater than the typical elastic limit. Then the thermal stress is

$$\sigma = Y\varepsilon = 4.5 \times 10^{9} \text{ dyne/cm}^2$$

and

$$\sigma\Omega = 0.045 \text{ eV}$$

On the other hand, $kT = 0.032$ eV at 100°C, so that we have $\sigma\Omega > kT$ in this case of a high-stress and low-temperature creep.

4.6 Diffusional Creep (Nabarro-Herring Equation)

In Fig. 4.4c, we showed that a square block under a shear stress s can be regarded as under a combination of tensile and compressive stresses. We now translate the picture to Fig. 4.6 and consider a hexagonal grain in a polycrystalline material which is under a shear stress. Again, we can imagine that the grain is acted upon by a combination of tensile and compressive stresses as shown. The effect of the elastic stress is to deform the grain from its original shape delineated by the solid lines, to that delineated by the broken lines. If the stress persists, the material can be deformed by transporting the part of material in the shaded area from the compressive region to the tensile region in order to release the stress. The transport is by atomic diffusion as indicated by the curved arrows. To analyze this problem, we shall follow the Nabarro–Herring creep model. In this model, it is assumed that grain boundaries are effective sources and sinks of point defects which mediate the mass transport, and that diffusion is via the point defects in the lattice.

In the tensile region very close to the grain boundary, the chemical potential deviates from the equilibrium value μ_0, due to the stress, by an amount

$$\mu_1 - \mu_0 = \sigma\Omega$$

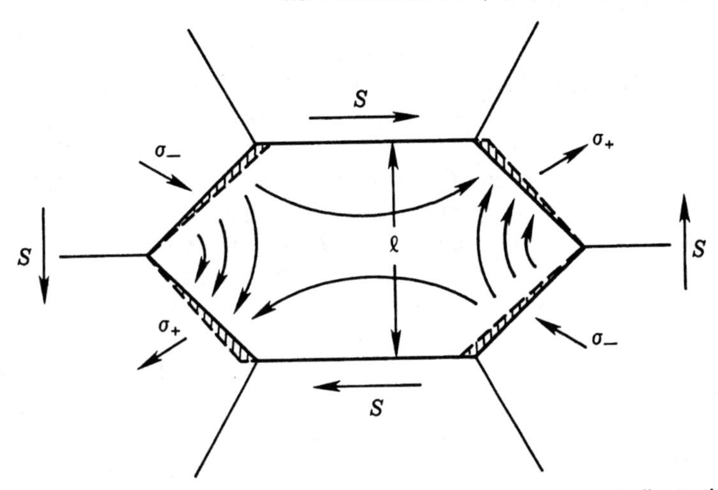

FIGURE 4.6 A grain under the shear stress *s*. The curved arrows indicate the atomic diffusion flux. The dimension of the grain is ℓ.

Similarly, in the compressive region, $\mu_2 - \mu_0 = -\sigma\Omega$, and thus the chemical potential difference in going from the compressive region to the tensile region is

$$\Delta\mu = \mu_2 - \mu_1 = -2\sigma\Omega \tag{4.26}$$

This potential difference will drive atoms to diffuse from the compressive regions to the tensile regions. The force acting on the diffusion atoms is

$$F = -\frac{\Delta\mu}{\Delta x} = \frac{2\sigma\Omega}{l} \tag{4.27}$$

where l is the grain size. The flux of the diffusing atoms, according to Eq. 3.5 in Chapter 3 is

$$J = \frac{C\nu\lambda^2}{kT} F = \frac{CD}{kT}\frac{2\sigma\Omega}{l} = \frac{2\sigma D}{kTl} \tag{4.28}$$

where $D = \nu\lambda^2$ and $C = 1/\Omega$ in a pure metal. The number of atoms transported by the flux in a period t and through an area A is $N' = JAt$, or the volume accumulated is

$$\Omega N' = \Omega JAt \tag{4.29}$$

The strain is then

$$\varepsilon = \frac{\Delta l}{l} = \frac{\frac{\Omega N'}{A}}{l} = \frac{\Omega Jt}{l} \tag{4.30}$$

so that the strain rate is

$$\frac{d\varepsilon}{dt} = \frac{\Omega J}{l} = \frac{2\sigma\Omega D}{kTl^2} \tag{4.31}$$

This is the well-known Nabarro–Herring creep equation. It was derived by considering the flux of atoms. Since creep occurs by having vacancies diffusing in the opposite direction, we should be able to obtain the same equation by considering the flux of vacancies. We shall illustrate that for a comparison, but first we deal with the concentration of vacancies in the tensile and the compressive regions.

We have argued that the chemical potentials in the tensile and the compressive regions have changed by the amounts $\sigma\Omega$ and $-\sigma\Omega$, respectively, from the equilibrium value. Since chemical potential is free energy per atom, this means that if we wish to remove an atom from these regions (i.e., to create a vacancy), the work needed to do so is changed by the same amounts. When we consider a vacancy in the stressed solid, the formation energy (which is actually the potential energy) is changed by $\pm\sigma\Omega$ assuming Ω is the volume of a vacancy. The positive and the negative signs are now reversed and refer respectively to the compressive stress and the tensile stress. In other words, the formation energy of a vacancy in the tensile regions is reduced by the amount $\sigma\Omega$, and in the compressive regions it is increased by $\sigma\Omega$. In a compressive region, it takes more energy to form a vacancy, and in a tensile region, it takes less, so that we have more vacancies in the tensile regions and fewer in the compressive regions at a given temperature. There is a gradient of vacancies between regions, and the vacancies will diffuse from the tensile region to the compressive region. According to Eq. (3.33), we can express the concentration of vacancies as

$$C_v^{\pm} = C \exp\,[(-\Delta G_f \pm \sigma\Omega)/kT] \tag{4.32}$$

where C_v^+ and C_v^- correspond to the vacancy concentrations in tensile and compressive regions, respectively. Assuming $\sigma\Omega << kT$, we have

$$C_v^{\pm} = C_v\left(1 \pm \frac{\sigma\Omega}{kT}\right) \tag{4.33}$$

where $C_v = C\exp(-\Delta G_f/kT)$ is the concentration of vacancies in the equilibrium state. Then the concentration difference is

$$\Delta C_v = C_v^+ - C_v^- = 2\sigma\Omega\,\frac{C_v}{kT} \tag{4.34}$$

The flux of vacancies going from the tensile to the compressive region is

$$J_v = -D_v\frac{\Delta C_v}{\Delta x} = \frac{-2\sigma\Omega D_v C_v}{kTl}$$

where D_v is the diffusivity of a vacancy. Then the atomic flux is

$$J = \frac{2\sigma\Omega DC}{kTl} = \frac{2\sigma D}{kTl} \tag{4.35}$$

where we have taken $DC = -D_vC_v$ since the atomic flux J is opposite to the vacancy flux J_v. Eq. (4.35) is the same as Eq. (4.28), so whether we consider the atomic flux or the vacancy flux, the creep equation is the same. We present both of them here because in Chapter 15, when we consider void formation we use vacancy flux, and for hillock growth we use atomic flux. The creep relation in Eq. (4.31) shows that the strain rate is linearly proportional to the stress σ, and inversely proportional to the square of grain size. It also shows that if we plot ln $(T\, d\varepsilon/dt)$ versus $1/kT$, we determine the activation energy of creep which is the same as the activation energy of lattice diffusion. Many high-temperature creep data for pure metals have been analyzed, and the measured activation energies agree well with those of lattice diffusion (see Fig. 4.7). However, lower-temperature creep data show a much smaller activation energy. This may be due to grain-boundary diffusion or to creep-induced dislocation motion. As shown in Fig. 4.6, the shaded volume in the compressive regions can be transported along the grain boundary to the tensile regions. In this

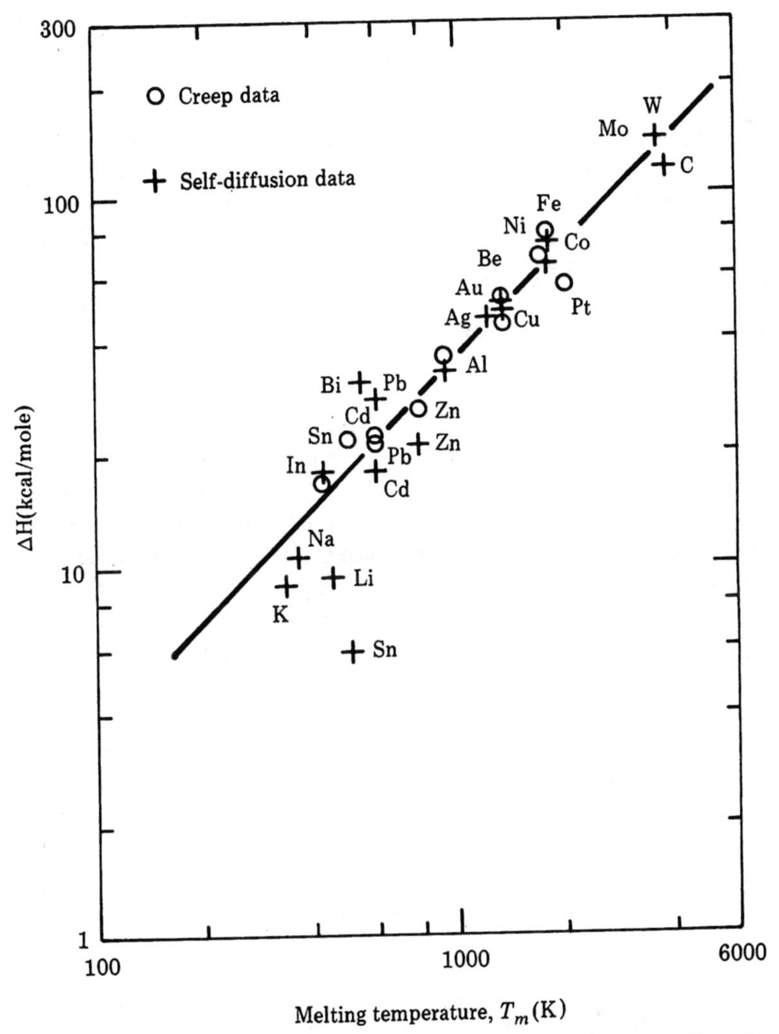

FIGURE 4.7 Activation energies for creep and self-diffusion as a function of melting point (from Chalmers, 1959).

case, the creep rate becomes

$$\frac{d\varepsilon_{gb}}{dt} = A\,\frac{\sigma\Omega D_{gb}\,\delta}{kTl^3} \tag{4.36}$$

where A is a constant and D_{gb} and δ are grain-boundary diffusivity and width, respectively. Comparing Eq. (4.36), to Eq. (4.31), we see that the difference is in replacing D by $D_{gb}\,\delta/l$. The factor $1/l$ can be regarded as the density of cross section of grain boundaries per unit area, hence δ/l is the cross-sectional area of grain boundaries per unit area. Creep by grain-boundary diffusion has a stronger dependence on grain size than does lattice diffusion. When both lattice diffusion and grain-boundary diffusion occur simultaneously, we have

$$\frac{d\varepsilon}{dt} = \frac{2\sigma\Omega D}{kTl^2}\left(1 + \frac{A}{2}\,\frac{D_{gb}\,\delta}{Dl}\right) \tag{4.37}$$

For a thin film on a substrate, diffusional creep leads to relaxation rather than deformation, yet if the relaxation is inhomogeneous (i.e., is localized), it induces void formation or hillock growth, which can be a serious reliability issue.

We have shown in Section 4.2 that the strain energy is unimportant in most of the chemical reactions such as silicide formation, and yet we show in this section that stress can influence vacancy concentration and affect diffusion. The difference lies in the period of time involved in the reactions; in silicide formation the reaction finishes typically in minutes or hours, whereas in creep it usually lasts for months. We ignore the long-term effect in short-term events.

4.7 Elastic Energy of a Misfit Dislocation

Since misfit dislocations release strain in epitaxial thin films, the elastic energy of an array of misfit dislocations at the interface between a film and its substrate is of interest. We show in Fig. 4.8 a schematic atomic picture of a misfit edge dislocation, where the atoms of the film and the substrate are the same. Such a misfit may occur in a doped layer of Si on an undoped Si substrate. Dopants can change the lattice constant of the film (slightly), giving rise to a film/substrate mismatch even though the system is basically homoepitaxial.

To form the misfit dislocation, we assume a simple cubic structure for film and substrate, having lattice parameters a_f and a_s, respectively. We further assume

$$(n + 1)a_f = na_s \tag{4.38}$$

and we define the misfit

$$f = \frac{a_s - a_f}{a_s} \tag{4.39}$$

Hence we have the spacing between neighboring misfit dislocations to be

$$na_s = \frac{a_f}{f} \simeq \frac{b}{f} \tag{4.40}$$

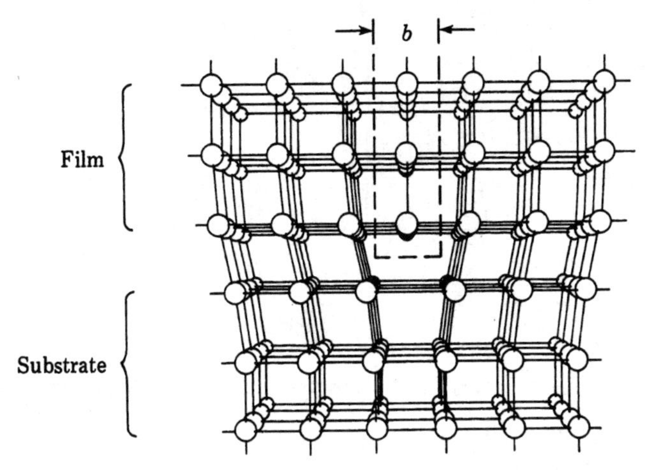

FIGURE 4.8 An edge-type misfit dislocation created by inserting an extra half-sheet of atoms.

where b ($\simeq a_f$) is the Burgers vector of the misfit dislocation. Clearly, to fit the film epitaxially without misfit dislocations on the substrate, we need to stretch the film by a strain of $\Delta l/l = 1/n$. For a thick film, it is energetically favorable to relax the film by misfit dislocations. We imagine that we make cuts in the film at a spacing of na_s, insert a single sheet of atoms in each cut as indicated by the broken lines in Fig. 4.8, and rejoin the atoms.

The elastic energy involved in forming a straight dislocation (an edge or a screw dislocation) has been treated in many textbooks. For a detailed discussion, see chapter 2 in the book by Hirth and Lothe. Here we shall present only a simple analysis. We consider in Fig. 4.9 a ring surrounding a misfit dislocation. We make a cut at the top of the ring, open the cut, and insert a sheet of atoms of width b. The strain and stress introduced into the ring are, respectively,

$$\varepsilon = \frac{b}{2\pi r}$$
$$\sigma = \frac{\mu b}{2\pi r} \tag{4.41}$$

where μ is the shear modulus. The elastic energy per unit length of the dislocation is

$$\begin{aligned} E_d &= \int_{r_1}^{r_2} \frac{1}{2} \frac{\mu b^2}{(2\pi r)^2} 2\pi r \, dr \\ &= \frac{\mu b^2}{4\pi} \int_{r_1}^{r_2} \frac{dr}{r} \\ &= \frac{\mu b^2}{4\pi} \ln \frac{r_2}{r_1} \end{aligned} \tag{4.42}$$

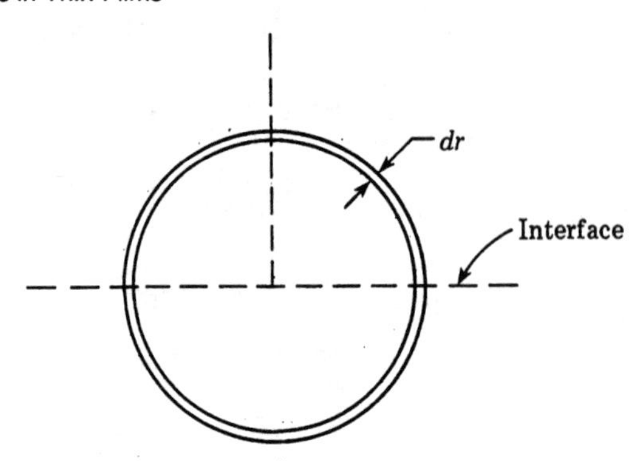

FIGURE 4.9 A ring of radius *r* and width *dr* surrounding an interfacial misfit dislocation.

We note that in a rigorous analysis for an edge dislocation, we must divide Eq. (4.42) by a factor of $(1 - \nu)$ where ν is Poisson's ratio. Now, what are r_1 and r_2? When we make the cut in the film, the cut should open up automatically since the film is under tension, so that it does not cause much strain to insert a sheet of atoms into the cut. Although the elastic field of a dislocation is long range and proportional to $1/r$, the elastic field produced by an array of misfit dislocations extends only a distance of $na_s/2$. The fields produced by neighboring dislocations cancel each other and we can take $r_2 = nb/2$. Then for r_1 it is clear that the dislocation core is a singularity, so that we have to take $r_1 > 0$ in the integration in order to avoid the singularity. Typically, $r_1 \simeq b$. Treatment of the actual dislocation core is a long-standing problem. The core energy is not infinite, because we know that the energy is finite for the case of a small-angle grain boundary which consists of a set of dislocations. This in turn means that the elastic field in the core of the dislocation does not go to infinity. In any case the interatomic distances between atoms in the core are finite.

If we take $r_2 = nb/2$ and $r_1 = b$, we have

$$E_d = \frac{\mu b^2}{4\pi(1 - \nu)} \ln \frac{n}{2} \simeq \frac{1}{2} \mu b^2 \tag{4.43}$$

for a typical misfit of 0.1% or less, and $n = 10^3$ to 10^4. Using Al as an example, we have $\mu \simeq 2 \times 10^{11}$ dyne/cm^2 and $b = 2.5 \times 10^{-8}$ cm.

Then,

$$E_d = \tfrac{1}{2} \times 2 \times 10^{11} \times (2.5 \times 10^{-8})^2$$

$$= 6.25 \times 10^{-5} \text{ ergs/cm of dislocation.}$$

If the misfit is 0.1% then there is a dislocation every 1000 lattice spacings ($\sim 2.5 \times 10^{-5}$ cm) or an (areal) energy density of 2.5 ergs/cm^2. For comparison purposes it is useful to evaluate this energy in eV/atom. A thin film of Al, thickness t(cm), contains $t \cdot 6 \times 10^{22}$ atoms/cm^2. A typical film of 100 nm then contains 6×10^{17}

atoms/cm^2. The dislocation energy averaged over all atoms in the film reduces to:

$$E_d = \frac{(2.5 \text{ ergs/cm}^2)(6.25 \times 10^{11} \text{ eV/erg})}{6 \times 10^{17} \text{ atoms/cm}^2},$$

or a value of 2.6×10^{-6} eV/atom. This value is comparable to the strain energy in a stressed solid discussed in Section 4.2. In both cases the stored energy is basically the displacement from an equilibrium position in an elastic solid. While the average energy/atom is small, the displacements and stored energy of atoms at the dislocation core can be large, of order 1eV/atom. Thus dislocations are rarely created by random thermal fluctuations. Stresses during thermal and mechanical processing are usually the cause of excess dislocations.

References

1. M. F. Ashby and D. R. H. Jones, *Engineering Materials I,* Pergamon Press, Oxford (1980).
2. B. Chalmers, *Physical Metallurgy*, Wiley, New York, (1959).
3. A. H. Cottrell, *Theory of Crystal Dislocations,* Gordon and Breach, New York (1964).
4. R. P. Feynman, R. B. Leighton, and M. Sands, *The Feynman Lectures on Physics* (Volume II, Chapter 38), Addison-Wesley, Reading, MA (1964).
5. C. Herring, *J. Appl. Phys.* **21**, 437 (1950).
6. J. P. Hirth and J. Lothe, *Theory of Dislocations*, McGraw-Hill, New York (1969).
7. R. W. Hoffman in *Physics of Thin Films* (Vol. 3, p. 211), edited by G. Hass and R. E. Thun, Academic Press, New York, (1964).
8. H. B. Huntington, "The Elastic Constants of Crystals," in *Solid State Physics,* Vol. 7, ed. by F. Seitz and D Turnbull (Academic Press, New York, 1958).
9. N. F. Mott and H. Jones, *The Theory of the Properties of Metals and Alloys,* Dover, New York (1958).
10. M. Murakami and A. Segmüller in *Analytical Techniques for Thin Films,* edited by K. N. Tu and R. Rosenberg, Vol. 27 in *Treatises on Materials Science and Technology,* Academic Press, Boston (1988).
11. A. S. Nowick and B. S. Berry, *Anelastic Relaxation in Crystalline Solids,* Academic Press, New York (1972).
12. J. Schell-Sorokin and R. Tromp, *Phys. Rev. Lett.* **64**, 1039 (1990).

Problems

4.1 Using the Lennard–Jones potential with $n = 8 \times 10^{22}$ atom/cm^3 and $\varepsilon_b =$ 0.6 eV/atom, $a_0 = (1/n)^{1/3}$

(a) Calculate the maximum force F_{max}.

(b) Assume the solid is in the linear elastic region and calculate Young's modulus Y.

(c) What is the elastic energy $E_{elastic}$ at F_{max}?

4.2 Show that a material maintaining constant volume during elastic deformation has Poisson's ratio $\nu = \frac{1}{2}$.

4.3 A 1 μm thick Al film is deposited without thermal stress on a 100 μm thick Si wafer at a temperature 100°C above ambient temperature. The wafer and film are allowed to cool to the ambient. Using the values provided in the table and assuming $\nu = 0.272$ for Si

(a) Calculate thermal strain and stress for $\Delta T = 100$°C.

(b) Calculate the radius of curvature.

	Expansion Coefficient α (10^{-6}/°C)	Young's Modulus Y (10^{11} N/m^2)
Al	24.6	0.7
Si	2.6	1.9

4.4 Consider Al and Cu at two-thirds their melting temperature T_m under a tensile strain of 0.2 percent. Using the data provided in the table, calculate for both materials

(a) The concentration of vacancies.

(b) The enhancement of the vacancy concentration caused by the strain.

	N ($\times 10^{22}$/cm^3)	Y ($\times 10^{11}$ N/m^2)	ΔH_f (eV)	T_m (°C)
Al	6.02	0.7	0.67	660
Cu	8.45	1.1	1.28	1028

4.5 For a 10^{-5} cm cubic grain of Cu held at two-thirds the melting temperature for 10 minutes, use the data in Table 3.1 and the Nabarro–Herring equations to calculate the volume and number of atoms accumulated. ($\varepsilon = 0.2\%$)

4.6 For Si, calculate the elastic energy per unit length of a misfit dislocation, and discuss whether dislocations would be formed by heating the crystal to 100°C. Use the Si parameters, $\mu = 7.5 \times 10^{10}$ N/m^2, $b = a/\sqrt{2}$, $a = 0.543$ nm, and $N = 5 \times 10^{22}$/cm^3.

4.7 A 200 μm diameter spherical balloon is made of 1 μm thick Al. At 20°C and 760 Torr, the balloon is fully inflated with no stress on the film. If the temperature is changed to 30°C and the pressure is constant, will the balloon burst? (Thermal expansion coefficient of Al is 2.5×10^{-5}/°C).

4.8 A 3000 Å oxide film is deposited on a 500 μm thick bare Si wafer that has a radius of curvature of 300 m. After deposition the radius is measured to be 200 m. A 6000 Å nitride film is now deposited on the oxide and the radius of curvature is measured to be 240 m. Calculate the dual film stress and the stress of the nitride film. ($\nu_{Si} = 0.272$, $Y_{Si} = 1.9 \times 10^{12}$ dyne/cm^2.)

4.9 The stress in films on wafers can be determined from the amount of bow of the substrate. The bow can be measured by a surface profileometer with an 18 cm scan length.

(a) Prove that Eq. (4.20) in the text is equivalent to the equation shown here.

$$\sigma = \left(\frac{\delta}{3\rho^2}\right)\left(\frac{Y}{1-\nu}\right)\left(\frac{t_s^2}{t_f}\right)$$

where $\delta \equiv$ the maximum bow height of the profileometer scan and $\rho \equiv$ half the scan length.

(b) Given a scan length of 5 cm and a bow of 20,000 Å calculate the stress for an unknown film 2 μm thick. The substrate is a 200 μm thick Si wafer ($Y/(1-\nu)$ for (100) Si $= 1.8 \times 10^{11}$ N/m^2).

CHAPTER 5

Surface Kinetic Processes

5.0 Introduction

The evolution of electronic materials progresses as we learn to form thin, planar solid films. The crucial requirement of film planarity has led to planar technology. State-of-the-art techniques can grow layers uniform to within one monolayer over 5 to 10 nm, the spatial extent of the electron wave function. The most striking examples of this mastery of layer growth are heterostructures, quantum wells, and superlattices which give direct evidence for control to tolerances of ± one monolayer as discussed in Chapter 7.

Monolayer control is a triumph of modern technology even though there are many physical reasons to suggest that a planar film does not represent the equilibrium state (lowest energy state) of a film/substrate combination. In this chapter we will discuss the tendency for deposited material to form three-dimensional clusters. We show that in many cases, depending on the surface energies of film and substrate, for sufficiently high temperatures and long periods of time, the clustered (nonplanar) film represents the equilibrium state. Planar technology has made advances which permit layer growth at low temperatures (nonequilibrium conditions), thus producing flat films.

Deposited atoms on surfaces seek atomic sites that minimize the total energy of the system. On a perfect surface such sites may be the relatively deep potential wells formed by the underlying substrate. An isolated atom on such a surface adds dangling (unsatisfied) bonds to the system, raising its total energy. A few atoms can form a cluster adding potential energy via dangling bonds in the atomic picture or adding surface area in the continuum picture.

In practice all surfaces are imperfect and have numerous defects. The simplest example of a surface defect is the surface step; an isolated atom with sufficient mobility will eventually adhere at a step site. If an atom sits alone on the surface it may have two unsatisfied bonds, whereas if it sits at a step, it may have only one. The step provides a favorable site for an atom to bond, and thus diffusing atoms may find the step a sink (Fig. 5.1). The step is a high-energy binding site and the existence of steps is important in thin film growth.

Now consider many deposited atoms on a surface. Instead of bonding at surface defects, they may combine, again reducing the number of free bonds. Additional

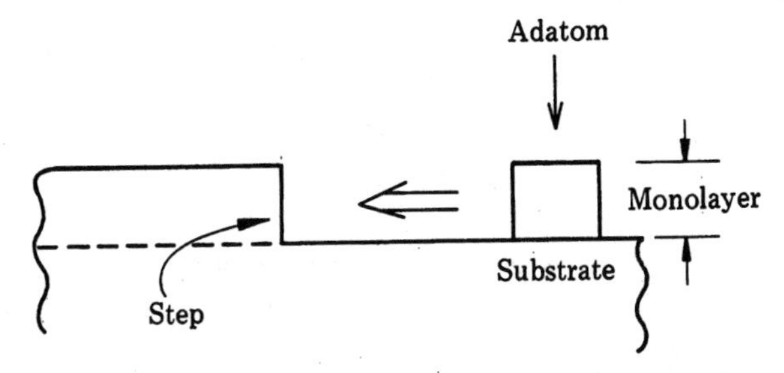

FIGURE 5.1 Schematic of a surface step one monolayer high with a deposited adatom migrating to the step (which is a favorable binding site).

deposited atoms may attach to the group to form a larger collection. If this collection of atoms grows two-dimensionally we refer to it as an *island*; three-dimensional collections are called *clusters*. Clearly, as islands or clusters grow they themselves can be thought of as surface defects providing high-energy binding sites for further atom attachment during deposition.

Planar growth is the desired state for technology. In this chapter however, we discuss growth of clusters, because under some circumstances the cluster is the lower energy state of deposited atoms on a solid. We have to understand clustering if we are going to understand thin film growth.

5.1 Atoms on a Surface

An atom impinging on a solid surface sees an array of binding sites or potential wells formed by substrate atoms. There is a finite probability that the atom diffuses along the surface by hopping from well to well. There is also a probability of an atom escaping from the well to vacuum (desorption). To characterize the surface diffusion and desorption, we assume that the surface atoms have a surface vibration frequency ν_s ($\nu_s \cong 10^{13}$ sec^{-1}). The frequency of desorption is defined as

$$\nu_{\text{des}} = \nu_s \exp\left(-\frac{\Delta G_{\text{des}}}{kT}\right) \tag{5.1}$$

where ΔG_{des} is the change in free energy associated with desorbing one atom. As in all such formulae, the frequency can be thought of as the number of attempts multiplied by the probability of success given by the exponential factor. The residence time τ_0 of an atom on the surface is

$$\tau_0 = \frac{1}{\nu_{\text{des}}} = \frac{1}{\nu_s} \exp\left(\frac{\Delta G_{\text{des}}}{kT}\right) \tag{5.2}$$

During the residence time, an atom moves from one surface site to a neighboring site (surface diffusion) with a frequency,

$$\nu_{\text{diff}} = \nu_s \exp\left(-\frac{\Delta G_s}{kT}\right) \tag{5.3}$$

where ΔG_s = energy required to move to a neighboring site. The surface diffusivity D_s is defined as

$$D_s = \lambda^2 \nu_{\text{diff}} = \lambda^2 \nu_s \exp\left(-\frac{\Delta G_s}{kT}\right) \tag{5.4}$$

where λ is the jump distance between two neighboring surface sites. In surface diffusion the sites are available, whereas in bulk diffusion a site must be generated by formation of a vacancy. Desorption is much less likely than surface diffusion; the difference lies in their activation energies. For the (111) surface of a face-centered-cubic noble metal, we typically have ΔG_{des} and ΔG_s to be about 1 eV and 0.5 eV, respectively. Desorption involves breaking bonds while surface diffusion consists of motion only with no net change in the number of paired bonds.

Using the concepts of desorption and surface diffusion, we can describe the behavior of an atom on a stepped surface where the mean spacing between steps is L_0. Atoms will not desorb if they arrive at a high-binding-energy site (a step) in times less than the residence time, τ_0. For growth we require that the atom diffuse to a step before desorption. So, for arrival at the step to occur before desorption

$$\sqrt{4D_s\tau_0} > L_0/2 \tag{5.5}$$

where $L_0/2$ is the largest distance to some step. The diffusion time t_D for an atom to reach the step is

$$t_D = L_0^2/16D_s \tag{5.6}$$

We define a ratio S_c to be

$$S_c = \frac{\tau_0}{\tau_D} \tag{5.7}$$

taking τ_D to be a "characteristic" diffusion time of L_0^2/D_s, $\tau_D = 16t_D$. For $S_c \geq 1$, an atom will have sufficient time to diffuse to the step and become bound on the surface. For $S_c < 1$, desorption can occur. From Eq. (5.7),

$$S_c = \frac{\lambda^2}{L_0^2} \exp\left(\frac{\Delta G_{\text{des}} - \Delta G_s}{kT}\right) \tag{5.8}$$

For a noble metal surface with $\lambda^2 = 10^{-15}$ cm^2, $L_0 = 10^{-5}$ cm, $\Delta G_{\text{des}} = 1$ eV, $\Delta G_s = 0.5$ eV, and $T = 293$ K, we have S_c appoximately equal to 10^3. This estimate shows that sticking is complete at room temperature for deposition on a noble metal surface. If we increase the substrate temperature to 600°K, we obtain $S_c \cong 10^{-1}$; the sticking is incomplete and desorption dominates. Thin film growth is, however, more complicated than a simple description of an atom on a stepped surface. Deposited atoms can react with the surface to form new chemical compounds, or they can interact with each other to form dimers and larger two dimensional clusters called *islands*. Surface morphologies can change as the result of surface energy variations from one material to another. Strain energy in heteroepitaxial growth can be a driving factor for clustering and defect propagation.

5.2 Vapor Pressure Above a Cluster

Clusters grow primarily in two ways: ripening and coalescence. An ensemble of clusters on a surface may undergo a ripening process with mass transport from smaller cluster to larger cluster by atomic surface diffusion (Fig. 5.2). *Ripening* is a process where larger clusters grow gradually at the expense of smaller clusters. At finite temperature there is a probability for atoms to escape a cluster, undergo surface diffusion and attach to another cluster. The flux of atoms from the cluster is dependent on the cluster size. There is some transport away from larger clusters, but the major motion is from smaller to larger with a net increase of average cluster size. The other cluster growth process is coalescence. *Coalescence* is simply whole clusters coming together. Two clusters combine to form a single larger cluster. The larger cluster represents a lower energy state than the two separated clusters.

The major driving force for ripening is a concentration gradient of material associated with the difference in vapor pressures over different size clusters. The vapor pressure over a small cluster is larger than the vapor pressure over a large cluster. As material diffuses across the surface to attach itself to a larger cluster, the vapor pressure must be maintained. Atoms from the smaller clusters keep resupplying the vapor. Atom by atom, the smaller cluster shrinks and the larger cluster grows as a function of time. The existence of Ga clusters on GaAs is shown in the electron micrograph of Fig. 5.3.

First consider the enhancement in pressure (or atomic concentration) above a cluster of radius r. The pressure difference across a curved surface is Δp. The work of expansion (of a bubble) must equal the increase in surface energy and

$$\Delta p \, dV = \gamma \, dA$$

For a sphere

$$\frac{dA}{dV} = \frac{8\pi r}{4\pi r^2} = \frac{2}{r}$$

so

$$\Delta p = \frac{2\gamma}{r} \tag{5.9}$$

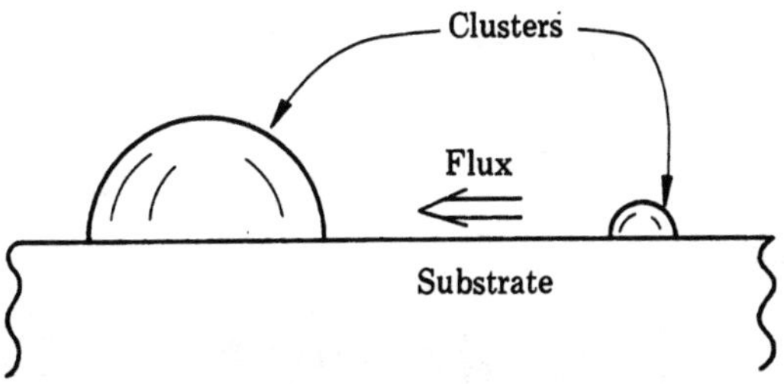

FIGURE 5.2 Schematic of a large and a small cluster on a surface with transport of atoms from the small to the large cluster.

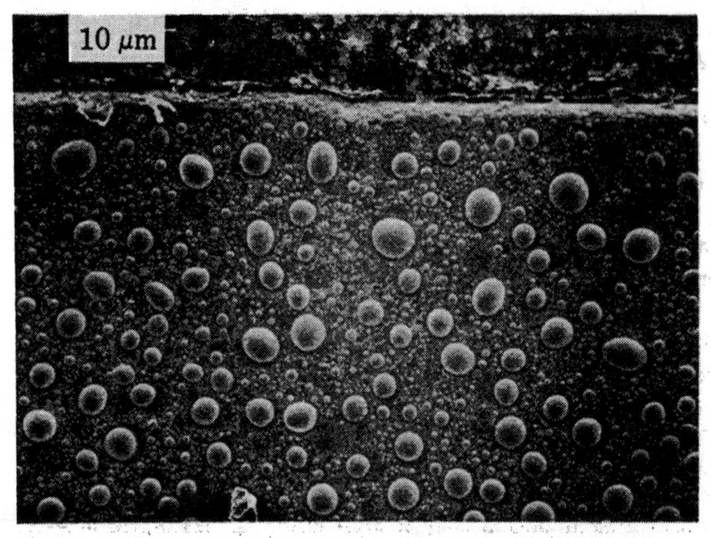

FIGURE 5.3 Clusters of Ga on a GaAs surface heated to 660°C for 5 minutes.

where dA is the increase in the surface area and γ is the surface energy per unit area. The increase in volume due to Δp results in work done. This work is limited by the increase in the surface energy; this is why bubbles do not continue to grow.

From the thermodynamic relations in Appendix B, the Gibbs free energy change is

$$dG = V\,dp - S\,dT$$

and

$$dG = V\,dp \tag{5.10}$$

for an isothermal process where $dT = 0$.

For an ideal gas,

$$pV = RT$$

and

$$p\Omega = kT \tag{5.11}$$

where $k = R/N_A$ and Ω is the volume per atom. For dG_A, the Gibbs free energy change per atom, Eq. (5.10) and Eq. (5.11) yield

$$dG_A = \Omega\,dp = kT\frac{dp}{p}$$

and by integration,

$$\frac{G_A}{kT} = \ln\,(p/p_0) \tag{5.12a}$$

On a curved surface, $G_A = \Omega\,\Delta p = \Omega\,2\gamma/r$, so that

$$\ln\,(p/p_0) = 2\gamma\Omega/rkT$$

TABLE 5.1 Values of p/p_0 in Eq. (5.12b) for $\gamma = 10^{15}$ eV/cm^2 and $\Omega = 20 \times 10^{-24}$ cm^3.

	p/p_0		
r (nm)	**300 K**	**600 K**	**900 K**
1	4.88×10^6	2.2×10^3	5.1
10	4.66	2.16	1.67
100	1.17	1.08	1.05

which we rewrite as

$$\frac{p}{p_0} = \exp\left(\frac{2\gamma\Omega}{rkT}\right) \tag{5.12b}$$

where p is the pressure above a cluster of radius r. The equilibrium pressure p_0 is the pressure if there was no cluster and this material was just a planar surface (i.e., $r \to \infty$). The relation in Eq. 5.12b is known as the Gibbs-Thomson equation.

For an estimate of the value of p/p_0, we consider a cluster of radius 1 nm at room temperature ($kT = 0.026$ eV). Then with a surface tension $\gamma = 10^{15}$ eV/cm^2 and an atomic volume $\Omega = 20 \times 10^{-24}$ cm^3, the factor in the exponential term in Eq. (5.12b) becomes:

$$\frac{2\gamma\Omega}{rkT} = \frac{2 \times 10^{15}\ \text{eV/cm}^2 \times 20 \times 10^{-24}\ \text{cm}^3}{10^{-7}\ \text{cm} \times 2.6 \times 10^{-2}\ \text{eV}} = 15.4.$$

yielding a very large enhancement in the effective pressure above a small cluster. As Table 5.1 shows, for cluster radii greater than 10 nm there are only minor enhancements. The enhanced vapor pressure over different-size clusters sets up concentration gradients which are the driving force for surface diffusion and cluster growth.

5.3 Growth of Clusters by Ripening

Thin film deposition often results in the formation of a distribution of discrete three-dimensional nuclei or particles. Annealing the film at temperatures below the desorption temperature results in particle growth. Large nuclei grow larger at the expense of smaller ones, thus reducing the total surface-to-volume ratio of the deposit. The total surface energy of a system is decreased when small particles combine to form larger clusters with no loss of mass. When nuclei agglomerate, we refer to the process as *coalescence*; it is discussed in Section 5.5. When clusters grow through single-atom processes we call the growth *ripening*. In this section we describe the kinetics of the ripening process. Energy minimization (Section 5.2) represents the driving force for cluster growth, while diffusion as outlined here describes the mechanism and the time-dependence (the kinetics) of the process.

We consider a cluster of adatoms on a surface containing free (unclustered) adatoms formed by an initial deposit. For simplicity we take the clusters to be hemispheres of radius r. The Gibbs-Thomson relation (5.12b) now represents an enhanced concentration of adatoms at the cluster radius r. A nucleus of radius r is in thermodynamic equilibrium with an adatom concentration N_r,

$$N_r = N_0 \exp (2\gamma\Omega/rkT) \tag{5.13}$$

where N_0 is the adatom concentration corresponding to the vapor pressure for the planar surface. Note that the units of N_0 and N_r are number of atoms per cm^2.

The mechanism of cluster growth is taken to be surface diffusion. We now solve the steady-state diffusion equation to obtain the growth rate as a function of time. Note that the "steady state" is an approximation here, as the cluster size is actually changing with time. However, the rates involved are sufficiently slow to allow a steady-state approximation.

The continuity equation for the concentration N of a diffusing species where the surface diffusion coefficient, D_s, is a function of the concentration may be written in three dimensions, based on relations in Chapter 3, as

$$\frac{\partial N}{\partial t} = \frac{\partial}{\partial x}\left(D_s \frac{\partial N}{\partial x}\right) + \frac{\partial}{\partial y}\left(D_s \frac{\partial N}{\partial y}\right) + \frac{\partial}{\partial z}\left(D_s \frac{\partial N}{\partial z}\right)$$

The corresponding two-dimensional equation in terms of cylindrical coordinates R and Θ is obtained by writing

$$x = R \cos \Theta \qquad \text{and} \qquad y = R \sin \Theta$$

and

$$\frac{\partial N}{\partial t} = \frac{1}{R}\frac{\partial}{\partial R}\left(RD_s \frac{\partial N}{\partial R}\right) + \frac{1}{R^2}\frac{\partial^2(D_s N)}{\partial \Theta^2}$$

In our case we consider a steady-state solution with $\partial N/\partial t = 0$. The picture consists of a cluster of radius r on a surface containing a concentration of adatoms. The steady-state diffusion equation in cylindrical polar coordinates, ignoring terms in Θ, is

$$\frac{1}{R}\frac{d}{dR}\left[RD_s \frac{dN}{dR}\right] = 0 \tag{5.14}$$

where R = polar coordinate, $N(R)$ = local concentration of adatoms/cm^2, and D_s = surface diffusion coefficient. The solution to this equation is of the form (Fig. 5.4)

$$N(R) = K_1 \ln R + K_2$$

where K_1 and K_2 are arbitrary constants. We impose the following boundary conditions:

$$N(R) = N_r \qquad \text{at } R = r$$

$$N(R) = N_0 \qquad \text{at } R = Lr$$

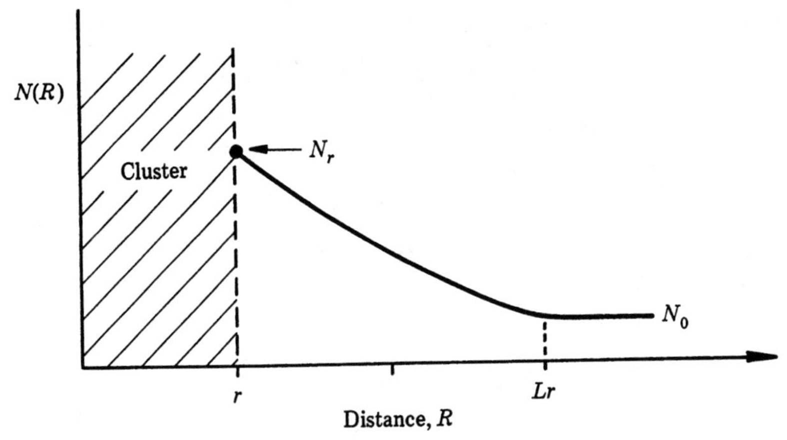

FIGURE 5.4 Sketch of values $N(R)$ versus distance for a cluster of radius r.

where L is a multiplier of the cluster radius r and measures the length (in units of r) in which the enhanced vapor pressure returns to N_0. Typically $L \sim 3$. These conditions imply that the concentration just at the cluster surface ($R = r$) is given by Eq. (5.13) and that the enhanced concentration returns to the equilibrium value at a distance Lr. Then,

$$N(R) = \frac{N_r \ln (Lr/R) - N_0 \ln (r/R)}{\ln (L)} \tag{5.15}$$

satisfies the differential equation.

In a diffusion problem the flux of particles is always given by

$$J_S = -D_S \frac{dN}{dR}$$

Note that with N in units of cm^{-2} and D_s in the usual units of cm^2/sec the units of J_s are particles/length/sec. This is the flux or number of particles/length/sec and is a unit length (rather than a unit area) because the problem is two dimensional.

The number of particles/sec diffusing in (or out) of the cluster of circumference $2\pi r$ is J where

$$J = -2\pi r D_S \left.\frac{dN}{dR}\right|_{R=r}$$

or

$$J = \frac{2\pi D_S}{\ln (L)} (N_r - N_0) \tag{5.16}$$

Noting that $N_r = N_0 \exp (2\gamma\Omega/rkT)$ and recognizing that a small-argument expansion is valid for $r \gtrsim 5$ nm we have

$$J = \frac{2\pi D_S}{\ln (L)} N_0 \frac{2\gamma\Omega}{rkT} \tag{5.17}$$

Now the number of atoms in a hemispherical cluster of radius r is Q where

$$Q = \tfrac{2}{3}\pi r^3 n = \tfrac{2}{3}\pi r^3/\Omega$$

and n is the number of atoms/unit volume. Then

$$\frac{dQ}{dt} = \frac{2\pi r^2}{\Omega}\frac{dr}{dt} \tag{5.18}$$

must be equal to the particle current J which is the divergence of J_s. So equating (5.17) and (5.18) and canceling 2π,

$$\frac{r^2}{\Omega}\frac{dr}{dt} = \frac{D_S}{\ln(L)} N_0 \frac{2\gamma\Omega}{rkT} \tag{5.19}$$

and

$$r^4 = \frac{N_0 8\gamma\Omega^2 D_S t}{kT \ln(L)} = Kt \tag{5.20}$$

where K contains the material properties. To obtain an estimate, we take $N_0 = 10^{13}/\text{cm}^2$, $\gamma = 10^{15}$ eV/cm^2, $\Omega = 20 \times 10^{-24}$ cm^3, kT at 525 K $= 0.046$ eV, $L = 3$, and $D_S = 10^{-8}$ cm^2/sec. Then in 20 minutes the cluster radius would be 9.3×10^{-6} cm.

Equation (5.20) corresponds to the *shrinkage* of the *small* cluster of radius r; this loss of mass is pictured as atom diffusion. The atoms attach themselves to *larger* clusters which *grow* at a rate $t^{1/4}$. It is a diffusion-limited process. The full treatment of loss and gain is considered in more complex treatments. Such treatments also consider the rate of atoms detaching themselves from the cluster (barrier passing) as the limiting process rather than diffusion. Depending on cluster size, temperature, and materials parameters, the surface diffusion or the detachment may actually limit the process. The time dependence then depends on the limiting mechanism. In addition, different cluster shapes will give rise to different time growth laws. For example, Eq. (5.18) is applicable for a hemispherical cluster, while a two-dimensional island will yield a different radial dependence for dQ/dt and hence a different time scaling. Nevertheless for all such processes small clusters shrink and large clusters grow. The different scaling laws are summarized in Table 5.2.

The general result for surface diffusion to hemispherical clusters is that the larger clusters form with the fourth power of the radius growing linearly with time. Experimental evidence confirms this power dependence (Fig. 5.5).

TABLE 5.2 Summary of $t^{1/n}$ Scaling (Surface Processes)

Cluster	Transport Mechanism	Scaling
3 Dimensional	Diffusion	$r \propto t^{1/4}$
2 Dimensional	Diffusion	$r \propto t^{1/3}$
3 Dimensional	Barrier Passing	$r \propto t^{1/3}$
2 Dimensional	Barrier Passing	$r \propto t^{1/2}$

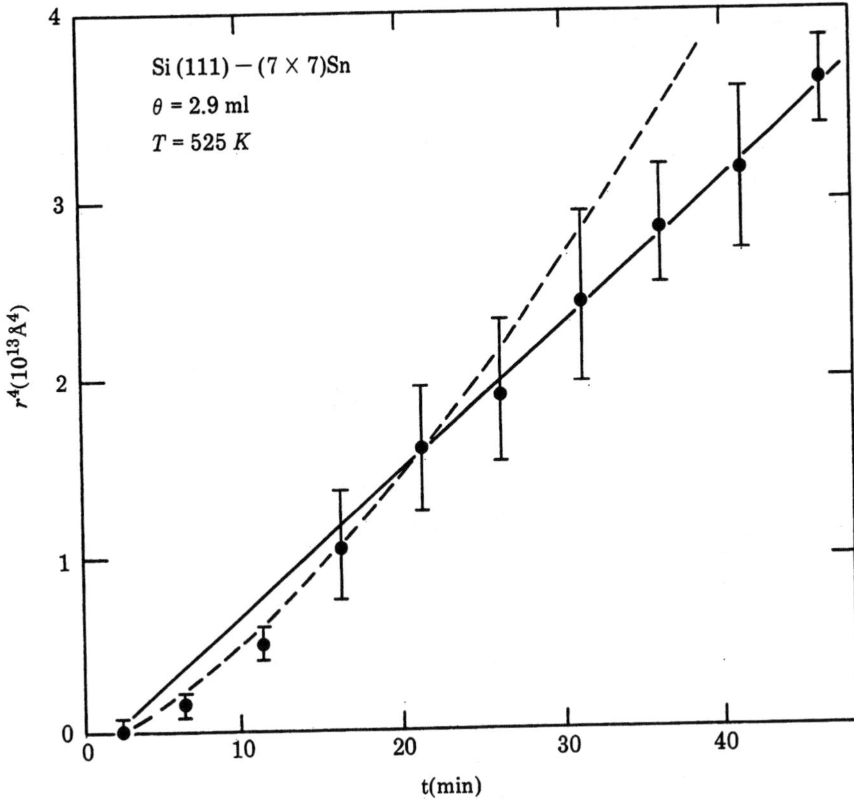

FIGURE 5.5 The cluster radius to the fourth power r^4 versus time for 2.9 monolayers of Sn on a (7 × 7) Si (111) surface at 525 K. The broken curve corresponds to $r^3 \alpha t$ for comparison. Data from Zinke-Allmang, 1987.

5.4 Activation Energies For Cluster Growth

For diffusion-limited processes the temperature dependence for the growth arises from the rate of atom diffusion on the surface. The higher the temperature the greater the surface diffusion and the faster the clusters grow.

Values for the cluster activation energy have been found from studies of Ga, Ge, and Sn cluster kinetics of overlayers on silicon. Rutherford backscattering techniques were used for (a) quantitative determination of deposited-atom overlayer coverages and (b) measurements of cluster growth by determination of depth profiles in relation to time at elevated surface temperatures (Feldman et al., 1990). Rutherford backscattering gives the mean height of the cluster distribution which can be used to calculate the mean radius when the number of adatoms on the surface is known. This analysis holds as long as the cluster distribution does not broaden significantly and the anneal times are sufficiently long that the diffusion length is greater than the intercluster distance on the surface.

The growth of the critical radius r_c is given by

$$r_c(t) = r_c(t_0)\,[1 + (t - t_0)/\tau]^{1/4} \tag{5.21}$$

The critical radius is the radius of the clusters which are in equilibrium and therefore neither grow nor shrink. The value $r_c(t_0)$ is the critical radius at a given time in the ripening regime. The constant τ in Eq. (5.21) is given by

$$\tau = r_c^4(t_0)/\beta$$

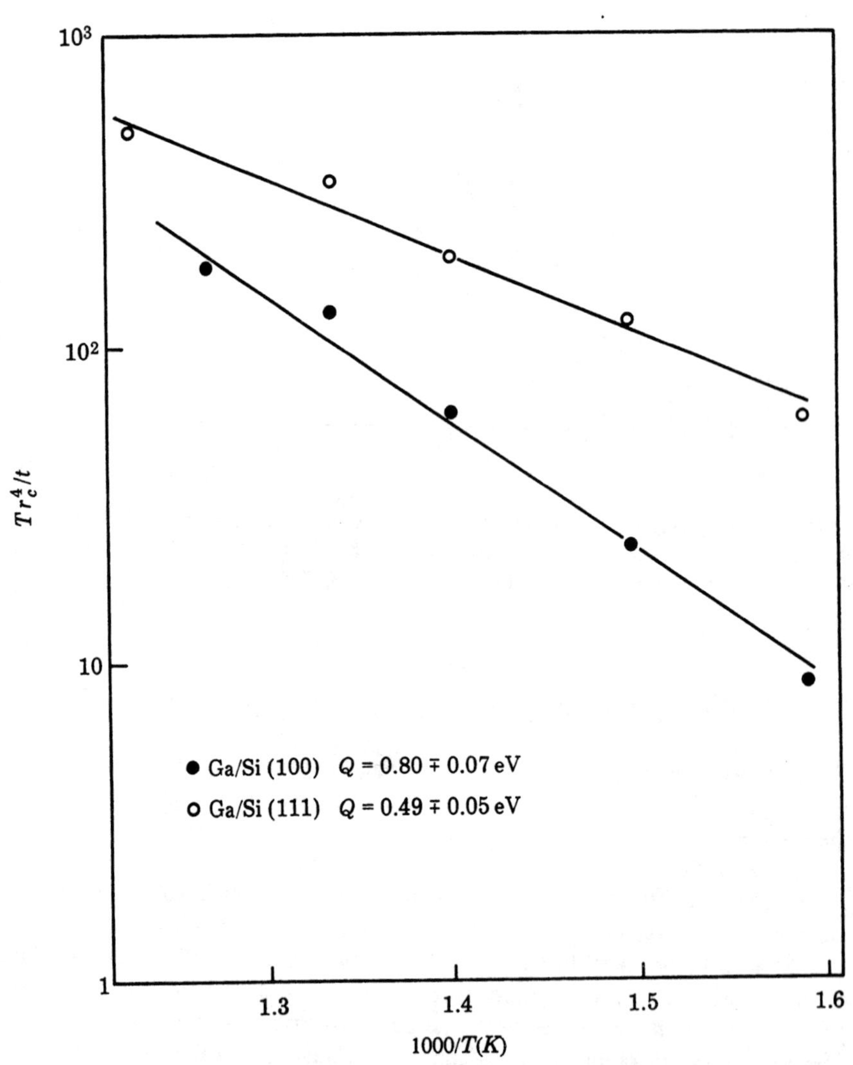

FIGURE 5.6 Arrhenius plot of growth of Ga clusters on Si (100) and Si (111) to determine activation energies for the clustering process. The additional temperature factor on the ordinate is due to the ripening model which predicts an $[\exp(-Q/kT)]/T$ dependence for the ripening.

TABLE 5.3 Activation Energies ΔG_S and Pre-exponential Factors for Cluster Growth on Si*

System	ΔG_S (eV)	D_0 (cm^2/s)	D_S (750°C) (cm^2/s)
Sn/Si (111)	0.32 ± 0.04	5×10^{-8}	3×10^{-10}
Sn/Si (100)	1.0 ± 0.2	10^{-4}	2×10^{-11}
Ge/Si (100)	1.0 ± 0.1	—	—
Ga/Si (111)	0.49 ± 0.05	8×10^{-9}	4×10^{-12}
Ga/Si (100)	0.80 ± 0.07	10^{-6}	6×10^{-12}
Ga/Si (100), 4° miscut	0.80 ± 0.07	4×10^{-6}	2×10^{-11}

*From Feldman, Nakahara, and Zinke-Allmang.

with

$$\beta = N_0 2\gamma\Omega^2 D_S/kT \tag{5.22}$$

as discussed for Eq. (5.20).

Activation energies for cluster formation follow from the temperature dependence of the growth rates, $\Delta r_c^4/\Delta t$. Figure 5.6 shows an Arrhenius type of representation of growth rates for Ga on Si (111) and Si (100). The solid lines fit the data with cluster activation energies of 0.80 ± 0.07 eV on Si (100) and 0.49 ± 0.05 eV on Si (111). Cluster growth parameters on Si surfaces with different adatoms are summarized in Table 5.3. Since diffusion is a limiting component, these activation energies are of the same order of magnitude as surface diffusion coefficients.

5.5 Coalescence

Another mechanism of cluster growth is coalescence, the combining of two clusters into a still larger cluster. Let us start by explaining why coalescence is favored, such as when two water drops come together to form a larger drop. We simply compare the total surface energy of two clusters to that of one cluster (Fig. 5.7). For two

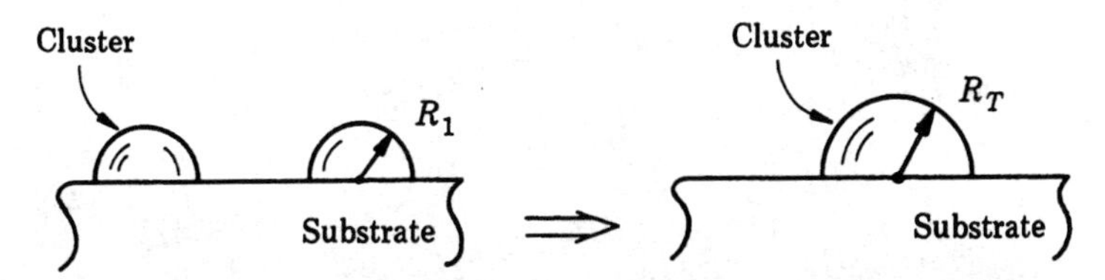

FIGURE 5.7 Schematic of a substrate with two clusters of radius R_1 compared to the same substrate with one cluster of radius R_T having the same total volume as the two smaller clusters.

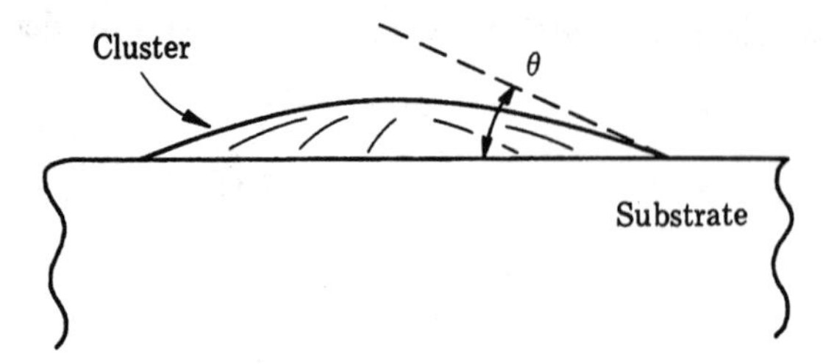

FIGURE 5.8 Schematic of a more realistic cluster configuration with wetting angle Θ.

clusters, hemispheres of radius R_1 in our approximation, the total surface energy is

$$E_S = 2 \times 2\pi(R_1)^2\gamma \tag{5.23}$$

This is the energy for the two clusters; γ is the surface energy per unit area between the cluster and the outside world, the vapor.

The combined cluster has a total volume which is twice the volume of the individual clusters,

$$V = 2 \times \tfrac{2}{3}\pi R_1^3 \tag{5.24}$$

The total volume may also be written as

$$V = \tfrac{2}{3}\pi R_T^3 \tag{5.25}$$

The radius R_T of the combined cluster is

$$R_T = (2)^{1/3}R_1 \tag{5.26}$$

The total surface energy for the single large cluster is

$$E_S(T) = 2\pi R_T^2\gamma = 2\pi(2^{1/3}R_1)^2\gamma \tag{5.27a}$$

and the ratio of total energies is

$$\frac{E_S}{E_S(T)} = \frac{4\pi R_1^2}{2\pi 2^{2/3}R_1^2} = \frac{2}{2^{2/3}} > 1 \tag{5.27b}$$

FIGURE 5.9 Sideview of Ga clusters on (111) GaAs for 13-monolayer deposition for 100 minutes at a temperature of 750 K.

The new surface area is less than the total surface area of the two smaller clusters and there is thus less total surface energy.

Not all clusters are hemispheres, as shown in Fig. 5.8 and Fig. 5.9. The angle of wetting the surface is related to the surface and interfacial energies. All the arguments that we have given about ripening and coalescence hold, even if the clusters are not hemispheres. When the wetting angle Θ equals 90 degrees, the cluster is a hemisphere.

5.6 Mass Transport on Patterned Surfaces

Diffusion of atoms along the surface was used to describe cluster formation where the driving force was associated with high vapor pressures above surfaces of large curvature. Macroscopic steps or grooves on the surface also represent large curvature and give rise to high concentration gradients. Mass transport can minimize surface energy by smoothing out stepped structures. Figure 5.10 shows a multilevel, stepped surface etched in InP and the resultant smooth surface obtained by mass transport.

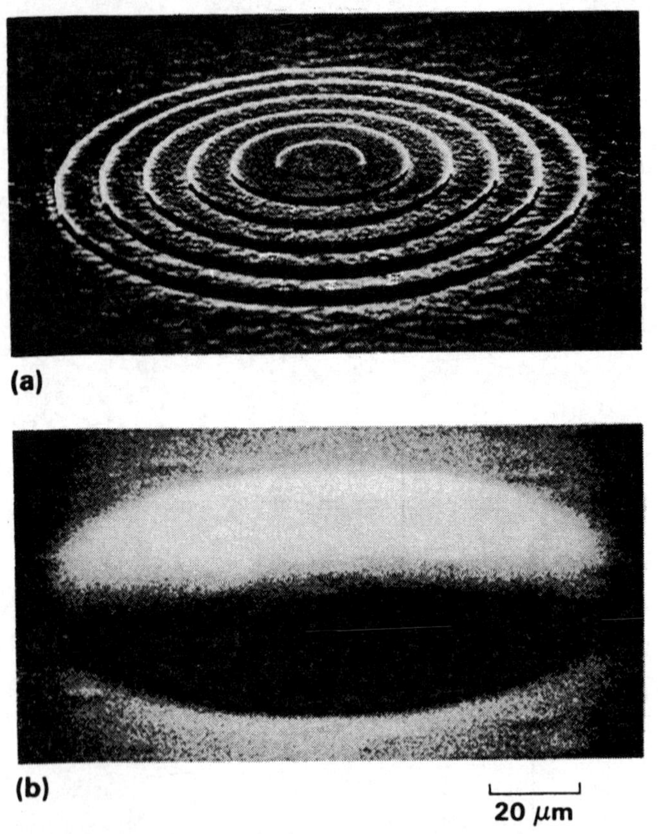

FIGURE 5.10 SEM photos showing the fabrication of a lenslet in InP: (a) multilevel mesa formed by etching and (b) surface after mass transport (from Liau et al., 1988a).

This mass transport of InP has been used to form buried heterostructure lasers (Liau and Walpole, 1985).

We now introduce a model of mass transport based on the work of Liau et 1988b. The basic concepts are similar to our discussion of cluster growth. Nan. the small radius of curvature associated with a sharp corner is a region of hi effective vapor pressure. Surface diffusion occurs as a result of this concentrati gradient smoothing out the surface and reducing the surface area. Addition c volume element dV to a curved surface (Fig. 5.11) results in an increase of surfac area dA. Since work is needed to create a new surface, there is additional energy (i.e., the surface energy) of an amount γdA, where γ is the surface tension. We define a flat surface as being the zero surface-energy state. Therefore the surface energy per atom is

$$E_{SO} = \frac{\gamma dA}{dV/\Omega} \tag{5.28}$$

where Ω is the atomic volume. For a cylindrical feature it is easily shown that

$$\frac{dA}{dV} = \frac{1}{r} \tag{5.29}$$

where r is the radius of curvature of the surface element.

For a profile such as shown in Fig. 5.12, the radius r is defined by the curvature $1/r = d\theta/ds$ which is

$$1/r = \frac{d^2z/dx^2}{[1 + (dz/dx)^2]^{3/2}}$$

For a slowly varying profile $\partial z/\partial x << 1$, where z and x are height and position, respectively, and the curvature is simply

$$\frac{1}{r} = \frac{\partial^2 z}{\partial x^2} \tag{5.30}$$

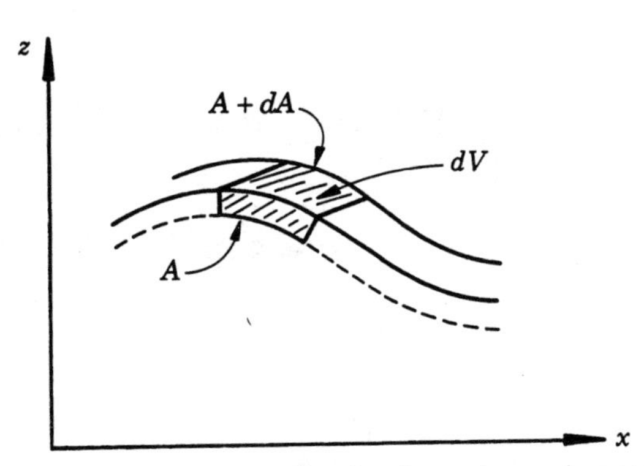

FIGURE 5.11 Surface profile showing volume element dV.

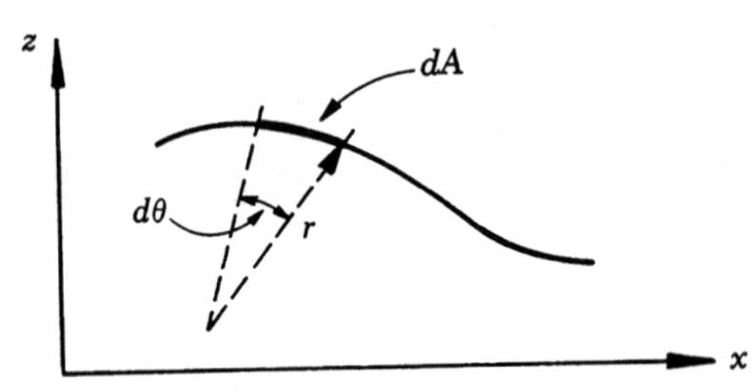

FIGURE 5.12 Schematic of a surface profile where $r = dS/d\Theta$.

Combining Eqs. (5.28)-(5.30), we have

$$E_{SO} = \gamma\Omega \frac{\partial^2 z}{\partial x^2} \tag{5.31}$$

The effect of surface curvature on surface smoothing and the resulting mass transport can now be considered.

The equilibrium concentration of adatoms on a flat surface, N_0, is N_s exp $(-E_{dis}/kT)$ where N_s is the number of surface sites and E_{dis} is the binding energy. For a curved surface, the extra surface energy per atom E_{SO} represents a decrease in the binding energy and the adatom concentration, N_{ad}, therefore becomes

$$N_{ad} = N_0 \exp(E_{SO}/kT) \tag{5.32}$$

The deviation in the surface concentration, $N' \equiv N_{ad} - N_0$, is then given by

$$N' = N_0 [\exp(E_{SO}/kT) - 1] \tag{5.33}$$

For the surface features of interest, $E_{SO} << kT$, and $N' \cong N_0 E_{SO}/kT$. Hence, by using Eq. (5.31), N' may be expressed as:

$$N' = \frac{\gamma\Omega N_0}{kT} \frac{\partial^2 z}{\partial x^2} \tag{5.34}$$

When there is a variation of curvature across the surface, free atoms will diffuse from regions of large curvature (small radius) where N' is large, to regions of small (or negative) curvature.

The conservation of mass (i.e., the continuity equation) can then be expressed as

$$D_S \frac{\partial^2 N'}{\partial x^2} \cong -\frac{1}{\Omega} \frac{\partial z}{\partial t} \tag{5.35}$$

where D_S is the surface diffusion coefficient and t is the time. This relation follows from the basic concepts of diffusion discussed in Chap. 3. At any point on the surface the net accumulation of material is given by the negative of the divergence of the flux: $-\frac{\partial}{\partial x}J = -D_s\partial^2 N'/\partial x^2$. The actual motion of material represents a change in the height z, equivalent to $\partial N'/\partial t = (1/\Omega)\,\partial z/\partial t$. Using Eq. (5.34) for N', Eq. (5.35)

can be written as

$$\frac{\gamma\Omega^2 D_S N_0}{kT}\frac{\partial^4 z}{\partial x^4} = -\frac{\partial z}{\partial t} \tag{5.36}$$

For convenience, we define a term similar to that in Eq. (5.22),

$$\Gamma = \frac{\gamma\Omega^2 D_S N_0}{kT} \tag{5.37}$$

and Eq. (5.29) becomes

$$\Gamma\frac{\partial^4 z}{\partial x^4} = -\frac{\partial z}{\partial t} \tag{5.38}$$

Solutions of Eq. (5.38) can be found by using the Fourier expansion of the initial profile $z(x,0)$. In the case of a periodic profile, a Fourier series can be used, such as

$$z(x,0) = \sum_{n=0}^{\infty} A_n \sin(2n\pi x/\lambda) \tag{5.39}$$

for a profile with odd symmetry about some point, where λ is the periodicity. Then, it can be shown that

$$z(x,t) = \sum_{n=0}^{\infty} A_n \exp(-t/\tau_n)\sin(2n\pi x/\lambda) \tag{5.40}$$

satisfies Eq. (5.38) provided that

$$\tau_n = \frac{1}{\Gamma}(2n\pi/\lambda)^{-4} \tag{5.41}$$

Note that τ_n is proportional to $(\lambda/n)^4$. This means that the higher n components (the high-frequency components) decay much faster than lower n ones. These components represent the small radius-of-curvature position of the surface. These small radius of curvature points correspond to the greatest values of $\partial^2 z/\partial x^2$ such as sharp features or the smallest clusters. This result suggests that a square-wave profile will quickly change into a sinusoidal one followed by an exponential decay of the latter. A perfect sinusoidal profile remains sinusoidal with an exponential decay of the amplitude. If $z(x, 0) = A \sin Kx$ initially then $z(x, t) = z(x, 0)\exp(-\Gamma K^4 t)$ under the mechanism of surface transport.

Demonstration of this decay of surface features was made by observing the temperature behavior of gratings etched in (100) oriented InP. Figure 5.13 shows a square-wave structure before and after mass transport at 780°C for 3.25 and 19.25 hours. As predicted the square wave developed into a sinusoidal shape with an amplitude decaying in time. The decay time τ follows a fourth power of the periodicity (Fig. 5.14), showing that close-spaced gratings decay faster than long-period ones. The λ^4 dependence of the sinusoidal decay time permits easy control of lens fabrication from stepped surface. The mass transport process will virtually stop when the short-period steps have been smoothed.

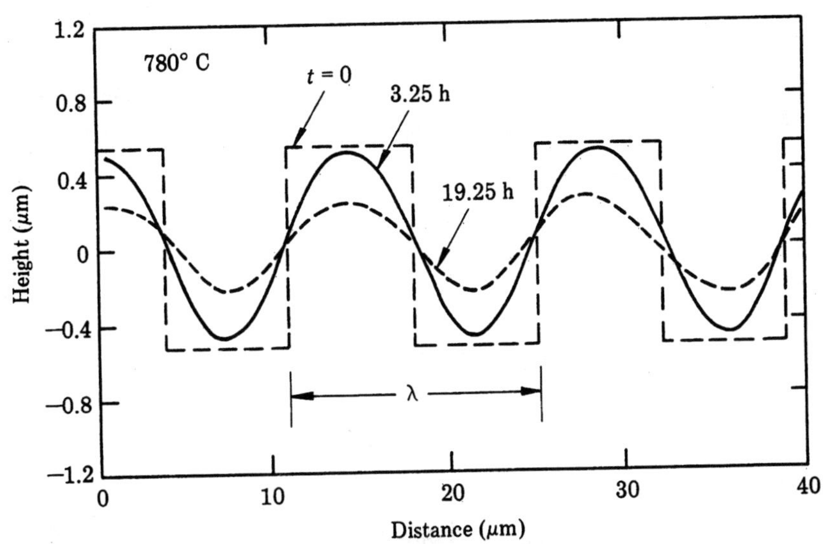

FIGURE 5.13 A square-wave mesa on (100) InP before and after mass transport at 780°C for 3.25 and 19.25 hours in a PH_3 ambient diluted by H_2 to a concentration less than 5% (Liau et al., 1988b).

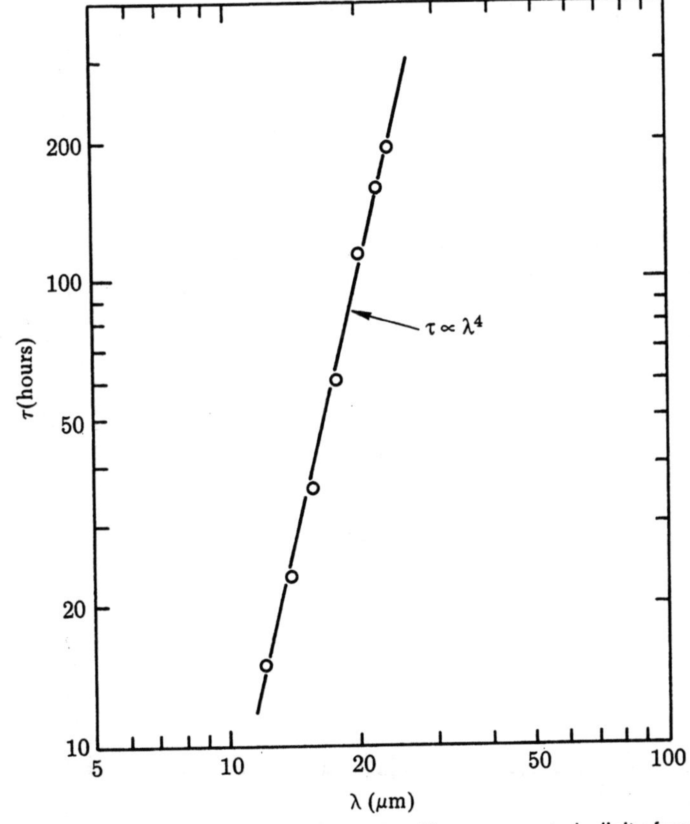

FIGURE 5.14 The decay time of sinusoidal profiles versus periodicity for various period structures on InP annealed at 780°C (Liau et al., 1988b).

We can also estimate a characteristic length, L, for mass transport (similar to the diffusion length in bulk diffusion) by using a trial solution

$$z(x,t) = A \exp\left[-(x + vt)/L\right] \tag{5.42}$$

and substituting in Eq. (5.38). The solutions are

$$\frac{\Gamma}{L^4} = \frac{v}{L} \tag{5.43}$$

Thus $v = \Gamma/L^3$. The trial solution can be written

$$z(x,t) = A \exp\left[-\left(\frac{x}{L} + \frac{\Gamma}{L^4}t\right)\right] \tag{5.44}$$

which for long times can be approximated as

$$z(x,t) \simeq A \exp\left(-\Gamma t/L^4\right) \tag{5.45}$$

This suggests that the characteristic length L_m for mass transport associated with time t has the form

$$L_m = (\Gamma t)^{1/4} \tag{5.46}$$

or

$$L_m = (\gamma \Omega^2 N_0 D_s t/kT)^{1/4} \tag{5.47}$$

Measurement of the transport length on an InP square step shows (Fig. 5.15) the expected $t^{1/4}$ dependence.

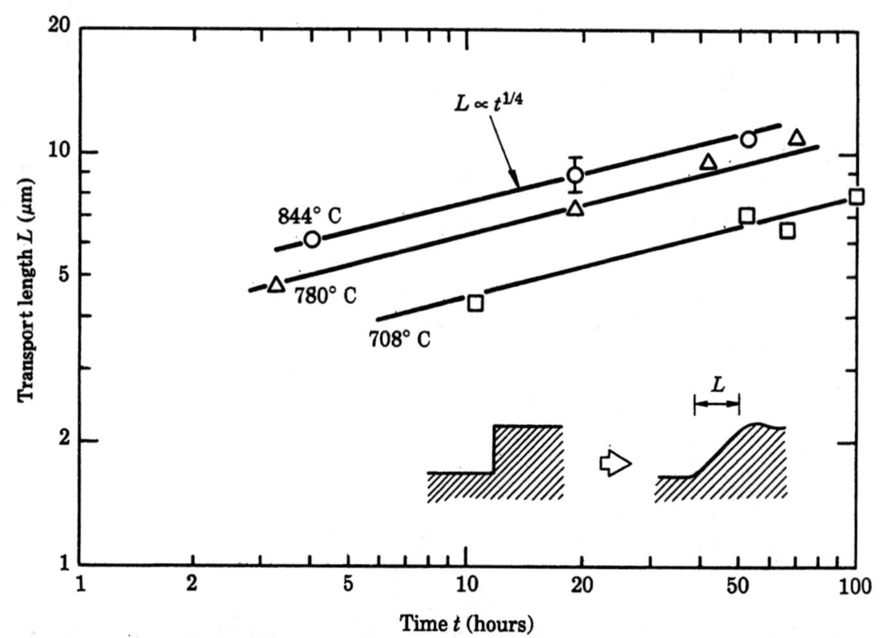

FIGURE 5.15 Transport length L determined from the decay of mesa edges in InP annealed at different temperatures (Liau et al., 1988b).

Mass transport on a curved surface is a material migration phenomenon driven by surface-energy minimization. It has applications to grain boundary grooving and the formation of buried structures and microlenses. It is a factor to consider for preserving the sharpness of gratings in microelectronic structures and for definition of features in semiconductor processing.

5.7 Nucleation of a Surface Step

This chapter has been concerned with the surface morphology formed by a collection of surface atoms. The structures can be generated by the agglomeration of deposited atoms (clusters) or by creation of a surface structure using modern lithographic techniques. The time evolution of the morphology changes have been driven by the concentration gradients set up at curved surfaces in the sense of decreasing the net surface area. In this section we consider the conditions for nucleating a surface feature on a solid. The feature we consider is of the form of a step since steps play such a critical role in film growth. They are the high binding sites for deposited atoms and act as a sink for mobile adatoms.

To create a step on a plane crystal surface, we assume that the equilibrium pressure on the plane surface is p_0. At equilibrium, the fluxes condensing on or desorbing from the surface are equal. For condensation, the flux from Eq. (1.4) is

$$J_c = nv_a \tag{5.48}$$

where n $(= p_0/kT)$ is the number of atoms per unit volume of the vapor, and v_a is the root mean square velocity of the atoms in the vapor. For desorption, we express Eq. 5.1,

$$J_c = J_0 = N_0 \nu_S \exp\left(-\frac{\Delta G_{\text{des}}}{kT}\right) \tag{5.49}$$

where N_0 is the number of adatoms per unit area of the plane surface, and ΔG_{des} is the activation energy of desorption.

To create a step on the plane surface, we consider the formation of a circular disk of atomic layer height a and radius r on the surface (see Fig. 5.16a). We assume that the disk is epitaxial to the plane surface, so that there is no extra interfacial energy. Surface energy per unit area of the circumference (the extra surface energy of the disk) is

$$E_d = 2\pi r a \gamma \tag{5.50}$$

This exerts a pressure on the disk,

$$p = \frac{1}{A}\frac{dE_d}{dr} = \frac{2\pi a \gamma}{2\pi r a} = \frac{\gamma}{r} \tag{5.51}$$

where A is the area of the perimeter of the disk. Under this pressure, the energy of each atom in the disk is increased by the amount

$$p\Omega = \frac{\gamma\Omega}{r} \tag{5.52}$$

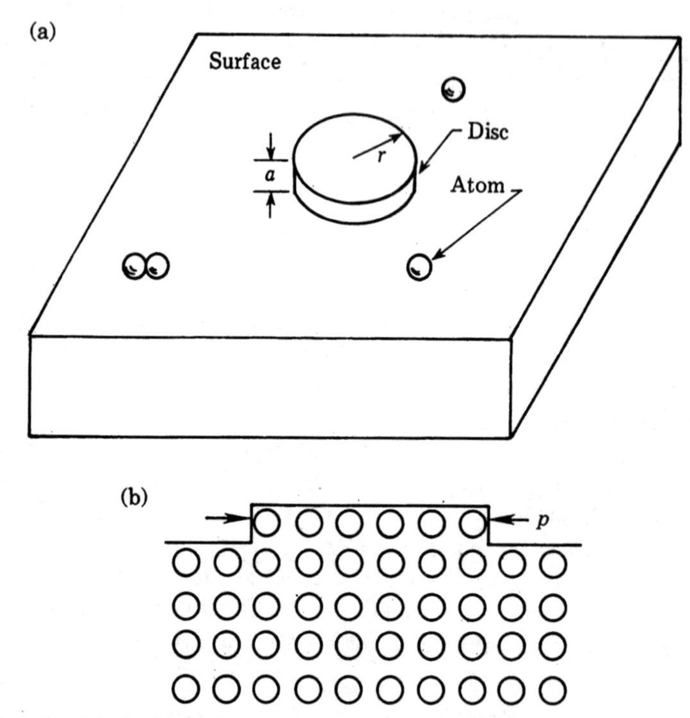

FIGURE 5.16 (a) Nucleation of a disc of radius *r* and step height *a* on a surface. (b) A schematic cross-sectional view of the disk. The circumferential surface exerts a compressive pressure on the disk.

where Ω is the atomic volume. Because of the increase in energy, all the atoms in the disk can sublime more easily than atoms in the plane surface, and the rate of sublimation is increased to

$$J_c' = N_0 \nu_S \exp\left(-\frac{\Delta G_{\text{des}}}{kT} + \frac{\gamma\Omega}{rkT}\right) \tag{5.53}$$

Then

$$\frac{J_c'}{J_c} = \exp\left(\frac{\gamma\Omega}{rkT}\right) \tag{5.54}$$

When the disk is small as in the nucleation stage, we can take this ratio to be equal to the ratio of vapor pressures on the disk and on the plane surface, since J_c and J_c' are directly proportional to their vapor pressures at a given temperature. Then

$$\frac{p}{p_0} = \exp\left(\frac{\gamma\Omega}{rkT}\right) \tag{5.55}$$

This equation is analogous to equation 5.12b aside from a factor of 2. This difference is simply attributed to the different shapes under consideration: For a hemisphere $dA/dV = 2/r$; for a cylinder of fixed height, $dA/dV = 1/r$. Since the atoms in the disk can desorb more easily, the result is a higher vapor pressure above the disk. The higher vapor pressure is due to a larger number of atoms per unit volume in the vapor (n'), which in turn increases the impingement flux (i.e., $J'_c = n'v_a$) of atoms on the disk surface. When these two processes reach equilibrium, the equilibrium vapor pressure on the disk is greater than the equilibrium vapor pressure on a flat surface by the ratio given in Eq. (5.54).

The equilibrium state of the disk depicted above is not stable; the disk will either grow or shrink in radius. The net change in energy in nucleating the disk of radius r and step height a is

$$\Delta E_d = 2\pi ra\gamma - \pi r^2 a\Delta E_S \tag{5.56}$$

where ΔE_S is the latent heat of sublimation per unit volume and γ is the surface energy per unit area of the circumference of the disk. The first term is the extra surface energy created while the second term is the decrease in energy associated with bonding free atoms. We define a critical radius r_{crit} such that when $r = r_{\text{crit}}$, we have

$$\frac{d\,\Delta E_d}{dr} = 0 \tag{5.57}$$

$$r_{\text{crit}} = \frac{\gamma}{\Delta E_S} \tag{5.58}$$

and the net change in energy of the disk having the critical radius is

$$\Delta E_{\text{crit}} = \pi r_{\text{crit}} a\gamma \tag{5.59}$$

which is half the surface energy of the critical disk. Fig. 5.17 shows that the critical disk is not stable because ΔE_{crit} is a maximum, so that a slight deviation from r_{crit} in either direction will lead to a decrease in energy. For example, if the disk at a pressure p has increased its radius, the sublimation rate is reduced but the condensation rate remains the same, and the disk will grow larger. Conversely, the disk will shrink if its radius is reduced below the critical value.

Since the disk has to achieve the critical size to grow, we regard ΔE_{crit} as the activation energy of nucleation and r_{crit} as the critical nucleus size. To calculate the number of nuclei, we note that the number of atoms on the unit area of the plane surface is equal to the product of the impinging flux and the residence time τ_0. So the number of critical nuclei per unit area is

$$\begin{aligned} N_{\text{crit}} &= J\tau_0 \exp\left(-\frac{\Delta E_{\text{crit}}}{kT}\right) \\ &= J\tau_0 \exp\left(-\frac{\pi r_{\text{crit}} a\gamma}{kT}\right) \end{aligned} \tag{5.60}$$

where

$$r_{\text{crit}} = \frac{\gamma\Omega}{kT \ln\left(\dfrac{p_{\text{crit}}}{p_0}\right)} \tag{5.61}$$

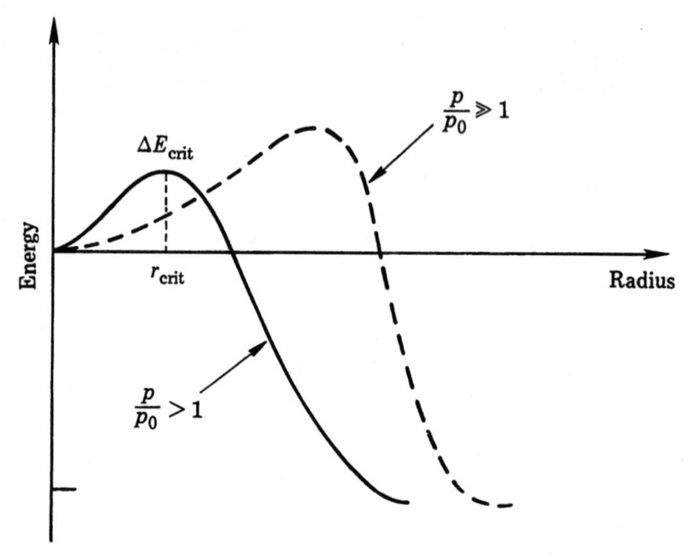

FIGURE 5.17 Activation energy of the critical nucleus as a function at supersaturation.

and p_{crit} is the pressure corresponding to the critical nucleus. This last result corresponds to Eq. 5.55. The quantity $\left[\left(\frac{p_{crit}}{p_0}\right) - 1\right]$ is generally defined as the supersaturation needed for nucleation.

It is clear that as p_{crit} approaches p_0 the critical nucleus size gets very large and the energy ΔE_{crit} becomes large. Spontaneous nucleation of clusters or steps is very unlikely. The nucleation generally requires a high supersaturation.

We shall estimate the value of p_{crit}/p_0 needed for the nucleation of a monolayer-thick disk of Si on a plane Si surface according to Eq. (5.61). We can grow a monolayer of Si in 10 seconds in molecular beam epitaxial growth. Since one monolayer needs 10^{15} atoms/cm^2, the minimum flux of Si needed for the growth is 10^{14} atoms/cm^2-sec. Using $J = n_1 v_a = 10^{14}$ atoms/cm^2 sec, we obtain $n_1 = 10^9$ atoms/cm^3 which corresponds to a pressure of 3×10^{-8} torr. If we maintain the Si substrate at 1223 K, the equilibrium vapor pressure is 3.2×10^{-7} Pa ($\sim 2.4 \times 10^{-9}$ torr) (see Table 6.1). The supersaturation is thus about 10. Indeed, such a growth condition is typical in molecular beam epitaxial deposition.

Now if we examine Eq. (5.60), we begin with

$$\Delta E_{crit} = \pi r_{crit} a \gamma = \frac{\pi \gamma^2 a \Omega}{kT \ln\left(\frac{p_{crit}}{p_0}\right)} \tag{5.62}$$

The product γa^2 is equal to the surface energy per atom, which for a plane Si surface is about 0.6 eV/atom as given in Table 2.3. However, the surface energy per atom at the circumference of the disk should be higher because of its curvature, and we shall take γa^2 to be about 1 eV/atom. For a substrate temperature of 1223 K, the value of kT is about 0.1 eV/atom. Then,

$$\Delta E_{\text{crit}} = \frac{10\pi}{\ln\left(\frac{p_{\text{crit}}}{p_0}\right)} \text{ eV/atom}$$

where $\Omega = a^3$. At a low supersaturation (e.g., $p_{\text{crit}}/p_0 = 2.7$), we have $\Delta E_{\text{crit}} =$ 31.4 eV/atom. This is unreasonably high, indicating that nucleation would not occur at low supersaturations.

A reasonable nucleation density for growth would be one site for each 1 μm^2. This is an appropriate number since the diffusion length is of this order. Using Eq. 5.60 this density of $10^8/\text{cm}^2$ would correspond to an unreasonably large value for p_{crit}/p_0.

Nucleation of a step on a "perfect" surface is referred to as *homogeneous* nucleation. It is an unlikely process at a low supersaturation due to the large value of ΔE_{crit}. On real surfaces there is always an ensemble of imperfections; structural defects such as vacancies and steps, or chemical defects such as impurities. These defects often act as efficient nucleation centers for the conglomeration of atoms. This process is known as *heterogeneous* nucleation, which is the dominant process in real thin film growth.

TABLE 5.4 Overview of Notation Used in Kinetic Processes Occurring on Surfaces

1. **Frequencies, ν, times, τ, and distances, λ;**
   - ν_S–surface vibration frequency
   - ν_{des}–frequency of desorption
   - τ_0–residence time on surface, $\tau_0 = 1/\nu_{\text{des}}$
   - λ_S–diffusion length on surface, $\lambda_s = (D_s\tau_0)^{1/2}$
   - λ–jump distance between sites
2. **Energies; E, G and W**
   - E_S/A–surface energy per unit area $= \gamma$
   - E_{S0}–surface energy per atom
   - ΔG_{des}–activation energy, desorption, $\nu_{\text{des}} = \nu_S \exp(-\Delta G_{\text{des}}/kT)$
   - ΔG_S–activation energy, diffusion, $D_S = \lambda^2\nu_S \exp(-\Delta G_S/kT)$
   - W_S–energy to remove atom from kink site
   - W–total energy to remove atom from surface, $W = W_S + \Delta G_{\text{des}}$
3. **Numbers per cm^2 of atoms on surfaces, N**
   - N_S–number of sites (or atoms) per unit area
   - N_0–equilibrium number of adatoms on flat surface, $N_0 = N_S \exp(-W_S/kT)$
   - N_r–number of adatoms on cluster at radius r, $N_r = N_0 \exp(2\gamma\Omega/rkT)$
   - N_{ad}–number of adatoms on arbitrary surface, $N_{ad} = N_0 \exp(E_{S0}/kT)$
4. **Pressures, p, and particle fluxes, J (atoms/cm^2·sec)**
   - p_0–equilibrium pressure on flat surface
   - p–pressure on curved surface, $p = p_0 \exp(2\gamma\Omega/rkT)$
   - J_0–equilibrium impinging flux, $J_0 = p_0(2\pi mkT)^{-1/2}$

The calculated value of supersaturation for nucleating a surface step is orders of magnitude greater than the value used in actual deposition. This is the dilemma of classical nucleation theory, which has failed to explain the extremely fast crystal growth rate at a low supersaturation. For this reason, the pre-existence of steps on a crystal surface was assumed and the Burton–Cabrera–Frank (BCF) theory of dislocation or step-mediated crystal growth was developed. The BCF theory is discussed in the following chapter.

For the convenience of the reader, Table 5.4 gives an overview of some of the notation used to describe surface processes.

References

1. J. C. Bean, *Physics Today* 2, 8 (October, 1986).
2. A. Y. Cho in *Molecular Beam Epitaxy and Heterostructures,* edited by L. L. Chang and K. Ploog, NATO ASI Series, *Series E,* No. 87, Martinus Nijhoff, Amsterdam (1985).
3. J. Crank, *The Mathematics of Diffusion,* 2nd edition, Clarendon Press, Oxford (1975).
4. L. Eckertova, *Physics of Thin Films,* 2nd edition, Plenum Press, New York (1986).
5. Z. L. Liau and J. N. Walpole, *Appl. Phys. Lett.* 40, 568 (1982).
6. Z. L. Liau, V. Diadiuk, J. N. Walpole, and D. E. Mull, *Appl. Phys. Lett.* 52, 1859 (1988a).
7. Z. L. Liau, H. J. Zeiger, and J. N. Walpole, *Solid State Research Report* 3, Lincoln Lab (1988b).
8. W. W. Mullins, "Sold Surface Morphologies Governed by Capillarity," *Metal Surfaces*, ASM, Metal Park, Ohio (1963) pp 17–66.
9. C. A. Neugebauer, Chapter 8 in *Handbook of Thin Film Technology,* edited by L. I. Maissel and R. Glang, McGraw Hill, New York (1970).
10. E. H. C. Parker, editor, *The Technology and Physics of Molecular Beam Epitaxy,* Plenum Press, New York (1985).
11. A Zangwill, *Physics at Surfaces*, Cambridge University Press, Cambridge (1988).
12. M. Zinke-Allmang, L. C. Feldman, and S. Nakahara, *Appl. Phys. Lett.* 51, 975 (1987).

Problems

5.1 The growth of Sn clusters on (111) Si at 525 K is shown in Fig. 5.5 with Q given as the number of atoms in a cluster of radius r. At t = 40 min, what is

dQ/dt and how does dQ/dt vary with r for $t \geq 40$ min? Assume a hemispherical cluster and a density of Sn $= 3.6 \times 10^{22}$ atoms/cm^3.

5.2 A 1 cm^2 clean surface is covered with 4×10^{10} hemispherical tantalum clusters with equal numbers that are either 10 or 30 nm in diameter. If ripening allows the large clusters to consume all of the small clusters, what will the reduction of the tantalum's surface energy be if $\gamma_{Ta} = 2890$ ergs/cm^2?

5.3 Find the surface energy E_S of two hemispherical Ti clusters ($\gamma = 1650$ ergs/cm^2) of radius $r = 40$ nm. How much energy will be gained if they coalesce? Ignore the contribution from the interface.

5.4 A deposited film forms clusters whose radius has a (time)$^{1/4}$ dependence ($r^4 \propto t$). The figure for this problem shows r^4 versus time at different temperatures. From this data determine the activation energy of this process.

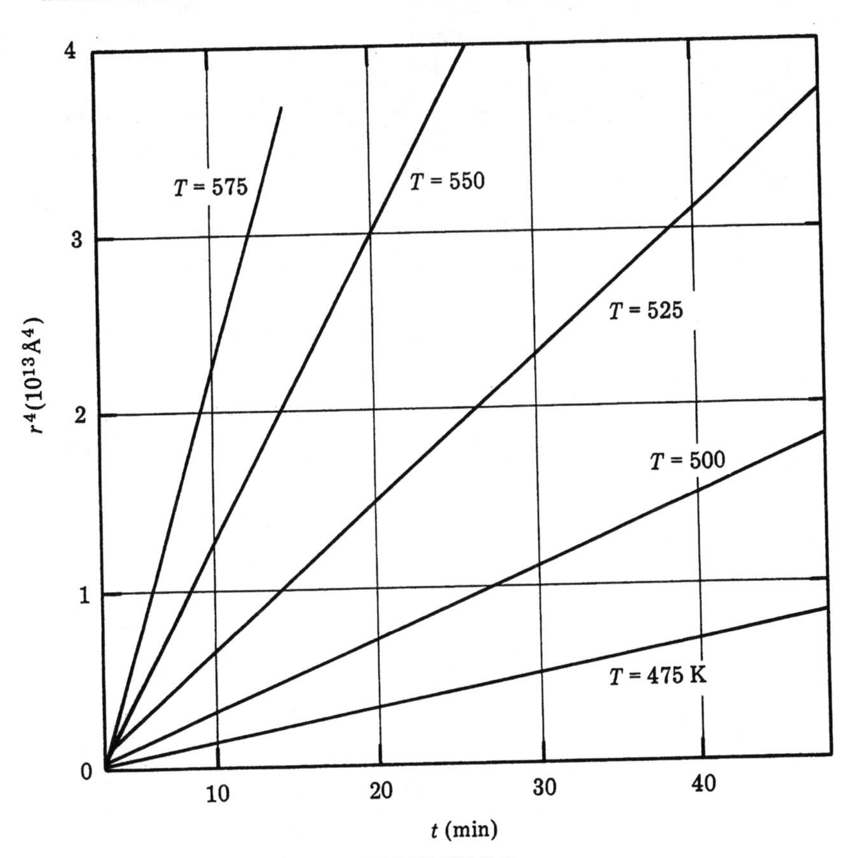

PROBLEM 5.4

5.5 From Eq. (5.20), calculate the cluster radius at 10 min for the case where the surface diffusion coefficient at 550 K is 10^{-9} cm^2/sec on a (100) surface where $N_0 = 10^{14}$/cm^2 and $\gamma = 10^{15}$ eV/cm^2. Assume $L = 3$ and $\Omega = 2 \times$

10^{-23} cm^3. If the hemispherical clusters are equally spaced on the sample surface at a distance $2r$ apart, how many atoms/cm^2 of material would be in the cluster?

5.6 You produce a square-well grating of InP with a periodicity λ of 1 μm. At what temperature would you heat the sample for 1 hour so that the transport length L equalled the periodicity? Use the data in Fig. 5.15 and assume that the surface diffusion coefficient D_S has an activation energy of 1.5 eV.

5.7 Given that at time $t = 1200$ sec, a cluster has radius $r = 1$ μm at temperature $T = 800$ K, at what time did $r = 0.5$ μm (at 800 K)? Assume that the activation energy for surface diffusion = 1.5 eV.

5.8 A hemispherical cluster is in the process of ripening. Calculate the time at which the pressure over the cluster will be 2.5 times the equilibrium pressure. $N_{eq} = 10^{15}$ cm^{-2}, $\gamma = 2890$ ergs/cm^2, $\Omega = 2.96 \times 10^{-23}$ cm^3, $T = 500$ K, $Lr = 10^{-6}$, $D_S = 10^{-10}$ cm^2/sec.

5.9 Determine the total surface energy difference between two raindrops whose radii are 0.1 mm and 0.2 mm. Assume that the drops are perfect hemispheres and that $\gamma = 10^{15}$ eV/cm^2 for water.

5.10 Suppose a cluster of 35 water molecules on a surface approximates the shape of a hemisphere. Calculate the vapor pressure over this cluster at 25°C given the following data: $\gamma = 73.05$ dyn cm^{-1}, vapor pressure $= 3.13 \times 10^{-2}$ atmospheres, density $= 0.99$ g cm^{-3}, molecular weight $= 18.02$ g mol^{-1}.

CHAPTER 6

Homoepitaxy: Si and GaAs

6.0 Introduction

Integrated circuits are formed on single-crystal semiconductors, usually silicon or gallium arsenide. In most modern semiconductor processing, these single crystals are formed from polished wafers cut from large ingots. A thin (typically 100 nm to 1 micron) epitaxial layer of the same semiconductor material is then grown on top of the wafer. This epitaxial layer of silicon (for example) on the single-crystal silicon substrate is grown so that the crystal structure has a smooth and continuous transition from substrate crystal to epitaxial layer, without disorder, impurities or misaligned atoms at the interface (dashed line in Fig. 6.1). Dopant atoms are added during epitaxial growth in order to form a layer with controlled conductivity that will have the correct electrical behavior for subsequent treatment leading to a finished device.

In terms of the discussion on morphology in Chapter 5, the epitaxial layer is required to be planar without clusters. In the present chapter we concentrate on homoepitaxy where by definition there is no lattice mismatch between layer and substrate. In Chapter 7 we consider the case of heteroepitaxy where lattice-mismatched materials are involved.

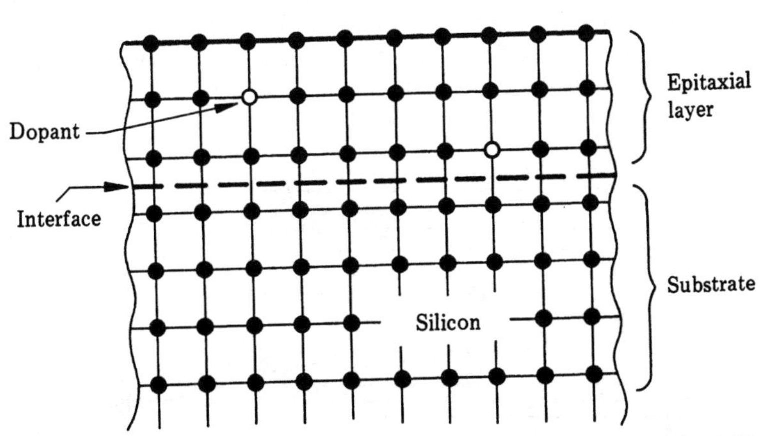

FIGURE 6.1 Schematic representation of the cross-section of a Si epitaxial layer with dopant atoms as grown on a silicon substrate.

6.1 Growth Techniques

Most commercial thin film semiconductor growth employs chemical vapor deposition (CVD). In this process, reactants are transported in the vapor phase to the substrate surface where they are adsorbed. A chemical reaction takes place on the surface leading to formation of the desired film and to reaction products which are desorbed and transported away from the surface. One method for silicon uses the silane reaction, in which silane gas reacts with silicon at ~900°C (Fig. 6.2):

$$SiH_4 + Si_{substrate} \rightarrow 2\ Si + 2H_2 \tag{6.1}$$

Dopants may be introduced also through the gas phase: PH_3 for phosphorus, AsH_3 for arsenic and B_2H_6 for boron. This is the most common way of growing a silicon thin film. Gallium arsenide can also be grown by CVD, in this case MOCVD, metal–organic chemical vapor deposition. A typical example would be trimethyl gallium, $Ga(CH_3)_3$, plus arsine (AsH_3) on a gallium arsenide substrate held at 550°C.

$$Ga(CH_3)_3 + AsH_3 \rightarrow GaAs + 3CH_4\ (gas) \tag{6.2}$$

These vapor deposition techniques are the most useful for large scale production. However, because of the complex chemistry they are difficult to describe microscopically in terms of surface diffusion, desorption, and so on. Furthermore, they are difficult to control on the monolayer scale which is necessary when extremely sharp interfaces and sharp doping profiles are necessary.

An alternative to CVD is molecular beam epitaxy (MBE). It requires ultrahigh vacuum, an atomically clean surface and an impinging beam of atoms. The MBE technique creates ultrathin films and is amenable to mathematical analysis; we can use kinetic theory to describe how the material grows. The name molecular beam epitaxy is used because the first applications of MBE were for gallium arsenide growth. In gallium arsenide MBE, one uses a gallium oven (Knudsen cell) and an arsenic oven in a vacuum system. The arsenic comes off as a molecule, As_2 or As_4, depending on the temperature. Hence, the name molecular beam epitaxy. A better name for silicon molecular beam epitaxy, would be silicon atomic beam epitaxy, because the silicon flux is an atomic beam.

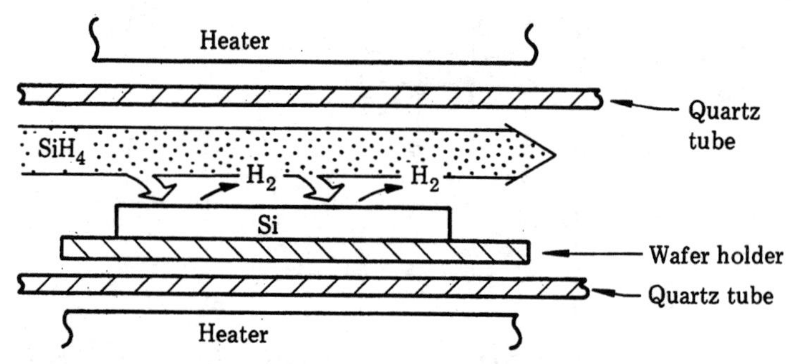

FIGURE 6.2 Sketch of a chemical vapor deposition chamber with reactant gas (silane) flowing over and reacting with a wafer of silicon.

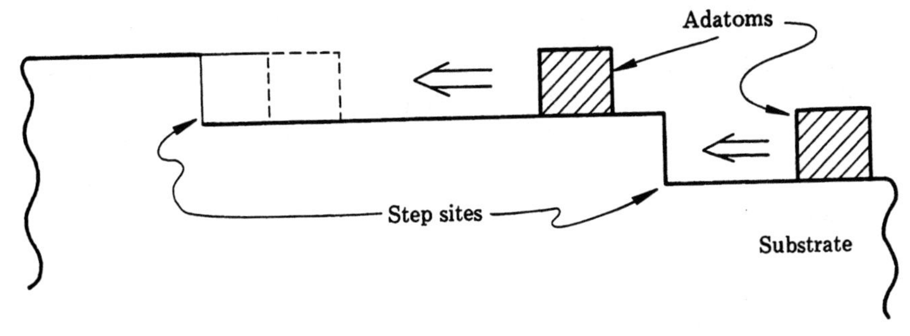

FIGURE 6.3 Representation of step-mediated growth where adatoms migrate to step sites with high binding energies.

Molecular beam epitaxy is pristine from an analytical point of view, for it employs a flux of atoms impinging on a clean surface. The picture we are going to develop is step growth associated with homoepitaxy. Every surface has steps, of course, and the steps come about most commonly because the surface is cut at a slight angle to the atomic plane. Atoms impinge on the surface from the elemental source. At high temperature, atoms will be adsorbed on the surface, undergo diffusion, and eventually arrive at the high-binding-energy site of a step (Fig. 6.3). In this picture, the step can be thought of as moving across the surface. We will calculate the velocity of steps due to the impinging beam. This is called the Burton–Cabrera–Frank model (BCF) of epitaxial growth. The discussion of this process follows that given by Allen and Kasper (1988).

We will also consider the lower temperature limits for epitaxial growth. Ultimate device designs require the lowest temperature processing possible for the sharpest dopant profiles. At low temperatures the picture of epitaxy is two-dimensional nucleation, monolayer by monolayer, since there is not sufficient thermal energy for diffusion to a step. In this low-temperature range the rate of deposition is also a critical parameter.

6.2 Fundamental Concepts

6.2.1 Solid Vapor Equilibrium

A solid in equilibrium at a given temperature has atoms leaving and returning to the surface; the fluxes are equal. The equilibrium flux density J_0 of incident atoms is proportional to the equilibrium vapor pressure p_0:

$$J_0 = \left(\frac{1}{2\pi MkT}\right)^{1/2} p_0 = 3.51 \times 10^{22}\, p_0/(MT)^{1/2} \tag{6.3}$$

where T is absolute temperature and M is the molecular weight. In this relation the flux is in units of $\text{cm}^{-2}\text{-sec}^{-1}$ and the pressure in units of torr. Equation (6.3) is a result of the derivation presented in Appendix A (see Eq. (A.11).

Typical vapor pressures of Si at different temperatures are given in Table 6.1; note that 1 pascal (Pa) $= 7.5 \times 10^{-3}$ torr.

TABLE 6.1 Values of Vapor Pressure p_0 and Flux Density J_0 for Silicon*

T(K)	823	923	1023	1123	1223
p_0(Pa)	2.5×10^{-16}	2.7×10^{-13}	7.1×10^{-11}	6.9×10^{-9}	3.2×10^{-7}
$J_0\left(\frac{1}{\text{cm}^2\text{-sec}}\right)$	4.4×10^{2}	4.3×10^{5}	1.1×10^{8}	1.0×10^{10}	4.6×10^{11}

*From Allen and Kasper (1988), p. 65.

One monolayer of material is about 10^{15} atoms/cm^2, and a reasonable growth rate is one monolayer/sec. Therefore, based on values in Table 6.1, the equilibrium vapor pressure of the material is always much smaller than the impinging flux. It is the impinging flux that is of importance here; the equilibrium between the vapor pressure and the substrate is upset by the impinging flux. Etching corresponds to upsetting the equilibrium in the opposite direction.

6.2.2 Characteristic Surface Diffusion and Binding Energies

Figure 6.4 shows a step on the surface and a kink site. Atoms can either bind at a kink site or diffuse along the surface. Such diffusing atoms are called *adatoms*. We consider the case where the incident flux is zero. The equilibrium density of atoms on the surface, N_0, is equal to the number of atomic sites, N_s (refer to Eq. 1.14) on the surface, times the probability of detachment

$$N_0 = N_s \exp(-W_s/kT) \tag{6.4}$$

where W_s is the amount of energy it takes to remove an atom from the kink site to the surface.

In MBE growth, an impinging flux of atoms on the surface upsets the equilibrium density of surface adatoms. The step is a sink, which results in a concentration gradient of adatoms. The atoms diffuse along the surface with a diffusion coefficient (described by Eq. 5.4)

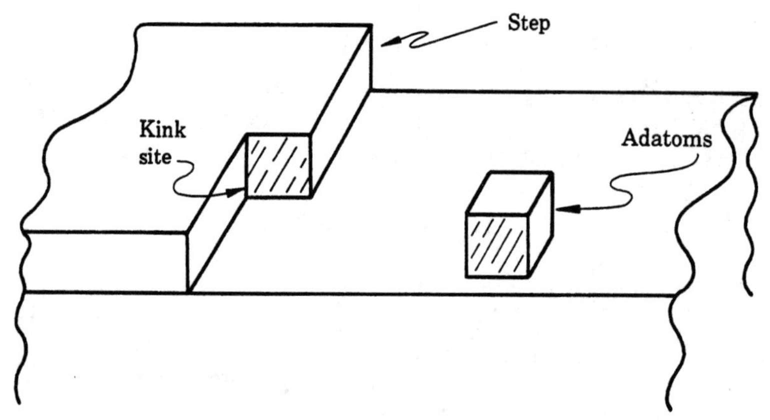

FIGURE 6.4 Schematic of a step on the surface with a kink site.

$$D_s = \lambda^2 \nu_s \exp(-\Delta G_s/kT) \tag{6.5}$$

where λ is the interatomic spacing, ν_s is the vibrational frequency, and ΔG_s is the activation energy for surface diffusion.

There is a probability for desorption and a mean residence time, τ_0 given by

$$\frac{1}{\tau_0} = \nu_s \exp(-\Delta G_{\text{des}}/kT) \tag{6.6}$$

where ΔG_{des} is the binding energy of the adatom to the surface. The process of evaporation may be thought of as atoms leaving the kink site and desorbing from the surface. The total energy required to take an atom from solid silicon is

$$W = W_s + \Delta G_{\text{des}} \simeq 4.5 \text{ eV} \tag{6.7}$$

This is the sum of the energy needed to remove the atom from the kink to the surface (W_s), and to remove it from the surface to the vapor (ΔG_{des}). We know the total energy W extremely well from vapor pressure measurements; it is 4.5 eV for silicon.

6.2.3 Path Length on the Surface

The diffusion length on the surface is limited by desorption and is given by

$$\lambda_s = (D_s \tau_0)^{1/2} = \lambda \exp\left[(\Delta G_{\text{des}} - \Delta G_s)/2kT\right] \tag{6.8}$$

The diffusion length is the mean path length due to diffusion before desorption.

6.2.4 Vapor Pressure

At equilibrium the desorbing flux is given by

$$\begin{aligned} J_0 &= \frac{N_0}{\tau_0} = N_s \exp(-W_s/kT)\, \nu_s \exp(-\Delta G_{\text{des}}/kT) \\ &= N_s\, \nu_s \exp(-W/kT) \end{aligned} \tag{6.9}$$

Measurement of the vapor pressure as a function of temperature is used to determine W. Values of J_0 are given in Table 6.1.

6.2.5 Supersaturation

In MBE, the impinging flux of atoms onto the surface creates an excess of adatoms, called the supersaturation σ. This is, by definition,

$$\sigma = \left(\frac{J_{\text{Si}}}{J_0} - 1\right) >> 1 \tag{6.10}$$

where J_{Si} is the impinging flux and J_0 the equilibrium flux.

For practical growth, the values of J_{Si} are about a monolayer per second or 10^{15} atoms per centimeter^2 per second. For $J_{\text{Si}} = 2 \times 10^{15}$, σ is always much greater than unity as shown in Table 6.2 (for a growth rate of 0.4 nm/sec on (100) Si where a monolayer has 6.78×10^{14} atoms/cm^2 and a height $h = a/4$ or 0.135 nm).

TABLE 6.2 Values of Supersaturation σ at $J_{Si} = 2 \times 10^{15}$/cm^2-sec*

T(K)	723	823	923	1023	1123	1223
σ	3×10^{16}	4.5×10^{12}	4.7×10^{9}	1.8×10^{7}	2×10^{5}	4.4×10^{3}

*From Allen and Kasper (1988), p. 65.

6.2.6 Characteristic Energies

Values of ΔG_s (activation energy for surface diffusion) and W_s (energy for removal from a kink site) are needed to calculate explicitly MBE growth rates. Such quantities cannot be measured directly and are estimated from simple surface models. We do know the value W, the total energy required to remove a Si atom from a solid surface. At this stage we can only estimate W_s by bond-breaking arguments. For a simple cubic lattice it is clear that an adatom has only one bond to the surface, that an atom near a step has two bonds, and that atoms at kinks have three or more bonds depending on the kink configuration. Thus we estimate W_s to be a fraction of the total bonding energy depending on the bond configuration. For a step associated with the perfect surface of silicon, such bond-breaking arguments suggest that $W_s = 3W/4$ for the (100) surface and a simple step configuration. The assumption of a perfect surface is clearly oversimplified, as we know that atoms on semiconductor surfaces are not in bulk-like positions. For $W = 4.5$ eV, $\Delta G_{\text{des}} = W - W_s = 1.12$ eV for the (100) surface. For ΔG_s, the range of experimental values of the activation energy for Si self-diffusion on Si (100) surfaces varies from 0.25 to 1.0 eV. For Si (111) the values are much higher, from about 0.8 to 1.6 eV. We shall use a value of 0.5 eV in our discussion of (100) surface step-mediated growth. Table 6.3 gives values for growth parameters on (100) Si.

6.2.7 Steps

There are three ways that a surface acquires steps:

6.2.7.1 A Bulk Dislocation Intercepts a Surface. A bulk defect extending through the material represents a misarrangement of atoms. The termination of this defect gives rise to a surface step. In the highest-quality electronic-grade material, the density of bulk dislocations is very small. It ranges between 1 and 1000 per cm^2. The best silicon can be made with less than 1 dislocation per square centimeter, and

TABLE 6.3 Growth Parameters on (100) Si with $W = 4.5$ eV, $\lambda = a = 0.543$ nm, $\nu = 10^{13}$/sec, $\Delta G_s = 0.5$ eV and $\Delta G_{\text{des}} = 1.1$ eV

	25°C	**300°C**	**500°C**	**700°C**
kT (eV)	0.0257	0.0494	0.0666	0.0838
D_s (cm^2/sec)	5.2×10^{-11}	5.9×10^{-7}	8.1×10^{-6}	3.8×10^{-5}
τ_0 (sec)	3.8×10^{5}	4.7×10^{-4}	1.4×10^{-6}	5.0×10^{-8}
λ_s (μm)	44.4	0.16	0.033	0.014

gallium arsenide generally has about 1,000 dislocations per square centimeter. The spacing between dislocation-formed steps is of the order of one over the square root of the bulk dislocation density. This spacing turns out to be about 0.03 centimeter or greater. This is a very large step spacing (or small step density) compared to that produced by other step-causing mechanisms.

6.2.7.2 Crystal Miscut. The major contributor to steps on a surface is referred to as *miscut* (see Fig. 6.5 for an illustration). Miscut occurs simply because a crystal is never cut exactly parallel to a crystal axis. For a small angle of miscut, the surface plane will be described by extended terraces of major planes and steps with a given periodic spacing. For a miscut of angle θ and steps of height h, the step spacing L_0 is given by (Fig. 6.5)

$$L_0 = \frac{h}{\tan \theta} \tag{6.11}$$

Typically, crystals are cut to a tolerance of 0.1°, corresponding to a step spacing of about 80 nm for monolayer height steps.

6.2.7.3 Nucleation of Clusters. For growth on a surface, there is a certain probability that atoms will cluster and stick before they bind at a step site. Such a two-dimensional cluster may be thought of as a step, in that a diffusing adatom can attach and become immobile. It is this kind of two-dimensional cluster growth that describes the epitaxial growth process at low temperature. Nucleation of such two-dimensional clusters, steps, is discussed in Section 5.7.

6.2.8 Natural Density of Steps

Defects occur in a solid at any finite temperature. For example, the number of vacancies in a solid is proportional to $\exp(-\Delta G_f/kT)$, where ΔG_f is the amount of energy required to create a vacancy. Similarly a step is a surface defect. However a step formed on a planar surface has an extremely high activation energy (see Chapter 5, Section 5.7); hence there is no natural density of steps on a surface. This is true

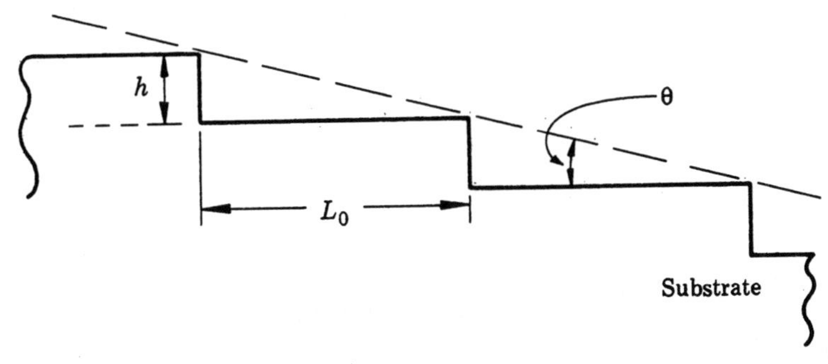

FIGURE 6.5 Cross section of terraces with monolayer steps and step spacing determined by the miscut angle θ.

until the temperature approaches the melting temperature, when the surface begins to get rough (produces a natural density of steps). Epitaxy occurs well below the roughening temperature.

6.3 Growth Modes of Homoepitaxy

The ability to grow an epitaxial layer depends primarily on surface diffusivity and on a variety of experimental factors (surface cleanliness, deposition rate, surface miscut, etc.). Generally our picture of semiconductor homoepitaxy is governed by substrate temperature. The different configurations of the deposited layers are illustrated schematically in Fig. 6.6. At the lowest temperatures (room temperature or below) impinging material will generally be deposited as an amorphous layer. There is little surface diffusivity at such a low temperature and deposited atoms get trapped into a conglomerate of noncrystalline sites. Heat treatment is necessary for atoms to make the necessary rearrangement to form an epitaxial layer. At higher deposition temperatures, the atoms have sufficient surface mobility to arrange themselves epitaxially on the surface. Epitaxial islands are formed. At yet higher temperatures growth occurs by lateral growth of steps. We indicate the transition temperatures in Fig. 6.6 with the understanding that the concept of a transition temperature applies to realistic vacuum conditions and growth rates. In principle, step-mediated growth

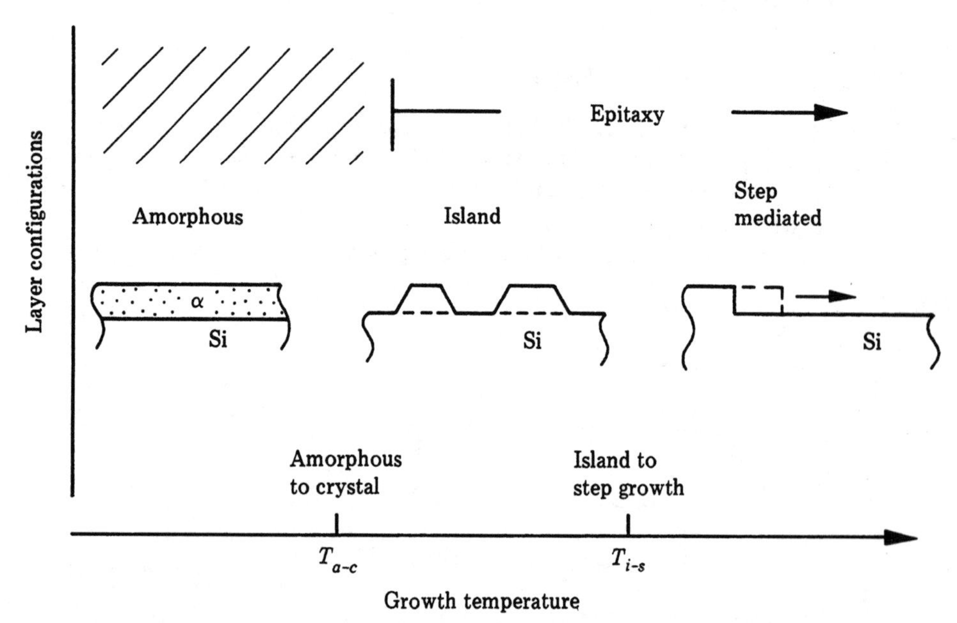

FIGURE 6.6 Configuration of deposited layer (amorphous to epitaxial) versus growth temperature showing the transition temperatures. The actual transition temperatures depend on surface preparation, deposition rate, substrate orientation and other parameters.

of epitaxial layers could occur at any temperature in a perfect vacuum if the deposition rates were slow enough.

The temperature dependence can be understood on a more quantitative basis by consideration of the surface diffusion coefficient. In general such diffusion coefficients are not well known, but there has been enough work on a few systems that we can make some estimates. From the previous discussion of surface diffusion coefficients we can estimate the time for an atom to make a single jump. If D_s is written as

$$D_s = \lambda^2 \nu_s \exp(-\Delta G_s/kT) \tag{6.12}$$

the mean time for a single jump, τ_s, is given by

$$\lambda^2 = D_s \tau_s \tag{6.13}$$

$$\tau_s = \frac{1}{\nu_s} \exp(\Delta G_s/kT)$$

As usual $\nu_s \simeq 10^{13}$/sec and we choose $\Delta G_s = 1.0$ eV, so at room temperature $\tau_s \simeq 10^4$ sec $\simeq 2.8$ hours. If the deposition rate is R monolayers/sec, other atoms will fall in the vicinity of the initial deposited atom in a time $1/R$. We need time for at least one jump before other atoms impinge. If we require at least one jump before completion of a monolayer, this implies a growth rate of a monolayer every 2.8 hours.

It is possible to do epitaxy at such low temperatures. In practice, such slow growth is limited by the quality of the vacuum system. Even at a pressure of 1×10^{-10} torr, a full monolayer worth of residual gases (O_2, H_2, N_2, etc.) impinges on the sample every 10^4 sec. If only a small fraction of impinging gas atoms stick, they contaminate the surface and ruin the epitaxial growth.

Because of the exponential dependence on temperature of the diffusion coefficient, the relevant times change dramatically with temperature. For example at 252°C (525 K) τ_s is $\simeq 3.9 \times 10^{-4}$ sec, and under typical conditions there are many possible jumps with deposition rates of 1 to 10 monolayers/sec. In this case growth is pictured as the assembly of two-dimensional epitaxial clusters. This argument is very sensitive to the choice of ΔG_s, the activation energy for surface diffusion. A value of $\Delta G_s =$ 0.5 eV implies $\tau_s = 3 \times 10^{-5}$ sec rather than 10^4 sec, a change of nine orders of magnitude. Nevertheless the general $\Delta G_s/kT$ scaling is correct and practical limits are set by experimental conditions.

Higher binding energy locations near surface steps are favored sites for deposited atoms. If atoms have sufficient mobility (diffusivity), they can diffuse to a step and reach the high-temperature regime associated with step-mediated growth. We can estimate the temperature required for this growth by assuming that steps are separated by ~100 lattice constants ($100a$) and asking what is the temperature required so that $\tau_{100a} = (100a)^2/D_s$. We take R as 10 monolayers/sec, a typical MBE rate. Then

$$\tau_{100a} < 1/R$$

$$\frac{10^4}{\nu_s} \exp(\Delta G_s/kT) < 0.1$$

yields $T > 270°C$. In fact this argument gives only a rough idea of the quantitative temperatures involved in epitaxy, but the sequence of growth modes is well established experimentally.

One type of experiment that gives information on growth modes is the dynamic electron diffraction known as RHEED, reflection high-energy electron diffraction. In this technique electrons are diffracted from the first monolayer of an ordered surface, and the diffraction intensity indicates the perfection of long-range order on the surface. One of the advantages of RHEED is that the diffraction intensity can be

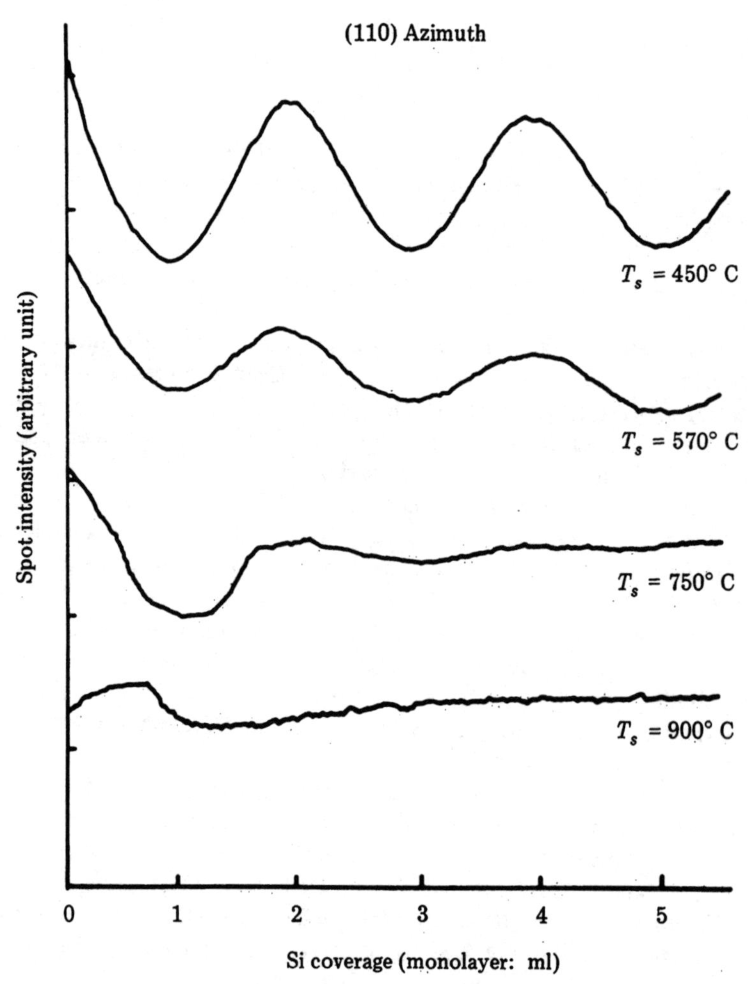

FIGURE 6.7 Reflection high-energy electron diffraction (RHEED) intensity oscillations versus silicon coverage in monolayers (ML) at different substrate temperatures (T_s) for the 2 × 1 reconstruction spot intensity during Si MBE (from Ichikawa, 1989).

measured during growth. The "two-dimensional island growth" mode gives a characteristic RHEED signal in the form of an oscillating intensity, with each oscillation corresponding to the deposition of one monolayer worth of material (Fig. 6.7). The maximum in the diffraction intensity occurs when the layer is complete while the minimum occurs at $\frac{1}{2}$ monolayer coverage, corresponding to maximum disorder in the sense of a random array of two-dimensional epitaxial clusters. As the substrate temperature is increased the RHEED oscillations decay and eventually vanish, signaling the onset of step-mediated growth. In this latter growth mode, the surface appears as a series of flat terraces at all times, and yields an essentially constant diffraction intensity.

A second experiment which illustrates the different Si MBE growth modes is described by Tung and Schrey (1989). In their technique, the surface step structure is imaged by formation of a low-temperature epitaxial silicide grown immediately following Si MBE. Each step on the original Si surface manifests itself as a dislocation in the epitaxial silicide. The dislocation network, easily seen with transmission electron microscopy (TEM), represents a visual image of the deposited atom configuration on the grown surface.

Figure 6.8 shows an array of three TEM micrographs for growth at different temperatures on Si (111). The highest temperature growth, at 750°C, clearly shows the flat terraces and step structure associated with step-mediated growth. Lower-temperature growth, at 650°C, has a disordered array associated with two-dimensional cluster formation.

In summary, homoepitaxy is associated with three different growth modes dependent primarily on the substrate temperature. At low temperature, there is amorphous layer deposition which requires subsequent thermal treatment for epitaxy. At intermediate temperature, structural epitaxy occurs via two-dimensional cluster formation. At higher temperature, there is epitaxy via step-mediated growth.

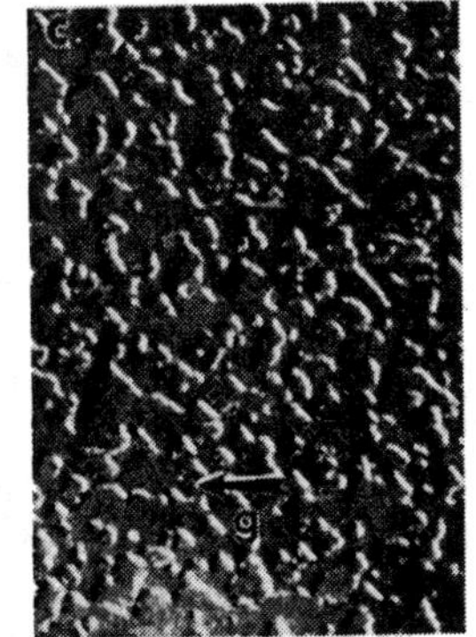

FIGURE 6.8 Plane-view transmission electron diffraction micrographs (TEM) of lattice-matched $CoSi_2$ grown on Si surfaces after Si MBE growth, for (a) as prepared surface, for (b) deposition temperature of 750°C, and for (c) deposition temperature of 650°C (from Tung and Schrey, 1989).

6.4 Step-Mediated Growth: The High-Temperature Regime

This model is an outgrowth of the Burton–Cabrera–Frank (BCF) theory of crystal growth. It describes diffusion of atoms to a step and considers the step as an infinite sink (Fig. 6.9). It includes the probability that atoms will desorb. The motion of atoms to the step is related to the velocity of steps. Every time a step moves past a particular point, another monolayer is grown. The rate of growth in the direction normal to the substrate surface is given by the product,

$$\text{Rate of growth} = \mathcal{N}_{\mathrm{L}} v h \tag{6.14}$$

where $\mathcal{N}_{\mathrm{L}}$ = number of steps per unit length, v = step velocity, and h = the height of the step.

6.4.1 Density of Atoms on the Surface

The net flux of atoms to the surface per unit area is

$$J_v = J_{\mathrm{Si}} - N_{ad}/\tau_0 \tag{6.15}$$

where J_{Si} is the incident flux of atoms, N_{ad} is the density of adatoms on the surface, and the second term is the loss due to the desorption (see Eq. (6.9)). In Chapter 5, Eq. (5.32), we defined N_{ad} with respect to a change of surface energy by curvature. Here we define N_{ad} with respect to supersaturation from an incident flux.

Surface diffusion is driven by the concentration gradient of atoms on the surface and the surface diffusion coefficient. The flux to the steps per unit length is:

$$J_s = -D_s \frac{\partial N_{ad}}{\partial y} \tag{6.16}$$

The boundary condition imposed on the flux equation is that the density of adatoms at the step is the equilibrium concentration of adatoms, N_0. The equation of continuity

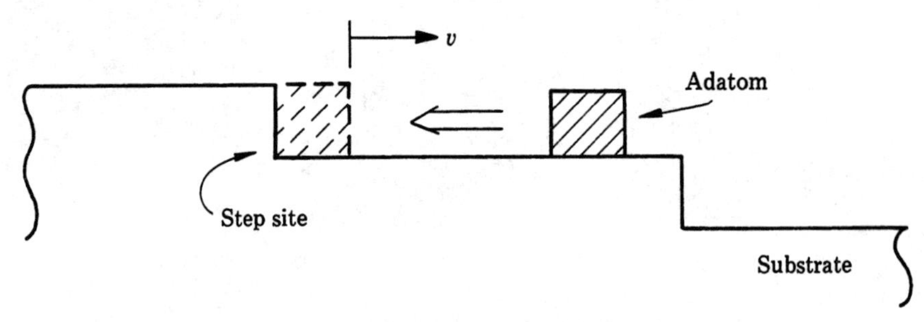

FIGURE 6.9 Representation of the migration of adatoms to steps leading to a lateral velocity v of steps across the surface.

relates J_s (per unit length) and J_v (per unit area) in steady state:

$$\frac{\partial J_s}{\partial y} = J_v \tag{6.17}$$

These two equations yield the differential equation:

$$-D_s \frac{\partial^2 N_{ad}}{\partial y^2} = J_{\text{Si}} - N_{ad}/\tau_0 \tag{6.18}$$

which can be rewritten using Eqs. (6.8) and (6.10)

$$-\lambda_s^2 \frac{\partial^2 N_{ad}}{\partial y^2} + N_{ad} = \tau_0 J_0(\sigma + 1) \tag{6.19}$$

This is the partial differential equation for diffusion of adatoms to steps on the surface. The atoms are supplied by the impinging flux, undergo surface diffusion, and arrive at the step. A single step solution is

$$N_{ad} = N_0 \{1 + \sigma[1 - \exp(-y/\lambda_s)]\} \tag{6.20}$$

where y is the distance from the step. The fact that this is a solution of Eq. (6.19) can be verified by the reader.

6.4.2 Periodic Array of Steps

A real surface has a finite miscut angle, with an array of steps spaced a distance L_0 apart. We treat this array as perfectly periodic, again with the condition that $N_{ad} = N_0$ at each step. The solution is

$$\frac{N_{ad}}{N_0} = 1 + \sigma\left[1 - \frac{\cosh(y/\lambda_s)}{\cosh(L_0/2\lambda_s)}\right] \tag{6.21}$$

This is for a periodic array of steps spaced a distance of L_0 apart with the origin ($y = 0$) at the midpoint between steps (Fig. 6.10).

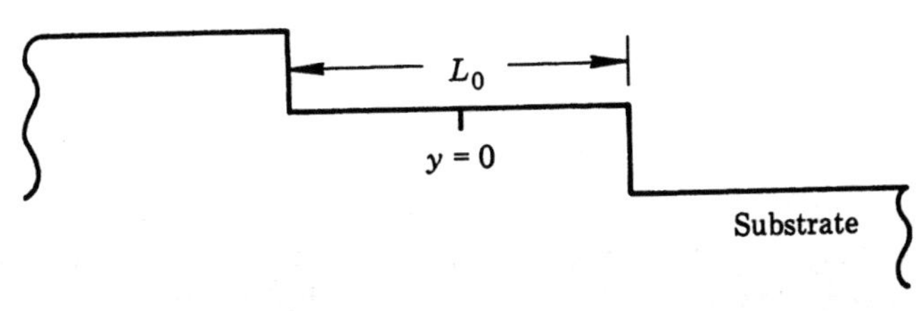

FIGURE 6.10 Geometry for step-mediated growth for a periodic array of steps spaced L_0 apart with $y = 0$ at the midpoint.

6.4.3 Growth Rate

In a diffusion problem the step velocity is related to the diffusion flux by

$$N_s v = J_s\Big|_{\text{step}} = -D_s \frac{\partial N_{ad}}{\partial y}\Big|_{\text{step}} \tag{6.22}$$

where N_{ad} is the density of adatoms on the surface and the flux is evaluated at the step edge. For a single step the origin is at the step, while for the periodic steps the origin is $y = L_0/2$. The relationship between the flux J_s and the velocity given by Eq. 6.22 is easily understood. The flux represents the number of atoms/cm/sec arriving at the step. If a is the surface lattice constant then the number of atoms/length of step is $1/a$ and the time interval Δt to accumulate this number of atoms is $\Delta t = (1/a)/J_s$. In this time the step moves a distance a so the velocity is given by $a/\Delta t$ or J_s/N_s, where $N_s = a^{-2}$. Then for a single step, using Eq. (6.20),

$$\frac{\partial N_{ad}}{\partial y} = \frac{\sigma N_0}{\lambda_s} \exp(-y/\lambda_s) \tag{6.23}$$

For $y = 0$, this yields a velocity of the step given by using Eq. (6.9) and Eq. (6.10),

$$v = \lambda_s(J_{\text{Si}} - J_0)/N_s \tag{6.24}$$

where the velocity of the step is in the opposite direction to that of the diffusional flux yielding a change in sign. In a typical MBE experiment J_{Si} is much greater than J_0. Therefore,

$$N_s v = \lambda_s(J_{\text{Si}} - J_0) \simeq \lambda_s J_{\text{Si}} \tag{6.25}$$

and the rate of crystal growth is

$$R = v\mathcal{N}_L h = \frac{\lambda_s J_{\text{Si}} \mathcal{N}_L h}{N_s} \tag{6.26}$$

It is interesting to compare this diffusion-limited growth rate with the rate if the silicon atoms just came down to the surface and stuck there. The rate of growth R_{ND} for the no diffusion model is

$$R_{\text{ND}} = \frac{J_{\text{Si}}}{N_s/h} \tag{6.27}$$

Then,

$$\frac{R}{R_{\text{ND}}} = \lambda_s \mathcal{N}_L = \frac{\lambda_s}{L_0} \tag{6.28}$$

The result is the ratio of path length (diffusion length before desorption) to step spacing. As long as this quantity is greater than one, atoms can reach a step, and step-mediated growth is possible. The factor is called the condensation coefficient η. For the single step, $\eta = \lambda_s \mathcal{N}_L$. For the periodic array,

$$\eta = \lambda_s \mathcal{N}_L \tanh(L_0/2\lambda_s) \tag{6.29}$$

The most critical parameter, the temperature dependence of the growth rate, is included in λ_s. This includes surface diffusion and surface desorption. Under typical MBE conditions, desorption is never very probable.

As a rule of thumb, our model requires atom diffusion to a step and excludes island formation resulting from two or more atoms. This condition is met by requiring that an atom diffuse a step length $1/\mathcal{N}_L$ in a time less than N_S/J_{Si}, the inverse of deposition rate. The time to diffuse a length $1/\mathcal{N}_L$ is $(1/\mathcal{N}_L)^2/2D_s$. Therefore,

$$\frac{(1/\mathcal{N}_L)^2}{2D_s} < N_S/J_{Si}$$

$$J_{Si} < 2D_S N_S \mathcal{N}_L^2$$

In terms of monolayers the relation may be written as $J_{Si}/N_S < 2D_S\mathcal{N}_L^2$. Estimating D_S from Eq. 6.12 with $\lambda = 3.84 \times 10^{-8}$ cm, $\nu_s = 10^{13}$/sec, $\Delta G_s = 0.5$ eV and $\mathcal{N}_L = 10^5$/cm, the maximum flux rate may be calculated as $\sim 10^5$ monolayers/sec at 550°C and < 1 monolayer/sec at 25°C. This imposes a temperature dependent limit on J_{Si}; if J_{Si} exceeds this critical quantity dimer and cluster formation is possible and less perfect growth may occur. In practice another mechanism takes over at this point. It is known as solid-phase epitaxial regrowth and is described in Chapter 10.

6.5 Step Periodicity in MBE

The basic concept in the step-growth picture is that deposited atoms diffuse along a terrace of crystal and become bound at a step (a strong bonding site). It is this concept that is central to the BCF model of growth. In this section we show that this growth mode can also lead to a more uniform distribution of step spacings.

A miscut sample will not necessarily provide a periodic terrace size distribution. While the average spacing of steps is given by the ratio of the step height to the miscut angle, it is clear from numerous surface science investigations that step spacings can vary considerably. This situation may be overcome through the growth of a buffer layer of the same material as the substrate. Assuming that the buffer layer itself grows in a step-growth mode, the surface step distribution may become more periodic. In a submonolayer deposition each terrace accumulates a fraction of atoms proportional to its size. The atoms diffuse to the terrace edge (the step) where they become part of the adjacent terrace. Larger terraces tend to shrink and smaller terraces tend to grow toward the average value. The mathematical description given in Appendix D provides an analytical description of this process to establish the approach to periodicity as a function of coverage and for different step-growth models. The calculation assumes a perfectly random starting distribution (the worst case scenario).

6.6 Low Temperature Epitaxy

It is natural to ask whether there is a lowest temperature for epitaxy. We discuss this question both with respect to practical considerations and in terms of fundamental surface parameters. Following Jorke (1989), we think of deposited atoms as being in epitaxial, covalent sites or nonepitaxial sites. We apply this dicussion to Si MBE because of its relative simplicity.

An atom acquires a covalent or epitaxial site according to the probability P, where P is given by

$$P = 1 - \exp(-\nu t) \tag{6.30}$$

and ν is the "hopping" rate for diffusion (Eq. 5.3),

$$\nu = \nu_s \exp(-\Delta G_s/kT) \tag{6.31}$$

In these relations t is the time after deposition, ΔG_s is the activation energy for surface diffusion, and ν_s is the surface frequency, about 10^{13}/sec.

We assume that surface diffusion essentially gets turned off when additional impinging atoms form a dimer or larger cluster with the initial atom. For a deposition rate R, this will happen in a time of $t_m = 1/R$. The probability for occupation of the epitaxial site is then $P(t_m) = 1 - \exp(-\nu t_m)$. Note that as the rate R goes to zero the probability for epitaxy goes to infinity. Thus in this picture there is no real lower temperature limit for epitaxial growth, as long as the rate of deposition is low enough.

From a practical point of view this very low rate may not be realizable, as the residual vacuum in the system will correspond to impurity species impinging and sticking. At 10^{-10} torr about one monolayer of impurities impinge every three hours (10^4 sec); thus lower deposition rates are not possible for the growth of high-quality material (barring substantial improvements in vacuum).

It is interesting to build on the Jorke model and describe the growth of a thin film. The nth layer of the film will have an epitaxial fraction f_n given by

$$f_n = f_{n-1}P(t_m,T) = f_0 P^n(t_m,T) \tag{6.32}$$

Clearly $P < 1$ so that each layer is less perfect during film growth. This analysis implies that for a given temperature there may be a "critical thickness" where epitaxy is no longer established. Actually the model implies a gradual deterioration of epitaxy, but we may define a thickness where $f_n = 0.5$, is the critical thickness for epitaxy. This yields an exponential dependence of the critical thickness Δ of form

$$\Delta = 0.5 \exp(-\nu t) \tag{6.33}$$

where we have expanded $(1 - e^{-\nu t})^n \simeq 1 - n \exp(-\nu t)$. At modest growth rates (~ 0.1 nm/sec) and high temperatures ($T > 550$°C), Δ becomes virtually infinite.

Two comments should be noted. First, Eq. (6.32) implies that epitaxial layers are always imperfect since f_n is, by definition, always < 1. This conclusion however, only applies in the temperature regime associated with two-dimensional island formation and growth. At higher temperatures the surface diffusion rate becomes so high that atoms can always find a step and an associated high-binding-energy, covalently bonded epitaxial site. Different mechanisms govern this high temperature regime and high-quality epitaxy is possible. Second, in this model the originally flat surface is pictured as forming facets and hills and valleys due to the low-temperature deposition. After some thickness the surface is simply too rough for successful growth.

6.7 GaAs Growth: MBE and MOCVD

Molecular beam epitaxy can best be described as a highly controlled deposition process. Epitaxial layers are grown by impinging thermal beams of molecules or atoms upon a heated substrate. The apparatus used for the MBE growth of GaAs and $Al_xGa_{1-x}As$ is shown schematically in Fig. 6.11. The principal components in this system are the resistance-heated source furnaces (called effusion or Knudsen cells), the shutters, and the heated substrate holder. There is surface analysis equip-

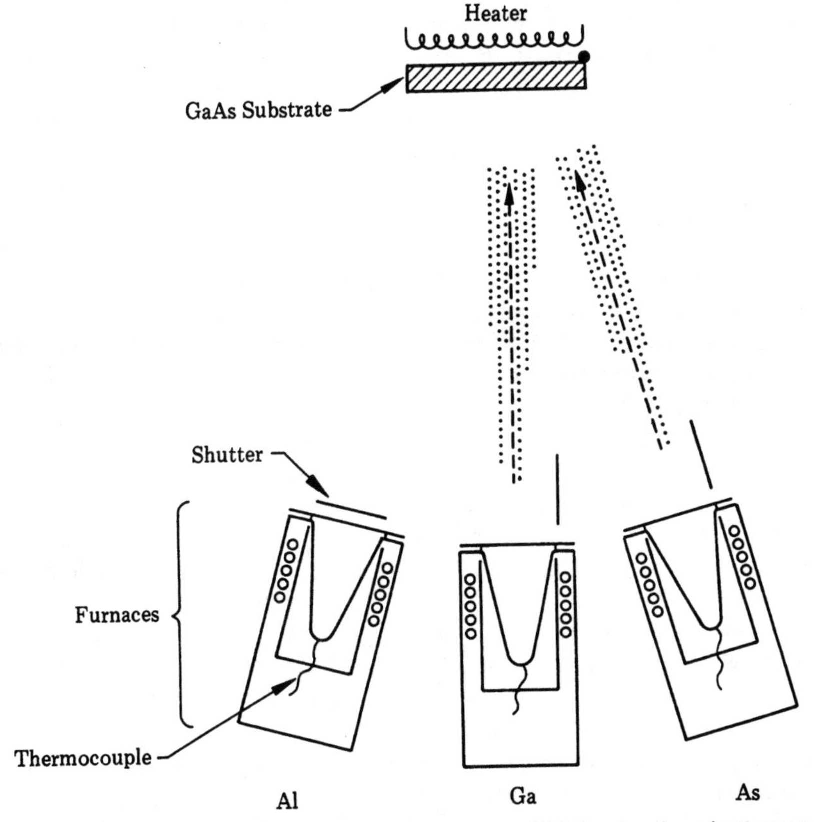

FIGURE 6.11 Schematic of molecular beam epitaxy (MBE), showing shutters open on the Ga and As Knudsen cells and the beams incident on the heated GaAs substrate.

ment attached to the growth chamber to monitor surface cleanliness and surface crystallography before and during growth. The effusion cells have small apertures (radius = 1 cm) for the vapor to escape in a molecular beam. The flux J_c or total number of atoms escaping through the aperture per unit area per second is (see Eq. (6.3))

$$J_c = \frac{p(N_A)^{1/2}}{(2\pi MRT)^{1/2}} = 3.5 \times 10^{22} \frac{p}{(MT)^{1/2}} \tag{6.34}$$

where p is the pressure (torr) inside the effusion cell, N_A is Avogadro's number, M is the molecular weight, R is the gas constant, and T(K) is the temperature of the cell. If the substrate is positioned at a distance from the aperture and directly in line with it, the flux J_s or total number of molecules per second striking the substrate of unit area is

$$J_s = 1.12 \times 10^{22} \frac{pA}{L^2(MT)^{1/2}} \frac{\text{molecules}}{\text{cm}^2\text{-sec}} \tag{6.35}$$

where A is the area of the aperture and L is the linear dimension of the substrate, and $J_cA = J_sL^2$. Take Ga as an example. At $T = 970$°C (1243 K) the vapor pressure is 2.2×10^{-3} torr, and $M = 70$. For $A = 5$ cm^2 and $L = 12$ cm, the value of J_s is 2.94×10^{15}/cm^2-sec. This arrival rate is about five monolayers of Ga or 1.5 nm (15 Å) per second. For the successful growth of GaAs, the arrival rate of arsenic as As_2 or As_4 must be about ten times greater than that of Ga. Arsenic atoms only stick to a Ga surface; therefore, the rate limiting factor in step growth is the arrival rate of the Ga atoms. At the Ga arrival rate of five monolayers per second (2.94×10^{15}/cm^2-sec), the growth rate of a 1 μm thick layer of GaAs takes less than 10 minutes. With a more typical arrival rate of 10^{12} to 10^{14}/cm^2-sec, it may take hours to grow a 1 μm thick GaAs layer.

The growth of GaAs by MBE occurs with fluxes of atomic Ga and molecular As dimers (As_2) or tetramers (As_4) as shown in Fig. 6.12. With growth from As_2, the reaction is one of dissociative chemisorption of As_2 molecules on single Ga atoms. The sticking coefficient of As_2 is proportional to the Ga flux. Excess As_2 is evaporated, leading to the growth of stoichiometric GaAs.

With growth from As_4 molecules, pairs of molecules react on adjacent Ga sites. Even when excess Ga is present, there is desorbed As_4 flux with a maximum sticking coefficient of 0.5. When the As_4 population is small compared to the number of Ga sites, the rate limiting step growth is that of encounter and reaction between As_4 molecules. In most growth conditions, however, the flux of incident As_4 molecules is much greater than the flux of incident Ga atoms, and arriving As_4 molecules will find adjacent sites occupied by other As_4 molecules. The desorption rate then becomes proportional to the number of As_4 molecules supplied. Growth thus proceeds by adsorption and desorption of As_4 with one As atom sticking for each Ga atom on the surface.

The growth of GaAs from As_4 and the resulting surface condition depend on the substrate temperature and the flux ratio of As_4 to Ga, as shown in Fig. 6.13. For growth on (100) and (111) surfaces, the crystal is formed with alternate layers of

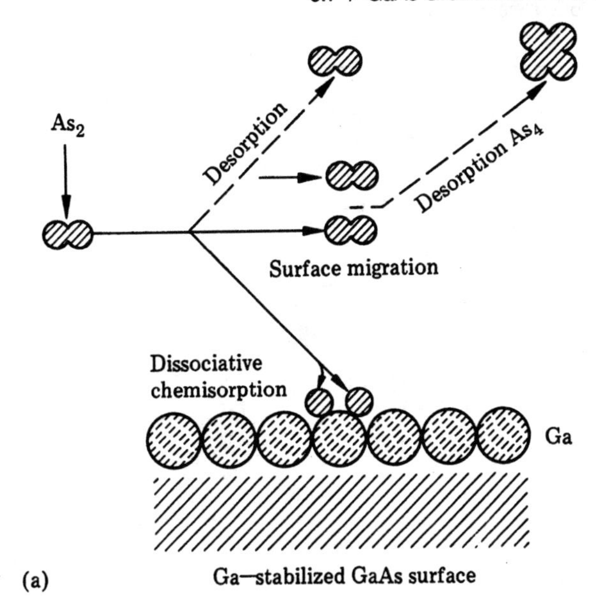

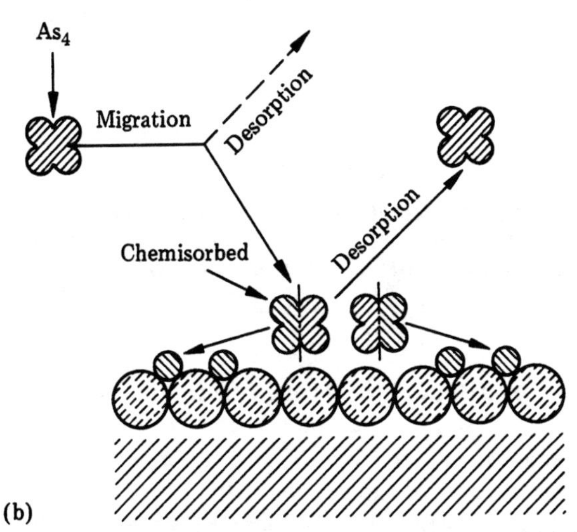

FIGURE 6.12 Model for growth of GaAs from incident fluxes of (a) As_2 and (b) As_4 (adapted from Grange, 1985).

Ga and As atoms. Stable surface structures rich in either Ga or As are formed. On the (100) surface, the Ga-stabilized surface has a centered c (8×2) structure and the As-stabilized surface has a c (2×8) structure. Analysis of the surface by electron diffraction shows that at low temperature and high As/Ga flux ratios an As-stabilized surface is formed, while at high temperature and low As/Ga ratios a Ga-stabilized surface results.

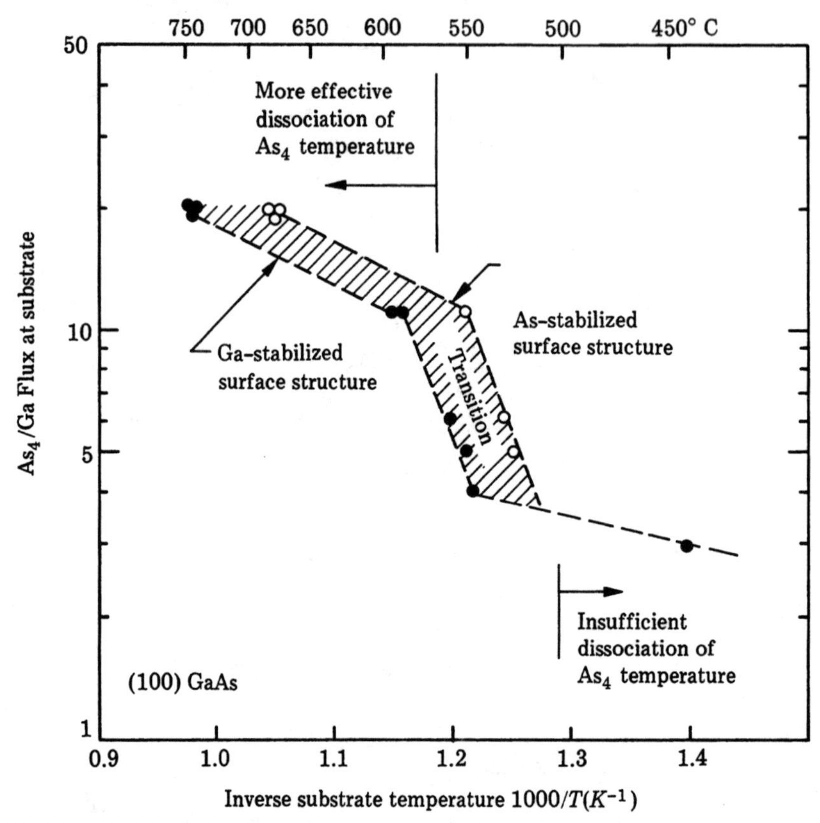

FIGURE 6.13 Molecular beam flux ratio of As_4/Ga at the substrate versus inverse substrate temperature, showing the transition between As-stabilized and Ga-stabilized surfaces on the (100) GaAs surface (from Cho, 1985).

For the growth of AlGaAs, the sticking coefficients of Al and Ga are both close to unity at the usual substrate temperature. Therefore, the Ga to Al ratio in the grown layer is the same as the ratio of Ga to Al flux at the substrate surface.

There are alternate methods of growth of III–V epitaxial layers. The growth of epitaxial materials in metal–organic chemical vapor depositon (MOCVD) or metal–organic vapor phase epitaxy (MOVPE) is typically accomplished by the co-reaction of reactive metal alkyls with a hydride of the nonmetal component. A diversity of chemical growth precursors and growth system designs has allowed for the successful growth of a large number of materials and structures, despite the complex nature of the growth process.

In the growth of III–V compounds, a general overall reaction is given by

$$\begin{gathered} MR_3 \text{ (a metal alkyl)} + XH_3 \text{ (a hydride)} = MX + 3RH \\ \text{where } RH = CH_3, C_2H_5, \ldots \end{gathered} \tag{6.36}$$

The most common example of this reaction is found in the growth of GaAs and

$Al_xGa_{1-x}As$:

$$Ga(CH_3)_3 + AsH_3 \rightarrow GaAs + 3CH_4 \tag{6.37}$$

or

$$xAl(CH_3)_3 + (1 - x)Ga(CH_3)_3 + AsH_3 \rightarrow Al_xGa_{1-x}As + 3CH_4 \tag{6.38}$$

A simplified description of the growth process occurring near and at the substrate surface is shown in Fig. 6.14 for the case of GaAs growth from $Ga(CH_3)_3$ and AsH_3 (Kuech, 1987). The growth ambient generally has a large excess of the Group V constituent over the metal alkyl. Several steps must occur in order for epitaxial growth to proceed. Mass transport of the reactants to the growth surface, their reaction at or near the surface, incorporation of the new material into the growth front, and removal of the reaction by-products must all take place. The slowest step in this sequence will determine the growth rate step. The substrate temperature is typically held at a value substantially higher than the pyrolysis temperature of the metal alkyl, ensuring its rapid decomposition at the growth surface. The growth rate is then limited by the mass transport of the Group III reactant to the growth surface. The surface and gas phase chemical reactions influence the material's properties. The MOVPE growth technique has proven capable of fabricating materials of high purity and excellent morphology.

Another growth technique combines the concepts of MBE and MOVPE. Metal–organic molecular beam epitaxy (MOMBE) can be carried out in an MBE chamber with a gas source of trialkyl Group III metals, $Ga(CH_3)_3$, replacing the Group III elemental sources. A source of As_2 or P_2 beam can be provided by thermal decomposition of the hydrides (AsH_3 or PH_3) in the furnace cells. The organometallic compounds, the alkyls, decompose after impinging on the substrate surface. The chemistry that takes place at the growing surface in MOMBE is more complex than with conventional MBE. Although MOMBE is carried out in an MBE chamber

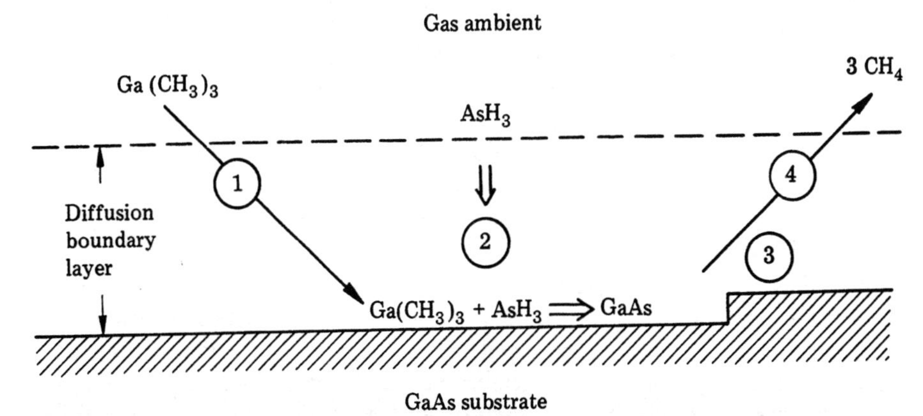

FIGURE 6.14 Surface reactions in metal–organic vapor phase epitaxy (MOVPE), showing transport through the boundary layer, surface reaction, and removal of reaction products (from Kuech, 1987).

(the dopants are supplied from conventional MBE effusion, or Knudsen, cells) the pressure is higher—between 10^{-3} and 10^{-4} torr. This is still in the regime of molecular gas flow and the minimum mean free path is greater than the source-to-substrate distance. The source of material is then in the form of a beam and can be controlled. This is in contrast to conventional chemical vapor deposition where the mean free path for collisions is small and the source is no longer "beam-like."

6.8 Semiconductor Junction and Electronic Potential

We dope semiconductors to make transistors by introducing electrically active impurities, *dopants*. The doping profile has to be nonuniform in depth. We use MBE to grow epitaxial layers to form heterojunction devices; the composition of the layers should change rather abruptly across the junction. The key materials feature of modern electronic devices is that the crystal structure of the active semiconductor is not at chemical equilibrium (i.e., its chemical composition is far from homogeneous). Indeed, it is the use of nonequilibrium structures which enables us to tailor the electrical potential in a semiconductor to produce unique device behavior. By ion implantation, rapid thermal annealing, molecular beam heteroepitaxial growth, and so on, we process nonequilibrium structures to make devices. A nonequilibrium structure is essential.

For a chemical system at equilibrium, we expect that the temperature, the pressure, and the atomic species are homogeneous in the system. Equilibrium is defined specifically by the use of chemical potential,

$$\mu = \frac{\partial F}{\partial N} \tag{6.39}$$

where F and N are Helmholtz free energy and number of atoms per unit volume, respectively. Chemical equilibrium of atoms means that each atom in the system has the same potential energy. The concept of having the same potential energy, however, can not be extended to electrons in equilibrium; it violates the exclusion principle. For electrons, we need to describe their energy and occupancy by a distribution function.

The energy spectrum of electrons obeys the Fermi–Dirac distribution,

$$F_{\mathrm{FD}}(E) = \frac{1}{\exp[(E - \mu)/kT] + 1} \tag{6.40}$$

where μ is defined as the electron chemical potential and $F_{\mathrm{FD}}(\mu) = \frac{1}{2}$; μ is also called the Fermi level. At 0 K, the Fermi level μ is equal to the Fermi energy E_F, which is the highest energy level in the electron distribution in a metal. At $T > 0$ K, μ changes slightly with temperature, and a small fraction of the electrons can have an energy higher than the Fermi level due to thermal excitation (Fig. 6.15a). The thermally excited electrons can be considered to obey Boltzmann's distribution above the Fermi level. In semiconductors containing a small concentration of ionized donor atoms, there are relatively few free electrons in the conduction band, and the

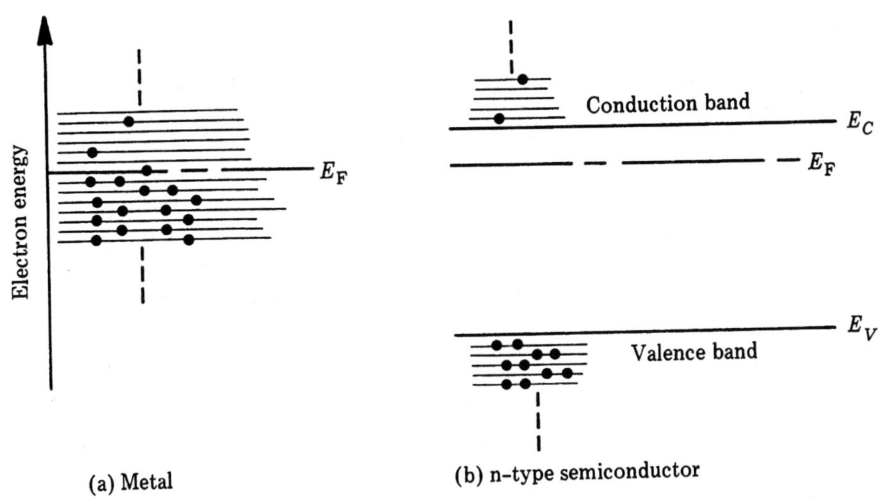

FIGURE 6.15 The location of the Fermi energy E_F at $T > 0$, in (a) metal and (b) *n*-type semiconductor. The spacings between the energy levels are greatly exaggerated to show electron occupation.

Fermi level is located in the energy gap between the conduction and valence bands (Fig. 6.15b). The energy gap is a region of energies in which there are no allowed electron energy levels; the density of states is zero between the conduction and valence bands. The Fermi energy describes the occupation of levels in the two bands. For silicon, Fig. 6.16 indicates the position of the Fermi level within the energy gap for a given concentration of ionized donors (n-type silicon) or acceptors (p-type silicon). As the temperature increases, the Fermi level approaches the intrinsic Fermi level where concentrations of electrons and holes are equal. The slight variation of the energy gap with temperature can also be seen in Fig. 6.16.

Electrical equilibrium in a system is defined as having the same Fermi level everywhere. Hence, when we bring two pieces of solids into contact with each other, electron and/or holes will flow between them to align their Fermi levels in order to reach electrical equilibrium. Then the immobile ions left behind will give rise to an electrical potential.

We must point out that becaue of the extremely high mobility of electrons, the rate of reaching electrical equilibrium is very fast, even at low temperatures. But this is not true for reaching atomic chemical equilibrium in general. It is a much slower process; the redistribution of atomic species depends on atomic diffusion which can be very sluggish—especially in a semiconductor, as we have discussed in Chapter 3. This difference in rates to equilibrium between electrons and atoms is essential in fabricating modern electronic devices. It is a basic concept in device physics that we design the electrical potential in a device by using inhomogeneous compositions: p–n junctions, heterojunction transistors, selectively doped manmade superlattices. To illustrate this point, we shall briefly consider a p–n junction.

A p–n junction in Si is formed by nonuniformly doping the semiconductor with two types of dopants. Across the junction, the concentrations of the donor and the

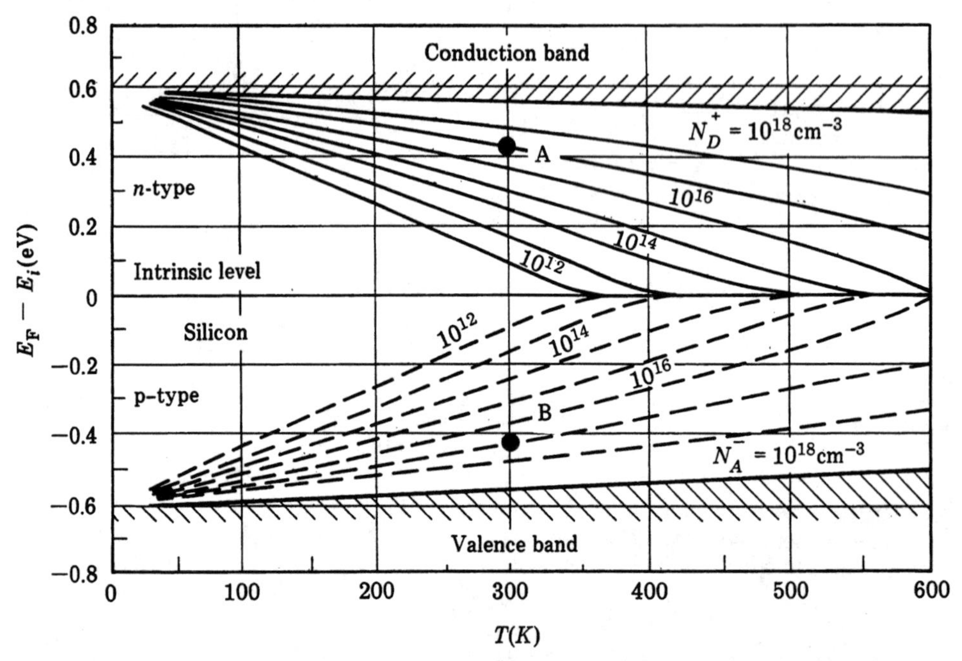

FIGURE 6.16 The Fermi level for *n*-type (solid lines) and *p*-type (dashed lines) silicon as a function of temperature and dopant concentration. The concentrations of ionized donors (N_D^+) and acceptors (N_A^-) per cm^3 are indicated (from Sze, 1981).

acceptor change abruptly. At 300 K, an n—type Si with the Fermi level shown at point A in Fig. 6.16 and a p-type Si with the Fermi level shown at point B in Fig. 6.16 are joined to form a p–n junction. The Fermi levels are aligned to achieve electrical equilibrium which results in the well-known energy band structure shown in Fig. 6.17. There is a built-in potential V_{bi} across the junction. The ionized dopant atoms are immobile in the Si and they sustain the built-in potential. Under an applied field, the potential manifests itself in the rectifying effect on current flow. If the

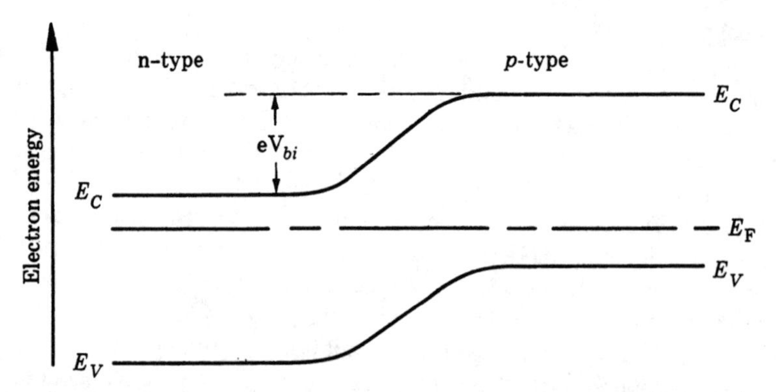

FIGURE 6.17 The energy band diagram of a p–n junction with a built-in potential V_{bi}.

applied field lowers the built-in potential, it enhances electron flow. If it adds to the built-in potential, it retards electron flow.

Clearly, we can homogenize the junction by high-temperature annealing. We achieve chemical equilibrium of the dopants by doing so, but we also destroy the junction. In thin film processes, we grow heteroepitaxial structures, and we react metal and semiconductor to form Schottky barriers. The effect of these structures on electrical potential will be discussed in Chapters 7, 8, and 9.

6.9 Direct and Indirect Band Structure

As an introduction to Chapter 7, we point out that Si, GaAs, and other III–V semiconductors are distinguished by having indirect (Si) or direct (GaAs) band gaps. The electron chemical potential and the Fermi level were introduced in the previous section to describe the occupation of an energy level. Here, we introduce the wave vector **k** to describe indirect and direct band gaps. For any electron there is a wave vector **k** parallel to the momentum **p** of the electron such that

$$\mathbf{p} = \hbar\mathbf{k} \tag{6.41}$$

where $\hbar = h/2\pi$. The $\hbar$ notation is standard in quantum mechanics. From classical mechanics where $E = mv^2/2 = p^2/2m$, we can express the energy of a free electron as

$$E = \frac{\hbar^2 k^2}{2m} \tag{6.42}$$

with spherical energy surfaces where $k^2 = k_x^2 + k_y^2 + k_z^2$.

The same treatment can be carried over to electrons in semiconductors; for low values of k, E is proportional to k^2. However, to account for the differences in mobilities between electrons in Si and those in GaAs, we introduce the concept of an effective mass m^* which is defined by

$$m^* = \hbar^2 \left(\frac{d^2E}{dk^2}\right)^{-1} \tag{6.43}$$

Under an applied force, such as an electric field, the electron can move faster (or slower) than a free electron depending on the curvature of the band. So it appears that the mass of the electron is lighter (or heavier) than the free electron mass. In the low-field region, the drift velocity v_d is given by

$$v_d = \mu\mathscr{E} = \frac{e\tau}{m^*}\mathscr{E} \tag{6.44}$$

where $\mathscr{E}$ is the electric field, μ is the carrier mobility in cm^2/volt-sec, e is the charge on the electron, τ is the average time between collisions and m^* is the effective mass of the carrier. For comparable collision times, the carrier mobility increases with a decrease in effective mass. The use of an effective mass reflects the fact that the carriers travel in real crystals with band curvature that differs from one material to the other. In silicon the effective mass of the electrons m_e^* can be close to that of the free electron, but in GaAs, $m_e^* = 0.067\ m$.

We denote m_e^* as the electron effective mass and m_h^* as the hole effective mass, so that the effective densities of states, N_C and N_V, in conduction and valence are given by

$$N_C = 2\left(\frac{2\pi m_e^* kT}{h^2}\right)^{3/2}$$

and (6.45)

$$N_V = 2\left(\frac{2\pi m_h^* kT}{h^2}\right)^{3/2}$$

Values or N_C and N_V are given in Table 6.4

There are further complications when we treat real crystals in more detail. The band structures for direct-gap (GaAs) and indirect-gap (Si) semiconductors are shown in Fig. 6.18 in the region near $k = 0$. Gallium arsenide is called a direct-gap semiconductor, because the minimum in the conduction band and the maximum in the valence band are located at $k = 0$. When holes and electrons are present, the electrons can make a direct transition to a hole state without a change in momentum.

In Si as well as in Ge and in some of the III–V semiconductors (such as GaP and AlSb), the minimum in the conduction band is located away from $k = 0$ (in Si along the $\langle 100 \rangle$ axes in k space), as shown in Fig. 6.18b. In thermal equilibrium, electrons occupy states at the two minima of the conduction band. An electron cannot make a direct transition to a hole (unoccupied state) near $k = 0$. In order to conserve momentum, there must be an interaction with a lattice atom so that a lattice vibrational wave (phonon) is generated.

Direct-gap semiconductors absorb and emit photons efficiently. Consequently, it is the III–V semiconductors with direct gaps, notably GaAs-based materials, that are most used for light-emitting diodes and semiconductor lasers. Efficient lasers are constructed by growing sandwich structures of smaller band-gap materials such as GaAs between larger gap materials such as $Al_xGa_{1-x}As$. To construct these devices requires heteroepitaxial growth (Chapter 7). The electronic properties of heterostructures are described in Chapter 8.

TABLE 6.4 Values for N_C, N_V, and E_G at 300 K

	Si	GaAs	Ge
E_G (eV)	1.12	1.42	0.66
N_c (cm^{-3})	2.8×10^{19}	4.7×10^{17}	1.04×10^{19}
N_v (cm^{-3})	1.04×10^{19}	7.0×10^{18}	6×10^{18}

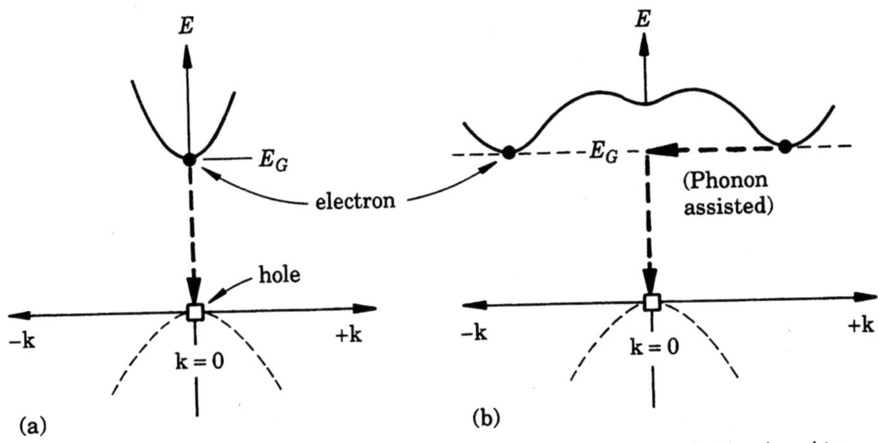

FIGURE 6.18 *E–k* diagrams of electron transitions from the conduction band to empty (hole) states in the valence band for (a) direct band-gap materials such as GaAs and (b) indirect band-gap materials such as Si.

References

1. F. Allen and E. Kasper, "Models of Silicon Growth and Dopant Incorporation," Chapter 4 in *Silicon Molecular Beam Epitaxy* (Vol. I, p. 65), edited by E. Kasper and J. C. Bean, CRC Press, Boca Raton, FL (1988).
2. W. K. Burton, N. Cabrera, and F. C. Frank, *Philos. Trans. Royal Soc. London,* Series A, 243–291 (1951).
3. A. Y. Cho in *The Technology and Physics of Molecular Beam Epitaxy,* edited by E. H. C. Parker, Plenum Press, New York (1985).
4. H. J. Gossmann, F. W. Sinden, and L. C. Feldman, *J. Appl. Phys.* 67, 745 (1990).
5. M. H. Grabow and G. H. Gilmer in *Semiconductor-Based Heterostructures* (pp. 3–19), edited by M. L. Green, J. E. E. Baglin, G. Y. Chin, H. W. Deckman, W. Mayo, and D. Narasinham, The Metallurgical Society, Warrendale PA (1986).
6. J. D. Grange, "The Growth of MBE III–V Compounds" in Chapter 3 of *The Technology and Physics of Molecular Beam Epitaxy,* edited by E. H. C. Parker, Plenum Press, New York (1985).
7. M. Ichikawa, "Crystallographic Analysis and Observation of Surface Micro-area Using Microprobe Reflection High-Energy Electron Diffraction," *Materials Science Reports* 4, 147–192 (1989).
8. H. Jorke, H.-J. Herzog and H. Kibbel, *Phys. Rev.* B40, 2005 (1989).
9. E. Kasper and J. C. Bean, "*Silicon-Molecular Beam Epitaxy,*" Vol. I and II. (CRC Press, Boca Raton, Florida, 1988).

10. T. F. Kuech, "Metal-Organic Vapor Phase Epitaxy of Compound Semiconductors," *Materials Science Reports*, Vol. 2, Number 1, 1987.

11. R. T. Tung and F. Schrey, *Phys. Rev. Lett.* 63, 1277 (1989).

Problems

6.1 On (100) Si ($a = 0.543$ nm) what is the surface density of atoms/cm^2, and what miscut angle along a [100] direction is required to produce a step length of 10^{-5} cm? of 10^{-6} cm?

6.2 Using the values in Table 6.3 for surface diffusion on (100) Si, calculate the values for D_S (cm^2/sec) and τ_0 (sec) at 800°C. How long would it take for an atom to diffuse 10^{-5} cm? 10^{-6} cm? Compare these values with τ_0.

6.3 An As-doped layer of Si is grown on Si(100) by MBE. As has a characteristic desorption time given by $\tau = \frac{1}{\nu_s} \exp(\Delta G_{des}/kT)$.

(a) Assuming that the As and Si coefficients are unity and Si desorption is negligible, show that the fraction of As left on the surface after deposition of one monolayer of Si is given by

$$N_{As} = R\,\tau\left[1 - \exp\left(-\frac{1}{R_{Si}\tau}\right)\right]$$

where N_{As} is the number (fraction) of As monolayers and R is the rate in monolayers/sec.

(b) Evaluate τ for 500°C and 600°C growth assuming $\nu_s = 10^{13}$/sec and $\Delta G_{des} = 0.8$ eV.

(c) Show that for typical MBE deposition rates of 1–10 monolayers/sec,

$$N_{As} = R_{As}\,\tau$$

(d) Assuming constant R_{As}, how should the Si rate at 600°C be adjusted to maintain a doping concentration of 5×10^{18}/cm^3 if the Si rate at 500°C is one monolayer per sec?

6.4 A simple cubic ($a = 0.3$ nm) substrate has been cut along the (100) at a 3° angle to create a stepped surface. If homoepitaxial growth is attempted at 600°C, will the diffusion length be greater than the distance between steps? Use $\Delta G_{des} = 1.0$ eV, $\Delta G_S = 0.5$ eV, and $\nu_S = 10^{13}$/sec.

6.5 Rutherford backscattering spectrometry was used to determine the number of Ga atoms/cm^2 on Si (100) surfaces that had been heated in vacuum for various times at three different temperatures. The Ga desorbed without forming clusters. The figure for this problem shows the number of Ga atoms/cm^2 on a ln scale versus time, with the starting point at 80 (arbitrary units). Calculate the activation energy for desorption.

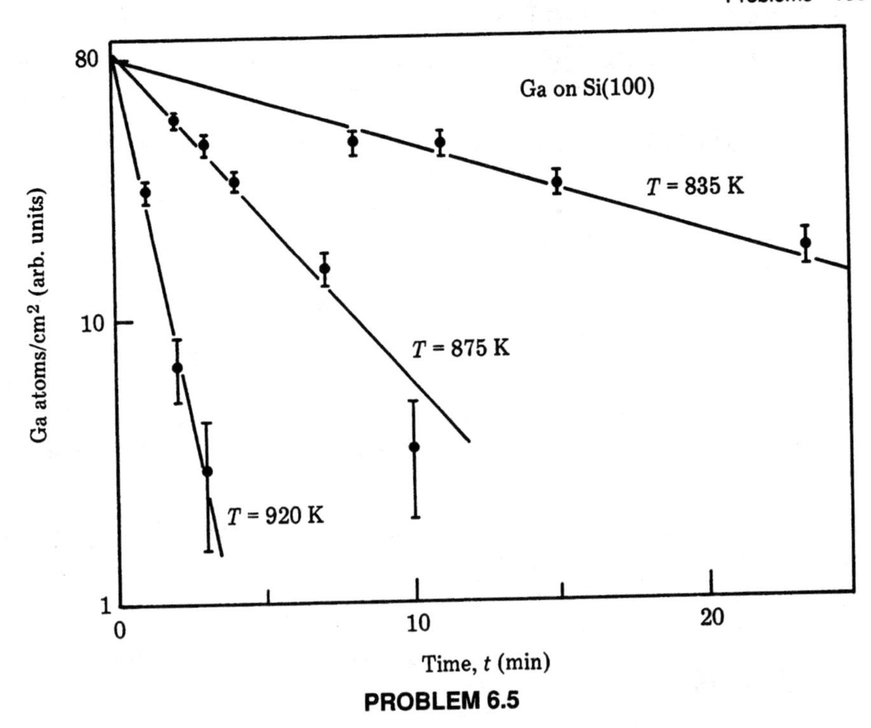

PROBLEM 6.5

6.6 RHEED oscillations provide a measure of surface diffusion as the oscillations disappear when the surface diffusion length λ_S exceeds the step length. For a step length of 33 nm on Si where $\Delta G_{des} = 2$ eV and $\Delta G_S = 1$ eV, at what temperature will the oscillations disappear?

6.7 You want to carry out Si MBE at a deposition rate of 1 monolayer per second with step-mediated growth on a sample where steps are separated by 50 nm. What growth temperature and what vacuum level are required if $\Delta G_S = 1.0$ eV or if $\Delta G_S = 0.5$ eV?

6.8 For diffusion-limited growth on (100) Si at 500°C with a spacing between steps of 50 nm, $6.8 \times 10^{14}/\text{cm}^2$ surface sites, and $\lambda_S = 0.033$ μm, what is the growth rate R for $J_{Si} = 10^{15}/\text{cm}^2\text{-sec}$? What is the condensation coefficient η for the single-step case?

6.9 For a substrate with a simple cubic lattice ($a = 0.3$ nm) homoepitaxial growth with $J = 5 \times 10^{15}$ atoms/cm² sec resulted in a growth rate $R = 3$ monolayers/sec. If the path length $\lambda_S = 15$ nm, estimate the average step length.

6.10 The following parameters are known for the desorption of an element from the (111) surface:

	850 K	1000 K
N_0 (eq) per cm²	10	0.08
τ_0 (sec)	5×10^4	2

Use these quantities to estimate W, the total energy required to remove an atom from the surface.

6.11 **(a)** In step-mediated growth in the high-temperature growth regime, Si atoms have a step velocity v (along the step) of 5×10^{-7} cm/sec. If the substrate was created by making a 0.69° miscut along the (100) plane, how long does it take to grow 1 μm of Si?

(b) Calculate the step velocity for a Si flux of 5×10^{15} atoms/cm^2 sec at 650°C on (100) Si. Use values from Problems 6.1 and 6.2, (Table 6.3).

6.12 You want to perform step-mediated homoepitaxial MBE growth on a simple cubic substrate with $a = 0.48$ nm. You know that the substrate has been miscut along the (100) plane at a .25° angle. If $\Delta G_S = 1.1$ eV, $\nu_s = 10^{13}$/sec, estimate the minimum temperature that would assume step-mediated growth at a rate of 5 monolayers/sec.

6.13 We want to grow $Al_{.30}Ga_{.70}As$. Assume a sticking coefficient of 1.0 for Al and Ga. Gallium growth rate is 1 μ/hr.

(a) Find the Al growth rate.

(b) Find the total growth rate.

(c) Assume a for AlAs = a for GaAs. Find the number of RHEED oscillations/sec for this growth rate.

(d) Assuming As_4 has a sticking coefficient of 0.5, what is the growth pressure assuming an initial background pressure of 1×10^{-10} torr and $T =$ 710°C?

CHAPTER 7

Heteroepitaxy and Superlattices

7.0 Introduction

Modern epitaxial growth offers the exciting possibility of creating materials with electronic properties tailored for particular applications. In the previous chapter, we discussed homoepitaxy, the growth of a thin single-crystal film on a single-crystal substrate of the same material. "Epitaxy" is in fact derived from Greek terms meaning "ordered upon." Homoepitaxy is a straightforward example of lattice-matched epitaxy, since film and substrate obviously have the same structure and lattice constant.

Heteroepitaxy refers to epitaxial growth of dissimilar materials on each other. The most common example in electronic thin film technology is the III–V system of AlGaAs on GaAs. When we speak of $Al_xGa_{1-x}As/GaAs$, we employ film/substrate notation; that is, AlGaAs is the film in this case, and x and $(1 - x)$ represent the relative amounts of Al and Ga. The two III–V compounds, AlAs and GaAs, have almost the same lattice constant—0.5661 nm (AlAs) and 0.5654 nm (GaAs)—and have a percentage lattice constant difference, or mismatch, of 0.1%. In practical applications the lattice mismatch is even less than the 0.1% given by the two pure materials, since one uses alloy compositions of the sort $Al_{0.5}Ga_{0.5}As/GaAs$, where the alloy has a lattice constant intermediate to the two pure materials. A superlattice is a periodic arrangement of these layers. An example would be a repeated structure consisting of 100 nm of $Al_{0.5}Ga_{0.5}As$ and 10 nm of GaAs on a GaAs substrate, $Al_{0.5}Ga_{0.5}As$ (100 nm)/GaAs (10 nm)/ . . . /$Al_{0.5}Ga_{0.5}As$ (100 nm)/GaAs (10 nm)/GaAs, where . . . indicates a repeat of the individual layers. Since the AlGaAs composition is periodic in the growth direction this structure is called a superlattice.

The combination of Ge and Si, Ge/Si, is an example of a system with large lattice mismatch. The lattice constants of the two pure elements differ by approximately 4% ($a_{Ge} = 0.5646$ nm and $a_{Si} = 0.5431$ nm), which is a huge mismatch in epitaxial systems. Here also, practical applications often involve the growth of the alloy, Ge_xSi_{1-x}/Si.

Heteroepitaxy allows the formation of artificial structures not available in nature. For semiconductor applications the choice of materials is governed by the band gap versus lattice constant chart shown in Fig. 7.1. Here the common semiconductors

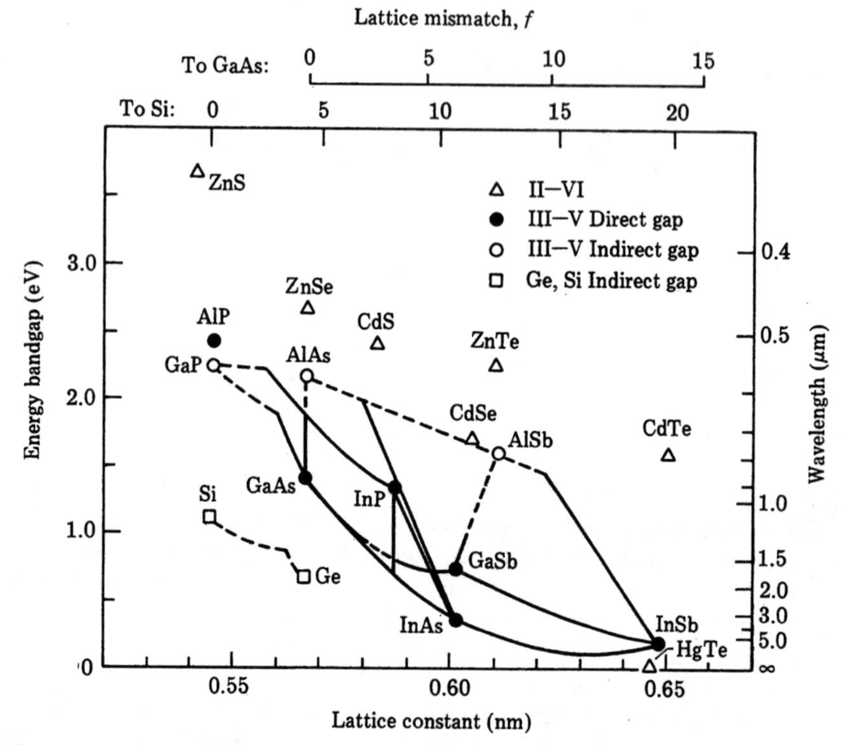

FIGURE 7.1 Energy gap versus lattice constant for III–V, II–VI, and IV semiconductors. Open symbols are indirect-gap, closed are direct-gap materials. The lines joining the III–V compounds give the ternary energy gap and lattice constant. The lattice mismatch f to Si and GaAs is shown at the top of the figure (after Cho, 1985).

(Groups IV, III–V, and II–VI) are listed by their lattice constants and energy gaps. Two materials on the same vertical line (same lattice structure and lattice constant) afford an opportunity for nearly lattice-matched heteroepitaxy. One can use the chart, for example, to see that an alloy composition of $In_xGa_{1-x}As$ will be a good lattice match to the important semiconductor InP.

7.1 Lattice Constants and Energy Gaps

The range of lattice constants for the common Group IV, III–V, and II–VI semiconductors, along with the percentage lattice mismatch relative to substrates Si, GaAs, and CdTe, are shown in Fig. 7.2. For strained-layer coherent epitaxy of films with practical thicknesses, lattice mismatch is usually less than 2%, which limits the choice of combinations. Heterostructures are often formed by growing ternary (three-element) alloys or quarternary (four-element) alloys on binary alloys with properly chosen composition in order to provide lattice matching and (more importantly) the desired energy gap and band offset. Figure 7.3 shows the lattice constant as a function

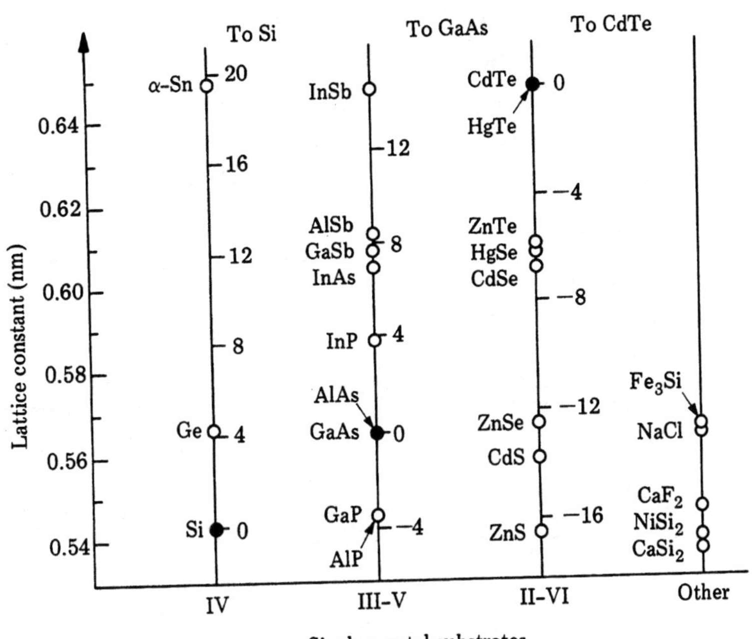

FIGURE 7.2 Lattice constants for common Group IV, III–V, and II–VI semiconductors and other cubic materials, and the percentage mismatch to Si, GaAs, and CdTe (closed symbols) (from Picraux et al., 1990).

of composition for ternary III–V solid solutions. Lattice constants for binary alloys are also charted. This diagram assumes that the lattice constant of a ternary alloy is a linear combination of the lattice constants of the pure materials. When the lattice-constant line for a ternary solution crosses the horizontal line for a binary compound, the composition at the intersection provides lattice matching at room temperature. For example, $Al_xIn_{1-x}As$ is lattice matched with InP at $x \cong 0.5$. At about this same x value, $Al_xIn_{1-x}P$ is lattice matched with GaAs. One can see that the line for AlAs almost coincides with that for GaAs, indicating that $Al_xGa_{1-x}As$ is closely lattice matched to GaAs for any value of x.

The energy gap of a ternary alloy depends on the composition. For $Al_xGa_{1-x}As$, the energy gap increases with Al content and changes from a direct to an indirect gap for x greater than about 0.4. Casey and Panish (1978) report $E_G = 1.424 + 1.24x$ with $0 < x < 0.45$; Kuech et al. (1987) give $E_G = 1.424 + 1.455x$ with $0 < x < 0.37$. The difference in the two determinations presumably lies in the measurement of the Al content. The compositional dependence of the energy gap in some III–V compounds at 300 K is listed in Table 7.1.

If one combines the information shown in Fig. 7.3 with the information listed in Table 7.1, energy gap versus lattice constant plots may be constructed as shown in Fig. 7.1. The lines joining the binary compounds give the energy gap as a function of lattice constant. Solid boundaries denote a direct band gap and dashed boundaries

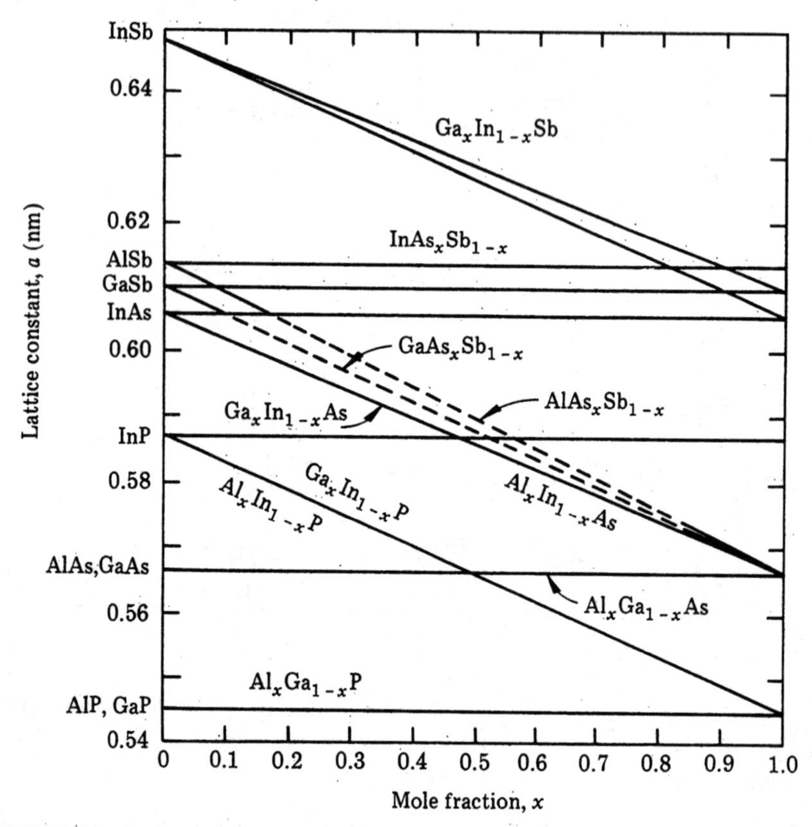

FIGURE 7.3 Lattice constant as a function of composition for ternary III–V solid solutions. The slanted lines connect binary compounds whose lattice constants differ by less than 0.05. The dashed lines show regions where miscibility gaps are expected (after Casey and Panish, 1978).

denote an indirect band gap. The area inside the boundaries denotes quaternary alloys. From this figure, one can see that GaAs is lattice matched to AlAs and to any ternary solid solution along the boundary between GaAs and AlAs. The ternary alloy can have energy gap from ~1.4 eV to ~2.2 eV. For the epitaxial growth of $In_xGa_{1-x}As$ on InP, lattice matching is possible only at $x = 0.53$ (E_G ~0.8 eV). Quaternary alloys (GaInPAs) along the GaAs–AlAs boundary but inside the area bounded by GaP–GaAs–InAs–InP are also lattice matched to GaAs. On the other hand a series of quaternary solid solutions, GaInAsP, inside the area bounded by GaP–GaAs–InP can be lattice matched to InP.

The requirement for lattice-matched epitaxy places severe restrictions on material combinations. Possible use of lattice-mismatched heteroepitaxy greatly increases the number of possible material combinations. Accompanying this flexibility in materials choice is the difficulty of growing planar (nonclustered), single-crystal layers without dislocations. Consequently this chapter will be devoted to the materials science issues associated with lattice-mismatched systems and superlattices. Electri-

TABLE 7.1 Compositional Dependence of the Energy Gaps of the Binary III–V Ternary Solid Solutions at 300 K*

Compound	Direct Energy Gap E_G (eV)
$Al_xIn_{1-x}P$	$1.351 + 2.23x$
$Al_xGa_{1-x}As$	$1.424 + 1.247x$[a]
	$1.424 + 1.455x$[b]
$Al_xIn_{1-x}As$	$0.360 + 2.012x + 0.698x^2$
$Al_xGa_{1-x}Sb$	$0.726 + 1.129x + 0.368x^2$
$Al_xIn_{1-x}Sb$	$0.172 + 1.621x + 0.43x^2$
$Ga_xIn_{1-x}P$	$1.351 + 0.643x + 0.786x^2$
$Ga_xIn_{1-x}As$	$0.36 + 1.064x$
$Ga_xIn_{1-x}Sb$	$0.172 + 0.139x + 0.415x^2$
GaP_xAs_{1-x}	$1.424 + 1.150x + 0.176x^2$
$GaAs_xSb_{1-x}$	$0.726 - 0.502x + 1.2x^2$
InP_xAs_{1-x}	$0.360 + 0.891x + 0.101x^2$
$InAs_xSb_{1-x}$	$0.18 - 0.41x + 0.58x^2$

*SOURCE: After Casey and Panish (1978).

[a] $(0 < x < 0.45)$.

[b] $(0 < x < 0.37)$; Kuech et al. (1987).

cal and optical properties of these heterostructures and superlattices are presented in Chapter 8.

7.2 Structure of Lattice-Mismatched Systems

A lattice-mismatched heterostructure can be represented by the schematic cross section shown in Fig. 7.4. Both the substrate and the film are cubic, with lattice constants a_s and a_f, respectively. The epitaxial layer is either strained or unstrained; the latter is usually called relaxed. Other phrases used to describe the strained system are *pseudomorphic* (i.e., there is a one-to-one correspondence between rows of atoms in the overlayer and those in substrate) and *commensurate* (in the sense that the atomic spacing of the overlayer is the same as that of the substrate). In both definitions, the lattice constant of the layer in the plane of the interface is equal to that of the substrate. Due to the constraint on the in-plane lattice constant, the unit cell will distort as allowed by Poisson's ratio, as illustrated in Fig. 7.5 for an epitaxial (epi) layer grown on the substrate. A cubic unit cell is distorted into a tetragonal cell. If the lattice constant of the layer material is smaller than that of the substrate, the cell must be stretched in the in-plane direction and its height will decrease.

The in-plane strain $\varepsilon_{\parallel}$ is defined as

$$\varepsilon_{\parallel} = \frac{a_{f\parallel} - a_f}{a_f} \qquad (7.1)$$

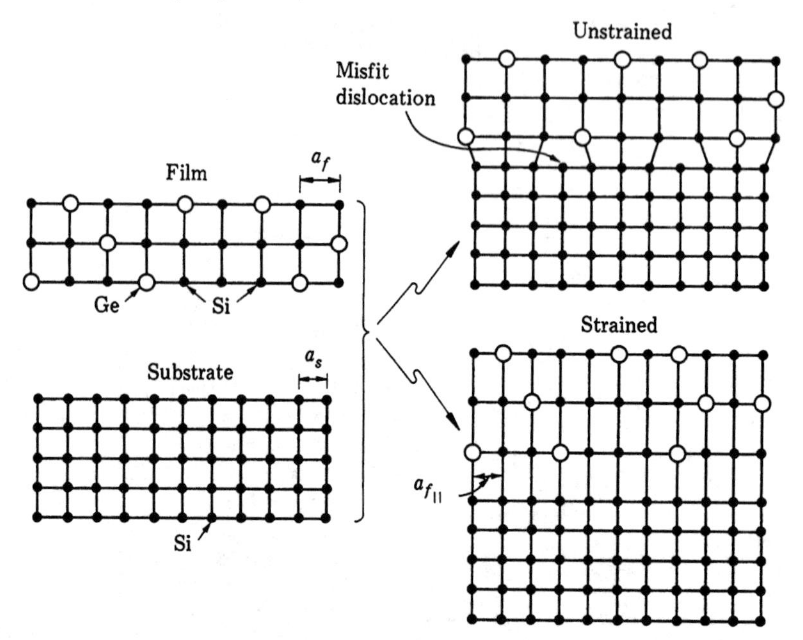

FIGURE 7.4 A Ge_xSi_{1-x} film and a Si single-crystal substrate are joined to form a lattice-mismatched heterostructure that is either strained or unstrained with misfit dislocations.

where $a_{f_{\|}}$ is the parallel-to-the-interface or in-plane lattice constant of the deposited film material, and a_f is the lattice constant of the film material in the bulk or unstrained state. For pseudomorphic material, $a_{f_{\|}} = a_s$ and the strain is equal to the lattice mismatch. The in-plane strain is often called the *coherency strain.*

We can also define the strain in the direction perpendicular to the interface.

$$\varepsilon_{\perp} = \frac{a_{f_{\perp}} - a_f}{a_f} \tag{7.2}$$

where $a_{f_{\perp}}$ is the lattice constant perpendicular to the surface. In general $a_{f_{\perp}} \neq a_{f_{\|}}$. Since most semiconductor crystals are cubic in the bulk state, strained-layer epitaxy results in conversion to noncubic structure with a tetragonal cell. The tetragonal distortion is defined as

$$\varepsilon_T = \frac{|a_{f_{\perp}} - a_{f_{\|}}|}{a_f} \tag{7.3}$$

Our picture of strained material is a one-to-one correspondence between atomic rows with lattice constants in the film modified as described above. Under certain conditions the deposited film is not strained (i.e., does not assume the in-plane lattice constant of the substrate). Unstrained epitaxial material corresponds to a structure where film and substrate retain their bulk lattice constants. Therefore, there cannot be an interfacial alignment on an atomic basis; the structure is not pseudomorphic.

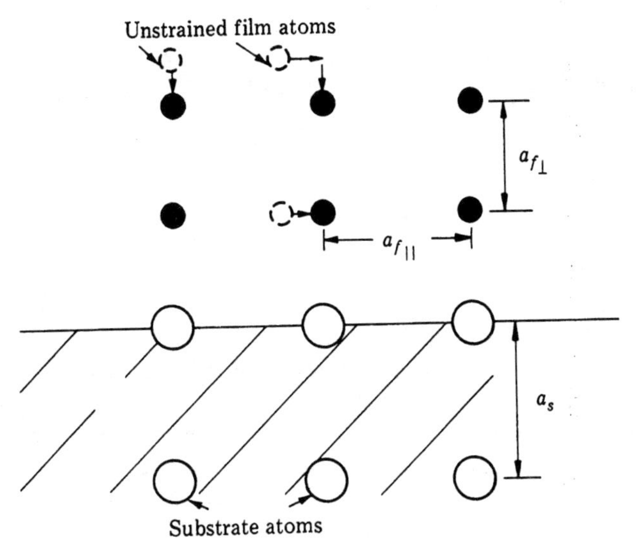

FIGURE 7.5 Atom positions for a strained layer on a thick substrate, illustrating the coherency strain and the Poisson effect for a commensurately grown layer (Picraux et al., 1990).

The mismatch is taken up by defects, called *misfit dislocations* (Fig. 7.4). The misfit f between substrate and film is defined as

$$f = \frac{|a_s - a_f|}{a_s} \tag{7.4}$$

Misfit dislocations (see Chapter 4) may be represented as edge dislocations uniformly spaced along the interface. The average spacing S of misfit dislocations depends on the mismatch f and is of the order of a_s/f. For $f = 1\%$ and $a_s = 0.4$ nm, $S = 40$ nm.

The relaxed or unstrained system maintains its cubic symmetry. In many cases, a film/substrate is not totally strained or relaxed but may contain a partial complement of dislocations.

From a materials science standpoint, epitaxy, strained or relaxed, refers to the crystallographic orientation alignment of film and substrate. From an electronic materials standpoint, the difference in electronic properties between strained and relaxed is enormous. Dislocations in electronic-grade material can have a strong adverse effect on charge transport. Present-day bulk Si can have dislocation spacings as large as one cm (dislocation density $< 1/\text{cm}^2$) and bulk GaAs can have spacings greater than 0.03 cm. For comparison, a relaxed film with 1% misfit f has 40 nm between dislocations at $a_s = 0.4$ nm. Since the mean free path of electrons in electronic-grade semiconductors is of the order of 100 nm, the dislocations in the relaxed film will adversely affect charge transport. Therefore the goal in lattice-mismatched semiconductors is to grow pseudomorphic films with minimal dislocation densities for electronic structures which require electronic transport through the heteroepitaxial interface.

7.3 Strain Energy in Heteroepitaxial Layers

In the following sections we consider the equilibrium configuration, pseudomorphic or relaxed, of strained-layer epitaxial systems. By equilibrium we mean that the film/substrate system is in its minimum-energy configuration. Energy minima are achieved by temperature treatment. In real systems and with modern epitaxial growth techniques, nonequilibrium film structures can be formed. Such systems are metastable. An example is a Ge_xSi_{1-x}/Si structure grown by MBE, which is highly strained although equilibrium arguments would suggest a relaxed state. The approach to equilibrium by dislocation generation and motion is sufficiently sluggish so that a metastable strained state is achieved.

In an elastic approximation, the strain energy is proportional to the product of stress and strain. In a cubic crystal the strain energy areal density E_ε can be written as

$$E_\varepsilon = \varepsilon^2 Bh \tag{7.5}$$

where ε is the in-plane strain ($\varepsilon = \varepsilon_\parallel$) and h is the film thickness. The quantity E_ε is quadratic in the strain (recall the energy of a strained spring)—a consequence of the elastic approximation—and linear in the thickness of the overlayer. The constant B is a complicated function of the elastic constants, depending on the symmetry of the crystal type and on the growth direction. For growth on (100), (111) or (110) surfaces

$$B = \left(\frac{C_{11} + C_{12}}{2}\right)\left[3 - \frac{(C_{11} + 2C_{12})}{C_{11} + 2(2C_{44} - C_{11} + C_{12})(l^2m^2 + m^2n^2 + n^2l^2)}\right] \tag{7.6}$$

where l, m, n are the indices of the growth surface; for (100) the factor $(l^2m^2 + m^2n^2 + n^2l^2) = 0$ while for (111) the factor has the value of 3. (See Appendix E for further details on the strain parameters.) There is a substantial simplification if we use the expressions for an elastically isotropic film. Semiconductors are not elastically isotropic, but there is little loss in conceptual understanding with such an approximation. The constant B is given for an elastically isotropic film as

$$B = 2\mu_f \frac{(1 + \nu)}{(1 - \nu)} \tag{7.7}$$

In this last equation μ_f is the shear modulus of the film and ν is Poisson's ratio for the film material.

The strain energy Eq. (7.5) is derived from the following relations originally discussed in Chapter 4. The total elastic energy per unit volume may be written as:

$$E_{\text{elastic}} = \int \sigma_x\, d\varepsilon_x + \sigma_y\, d\varepsilon_y + \sigma_z\, d\varepsilon_z$$

which is a generalization of Eq. (4.9). For a thin, isotropic film or a (100) oriented single crystal $\sigma_z = 0$, $d\varepsilon_x = d\varepsilon_y$, and $(\sigma_x + \sigma_y) = 2Y\varepsilon/(1 - \nu)$ from Eq. (4.11). Then for a film of thickness h, the energy/area is

$$E_\varepsilon = \frac{Y\varepsilon^2 h}{1 - \nu} = \frac{2(1 + \nu)}{(1 - \nu)} \mu_f \varepsilon^2 h = \varepsilon^2 B h \tag{7.5}$$

using Eq. 4.8.

The strain associated with a one monolayer film ($h = 0.141$ nm $= a/4$ on the (100) face) of Ge on Si (100) is

$$E_\varepsilon = (0.04)^2 \times 2.31 \times 10^{11} \frac{\text{newton}}{\text{m}^2} \times 1.41 \times 10^{-10} \text{ m}$$

$$= 5.2 \times 10^{-2} \text{ Joule/m}^2$$

where in this case $\varepsilon = 0.04$ and $B = 2.31 \times 10^{11}$ newton/m^2 ($\mu_f = 0.671 \times 10^{11}$ N/m^2 and $\nu = 0.272$). For the (100) direction there are 6.8×10^{18} atoms/m^2, so the strain energy in units of eV/atom is

$$E_\varepsilon = \frac{5.2 \times 10^{-2} \text{ J/m}^2 \times 6.25 \times 10^{18} \text{ eV/J}}{6.8 \times 10^{18} \text{ atoms/m}^2} = 0.047 \text{ eV/atom}$$

Note that this value is significantly greater than that calculated in Chapter 4, section 4.2, because here the strain and modulus are each greater by one order of magnitude.

7.4 Stability of Strained Layers

The previous section dealt with the calculation of the strain energy of a uniform film. In this section we compare this strain energy to the energy of a relaxed, clustered film. In effect we are determining if the film is metastable against cluster formation.

The ratio of the energy of the strained film to the energy of the relaxed, clustered film is denoted by R:

$$R = \frac{\text{energy of strained film}}{\text{energy of clustered film}} \tag{7.8}$$

The strain energy/area for a monolayer of thickness h is given by Eqs. (7.5) and (7.7)

$$E_\varepsilon = 2\mu_f \left(\frac{1 + \nu}{1 - \nu}\right) \varepsilon^2 h \tag{7.9}$$

The total energy of the strained film of area A and thickness Lh is the sum of the surface energy ($A\gamma$) and the strain energy ($LE_\varepsilon A$). For a clustered film, we assume that the clusters are cubic in shape; see Fig. 7.6. There are m clusters of individual volume X^3 with an atomic density of n atoms/cm^3. The strained layer has L monolayers with N_S atoms/cm^2 in each monolayer. From mass conservation,

$$mnX^3 = LN_S A$$

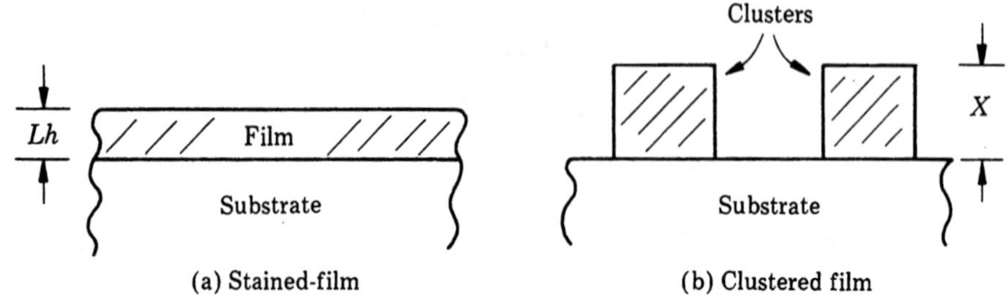

FIGURE 7.6 Schematic of (a) strained film of thickness *Lh* composed of *L* monolayers of height *h* and (b) a clustered film with cubic clusters of side *X*.

and each cluster has a vertical exposed surface area of $4X^2$ which is related to L and m from the foregoing by

$$4X^2 = 4LN_SA/mnX$$

The energy of the clustered film is made up of contributions from the exposed substrate, from the top of clusters, and from the vertical walls of the clusters (ignoring the cluster/substrate interface). Setting the surface energy per unit area of the film to γ_f, we have

$$A\gamma_f + \frac{4LN_SA\gamma_f}{Xn}$$

for the total surface energy of the clustered film, where the last term corresponds to the additional surface area exposed by the four sides of the clusters. In this analysis we will assume that the surface energy per unit area γ is the same for film and substrate. Then the ratio R is

$$R = \frac{A\gamma + LE_\varepsilon A}{A\gamma + \dfrac{4LN_S\gamma A}{Xn}} \tag{7.10}$$

Of course all the areas cancel out; this argument cannot depend on the starting area. Then

$$R = \frac{1 + \dfrac{LE_\varepsilon}{\gamma}}{1 + \dfrac{L(4N_S/n)}{X}} \tag{7.11}$$

We consider the critical cluster size X_c such that R is 1. In that case

$$X_c = \frac{4N_S/n}{E_\varepsilon/\gamma} \tag{7.12}$$

This is the critical dimension where a clustered configuration has an energy equivalent to the strain energy of a uniform film. When clusters are larger than X_c the cluster configuration will be favored, and when X is smaller than X_c, the strained film will be favored. For Ge on (100) Si, there is a 4% mismatch so the strain ε is 0.04. For one monolayer the strain energy is 0.05 eV/atom. The surface energy is 0.8 eV/atom (see Chapter 2). For Ge, $n \simeq 4.4 \times 10^{22}$ atoms/cm^3, and for the Si (100) monolayer, $N_S \simeq 6.8 \times 10^{14}$ atoms/cm^2, so that $X_c \simeq 5.7 \times 10^{-6}$ cm. When the film forms relaxed clusters larger than about 6×10^{-6} cm, these clusters are the lower energy state and R is greater than one.

There is always a value of X for which the clustered system is the lower energy state. Given sufficient temperature and sufficient time we could always form a relaxed, clustered configuration with this property. The art of thin film epitaxy is learning to grow under conditions low enough in temperature so that clusters do not form. Strained layer epitaxial films grown thus are metastable against clustering.

The estimate given by Eq. (7.12) can be improved in a number of ways. First of all, clusters do not exist as cubes but have hemispherical or faceted shapes. Second, we have ignored the interfacial energy between the film and the substrate. We have also assumed that the surface energy per unit area of the substrate is the same as that for the overlayer. This is a poor approximation if we are comparing a metal or an insulator to a semiconductor; however, when we compare one semiconductor to another it is a reasonable approximation.

There are growth modes, called Stranski–Krastanov (S–K) growth modes, where the first layer or layers remain smooth and clusters form on top. That is a case where the film–substrate interfacial energy is strong and the film spreads out over the entire surface. Clusters of the film material then form on top of the first few monolayers of the deposited film.

The three major growth modes are shown in Fig. 7.7 in a plot of relative energy potential W of substrate to film versus lattice mismatch. Under the proper conditions, layer-by-layer growth (Frank–Van der Merwe mode) can be achieved where the lattice mismatch is zero. In homoepitaxy, the surface energy ratio is, of course, unity. If the film has a high surface energy per unit area compared to the substrate, clusters form in the Volmer–Weber mode. With increased lattice mismatch, clusters form even with $W > 0$. In the intermediate Stranski–Krastanov mode, a few monolayers grow before clusters are formed. The calculation of cluster formation given above is directly applicable to the S–K morphology since the exposed surface and the clusters are the same material (i.e., $\gamma_f = \gamma_s$ is a good approximation).

In general E_ε is inversely proportional to the square of the mismatch f. The larger the mismatch, the smaller the critical cluster size, which is one reason lattice-matched epitaxy is more easily accomplished than lattice-mismatched epitaxy. As the strain gets larger, the critical cluster size gets smaller as the square of the mismatch.

7.5 Dislocation Energy

The energy per unit length E_d of a single edge dislocation has been derived in Chapter 4

$$E_d = \frac{\mu b^2}{4\pi(1 - \nu)} \left[\ln\left(\frac{r}{b}\right) + B^1 \right] \tag{7.13}$$

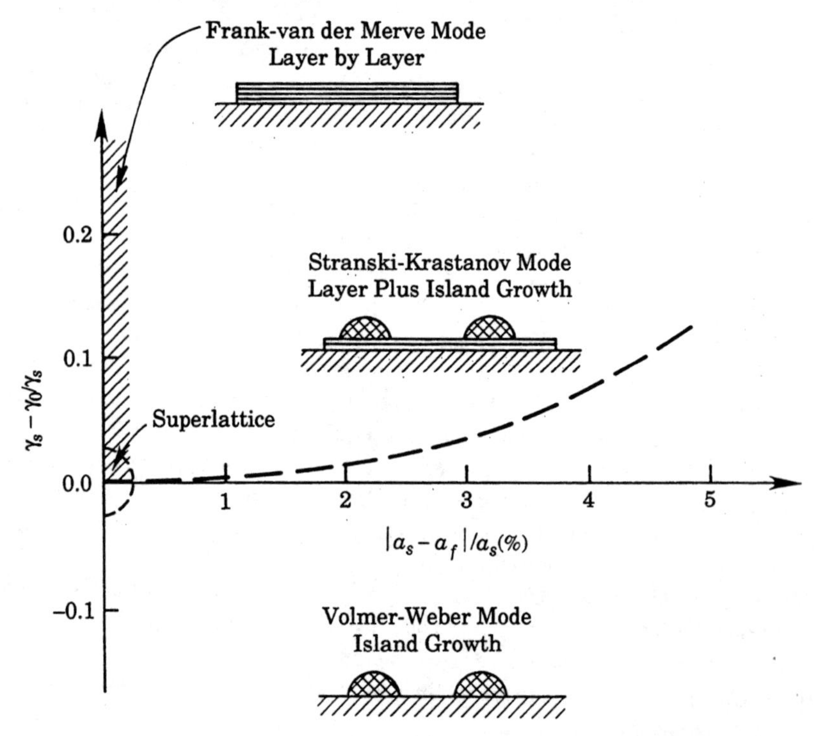

FIGURE 7.7 Energy ratio W (substrate to film) versus lattice mismatch, showing the three major growth modes. W is defined as $[(\phi_{fs}/\phi_{ss}) - 1]$ where ϕ is the film-substrate or substrate-substrate interatomic potential (adapted from Grabow and Gilmer 1986).

where b is Burgers vector of the dislocation, r is the distance of the elastic stress field, and B^1 is a factor which measures the core energy of the dislocation. The energy per unit length of a dislocation can be estimated to the first order if we ignore the logarithm term and set $b = a$, the lattice spacing. In Si, the shear modulus $\mu = 0.67 \times 10^{11}$ N/m^2, Poisson's ratio $\nu = 0.272$ and $a = 5.43 \times 10^{-10}$ m, so that

$$E_d \simeq \frac{\mu b^2}{4\pi(1 - \nu)} = \frac{0.67 \times 10^{11} \times (5.43 \times 10^{-10})^2}{4\pi(1 - 0.272)}$$

$$= 2.15 \times 10^{-9} \text{ Joule/m}$$

$$= 1.35 \times 10^{10} \text{ eV/m} = 13.5 \text{ eV/nm}$$

It is customary to make approximations to the form of B^1 and to write Eq. (7.13) as

$$E_d = \frac{\mu b^2}{4\pi(1 - \nu)} \left[\ln\left(\frac{r}{b}\right) + 1 \right] \tag{7.14}$$

where we have taken the core energy as $\mu b^2/4\pi(1 - \nu)$ corresponding to $B^1 = 1$.

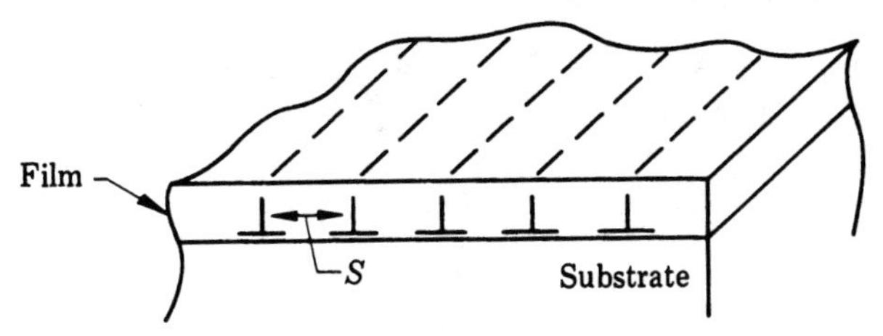

(a) One dimensional dislocation array

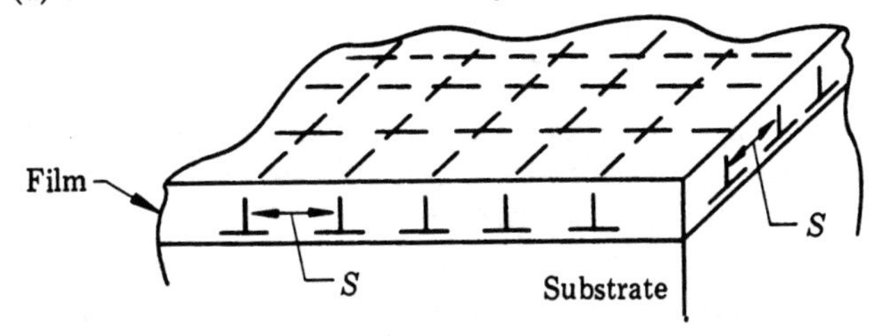

(b) Two dimensional dislocation network

FIGURE 7.8 Arrays of misfit dislocations with spacings between dislocations on (100) oriented substrates (the symbol ⊥ indicates the edge dislocation) for (a) one-dimensional array and (b) two-dimensional network.

In a lattice-mismatched system which is totally relaxed, the number of dislocations per unit length is $1/S$ where S is the spacing between dislocations (Fig. 7.8a). The dislocation energy per unit area E'_d for a dislocation array of one orientation is,

$$E'_d = \frac{E_d}{S} = \frac{fE_d}{b} \tag{7.15}$$

where $S = b/f$ and f is the misfit. Then

$$E'_d = \frac{f\mu_f b}{4\pi(1 - \nu)}\left[\ln\left(\frac{r}{b}\right) + 1\right] \tag{7.16}$$

for the single array shown in Fig. 7.8a with μ_f the shear modulus of the film material. The energy density is then $0.248\ \mathrm{eV/nm^2}$ for $f = .01$, ignoring the logarithm term. In a partially relaxed film, the separation between misfit dislocations is $b/(f - \varepsilon_{\|})$. The interpretation of this is as follows: If $\varepsilon_{\|} = f$, the system is totally strained, pseudomorphic, and the dislocation spacing is infinite. If the system is totally relaxed, $\epsilon_{\|} = 0$, and the dislocation spacing is given by $S = b/f$ as before. If the system is partially relaxed the spacing is intermediate between the two extremes and is given by $b/(f - \varepsilon_{\|})$. Then the areal energy density of dislocations for the single array is given by

$$E'_d = \frac{\mu_f b}{4\pi(1 - \nu)}(f - \varepsilon_{\|})\left[\ln\left(\frac{r}{b}\right) + 1\right] \tag{7.17}$$

In real systems, one observes a cross-hatched array of dislocations, shown in Fig. 7.8b for the (100) orientation. This cross-hatching is due to the equivalent directions in the plane of the interface and to strain relief in two independent directions within the plane. The energy density of an array of *two* noninteracting, perpendicular edge dislocations, each with spacing S, is given by

$$E_{\perp} = 2E'_d = \frac{\mu_f b(f - \varepsilon_{\parallel})}{2\pi(1 - \nu)} \left[\ln\left(\frac{r}{b}\right) + 1 \right] \tag{7.18}$$

7.6 Critical Thickness

Dislocations represent another mechanism (as opposed to strain) for accommodating mismatched materials. In the following section we compare the energy of a film with dislocations to that of a totally strained film.

The critical thickness h_c for pseudomorphic epitaxy is derived by considering the thickness dependence of the strain energy and dislocation energy, and by minimizing the total energy. If the dislocation spacing is greater than the film thickness h, the extent r of the strain field is determined by h: $r \approx h$, as indicated schematically in Fig. 7.9. For arbitrary strain $\varepsilon_{\parallel}$, the combined strain energy and dislocation energy E_{tot} is given by

$$E_{\text{tot}} = \varepsilon_{\parallel}^2 Bh + \frac{\mu_f b}{2\pi(1 - \nu)} (f - \varepsilon_{\parallel}) \left[\ln\left(\frac{h}{b}\right) + 1 \right]$$

where we maintain the approximation to the core energy, $B^1 = 1$. The strain that minimizes the total energy is obtained by setting $dE_{\text{tot}}/d\varepsilon_{\parallel} = 0$, to yield the critical strain, $\varepsilon_{\parallel}^*$. That is, for

$$\frac{dE_{\text{tot}}}{d\varepsilon_{\parallel}} = 2\varepsilon_{\parallel}^* Bh - \frac{\mu_f b}{2\pi(1 - \nu)} \left[\ln\left(\frac{h}{b}\right) + 1 \right] = 0 \tag{7.19}$$

Then

$$\varepsilon_{\parallel}^* = \frac{\mu_f b}{4\pi(1 - \nu)hB} \left[\ln\left(\frac{h}{b}\right) + 1 \right] \tag{7.20}$$

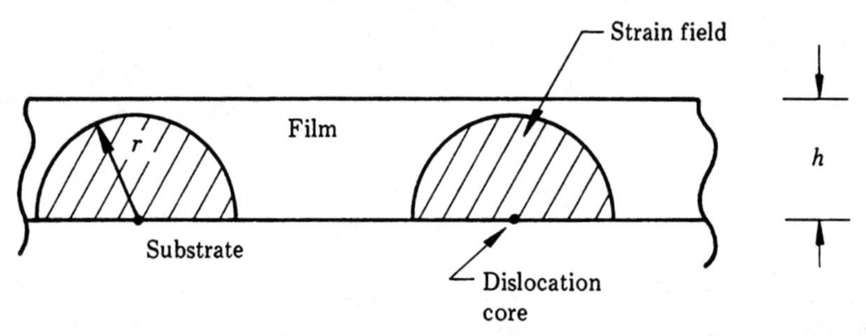

FIGURE 7.9 Side view of an epitaxial film of thickness h containing misfit dislocations with strain field extending a distance r from the core of the dislocation.

Using Eq. (7.7) we obtain

$$\varepsilon_{\parallel}^{*} = \frac{b}{8\pi(1 + \nu)h}\left[\ln\left(\frac{h}{b}\right) + 1\right], \qquad h > h_c$$

and (7.21)

$$\varepsilon_{\parallel}^{*} = f, \qquad h < h_c$$

with h_c defined by Eq. 7.22.

The largest possible value of $\varepsilon_{\parallel}^{*}$ is the misfit f. If the value of $\varepsilon_{\parallel}^{*}$ is equal to f, the film will be strained to match the substrate precisely. Of course, $\varepsilon_{\parallel}^{*}$ can never be greater than f, and the thickness h_c at which it becomes energetically possible for the first misfit dislocation to be made, is obtained by setting $\varepsilon_{\parallel}^{*} = f$. Then this critical thickness h_c is given by

$$h_c = \frac{b}{8\pi(1 + \nu)f}\left[\ln\left(\frac{h_c}{b}\right) + 1\right] \tag{7.22}$$

In many (100) semiconductor systems, b is given by $a/\sqrt{2}$; then for $f = 0.01$, $\nu = 0.3$, we have

$$h_c = \frac{a}{8\pi(2)^{1/2}(1.3)(0.01)}\left[\ln\left(\frac{h_c\sqrt{2}}{a}\right) + 1\right]$$

$$h_c = \frac{a}{0.46}\left[\ln\left(\frac{h_c\sqrt{2}}{a}\right) + 1\right]$$

This equality is closely satisfied by $h_c/a = 7.2$. With $a = 0.543$ nm (for Si)

$$h_c = 7.2a = 3.9 \text{ nm}$$

This value of h_c is 29 monolayers for a misfit of $f = 0.01$ on Si (100) substrates. Fig. 7.10 shows the critical thickness versus misfit f based upon Eq. (7.22) with $b = a/\sqrt{2}$ and $\nu = 0.274$.

For growth on (100) diamond-cubic semiconductors, the most common type of misfit dislocation is the 60° type, whose direction and Burgers vector are both along inclined $\langle 110 \rangle$ directions, 60° from each other. At very highly strained (>2%) interfaces, however, pure edge dislocations tend to form. These misfit dislocations are illustrated in Fig. 7.11. The greatest reduction in strain energy is achieved by locating the dislocations at the layer–substrate interface.

The case where the dislocations are not constrained to lie on the interface plane has been analyzed by Matthews. Here one obtains a formula for the critical strain $\varepsilon_{\parallel}^{*}$ as follows:

$$\varepsilon_{\parallel}^{*} = \frac{(1 - \nu)b}{2\mu_f h(\cos\lambda)(1 + \nu)}\left[\frac{\mu_f\mu_s(1 - \nu\cos^2\Theta_{db})}{2\pi(\mu_f + \mu_s)(1 - \nu)}\right]\left[\ln\left(\frac{h}{b}\right) + 1\right] \tag{7.23}$$

where Θ_{db} is the angle between the dislocation line and its Burgers vector, and λ is the angle between the slip direction and that line in the interface plane which is

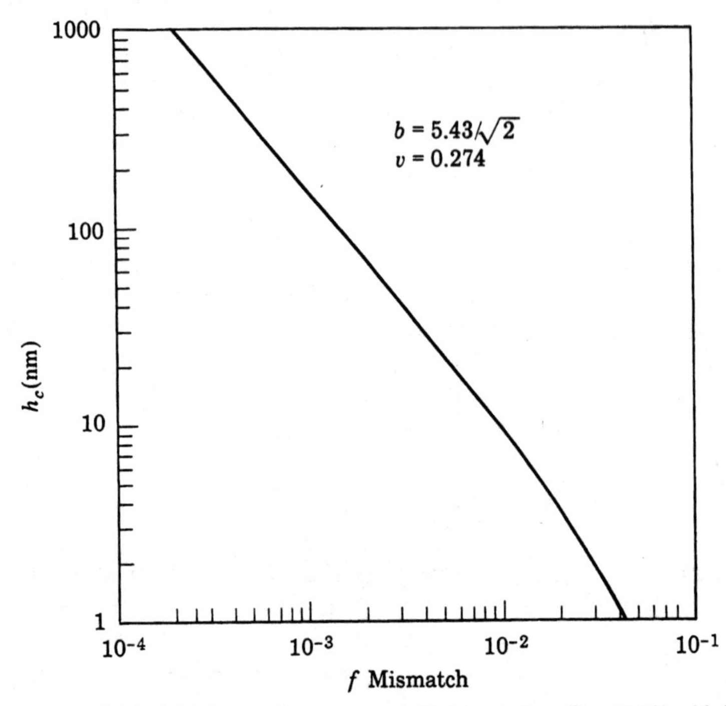

FIGURE 7.10 Critical thickness h_c versus misfit f based on Eq. (7.22) with $b = a$ and $\nu = 0.3$.

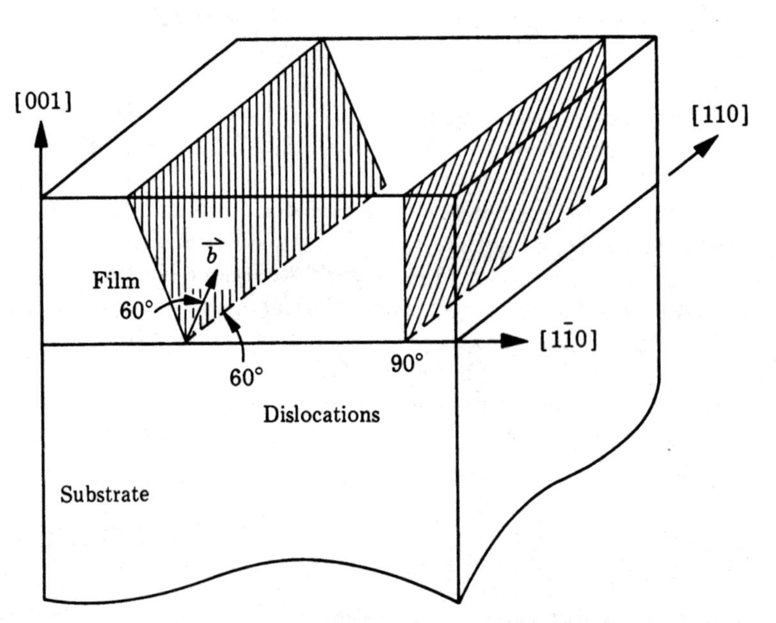

FIGURE 7.11 Diagram of the geometry of 60° and 90° misfit dislocations (from Picraux et al., 1990).

normal to the line of intersection between the slip plane and the interface. For growth on (100) substrates, $b = a/\sqrt{2}$ and $\Theta_{db} = \lambda = 60°$.

The critical thickness is again obtained by setting $\varepsilon_{\parallel} = f$ so that

$$h_c = \frac{b(1 - \nu \cos^2\Theta_{db}) \left[\ln\left(\frac{h_c}{b}\right) + 1 \right]}{8\pi(1 + \nu)f(\cos \lambda)} \tag{7.24}$$

using the approximation $\mu_f = \mu_s$. Letting $\lambda = 0°$ and $\Theta_{db} = 90°$ returns Eq. (7.22).

Fig. 7.12 shows the critical thickness versus Ge fraction for Ge_xSi_{1-x} films on Si (100). Changing the Ge fraction is equivalent to changing f; for $x = 0$, $f = 0$, and for $x = 1$, $f = 0.042$ at room temperature. The curve is based on Eq. (7.24) and the data points are the experimental measurements. Values (data points) which lie above the curve represent metastable structures which have not relaxed to their equilibrium positions.

As discussed, the derivation of the critical thickness relies on the energy balance between strain energy and dislocation energy. Since the contribution to the dislocation energy from the heavily disordered core is somewhat uncertain, it follows that the formula for the critical thickness, Eq. (7.22), should be considered only as a good estimate. One may obtain variations by factors of two when using somewhat different theoretical approaches. Nevertheless, the dependences on the elastic parameters and the mismatch are well obeyed. Large differences between experiment and

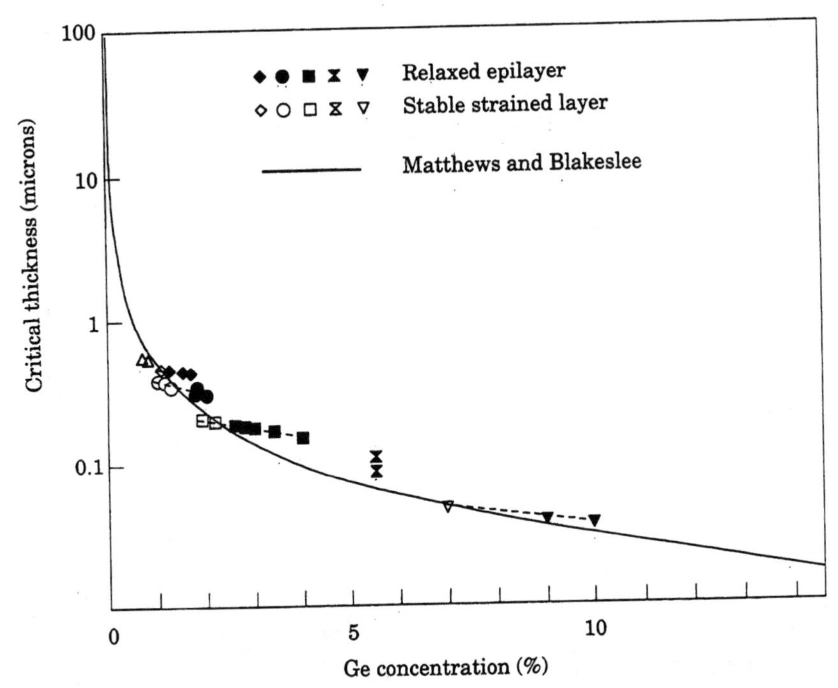

FIGURE 7.12 Critical thickness h_c versus Ge fraction for Ge_xSi_{1-x} films grown on Si (100). Experimental values indicate as-grown films (from Houghton et al., 1990).

theory may be observed indicating that the experimental films are metastable. The energy balance approach described here is considered an "equilibrium model" which considers the minimum energy state of the film. The fact that the experimental data may be different than expected from the equilibrium theory indicates that the films are not in equilibrium. That is, there has not been enough atomic motion (i.e., time and temperature) to allow formation of the relaxed state (large dislocation density). Such films may reach the equilibrium state upon annealing at high temperature. Since they are not in their lowest energy state they are considered metastable.

7.7 Reduced Strain

Equations (7.20) and (7.23) indicate that the critical strain $\varepsilon_{\parallel}^{*}$ is a function of the thickness of the film. The critical strain is that which minimizes the total energy. For $h < h_c$, no dislocations are expected. For $h > h_c$, the film may contain a dislocation density which partially or totally relaxes the strain.

A useful way to illustrate the strain versus thickness relationship is in reduced units with

$$\varepsilon_{\text{red}} = \frac{\varepsilon_{\parallel}^{*}}{\varepsilon_{\max}} \tag{7.25}$$

and

$$h_{\text{red}} = \frac{h}{h_c} \tag{7.26}$$

where $\varepsilon_{\max}$ is the maximum strain ($\varepsilon_{\max} = f$, the misfit).

In these reduced units,

$$\varepsilon_{\text{red}} = 1 \qquad h < h_c$$

$$\varepsilon_{\text{red}} = \frac{1}{h_{\text{red}}} \frac{\ln\left[\left(\frac{h}{b}\right) + 1\right]}{\ln\left[\left(\frac{h_c}{b}\right) + 1\right]} \qquad h > h_c \tag{7.27}$$

This formula holds for the different forms of the strain and critical thickness formulae derived above. The usefulness of this reduced-variable approach is the cancellation of the leading factors in the critical thickness formulae.

The strain versus thickness in reduced units is shown in Fig. 7.13. The strain assumes its maximum value, $\varepsilon_{\text{red}} = 1$, up to $h_{\text{red}} = 1$. Beyond this point the critical strain decreases approximately as $1/h_{\text{red}}$. This is a useful curve for the compilation of experimental results, since most experiments measure strain for a given thickness and sample configuration. It is difficult to measure h_c directly.

An experimental point that falls above the theoretical curve for any $h_{\text{red}} > 1$, represents a metastable condition. That is, the balance between strain energy and dislocation energy has not reached the minimum. Usually this is a kinetic limitation that occurs at low process temperatures where dislocation formation is inhibited.

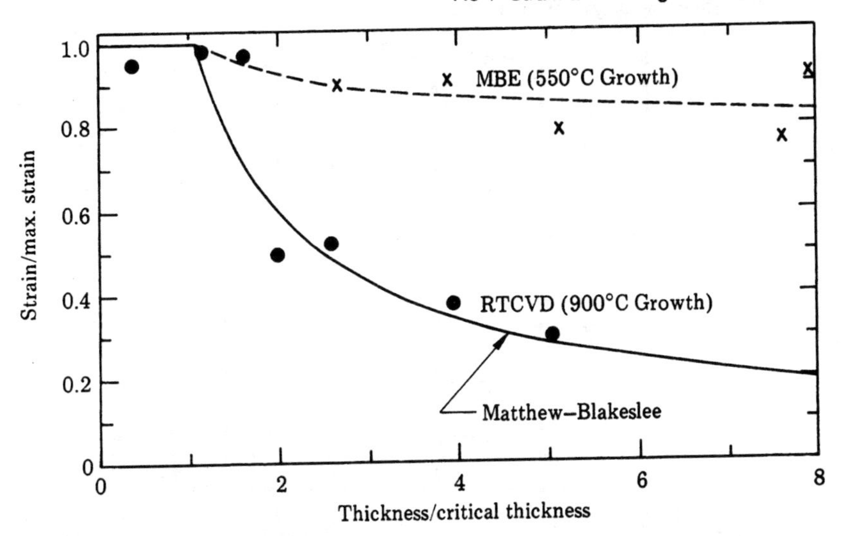

FIGURE 7.13 Reduced strain versus reduced thickness in Ge_xSi_{1-x} films on (100) Si. The Matthew-Blakeslee curve is a theoretical prediction of the same form as Eq. 7.27.

This limitation is illustrated in Fig. 7.13 which compares low temperature MBE growth to high temperature CVD growth of GeSi/Si structures.

7.8 Strain and Tetragonal Distortion

Lattice-mismatched epitaxial films are strained, leading to the generic term strained-layer epitaxy. There are two principal methods of measuring this strain: x-ray diffraction and ion channeling. In x-ray diffraction, one measures the $a_\perp$ and $a_\parallel$ lattice constants of the film independently (Fig. 7.14a). In ion channeling, one measures the deviation $\Delta\Theta$ in the angle between that expected for the cubic (relaxed) system and that found in the strained system (Fig. 7.14b). The relationship between this angle and the lattice constant follows from geometrical considerations.

Consider a small change in angle Θ due to a small variation in both x and y. From $\tan \Theta = y/x$, a small change in y and x correspond to a change in Θ, $\Delta\Theta$, given by

$$\frac{\Delta\Theta}{\cos^2 \Theta} = \frac{x\Delta y - y\Delta x}{x^2} \tag{7.28}$$

where we have used $d(\tan \Theta)/d\Theta = 1/\cos^2 \Theta$.

But

$$\cos \Theta = \frac{x}{\sqrt{x^2 + y^2}} \quad \text{and} \quad \sin \Theta = \frac{y}{\sqrt{x^2 + y^2}}$$

so that

$$\Delta\Theta = \frac{x\Delta y - y\Delta x}{x^2 + y^2}$$

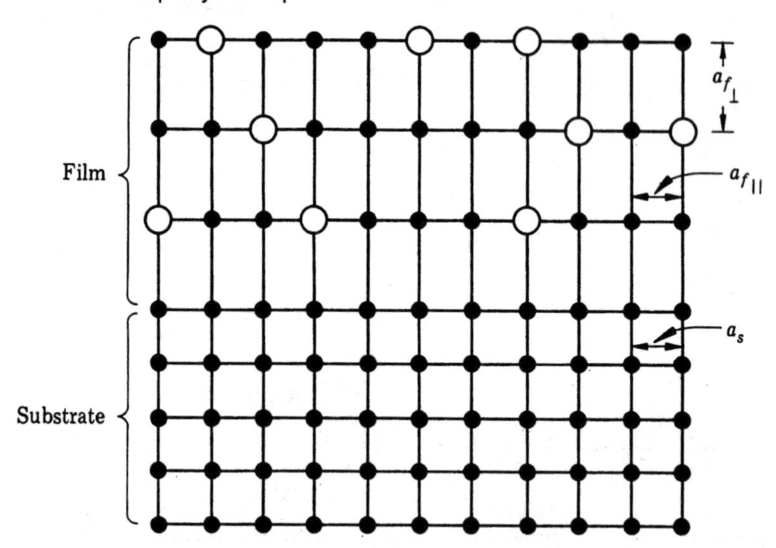

FIGURE 7.14a Strained film grown epitaxially on a substrate showing lattice constants of the film perpendicular ($a_{f_\perp}$) and parallel ($a_{f_\parallel}$) to the interface with the substrate where $a_s = a_{f_\parallel}$.

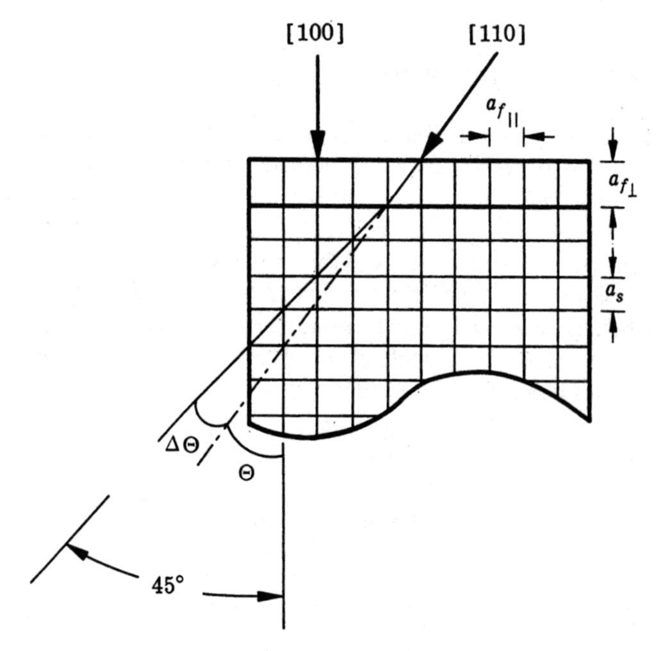

FIGURE 7.14b A strained film with lattice constants $a_{f_\parallel}$ and $a_{f_\perp}$ grown on a substrate with lattice constant a_s. An ion beam incident along the [110] of the film at angle Θ, will be at an angle $\Theta + \Delta\Theta$ to the substrate.

and

$$\Delta\Theta = \frac{(x\Delta y - y\Delta x)}{xy} \sin\Theta \cos\Theta$$

or

$$\frac{\Delta\Theta}{\sin\Theta \cos\Theta} = \frac{\Delta y}{y} - \frac{\Delta x}{x} = \frac{y + \Delta y}{y} - \frac{x + \Delta x}{x}$$

Associating Δy with $\Delta a_{f\parallel}$, Δx with $\Delta a_{f\perp}$, $(y + \Delta y)$ with $a_{f\parallel}$ and $(x + \Delta x)$ with $a_{f\perp}$, we have

$$\frac{\Delta\Theta}{\sin\Theta \cos\Theta} = \frac{a_{f\parallel} - a_{f\perp}}{a_f} \tag{7.29}$$

where a_f is the lattice constant of the relaxed film. In channeling measurements one measures $\Delta\Theta$ directly and therefore Eq. (7.29) relates $\Delta\Theta/(\sin\Theta \cos\Theta)$ to the difference in the parallel and perpendicular lattice constants.

Strain measurements, both x-ray and ion channeling, are often expressed in terms of a tetragonal distortion. The relation between the tetragonal distortion and the difference between $a_\perp$ and $a_\parallel$ is given below. The basic equations which relate stress σ and strain ε in a three-dimensional isotropic system are given in Chapter 4, Eq. (4.10). In thin film epitaxial growth there is stress within the plane of the film (x and y) but no stress in the growth direction, $\sigma_z = 0$ (biaxial stress, Section 4.4). In two-dimensional isotropic systems where $\varepsilon_x = \varepsilon_y$,

$$\varepsilon_z = -\frac{2\nu}{1 - \nu}\varepsilon_x \tag{7.30}$$

The tetragonal distortion ε_T is defined in Eq. (7.3) as

$$\varepsilon_T = \frac{|a_{f\perp} - a_{f\parallel}|}{a_f}$$

which can also be written as

$$\varepsilon_T = (1 + \varepsilon_z) - (1 + \varepsilon_x) = \varepsilon_z - \varepsilon_x \tag{7.31}$$

where we have treated $a_{f\perp} = a_f + \Delta a_{f\perp}$ and $\varepsilon_z = \Delta a_{f\perp}/a_f$, and similarly for $a_{f\parallel}$. Then the tetragonal distortion from Eq. (7.31) is

$$\varepsilon_T = \varepsilon_z - \varepsilon_x = \varepsilon_x\left(1 + \frac{2\nu}{1 - \nu}\right) = \left(\frac{1 + \nu}{1 - \nu}\right)\varepsilon_x \tag{7.32}$$

with the sign of ε_T positive. Therefore, the tetragonal distortion is,

$$\varepsilon_T = \frac{(1 + \nu)}{(1 - \nu)}\frac{(a_{f\parallel} - a_f)}{a_f}$$

In the case of pseudomorphic epitaxy $a_{f_{\|}} = a_s$ and

$$\varepsilon_T = \frac{(1 + \nu)}{(1 - \nu)} \frac{(a_s - a_f)}{a_f} = \frac{(1 + \nu)}{(1 - \nu)} f \tag{7.33}$$

For $f = 0.01$ and $\nu = 0.3$, $\varepsilon_T = .018$ and $\Delta\Theta \simeq 0.5°$.

7.9 Strain Measurements

The equilibrium critical layer thickness is a parameter associated with pseudomorphic growth of lattice-mismatched structures. For layers exceeding this critical thickness, strain relief occurs by the generation of misfit dislocations. In this section, we describe a measurement of strain in AlSb/GaSb (100). This is a III–V system with a 0.7% lattice misfit corresponding to AlSb lattice constant 0.6138 nm and GaSb lattice constant 0.6095 nm.

The channeling measurement determines the angular displacement $\Delta\Theta$, for the [110] off-normal axis ($\Theta = 45°$), as shown in Fig. 7.15 for the pseudomorphic layer on (100) GaSb. Channeling data in Fig. 7.16 clearly show the angular shift in the channeling minimum between film (Al signal) and substrate (Sb signal). Using Eqs. (7.29) and (7.33), the tetragonal distortion corresponds to 2.3% when $\nu = 0.331$.

An established way of determining stress in thin films is by measurement of the radius of curvature induced by the different stresses in the thin film and the substrate. This was discussed in detail in Chapter 4 where Eq. 4.20a was derived giving the relationship between the radius of curvature and the film thickness in the case of biaxial stress in isotropic thin films. Using the same level of approximation the formula is easily modified to express the curvature as a function of strain. The in-plane stress in one dimension is related to the mis-match strain by:

$$\sigma_f = \left(\frac{Y}{1 - \nu}\right)_f \varepsilon$$

where f refers to the film.

In terms of Eq. 4.20 this yields a relation between the expected curvature and the strain as:

$$r = \left(\frac{t_s^2}{6t_f\varepsilon}\right) \frac{\left(\frac{Y}{1 - \nu}\right)_s}{\left(\frac{Y}{1 - \nu}\right)_f}$$

where t_f is the film thickness and t_s is the substrate thickness. As a rough estimate we take the elastic constants as equal to show that a 10 nm film with 1% mismatch on a substrate of 500 μm results in a radius of curvature of 4.2×10^4 cm. A geometrical analysis will show that this corresponds to a deflection of the order of 3.0 microns at the center of a 4″ wafer. Although small, this is still sufficient to interfere with standard processing and lithography when sub-micron devices are desired.

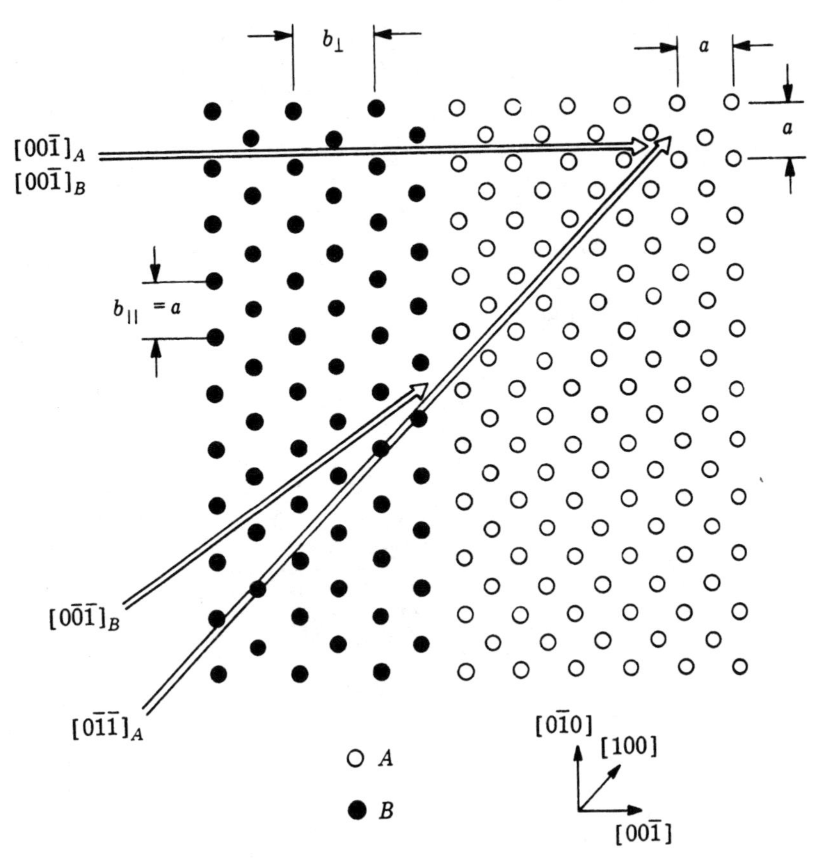

FIGURE 7.15 Schematic of AlSb grown on (100) GaSb showing the ion channeling directions along the $[00\bar{1}]$ and $[0\bar{1}\bar{1}]$ directions where A denotes AlSb and B, GaSb (from Gossmann et al.).

Strain may also be induced by differences in the thermal expansion coefficient of two different materials. This is discussed explicitly in Chapter 4. In this case the strain is given by the thermally induced lattice mismatch, usually expressed as $(\alpha_2 - \alpha_1)\Delta T$. Here α_2 and α_1 are the thermal expansion coefficients of the two materials and ΔT the temperature difference. This issue arises with even nominally lattice matched materials. Due to differences in their expansion coefficients, matching at the elevated growth temperature may not correspond to matching at room temperature where the device is to be used.

7.10 Superlattices

The superlattice was originally envisaged as a one-dimensional periodic structure consisting of alternating ultrathin semiconductor layers with its period less than the electron mean free path (Esaki and Tsu, 1970). It is a *superlattice* because there is a period of the lattice determined not by the lattice constant of the materials ($a \simeq$ 0.5 nm) but by the period of the layers with a repeat spacing of about 10 nm (for

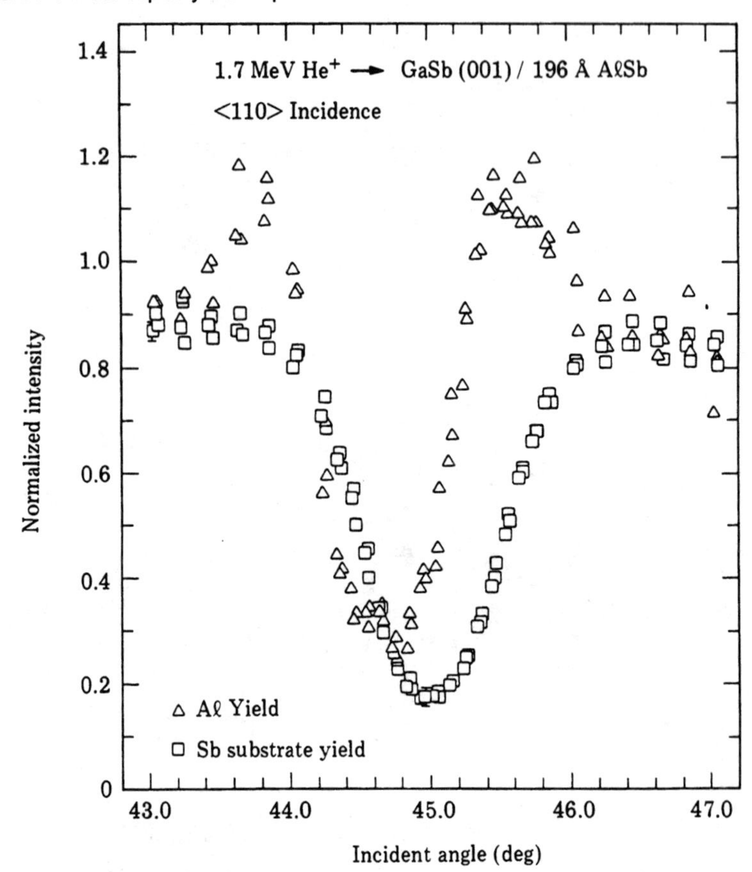

FIGURE 7.16 Channeling angular yield measurements along the ⟨110⟩ axis of the Al yield from the AlSb film and of the Sb yield from the GaSb substrate, showing the angular shift between the two yields (from Gossmann et al.).

example). Such a superlattice is shown schematically in Fig. 7.17. The electron mean free path relative to the superlattice period is the important parameter for the observation of quantum effects. If the superlattice period is less than the mean free path, the electron system enters into a quantum regime. The condition is easily reached as electron mean free paths can easily exceed 10 nm in well-ordered systems and epitaxial growth can be achieved on the monolayer scale.

Superlattices can be made either by depositing layers of the same material but with alternating levels of dopant concentration, *a doping superlattice*, or by depositing layers of different composition such as Ge_xSi_{1-x} on Si, *a compositional superlattice*. The latter is shown in Fig. 7.18. Superlattice effects may be produced by periodic variation of dopants or composition during epitaxial growth. The compositional superlattice involves growth processes which follow naturally from our discussion of heteroepitaxy.

Many applications of epitaxy in electronic materials require a periodic variation of the band gap. This periodic change in the potential seen by the electron can

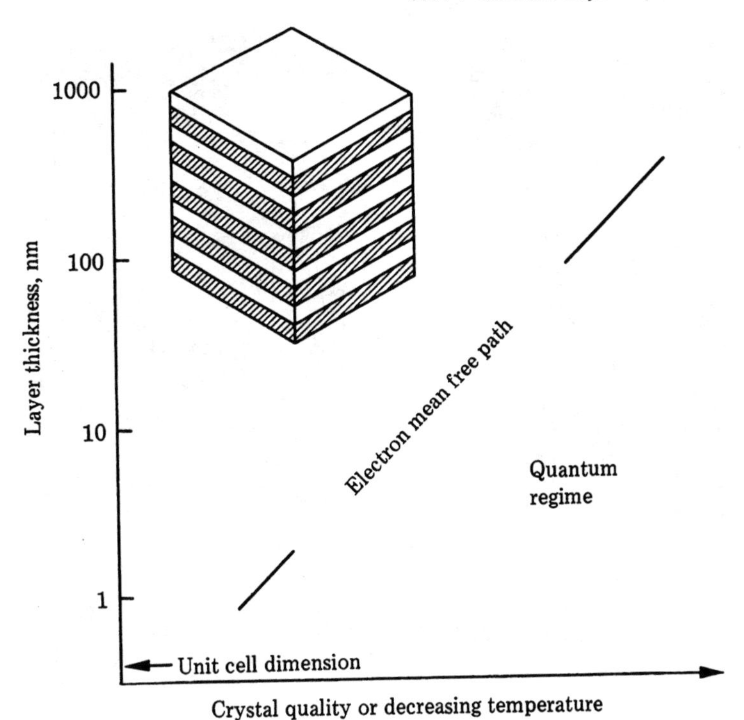

FIGURE 7.17 Schematic illustration defining the quantum regime with a superlattice in the insert (Esaki, 1985).

be achieved by alternating layers of two different materials with periodicities of 5 to 100 nm in a *compositional superlattice*.

As in single-layer heteroepitaxy, superlattices may be nearly lattice-matched such as AlGaAs/GaAs ... AlGaAs/GaAs where the "..." indicates a periodic array. The superlattices may also be formed from non-lattice-matched materials and are then called *strained-layer superlattices*. In the following we consider the strain associated with the superlattice. Figure 7.19 indicates the notation for an A/B ... A/B superlattice.

For structures with no dislocations, the elastic strain energy of the superlattice configuration can be written:

$$E_{\text{strain}} = n(k_A d_A \varepsilon_A^2 + k_B d_B \varepsilon_B^2) \tag{7.34}$$

where n is the number of superlattice layers, d_A and d_B are the layer thicknesses, k_A and k_B are the relevant elastic constants, and ε_A and ε_B are the strains. This assumes a superposition of the strains proportional to ε^2 in the individual layers.

7.11 Strained-Layer Superlattices

We will consider two superlattices: one grown with lattice-matched materials and the other with lattice-mismatched materials (a strained-layer superlattice). In a

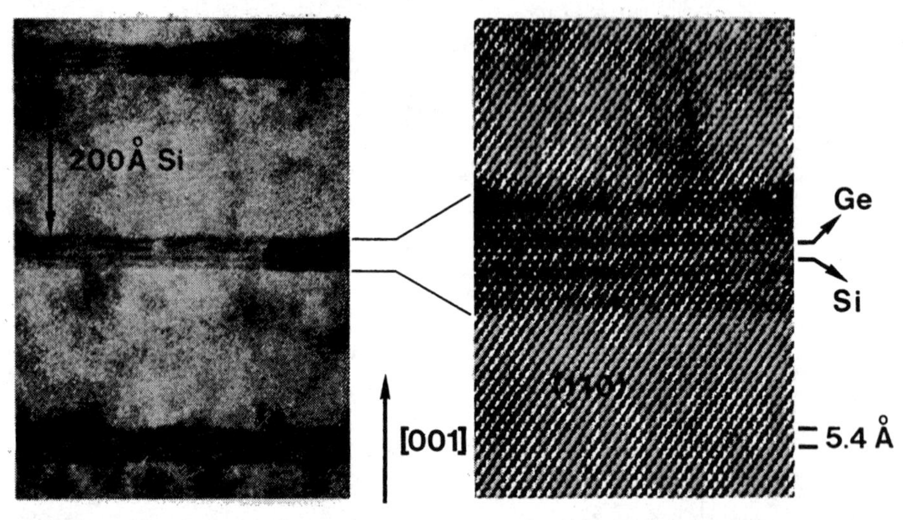

FIGURE 7.18 Cross-sectional transmission electron microscope lattice image of a double Ge/Si superlattice consisting of alternating layers of (4 ml Ge + 4 ml Si) spaced by 200 Å layers of Si: (a) low-magnification image showing overall structure and (b) high-magnification image highlighting the 4 × 4 superlattice (from Bevk et al., 1987).

strained-layer superlattice we grow layered structures with slightly different lattice constants in a fully commensurate fashion. The lattice mismatch is taken up by the biaxial strain in the plane of the two layers as shown in Fig. 7.20. A single strained layer is the building block of the strained-layer superlattice. The layers a and b with thickness t_a and t_b must have the same in-plane lattice constant as the substrate (or buffer layer). The alignment of the atoms in the plane of the interface is the condition for commensurate growth (i.e., the lateral positions of the atoms do not change across the interface). The in-plane strain of the epitaxial layer is called the coherency strain.

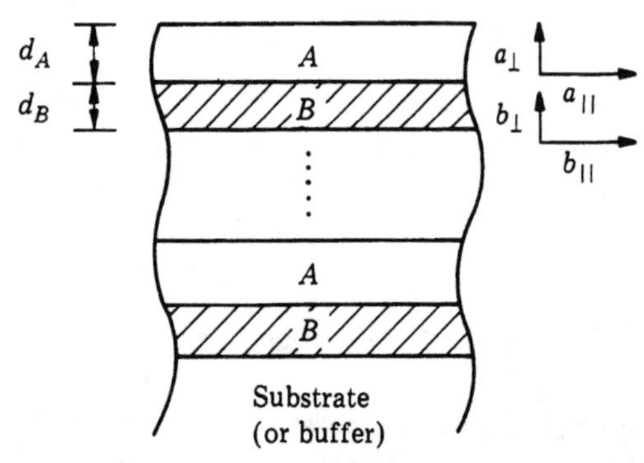

FIGURE 7.19 Schematic of the notation for an A/B superlattice.

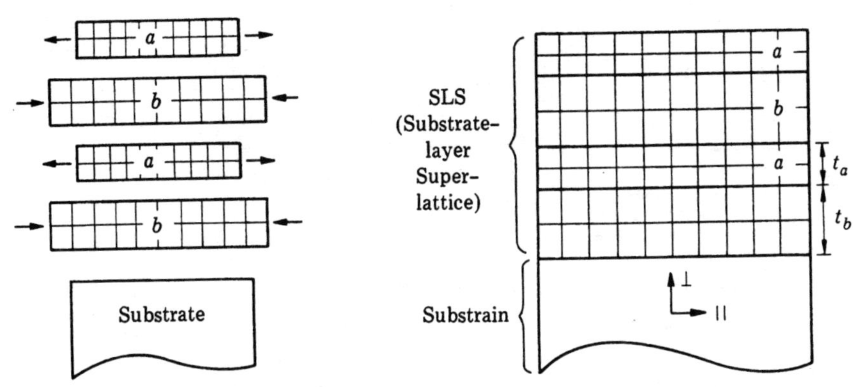

FIGURE 7.20 Schematic diagrams of (a) isolated (unstrained) layers and substrate and (b) commensurate superlattice with alternating layers in tensile and compressive stress (Picraux et al., 1990).

As discussed in Section 7.5, misfit dislocations can be generated to relieve strain. In the extreme case of perfect incoherence, misfit dislocations are so numerous as to relieve all the strain in the epitaxial layer. This implies a misfit dislocation spacing of $S = b_{\|}/f$, where $b_{\|}$ is the magnitude of the component of Burgers vector parallel to the plane of the interface and f is the lattice mismatch. The substrate lattice constant is usually fixed by available high-quality crystals (Si, GaAs, InP), so that partially relaxed alloy buffer layers are often grown to establish a given lattice constant prior to strained-layer superlattice growth.

The role of the substrate or buffer layer in establishing the strain distribution in coherent superlattices is illustrated schematically in Fig. 7.21. When a buffer layer is used prior to superlattice growth, the distribution of strain between layers a and b can be balanced. For example, by setting the lattice constant of buffer c to the in-plane lattice constant for the freely floating a/b superlattice, we can ensure that there will be no net accumulation of excess stress in the superlattice.

In many cases one of the layers of the superlattice will be the same material as the substrate, so that for a coherent structure the a layers will be unstrained, with the strain confined to the b layers. This configuration is usual for single quantum wells or for a few strained layers. However, the excess stress in the structure will increase with thickness, so that superlattices with many periods may exceed critical values and exhibit relaxation by dislocations.

The influence of the choice of the buffer layer on the total strain can be seen in the following way. Consider a superlattice made up of two components A and B with lattice constants a_A and a_B and equal individual thicknesses, $d_A = d_B = d$. Suppose we choose a buffer layer with a lattice constant just in-between these two values, $(a_A + a_B)/2$. Then the total strain energy of the (pseudomorphic) superlattice is given by Eq. 7.34 in the form of:

$$E^{b}_{\text{strain}} = kd[n_A(\varepsilon^{b}_{A})^2 + n_B(\varepsilon^{b}_{B})^2] \tag{7.35}$$

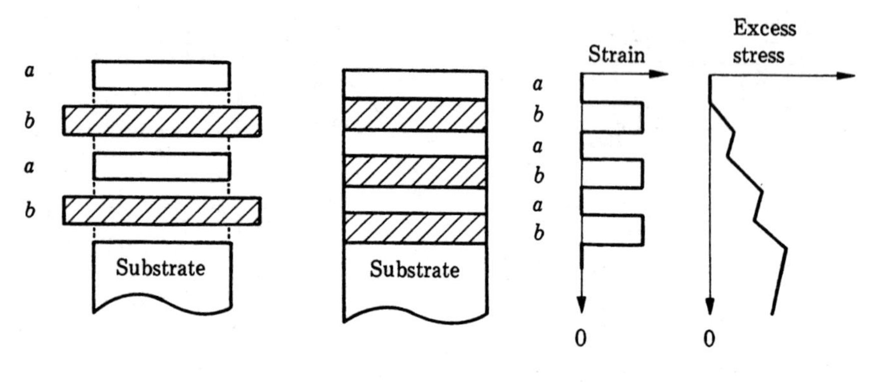

(a) No buffer layer

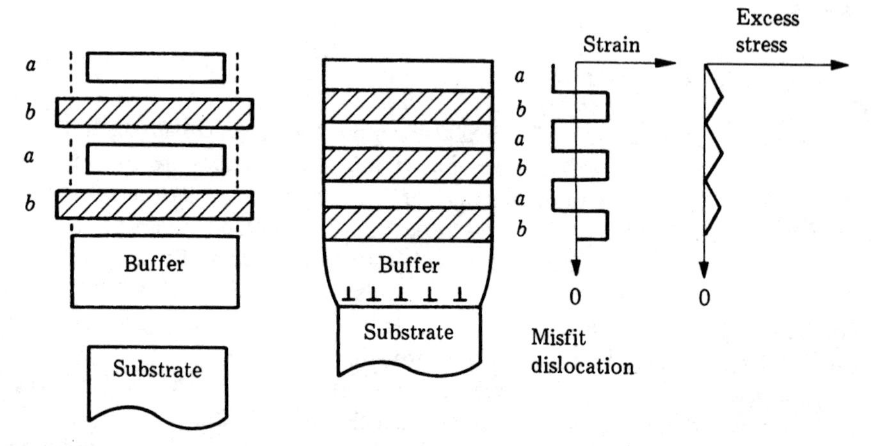

FIGURE 7.21 Schematic diagram of a strained-layer superlattice illustrating the layer stress for (a) no buffer layer and the substrate corresponding to one of the superlattice layers and (b) a buffer layer with lattice constant matched to the in-plane lattice constant of the free-standing superlattice (Picraux et al., 1990).

where ε_A^b and ε_B^b are the strains associated with the different lattice constant in the strained layers. In this expression we have used the simplifying assumptions of equal layer thickness, and equal elastic constants. The quantity E_{strain}^b, with superscript b, indicates the strain energy for a layer grown on a suitably chosen buffer layer. This strain energy should be compared to the identical compositional superlattice grown on a substrate of one of the individual layer materials, for example the material A. Then the total strain energy, E_{strain}^u, for this unbuffered system is

$$E_{\text{strain}}^u = kdn_B\,(\varepsilon_B^u)^2. \tag{7.36}$$

If $a_B = a_A + \Delta a$ then $\varepsilon_B^b = \Delta a/2a_B$, $\varepsilon_A^b = \Delta a/2a_A$, $\varepsilon_B^u = \Delta a/a_B$ and

$$E_{\text{strain}}^b = kdn_B\,(\Delta a)^2/2a_B^2 \tag{7.37}$$

where the last equation makes use of the fact that $a_A \cong a_B$ and $n_A = n_B$. This should be compared to $E_{\text{strain}}^u = n_B kd(\Delta a/a_B)^2$ for the unbuffered system. Thus the unbuffered system contains twice the strain energy for the same superlattice. Because of the quadratic dependence of the strain energy it is favorable to have all of the material partially strained than one half the material at maximum strain. As we have seen throughout this chapter the thickness of strained layer materials is limited by the total strain energy; thus it is desirable to choose a substrate that minimizes this quantity. A buffer layer can be chosen to minimize the energy.

7.12 Threading Dislocations

A basic concept in dislocation theory is that a dislocation line can never terminate within a crystal (Hull, 1975). Dislocations must form either closed loops, branch into other dislocations, or terminate at a grain boundary or crystal surface. This concept is implicit in our definition of the dislocation energy which has dimensions of energy/length; the total energy of the structure depends on defining the length of the dislocation.

As we have seen, the edge or misfit dislocation plays a critical role in the theory of heteroepitaxy. It is the mechanism for relief of the strain in lattice mis-matched systems. The formulation of the critical thickness expression has assumed that the misfit dislocations are extended throughout the crystal, lying in the plane of the interface and located at the solid-solid interface formed by the two materials. In practice it is common to find a misfit as part of a dislocation network, the network working its way through the thin overlayer and terminating at the (top) growth surface. Part of the dislocation network is envisioned as threading its way through the film; such dislocations are known as threading dislocations (Fig. 7.22).

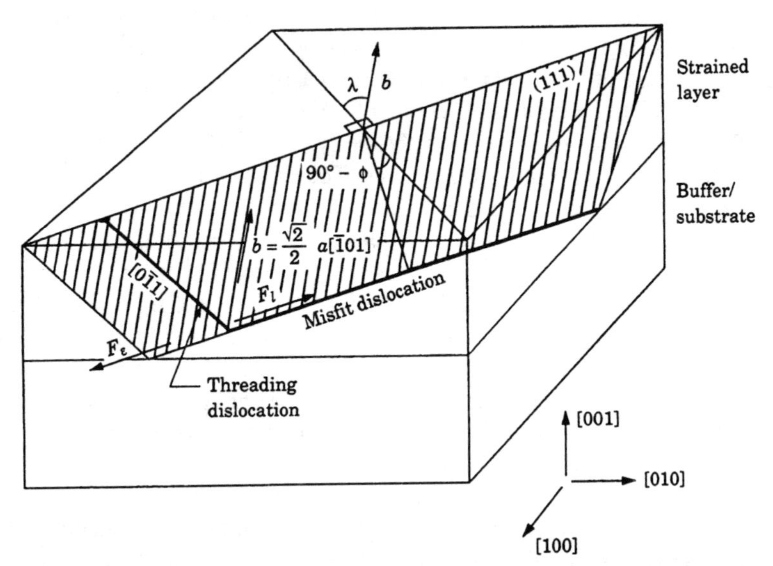

FIGURE 7.22 A strained layer epitaxial structure illustrating a misfit dislocation lying in the plane at the film/substrate interface connected to a threading dislocation lying in a (111) plane which intersects the surface.

The threading dislocation is associated with the nucleation mechanism for dislocation formation. Under certain circumstances equilibrium theory may predict a lowest energy state with a certain density of misfit dislocations. Growth at low temperature inhibits the atomic motion necessary to form the dislocations. Such a situation corresponds to a configuration which is metastable against dislocation formation. In other words the physical situation has inhibited the nucleation of the required dislocation density. As the temperature is increased the nucleation process will often occur at a free surface. If the nucleation originates at the growth surface and extends to the interface, a threading dislocation is created. A common configuration may correspond to a dislocation half-loop with a misfit dislocation extending along the buried interface terminated by two threading dislocations reaching the growth surface. In almost all applications of heteroepitaxy it is desirable to achieve the lowest threading dislocation density possible. While misfit dislocations are an intrinsic part of the structure and cannot be avoided in an equilibrium configuration, threading dislocations are a function of the experimental situation (i.e. purity, surface cleanliness, growth rate etc.) and can be varied by large factors.

The density of threading dislocations is usually measured by one of the following methods: (1) Etch pit decoration with optical microscopy; (2) scanning electron microscopy, often combined with electrical measurements; and (3) planar transmission electron microscopy. Etch pit decoration makes use of the fact that chemical etching can be enhanced at a dislocation which penetrates the free surface. The micron sized etch pits enlarge the original (nanometer size) dislocation and are easily observed and counted under an optical microscope. Since the pits have dimensions of about a micron the technique is most useful at dislocation densities of less than $10^6/cm^2$ to avoid overlap. Scanning electron microscopy techniques are useful in the range of 10^6 cm^{-2} to 10^8 cm^{-2}, comparable to a dislocation separation of one to ten microns. Transmission electron microscopy has excellent resolution but a rather limited field of view, less than a $1 \mu m$ by $1 \mu m$. Thus to have a dislocation in the field requires a density well in excess of 10^8 cm^{-2}.

In all of these techniques one measures the density of dislocations which intercept the surface, such as threading dislocations. The measured density is expressed as dislocation/cm^2. As we have seen the misfit dislocation density is expressed as misfit dislocations/cm, a linear measure. The ratio of these two quantities, misfits/cm to threading dislocations/cm^2, estimates the average length of the misfit dislocation. For example for pure GaAs on Si the misfit is about 4%, corresponding to a misfit dislocation density of 1 misfit/25 lattice constants or about 10^6/cm. A wide range of values of threading dislocation densities have been reported for this GaAs/Si system. For illustrative purposes we will use the value of $10^6/cm^2$. Then the average length of misfit dislocation is about 1.0 cm. From a crystallography point of view this represents a substantial length of good atomic order. On the other hand from a semiconductor point of view a dislocation density of $10^6/cm^2$ represents poor quality, with sufficient electron traps to result in unusable material. (Remember that the dislocation density of excellent bulk Si can be about $1/cm^2$ and good GaAs about $10^3/cm^2$). A phenomenological expression which relates the misfit dislocation density, $1/S$, the threading dislocation density, ρ_t, and the average length of the misfit dislocation, $\bar{l}$, is

$$\rho_t = \frac{2(1 - \bar{l}/z)}{S\bar{l}} \qquad (7.38)$$

where z represents the linear dimension of the sample. Since $\bar{l} \leq z$, the expression indicates a low threading dislocation density when $\bar{l}$ is close to the length of the sample.

Threading dislocations are intimately related to misfit dislocations. Therefore they are of particular significance in configurations of heteroepitaxy in which misfits are part of the design configuration. One example is the use of a specialized buffer layer grown to minimize the strain energy in a superlattice (Sec. 7.11). In this case the buffer is grown on a substrate of different lattice constant. The buffer is intended to be entirely relaxed with a complete complement of misfit dislocations at the substrate/buffer interface to entirely relax the strain. The large density of misfit dislocations is a driving force for the existence of threading dislocations which can propagate throughout the grown superlattice. The fact that a grown film replicates the dislocations in the substrate or buffer layer also indicates the necessity for the starting substrate to be of the highest quality. A second configuration in which misfit dislocations are "built-in" to the design are systems such as GaAs on Si. The overall device interest here is to combine the semiconductor advantages of Si, the most useful group IV semiconductor, with the advantages of GaAs, the most useful III-V material. The large lattice mismatch between these materials, about 4.2%, predicts a very small critical thickness, precluding the growth of pseudomorphic layers with a workable thickness, that is, thick enough for device processing. An alternate scheme requires non-pseudomorphic epitaxy, that is, a thick GaAs layer deposited on Si, totally relaxed, with misfit dislocations at the buried interface. Devices made in the upper layers of the GaAs would not be affected by the buried dislocations. However threading dislocations do exist at the upper surface, and successful growth of "electronic grade material" requires the minimization of this threading dislocation density.

References

1. J. C. Bean in *High Speed Semiconductor Devices,* edited by S. M. Sze, Wiley–Interscience, New York (1990).

2. J. Bevk, B. A. Davidson, L. C. Feldman, H.-J. Gossmann, J. P. Mannaerts, S. Nakahara, and A. Ourmazd, *J. Vac. Sci. Technol.* B5, 1147 (1987).

3. H. C. Casey, Jr. and M. B. Panish, *Heterostructure Lasers* (Part A: "Fundamental Principles" and Part B: "Materials and Operating Characteristics"), Academic Press, Orlando FL (1978).

4. L. L. Chang, L. Esaki, W. E. Howard, R. Ludeke and G. Schul, *J. Vac. Sci. Technol.*, 10, 655 (1973).

5. A. J. Cho in *Molecular Beam Epitaxy and Heterostructures,* edited by L. L. Chang and K. Ploog, NATO ASI Series, Series E, No. 87, Martinus Nijhoff, Amsterdam, The Netherlands (1985).

6. L. Esaki and R. Tsu, *IBM J. Res. Dev* 14, 61 (1970).

7. L. C. Feldman, J. W. Mayer, and S. T. Picraux, *Materials Analysis by Ion Channeling*, Academic Press, New York (1982).

8. H.-J. Gossmann, B. A. Davidson, G. J. Gualtieri, G. P. Schwartz, A. T. Macrander, and S. E. Slusky, *J. Applied Phys.* 66, 1687 (1989).

9. M. H. Grabow and G. H. Gilmer in *Semiconductor-Based Heterostructures* (pp. 3–19), edited by M. L. Green, J. E. E. Baglin, G. Y. Chin, H. W. Deckman, W. Mayo, and D. Narasinham, The Metallurgical Society, Warrendale PA (1986).

10. D. C. Houghton, C. J. Gibbings, C. G. Tuppen, M. H. Lyns, and M. A. G. Halliwell, "Equilibrium critical thickness for $Si_{1-x}Ge_x$ strained layers on (100) Si," *Appl. Phys. Lett.* 56, 460–462 (1990).

11. D. Hull, *Introduction to Dislocations*, 2nd Edition. Pergamon Press, Oxford (1975).

12. E. Kasper and J. C. Bean, editors, "*Silicon Molecular Beam Epitaxy*" (Vols. I and II), CRC Press, Boca Raton FL (1988).

13. T. F. Kuech, D. J. Wolford, R. Potoemski, J. A. Bradley, K. H. Kellecher, D. Yan, J. P. Farrell, P. M. S. Lesser, and F. H. Pollock, *Appl. Phys. Lett.* 51, 505 (1987).

14. J. W. Matthews, editor, *Epitaxial Growth* (Part A and Part B), Academic Press, New York (1975).

15. J. W. Matthews, "Coherent Interfaces and Misfit Dislocations," Chapter 8 in *Epitaxial Growth,* edited by J. W. Matthews, Academic Press, New York (1975).

16. J. W. Mayer and S. S. Lau, *Electronic Materials Science for Integrated Circuits in Si and GaAs*, Macmillan, New York (1990).

17. G. C. Osbourne, "Electronic Properties of Strained-Layer Superlattices," *J. Vac. Sci. Technol.* B1, 379 (1983).

18. E. H. C. Parker, editor, *The Technology and Physics of Molecular Beam Epitaxy*, Plenum, New York (1985).

19. T. P. Pearsall, editor, *Strained Layer Superlattices: Science and Technology*, Vol. 33 in the series *Semiconductors and Semimetals,* edited by R. K. Willardson and A. C. Beer, Academic Press, New York (1990).

20. S. T. Picraux, B. L. Doyle, and J. Y. Tsao, "Structure and Characterization of Strained Layer Superlattices", In Pearsall (1990).

21. R. A. Swalin, *Thermodynamics of Solids,* 2nd edition, Wiley–Interscience, New York (1972).

22. K. N. Tu and R. Rosenberg, *Preparation and Properties of Thin Films*, Vol. 24 in *Treatises on Materials Science and Technology*, Academic Press, New York (1982).

Problems

7.1 Calculate the ratio R of the energy of a strained film to that of a clustered film by assuming there are m hemispherical (rather than cubic) clusters of radius r. Treat the surface energies per unit area of film and substrate as equal. If $R = 1$, what is the critical radius r_c?

7.2 Calculate the ratio R (Eq. 7.8) for cubic clusters if the surface energy per unit area of the film γ_f is 20% greater than that of the substrate. Compare the analytical formula for X_c (i.e. $R = 1$) to the $\gamma_F = \gamma_s$ case in the text. Calculate X_c for the case of Ge on Si using the following values: $n_{Ge} = 4.4 \times 10^{22}$ at./cm^3, $N_{Si} = 6.8 \times 10^{14}$ atoms/cm^2, $E_{st} = 0.5$ eV/atom, $\gamma_f = 2.0$ eV/atom, $\gamma_s = 1.6$ eV/atom and $n = 6$ monolayers.

7.3 For Ge on (100) Si, calculate the critical cluster size X_c if $\gamma_{Ge} = 1.9$ eV/atom and $\gamma_{Si} = 2.0$ eV/atom. Remember if $\gamma_{overlayer} < \gamma_{substrate}$, we have Stransi-Krastanov (S-K) growth, an epitaxial wetting layer is formed of Ge on Si. The problem requires a derivation of X_c in the S-K situation.

7.4 A 20 nm film of pure Si is deposited on a (100) Ge_xSi_{1-x} substrate where the lattice mismatch is 1.9%.

(a) The lattice parameters of Si and Ge are 0.543 and 0.566 nm, respectively. What is the value of x in Ge_xSi_{1-x}?

(b) Dislocations are formed with Burgers vector b oriented along the [110] direction. For silicon, Poisson's ratio = 0.272 and the shear modulus = 0.67×10^{11} N/m^2. Assume the elastic stress field radius $r = 20$ nm (the film thickness) and the core energy parameter $B^1 = 1$. Calculate:

(1) Burgers the vector b and spacing S.

(2) The energy per unit length E_d of a single dislocation.

(3) The energy density $E_\perp$ of an array of noninteracting edge dislocations.

7.5 For an isotropic, two-dimensional system where the film is grown on a (100) plane, what is the value of the tetragonal distortion ε_T for $f = 0.02$ and Poisson's ratio $\nu = 0.27$. For channeling measurements along a $\langle 110 \rangle$ channel, what is $\Delta\Theta$?

7.6 The figure for this problem is a plot of ion backscattering yield as a function of tilt angle Θ, measured down a $\langle 111 \rangle$ channel with respect to the [100] surface normal, for a 100Å GeSi film on a Si substrate. The value of $|\Delta\Theta| = 1.0°$. (Figure from Fiory et al., *J. Appl. Phys.* 56 (1984), 1227.)

(a) Compute the tetragonal strain.

(b) What would the $\Delta\Theta$ of the minimum yield be for each material if the backscattering yield were measured down a $\langle 110 \rangle$ channel, rather than a $\langle 111 \rangle$ channel?

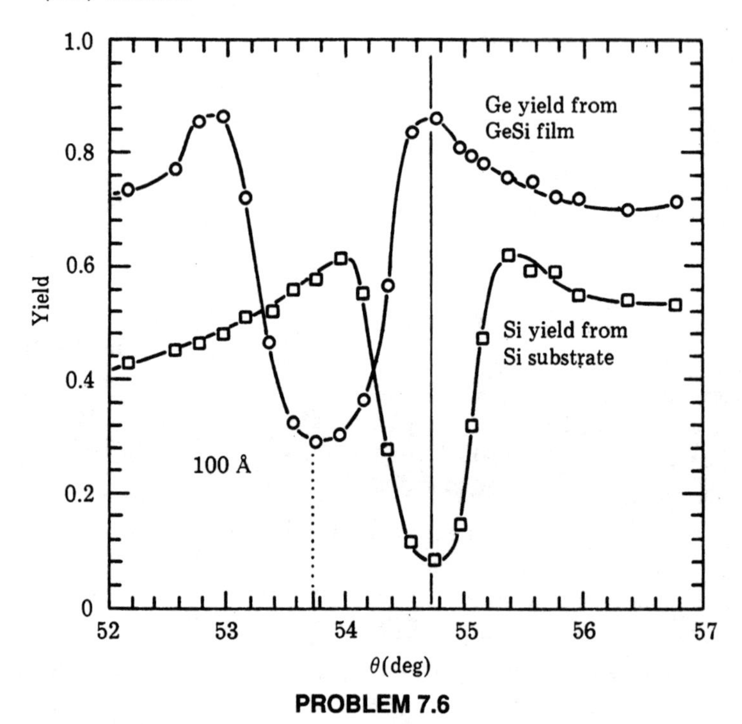

PROBLEM 7.6

7.7 **(a)** For a film of cubic material A ($a_f = 0.5513$ nm) grown on a substrate B ($a_s = 0.548$ nm), what is the maximum strain, $\varepsilon_{\parallel}$(max), in the film?
(b) If the critical thickness $h_c = 10$ nm for film A on substrate B, use the reduced strain figure to obtain a value of $\varepsilon_{\parallel}$ for a 20 nm film.

7.8 Consider a film of GaP ($a = 0.545$ nm) on (100) GaAs ($a = 0.565$ nm) where the film is relaxed with dislocations.
(a) In the simple approximation that the edge dislocations are perpendicular to the interface, what is the spacing S between dislocations?
(b) For the case that the dislocations are 60° dislocations which lie on the (111) plane in the $\langle 110 \rangle$ direction, what is the magnitude of Burgers vector and the spacing between dislocations?

7.9 For a GeSi film on a silicon crystal, rocking curve measurements for glancing angle x-ray diffraction, CuK_α, on (022) planes were used to find the strain ($\varepsilon_{\parallel}$) parallel to the interface, where we define $\varepsilon_{\parallel} = \Delta d_{022}/d_{022}$ when d is the interplanar spacing. A first-order Bragg reflection was obtained at 23.656° (0.4128 radians) corresponding to Si (022) planes. On rocking the sample, another (022) reflection was found at 23.196° (0.4048 radians).
(a) Find $\varepsilon_{\parallel}$.

(b) For a Burgers vector of 0.55 nm, find the average spacing of misfit dislocations.

7.10 Consider a thin epitaxial film of GaAs (100) with a maximum strain of 0.01. Estimate the dislocation spacing in GaAs if the strain is 0.99, 0.90, 0.50, 0.10 of the maximum value (the lattice constant of GaAs is 0.56 nm). At what fraction of the maximum strain is the dislocation spacing equal to the mean free path ($\lambda_{mfp} \simeq 100$ nm) of electrons in GaAs? Note that when the dislocation spacing is less than the mean free path, electronic transport in the film is substantially degraded.

7.11 If the maximum strain energy allowed in a film 20 nm thick is 0.1 eV/atom, what semiconductors could be grown on (100) Si and on (100) GaAs (calculate f and use Figs. 7.1 and 7.2)? Assume Poisson's ratio = 0.3, the shear modulus $\mu = 0.7 \times 10^{11}$ N/m^2, and there are 6×10^{18} atoms/m^2 on the (100) interface.

7.12 A strained-layer superlattice photo-detector is to be constructed with equal periods (10 nm) of GaP and $GaAs_xP_{1-x}$. The detector is to be grown on a GaP substrate with a $GaAs_yP_{1-y}$ buffer layer, as shown in the figure for this problem.

(a) Choose the superlattice mole fraction x such that the detector will be responsive to $\lambda \leq 0.58$ μm. Assume that the band gap of $GaAs_xP_{1-x}$ varies linearly from GaP (2.26 eV) to GaAs (1.42 eV) as x varies from 0 to 1, and that the band gap of the $GaAs_xP_{1-x}$ layers alone governs the optical response.

(b) Choose the buffer layer mole fraction y such that the net strain in the superlattice is zero.

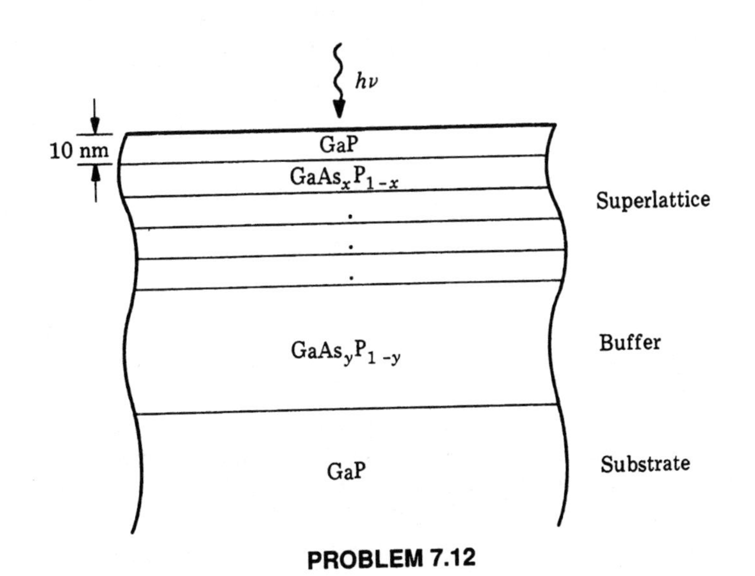

PROBLEM 7.12

(c) If the optical absorption coefficient α for the superlattice is $1 \times 10^4\ cm^{-1}$, how many layers (at 10 nm per layer) would be required to absorb 30% of the incident photons?

7.13 An epitaxial layer of $Al_xIn_{1-x}As$ is grown on an InP substrate.
(a) Find x for the lattice-matched condition.
(b) If $x = 0.25$, what would be the misfit and approximate critical thickness if $\nu = 0.3$ and $b = a/\sqrt{2}$?

7.14 The purpose of this problem is to show that in a thin film-thick substrate system essentially all of the strain is in the thin film. This is an underlying assumption used throughout this chapter and in the derivation of Eq. 7.22.

In general the strain energy can be written as $E_\varepsilon = \varepsilon^2 Bh$ where h is the thickness of the film. Consider a system of film, (thickness t_f, bulk lattice constant a_f) and substrate (thickness t_s, bulk lattice constant a_s) with the same elastic constants ($B_f = B_s = B$) and where $t_f << t_s$ and $a_f \geq a_s$. In a pseudomorphic state both the film and substrate have the same lattice constant a, where we may intuitively expect $a_f \geq a \geq a_s$.

(a) Show that the total strain energy of the system, E_{tot}, is given by:

$$E_{tot} = \frac{Bt_f}{a_f^2}(a_f - a)^2 + \frac{Bt_s}{a_s^2}(a_s - a)^2$$

(b) Find the absolute minimum of $E_{tot}(a)$ to show that the corresponding lattice constant, a_m, is given by

$$a_m = \frac{\dfrac{a_s^2}{a_f^2}\dfrac{t_f}{t_s}a_f + a_s}{\dfrac{a_s^2}{a_f^2}\dfrac{t_f}{t_s} + 1}$$

Noting that $a_f \sim a_s$ and $t_f << t_s$ show that $a_m \simeq a_s$. This proves the approximation that for a thin film/thick substrate system pseudomorphic growth corresponds to the film acquiring the lattice constant of the substrate.

(c) Consider the system of an InAs film (10nm) grown on a GaAs substrate of thickness 20 mils $\simeq 500\mu$. Using the formulas derived above calculate the difference $(a_s - a_m)/a_s$ for the system thus showing that the infinite substrate approximation is excellent.

7.15 This problem compares changes in lattice constants of semiconductors due to applied pressure to changes induced by pseudomorphic epitaxy. The "bulk modulus" of a solid is defined as $-Vdp/dV$ where V is the volume and p the applied pressure. The volume compressibility is the reciprocal of the bulk modulus. For a cubic material the volume compressibility is given in terms of the elastic constants as $3(C_{11} + 2C_{12})$. Using values given in Appendix E, estimate the change in lattice constant for Si by applying a pressure of 12 GPa. (Section 1.2 contains definitions of Pascal and conversion coefficients.) The value of

12 GPa is chosen here since recent experiments have shown that Si undergoes a phase transition to a different crystal structure at this pressure; the very open diamond structure converts to the more close-packed structure of beta-Sn. The highest pressures currently applied in the laboratory are closer to 400 GPa.

The change in lattice constant is a linear measure which may also be expressed as the linear compressibility, equivalent to one-third of the volume compressibility for cubic materials. Compare the change in Si lattice constant under 12GPa to the change in $a_{\parallel}$ and $a_{\perp}$ for pseudomorphic growth of Si on a Ge substrate. (Information on the pressure dependence of Si can be found in S. Duclos, Y. K. Vohra and A. L. Ruoff, *Phys. Rev. B.*, Vol. **41**, p. 12021 (1990).

CHAPTER 8

Electrical and Optical Properties of Heterostructures, Quantum Wells, and Superlattices

8.0 Introduction: Materials by Design

Higher operating speeds and more efficient data processing have been the driving forces for innovation in the electronics industry. Advances have been made through the control of size (submicron technology) and through the growth of new materials and materials combinations (band-gap engineering). On the materials side, the two principal approaches have been (1) the design of semiconductor thin film structures to attack basic questions of ultimate speeds in electron transport and (2) the use of optically active materials to attain the advantages associated with optical processing. Both have employed the most sophisticated aspects of materials design to control device fabrication at the monolayer level. Fundamental parameters associated with electron/photon transport (mobility, index of refraction, etc.) can be manipulated to produce active and passive devices of unprecedented dimensions and quality.

Successful growth of suitable materials requires special techniques able to produce sharp (1 nm) material interfaces with minimal width dopant profiles. As pointed out in Chapter 6 and 7, different processes, such as molecular beam epitaxy (MBE) and metal–organic chemical vapor deposition (MOCVD), have been employed in the growth of device structures. Although not discussed in this text, developments are underway to enhance the growth process. Usually these invoke a directed energy deposition process such as laser irradiation during growth or energetic ion beam deposition.

In current solid-state laser design, material combinations are formed not only to modify the electronic band gap but also to control the optical waveguiding properties of the material. The different indices of refraction of the semiconductors are used to control the optics, while the different band gaps are used to control the electronics.

Numerous issues in materials fabrication form the frontier in this field. Most devices currently consist of AlGaAs/GaAs combinations. These two materials have a close lattice match, and this match is one reason for successful fabrication. There is, however, an increasing need to combine different materials with large lattice mismatches in order to optimize device performances. Such would be the combination of germanium and silicon; in their natural form, they have a mismatch of 4%. As discussed in Chapter 7, manipulation of these lattice-mismatched structures is carried out both to ascertain limits of growth and to devise new low-temperature schemes to form metastable structures.

Our interest in this chapter is in heterostructures and in structures with two-dimensional periodicity on the 10 nm scale: quantum wells and superlattices. The requirement of 10 nm can be easily understood in terms of simple quantum mechanics. As we will show, the energy level associated with an infinite square well of width 10 nm the order of 1 eV, large compared to room temperature thermal energies and necessary if device structures are to operate at room temperature.

Dopant configurations ultimately determine the electronic properties of semiconductor devices. Layer-by-layer growth permits the precise control of dopants to achieve configurations unattainable by other techniques such as thermal diffusion or ion implantation. Epitaxial growth also permits the possibility of ordered dopant configurations. Such structures promise improvements in ultimate speeds for otherwise conventional devices. In this chapter we present an overview of the electrical and optical behavior of layered structures to provide insight into the applications of heteroepitaxial growth. We present only a small subset of device configurations. These are chosen to highlight the effect of controlled growth of unique thin film configurations on electronic and optical behavior.

8.1 Heterostructure Energy Band Diagram

The Si p–n junction has an energy barrier or potential offset due to a change in dopants from donors (n-type Si) to acceptors (p-type Si) across the junction region. A potential offset can be produced by a change in composition as well as by a change in dopants. Figure 8.1 shows the band diagram for the intrinsic ternary III–V compound AlGaAs grown on lightly doped n-type GaAs. The discontinuities in the conduction band and in the valence band are immediately obvious.

This choice of AlGaAs on GaAs is the classic example of heteroepitaxy in electronic materials. Heteroepitaxy is the epitaxial growth of one lattice on another. In this case, the two lattices have nearly the same lattice constant—this is called lattice-

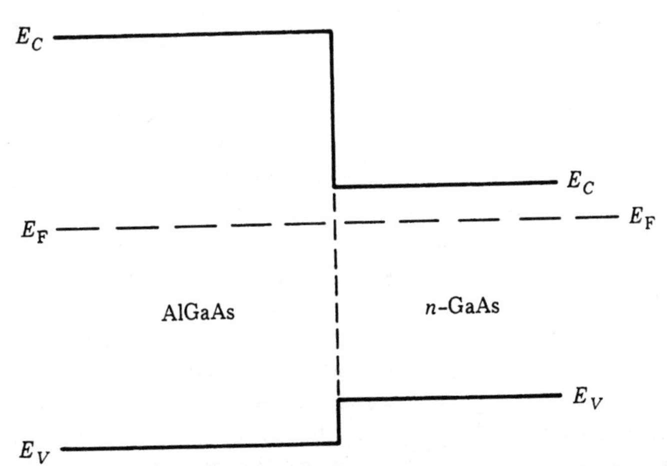

FIGURE 8.1 Energy-band diagram for AlGaAs grown on lightly doped n-type GaAs, showing the conduction band and valence band discontinuities.

matched heteroepitaxy—but the band-gap difference is substantial. The energy gap E_G of GaAs is 1.424 eV and that of $Al_{0.3}$ $Ga_{0.7}$ As is 1.861 eV. The offset is easy to visualize, as the energy-band lineup is straddling as opposed to staggered or broken (Figs. 8.2a, b, c, respectively). The lineup can be changed by changing the doping.

For device applications, the two heteroepitaxial semiconductors are often oppositely doped, as in the case of the homojunction. Figure 8.3a shows the Fermi level offset in p-type AlGaAs adjacent to n-type GaAs. In contact with each other, the two materials come to electrical equilibrium and their Fermi levels are aligned. This produces a large potential barrier in the conduction bands and a spike in the valence bands (Fig. 8.3b).

One obvious advantage of the band offset is carrier confinement. Electrons in GaAs in Fig. 8.3b have an additional potential barrier, ΔE_C, over and above that present if the n-type GaAs were in contact with p-type GaAs (the homojunction configuration). The confinement of carriers is exploited in double heterojunction structures (quantum wells) as shown in Fig. 8.4. The electrons—their position probability density, $|\phi_1|^2$, is shown by the dashed line in Fig. 8.4—can be confined in a quantum well of width L.

8.2 Electronic States in Two-Dimensional Structures

Interest in semiconductor superlattices arises primarily because of new device properties expected from these structures. The superlattice structure can affect both the motion of electrons in the solid and the interaction of photons with the solid. In both cases the material can be "engineered" to provide properties which permit conceptually new and novel devices. Since the motion of electrons in a semiconductor is usually described in terms of a band gap, the implementation of superlattices in electronic devices has become known as "band-gap engineering."

In this section we describe some of the basic quantum mechanics illustrating the modified motion of electrons in a quantum well or in a superlattice. We consider square-well superlattices since they are the predominant structure.

The quantum mechanics required to understand electron transport in a semiconductor superlattice is simple and straightforward and can be found in many standard textbooks (e.g., see Jaros, 1989). The problem can be divided into three cases: (1) motion of an electron in a single quantum well of infinite height; (2) motion of an electron in a single quantum well of finite height; (3) motion of an electron in a periodic array of quantum wells of finite height.

The two materials in a superlattice are grown with a sharp interface and they are lattice matched such that strain does not affect the band gaps. Then, in the absence of doping, the potential is formed by two potential steps in the back-to-back position. (If the materials were doped, the band bending must be taken into account.) We further assume that each layer is thick enough (>3 nm) that the material has the properties characteristic of bulk material and thin enough (<50 nm) to ensure that electrons can traverse the well without undergoing a collision. Furthermore, to actually draw the potential well, the relative placement of the two band gaps must be

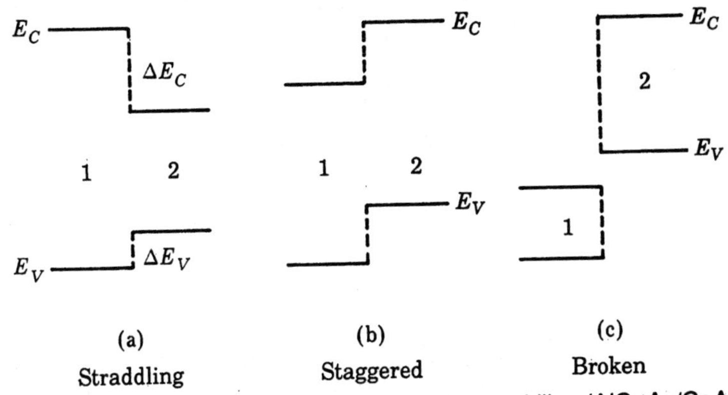

FIGURE 8.2 Types of energy-band lineups: (a) straddling (AlGaAs/GaAs), (b) staggered (InP/InSb), (c) broken gap (InAs/GaSb) (after Kroemer, 1985).

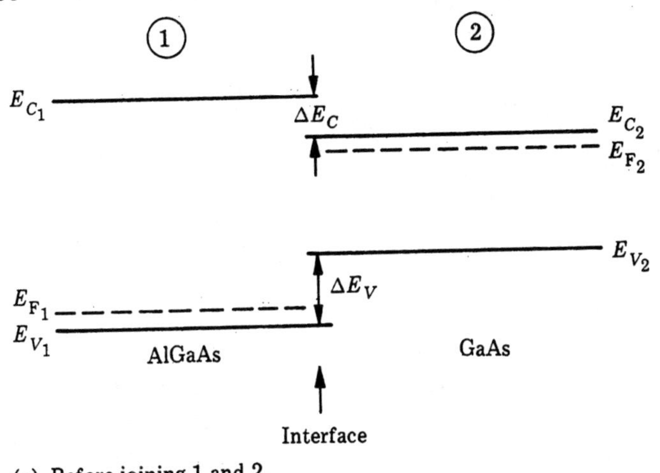

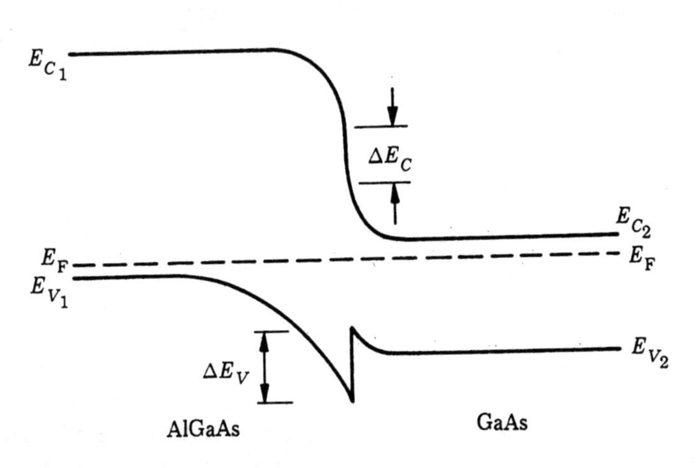

FIGURE 8.3 Energy-band diagram for p-type AlGaAs on n-type GaAs (a) before and (b) after joining (equilibrium), showing the band-gap discontinuities.

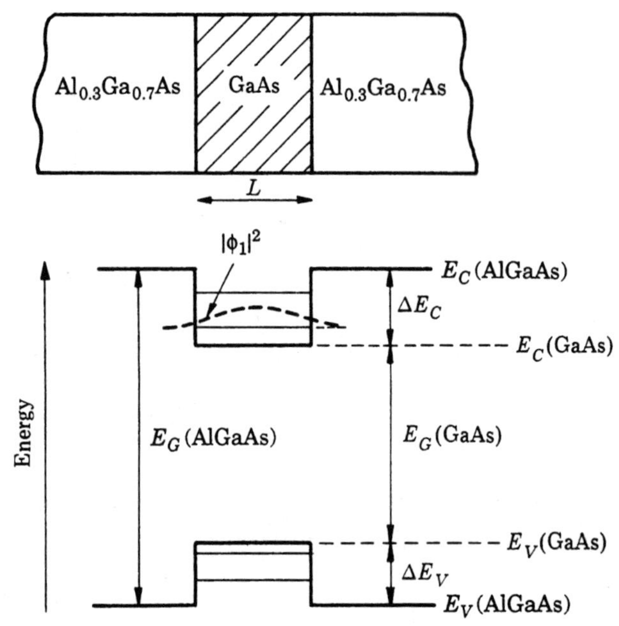

FIGURE 8.4 GaAs quantum well: (a) schematic of structure and (b) energy-band diagram showing the energy levels in the well and the electron occupation (dashed line).

known. This is illustrated in Fig. 8.4 for a $Al_{0.3}Ga_{0.7}As$ and GaAs structure. Explicitly, the conduction band offset ΔE_C or valence band offset ΔE_V must be known to solve the real problem.

The structure shown in Fig. 8.4 depicts the confinement of electrons in square wells consisting of the conduction band of the GaAs and $Al_xGa_{1-x}As$ structure. Another system might be a very large band-gap material in combination with a small band-gap material, such as InSb (E_G = 0.2 eV) sandwiched between CdTe (E_G = 1.4 eV) layers.

8.2.1 Electron in a Single Well

The electron motion is described by the solution to the Schrödinger equation:

$$\left[-\frac{\hbar^2}{2m^*}\left(\frac{\partial^2}{\partial x^2}+\frac{\partial^2}{\partial y^2}+\frac{\partial^2}{\partial z^2}\right)+V\right]\phi = E\phi \qquad (8.1)$$

where the growth direction (the superlattice) is taken as the z-direction, m^* is the effective mass of the electron in the semiconductor, h is Planck's constant, and $\hbar = h/2\pi$. The solutions in the x- and y-directions are free-electron-like, and the solution in the z-direction is of immediate interest. In particular we consider a well of width L and depth V. If the barrier seen by a particle in the well is very high ($V \to \infty$), the amplitude of the wave function must be zero at the well boundaries ($z = 0$ and $z = L$). Inside the well, the potential is a constant and the solution of the Schrödinger equation is of the form

$$\phi = A \exp(ikz) + B \exp(-ikz) \tag{8.2}$$

where A and B are determined from the boundary conditions $\phi(0) = \phi(L) = 0$. Hence

$$\phi = \text{constant} \times \sin\left[\frac{n\pi z}{L}\right] \qquad n = 1, 2, \ldots \tag{8.3}$$

The energy is obtained from

$$\left\{-\frac{\hbar^2}{2m^*}\frac{d^2}{dz^2} + V\right\}\phi = E\phi \tag{8.4}$$

Substitute for ϕ from Eq. (8.3), multiply the equation from the left by ϕ, and integrate over z to obtain the energy measured from $V = 0$, that is,

$$E_n = \frac{\hbar^2}{2m^*}\left(\frac{\pi n}{L}\right)^2 \qquad n = 1, 2, \ldots \tag{8.5}$$

In these equations $\hbar^2 k^2/2m^* = V - E$. The wave functions ϕ and energy levels E_n are illustrated in Fig. 8.5. The essential results from this simple calculation which are applicable to real systems are (1) electrons are confined within the well, (2) energy-level spacings vary like n^2, and (3) energy levels are quantized and proportional to $1/L^2$. This last fact is most exciting; electron energy levels of the semiconductor can be changed by simply changing the geometry!

8.2.2 Two-Dimensional Density of States

In addition to energy quantization an important property of the electronic structure of thin films is a new density of states. We consider the motion in the z-direction as quantized and in the other two dimensions as free.

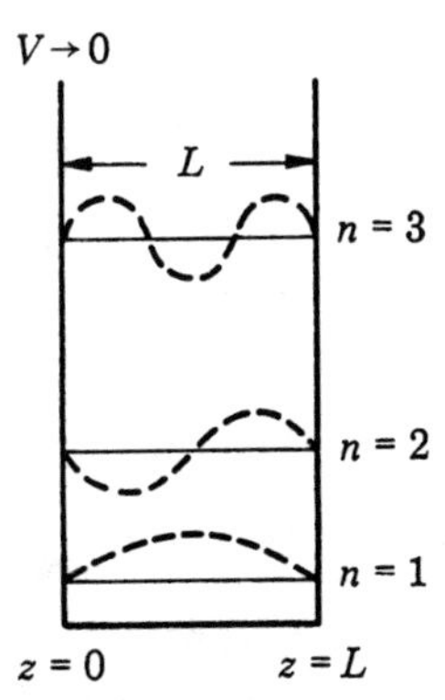

FIGURE 8.5 Energy levels of a particle in a one-dimensional potential well of infinite depth and width L.

For a given quantum state,

$$\rho_{2D}(E)\,dE = \rho_{2D}(k_\perp)\,dk_\perp \tag{8.6}$$

where ρ_{2D} is the energy or momentum density of states. In a parabolic (free-electron) approximation,

$$E = \hbar^2 k_\perp{}^2/2m^*$$

Using $dk_\perp = 2\pi k_1\,dk_1$, and adding a factor of 2 for the two spins and a factor of $1/(2\pi)^2$ from the usual calculation of k states allowed between E and $E + dE$, we have

$$\rho_{2D}(E) = m^*/\pi\hbar^2 \tag{8.7}$$

For a given quantum state E_n the density of states is constant in energy, independent of layer thickness and increases by an amount $m^*/\pi\hbar^2$ for each occupied energy (Fig. 8.6). Also shown in Fig. 8.6 is the usual three-dimensional density of states ρ_{3D}:

$$\rho_{3D}(E) = \frac{\sqrt{2}m^{*3/2}\,E^{1/2}}{\pi^2\hbar^3} \tag{8.8}$$

The total density of states just above the energy E_n is then $nm^*/\pi\hbar^2$ and the density shows a discontinuous step function at each E_n value. The effect of this difference in density of states is to strongly modify the basic semiconductor interactions which govern photon absorption or electron transport. Note that the density of states is finite even at the bottom of the conduction band of the two-dimensional level (which begins at E_1), whereas it tends to zero in the three-dimensional system.

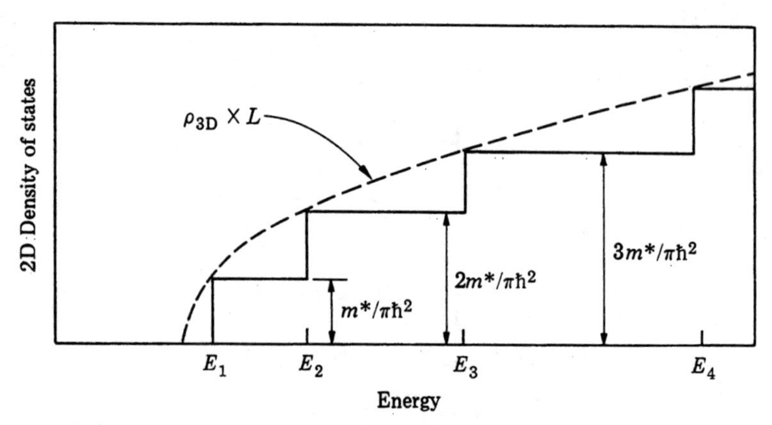

FIGURE 8.6 The two-dimensional (2D) density of states (DOS) and comparison with 3D density of states (after Weisbuch, 1987).

8.3 Excitons

The lowest energy required to induce a transition of an electron from the valence band to the conduction band in a quantum well is

$$E_G(\text{QW}) = E_G(\text{GaAs}) + E_e + E_h \tag{8.9}$$

where E_G is the band gap of the semiconductor material (GaAs) and E_e, E_h are the lowest-lying states of electrons and holes within the well. Optical spectroscopy, however, often shows a strong absorption at a slightly lower energy due to the exciton.

Emission spectra of superlattices, and of most semiconductors, are governed by the exciton. An *exciton*, to put it simply, is a bound electron–hole pair. In three dimensions the energy E_n^{ex} of such a pair is given by

$$E_n^{\text{ex}} = -13.6\,\frac{m_{\text{ex}}^*}{m}\frac{1}{\epsilon_r^2 n^2}\ (\text{eV}) \tag{8.10}$$

where m_{ex}^* is the reduced mass of the exciton and ϵ_r is the relative dielectric constant of the material. The emitted photon energy is then

$$h\nu = E_G - E_n^{\text{ex}} \tag{8.11}$$

Figure 8.7 shows excitonic transition luminescence peaks in GaInAs/InP quantum wells. For the 1 nm well, the peak is at 1.17 μm and is shifted 0.40 μm from the bulk GaInAs reference at 1.57 μm. The energy reflects the energy-level shift of the quantum well of varying thickness. The transition is primarily through the exciton via Eq. 8.11.

The exciton can be thought of as a bound electron–hole pair in a semiconductor. In three dimensions its properties are described by the same physics which is used to describe the Bohr atom, although the mass is now given by a reduced mass associated with the electron–hole pair, and the coulomb attraction is reduced by a factor of $1/\epsilon_r$ where ϵ_r is the dielectric constant of the material (e.g. see chapter 3 in Morrison et al., 1976).

Keeping in mind the Bohr model of the hydrogen atom, we can define the "three-dimensional exciton radius" as

$$a_{\text{ex}}^{3\text{D}} = \frac{\hbar^2\epsilon_r}{m_{\text{ex}}^* e^2} \tag{8.12}$$

corresponding to the ground state of the hydrogen atom. Note that ϵ_r is of the order of 10 in most semiconductors and that $m_{\text{ex}}^* \sim 0.05\ m_e$, so that $a_{\text{ex}}^{3\text{D}} \sim 5$ to 10 nm. As might be expected, the properties of the exciton will change when confinement forces the dimensions to be less than the three-dimensional radius.

In the limiting case, when the well width is narrow, we have

$$E_{n,\text{ex}}^{2\text{D}} = E_{\text{ex}}^{3\text{D}}\,[1/(n + 1/2)^2] \qquad n = 0, 1, 2, \ldots \tag{8.13}$$

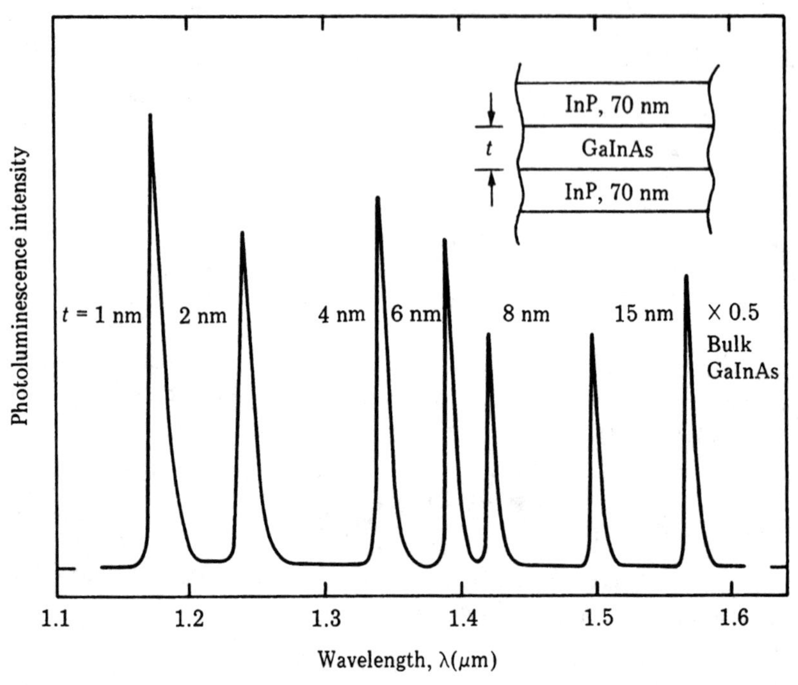

FIGURE 8.7 A typical photoluminescence spectrum from a stack of quantum wells with different thickness separated by 700 Å InP barriers at 2 K. The pumping power is 1 μW and the pumping area is of ~50 μm diameter (Tsang and Schubert, 1986).

where $E_{ex}^{3D} = E_1^{ex}$ given in Eq. 8.10. The $n = 0$ case corresponds to $E_{1,ex}^{2D} = 4E_{ex}^{3D}$. For intermediate confinement wells the exciton binding energy varies between E^{2D} and E^{3D}. The exciting aspect of this is that we can change this most fundamental semiconductor interaction by changing the film thickness once we are at the quantum scale of approximately 10 nm.

The derivation of the energy levels for the two-dimensional exciton is solved in a manner analogous to that for the three-dimensional case, basically "hydrogen-atom-like" with effective mass m^* (Shinada and Sugano, 1966). In two dimensions the relevant Hamiltonian and Schrödinger equation is given by

$$-\frac{\hbar^2}{2m^*}\left(\frac{\partial^2}{\partial x^2} + \frac{\partial^2}{\partial y^2}\right)\phi - \frac{e^2}{\epsilon_r(x^2 + y^2)^{1/2}}\phi = E\phi \tag{8.14}$$

where E is the eigenvalue associated with the energy levels measured from the zero of energy, ϵ_r is the dielectric constant of the material, and ϕ is the associated wave function. Using polar coordinates r, φ, the wave function can be separated into a radial part and an angular part with

$$\phi(r, \varphi) = \frac{1}{\sqrt{2\pi}} R(r) \exp(im\varphi) \qquad m = \text{integer} \tag{8.15}$$

and $R(r)$ satisfying the equation

$$\left[-\frac{\hbar^2}{2m^*} \left\{ \frac{1}{r}\frac{d}{dr}\left(r\frac{d}{dr} \right) - \frac{m^2}{r^2} \right\} - \frac{e^2}{\epsilon_r r} \right] R(r) = ER(r) \tag{8.16}$$

which can also be expressed as

$$\rho \frac{d^2F}{d\rho^2} + (1 - \rho)\frac{dF}{d\rho} + \left(2\lambda - \frac{1}{2} - \frac{m^2}{2} \right) F = 0 \tag{8.17}$$

with $R = F \exp(-\rho/2)$, $\rho = r/a_B\lambda$, $W = E/R_y$ and $\lambda^{-2} = -4W$. In these substitutions $a_B = \epsilon_r\hbar^2/m^*e^2$ and $R_y = m^*e^4/2\hbar^2\epsilon_r^2$ are the solid-state analogs of the usual atomic parameters: the Bohr radius a_B and the hydrogen energy level of the ground state R_y.

For $E < 0$, λ is real and the solutions to the differential equation give the discrete energy spectrum associated with bound states. To put the equation in standard form we make one more substitution, $F = \rho^{|m|}L$, to obtain

$$\rho \frac{d^2L}{d\rho^2} + (2|m| + 1 - \rho)\frac{dL}{d\rho} + \left(2\lambda - \frac{1}{2} - |m| \right) L = 0 \tag{8.18}$$

(This form of the equation is very analogous to the equivalent reduced equation for the three-dimensional hydrogen atom.)

The differential equation for L can be expressed in the general form

$$\rho \frac{d^2L_q^p}{d\rho^2} + (p + 1 - \rho)\frac{dL_q^p}{d\rho} + (q - p)L_q^p = 0 \tag{8.19}$$

with $p = 2|\mathrm{m}|$ and $q = 2\lambda - \frac{1}{2} + |\mathrm{m}|$, and the solution is the associated Laguerre polynomial:

$$L(\rho) = L_{n+|m|}^{2|m|}(\rho) \qquad |m| \le n \tag{8.20}$$

where n is zero or a positive integer and is related to λ via

$$n = 2\lambda - \tfrac{1}{2} \tag{8.21}$$

Consequently, the energy levels are given by

$$E_n = -R_y/(n + \tfrac{1}{2})^2 \qquad n = 0, 1, 2, \ldots \tag{8.22}$$

The major point is that $E_1(2\mathrm{D}) = 4E_1(3\mathrm{D})$.

Physically, this higher binding energy comes about due to the extra degree of confinement which results in a smaller separation, on average, between electron and hole, and hence a higher binding energy for the ground state. It is also clear that this two-dimensional derivation is only applicable when the film thickness $<< a_B \sim 5$ nm, where a_B is the Bohr radius of an electron in a dielectric medium. Intermediate thicknesses will result in intermediate exciton energies varying from the two-dimensional to the three-dimensional limit.

The change in the exciton binding energy in going from three-dimensional to two-dimensional geometry can have profound effects on photon semiconductor interactions. This is best illustrated through a numerical example: in bulk GaAs the exciton energy is a few meV below the transition energy corresponding to the minimum of band-to-band transitions. At room temperature (~300 K), the chance of observing bulk excitonic decay is very small, since the probability that the exciton is thermally ionized is given by a Boltzmann factor, $\exp(-E^{ex}/kT)$, and is large. For a quantum well, the ratio of ionization factors is $\exp(-4E_x^{ex}/kT)/\exp(-E^{ex}/kT)$, yielding a much larger probability for exciton preservation and eventual radiative decay. The optical emission spectrum of a direct-gap semiconductor is easily observed and has become a measure of material quality. The energy positions are dependent on the precise well thickness and the width of the emission lines yields information on the uniformity of the structure. Precise interpretation of these spectra depends on understanding these excitonic effects.

8.4 Photon Emission and Transport

Light-emitting devices operate by the injection of high concentrations of electrons and holes into the active region where electrons recombine with holes in radiative transitions. This requires the use of direct-band-gap materials which have strong radiative recombination cross sections. High-quality material is necessary to minimize nonradiative recombination caused by defects in the bulk or at interfaces.

The injection of electrons and holes is accomplished by use of forward bias to reduce the potential barrier between the n-type and p-type regions. Under forward bias the junction region is the high resistance element, and the applied voltage V_F

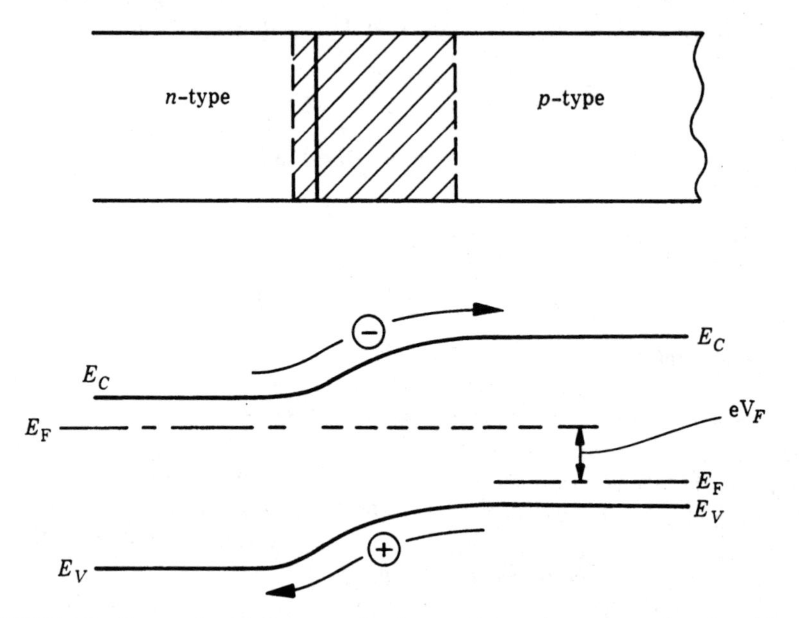

FIGURE 8.8 Energy-band diagram of a forward-biased n–p homojunction. The arrows represent the injection of electrons into the p-type and holes into the n-type material.

(positive voltage applied to p-type region) reduces the potential drop (Fig. 8.8) across the barrier to a value

$$V_T = V_{bi} - V_F \tag{8.23}$$

where V_{bi} is the potential barrier in thermal equilibrium ($V_F = 0$). With a reduction in the barrier height, more electrons and holes can get over the barrier than in the case of thermal equilibrium.

The number of electrons n_p injected into the p-type region (stated in other terms, the number of electrons that can surmount the energy barrier) is given by

$$n_p = n_n \exp\left(-\frac{eV_{bi} - eV_F}{kT}\right) \tag{8.24}$$

$$n_p = n_{p0} \exp\left(\frac{eV_F}{kT}\right) \tag{8.25}$$

where n_{p0} represents the number of electrons in the p-type material under thermal equilibrium conditions (the subscript zero is often omitted when describing carrier concentrations in bulk materials). Similar relations hold for holes injected into the n-type region.

8.4.1 Heterojunction Light-Emitting Diodes

In heavily doped GaAs n–p junctions operated under forward bias (Fig. 8.9a) electrons are injected at concentrations up to $10^{18}/\text{cm}^3$ into the p-type material. The electrons have a lifetime τ_n of about 10^{-9} sec in heavily doped material, but because of the high electron mobility, the diffusion length $L_n = (D_n\tau_n)^{1/2}$ can be 1 to 3 μm. In this case, we have treated the injected electrons as diffusing species in the p-type material because the electric field in the heavily doped p-type region is essentially zero as the applied voltage appears primarily across the potential barrier between n-type and p-type material. Consequently, we can apply the diffusion relations developed in Chapter 3. The diffusion coefficient D_n in this case can be determined from the mobility μ_n of electrons by use of the Einstein relation that

$$D_n = \mu_n \frac{kT}{e} \tag{8.26}$$

where the value of kT/e at room temperature is about 0.025 volts. The rate of photon emission is proportional to the product np of the electron and hole concentration. To achieve greater efficiency, it is desirable to confine the injected carriers. To optimize the confinement of the electrons and holes, double heterojunction (DH) structures are used with a 0.1 to 0.2 μm layer of GaAs—the active region—interposed between layers of AlGaAs (Fig. 8.9). The spacing between the heterojunctions is less than the diffusion length, and hence the carriers are spread uniformly throughout the active layer. The discontinuities ΔE_C and ΔE_V confine the injected electrons and holes within the active region.

The basic concept in the DH structure is to have the recombination region with a direct-band-gap material bracketed on either side by two layers having a higher

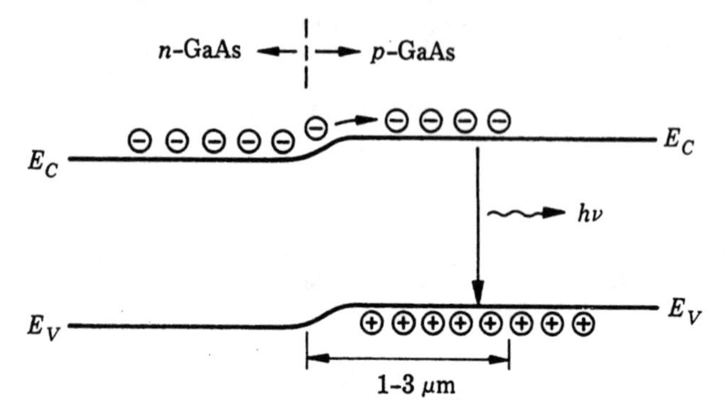

(a) GaAs homojunction laser

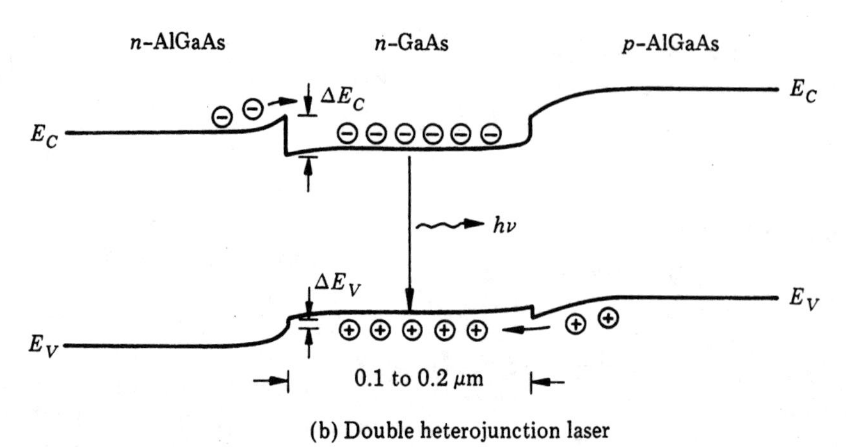

(b) Double heterojunction laser

FIGURE 8.9 Energy-band diagrams of laser diodes for (a) GaAs homojunction and (b) AlGaAs/GaAs/AlGaAs structure.

energy band-gap, to ensure carrier confinement. It is important to have high quality epitaxy. Misfit dislocations have nonradiative centers at the dislocation core and hence degrade the recombination efficiency of the device. For a 1% lattice mismatch at a relaxed interface, a dislocation will be generated every 100 atomic planes or every 30 to 50 nm. The long diffusion lengths of ~ 1 μm mean that carriers can easily find such defects if they exist.

8.4.2 Laser Diode

The most common example in this field is the GaAs/AlGaAs solid-state laser. This double-heterostructure device consists of a semiconductor sandwich of $Al_xGa_{1-x}As/GaAs/Al_xGa_{1-x}As$ layers grown epitaxially on a GaAs substrate. The AlGaAs layers (~ 100 nm) are doped n-type and p-type, respectively, to form a diode structure; the GaAs layer may be as thin as 10 nm. Under forward bias electrons and holes are injected to create a "population inversion"; the carriers then recombine with the emission of a photon equal in energy to the band gap of the GaAs well. For layers as thin as 10 nm, the electron wave functions and energy levels are quantum shifted resulting in the possible tuning of the emission wavelength. The device is a single crystal, all-epitaxial structure of high perfection as required for useful electron transport and efficient optical emission. Such a laser diode is a simple example of "band-gap engineering," the phrase coined to describe the materials design concept of the use of different semiconductor materials to engineer and control the energy-width of the band gap.

Laser and light-emitting diodes for optical communication use GaAs and AlGaAs for the 0.68 to 0.9 μm spectral region and InGaAs or GaInAsP for the 1.0 to 1.5 μm spectral region. In the $Al_xGa_{1-x}As$ system, the lattice constant shows little variation as x goes from zero to unity. Heterojunctions in the AlGaAs/GaAs system have negligible strain-induced defects. The GaInAsP system is usually designed to lattice match with InP to produce a InP/GaInAsP/InP laser structure. InP, E_G = 1.34 eV, is used as a substrate and a close lattice match is obtained with $Ga_xIn_{1-x}As_yP_{1-y}$ when $y \simeq 2.16x$. In this composition range, the emission wavelength varies from 0.9 μm ($x = y = 0$) to 1.5 μm ($x = 0.47$, $y = 1.0$).

8.4.3 Bragg Reflector Superlattice

The drive for high speed electronic devices has resulted in an increasing emphasis on photonics. Solid-state sources (lasers and light-emitting diodes), photodetectors, and optical fibers are combined to provide all-optical communications systems with unprecedented speeds. The emphasis on optics has placed an increasing demand on the use of thin film epitaxy to control photon transport in solid-state semiconductor devices. In this section we illustrate the basic aspects of optical design using a distributed Bragg reflector as an example.

Optical design of layered film structures is a well-understood subject and is applied in numerous technologies such as glass (window) design and solar cell fabrication. The application to a semiconductor structure follows the traditional format but is specific to epitaxial configurations. The growth of an epitaxial *optical* component allows subsequent integration with an epitaxial *electronic* element to form all-epitaxial optoelectronic integrated circuits (OEICs).

Photon transport in layered material is governed by the optical index of refraction of the individual media. In general the larger the difference in index, the larger the optical effects. Also, the larger the number of periods, the larger the optical effects. Small changes in index can be compensated for by increases in the number of layers. The index of refraction for GaAs is 3.62, for AlAs it is 3.18, and for layers of $Al_xGa_{1-x}As$, the index is given by the appropriate linear combination.

Transport of an electromagnetic wave (photons) through a layered medium is governed by Maxwell's equations with dielectric constant ϵ and magnetic permeability μ_M which are functions of z, the growth direction. The index of refraction n is related to both parameters through $n^2 = \epsilon\mu_M$ and here we assume that the materials are nonmagnetic, or $\mu_M = 1$. The transport behavior involves the usual solution to a partial differential equation with imposed boundary conditions (Born and Wolf, 1980).

Consider first the optics of a single dielectric film shown in Fig. 8.10. The reflection and transmission coefficients at the "12" interface are given by:

$$r_{12} = \frac{n_1 \cos \Theta_1 - n_2 \cos \Theta_2}{n_1 \cos \Theta_1 + n_2 \cos \Theta_2}$$

and (8.27)

$$t_{12} = \frac{2n_1 \cos \Theta_1}{n_1 \cos \Theta_1 + n_2 \cos \Theta_2}$$

where Θ_1 and Θ_2 are related by the usual law of refraction, Snell's law ($n_1 \sin \Theta_1 = n_2 \sin \Theta_2$). Similar expressions can be written for r_{23} and t_{23}. The reflection and transmission coefficients of the film (both interfaces) become

$$r = \frac{r_{12} + r_{23} \exp 2i\beta}{1 + r_{12} r_{23} \exp 2i\beta}$$

and (8.28)

$$t = \frac{t_{12} t_{23} \exp i\beta}{1 + r_{12} r_{23} \exp 2i\beta}$$

where $\beta = (2\pi/\lambda_0)\, n_2 h \cos \Theta_2$, with λ_0 the wavelength of light and h the film thickness. The factor $\exp i\beta$ takes into account the phase change or difference in optical path length due to the thin film.

These formulae arise by considering the multiple reflections of an incident electromagnetic field at each interface. The picture is of an incident wave, $\tilde{E}_0 = E_0 \exp(i\omega t)$, having a fraction r_{12} reflected at the first 12 interface, the transmitted fraction partly reflected at the 23 interface, and the subsequent transmission of this remaining component at the 21 interface. The total intensity of the reflected wave is determined by summing all of the back-reflected intensities. The different contri-

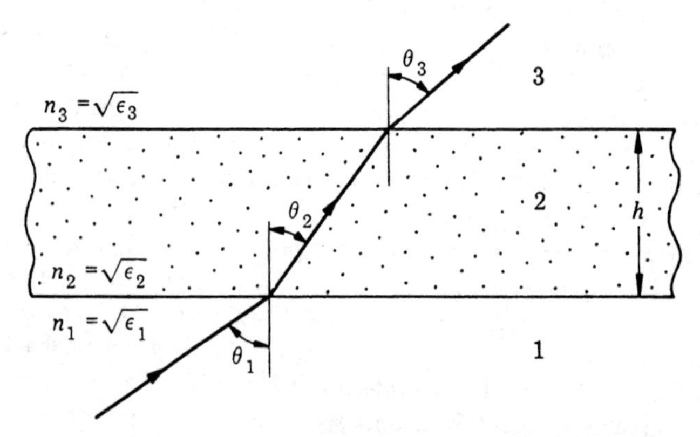

FIGURE 8.10 Propagation of an electromagnetic wave through a homogeneous film.

butions to the sum can be written as:

$$\begin{aligned}
\tilde{E}_0 &= E_0 e^{i\omega t} \qquad \text{(incident wave)} \\
\tilde{E}_1 &= r_{12}\tilde{E}_0 \\
\tilde{E}_2 &= r_{23}(1 - r_{12}^2)\tilde{E}_0 e^{-i\Delta\varphi} \\
\tilde{E}_3 &= r_{23}(1 - r_{12}^2)(r_{23}r_{21})\tilde{E}_0 e^{-2i\Delta\varphi} \\
\tilde{E}_4 &= r_{23}(1 - r_{12}^2)(r_{23}r_{21})^2\tilde{E}_0 e^{-3i\Delta\varphi} \\
&\vdots
\end{aligned}$$

The equation for $\tilde{E}_2$, for example, comes from the following factors: Of the incident wave $(1 - r_{12})\tilde{E}_0$ is transmitted, $r_{23}(1 - r_{12})\tilde{E}_0$ is reflected at the 23 interface and $(1 - r_{21})r_{23}(1 - r_{12})\tilde{E}_0$ is transmitted (reemerges) at the 21 interface. Using the fact that $r_{21} = -r_{12}$ gives the result for $\tilde{E}_2$. The reflected sum, $\tilde{E} = \Sigma_{m=1}^{\infty} \tilde{E}_m$ can be manipulated to yield Eq. 8.28 with $\Delta\varphi = 2\beta$.

The total reflectivity R and transmissivity τ are given by

$$R = |r|^2$$

and

$$\tau = |t|^2 \frac{n_3 \cos\theta_3}{n_1 \cos\theta_1}$$

so that

$$R = \frac{r_{12}^2 + r_{23}^2 + 2r_{12}r_{23}\cos 2\beta}{1 + r_{12}^2 r_{23}^2 + 2r_{12}r_{23}\cos 2\beta} \tag{8.29}$$

with a corresponding formula for $\tau = 1 - R$. We first note that R and τ remain the same when the film thickness h goes into $h + \Delta h$ where $\Delta h = \lambda_0/2n_2\cos\Theta_2$. Films which differ by an integral multiple of $\lambda_0/2n_2\cos\Theta_2$ have the same reflectivity and transmissivity.

For convenience we set the optical thickness of the film to be $H = n_2 h$ and seek the minimum and maximum values of R. The derivative dR/dH equals zero when $\sin 2\beta = 0$, so that extremum values of H correspond to

$$H = \frac{m\lambda_0}{4\cos\Theta_2} \qquad m = 1, 2, 3, 4, \ldots \tag{8.30}$$

A further analysis shows that the nature of the extremum (maximum or minimum) depends on the parity of m (odd or even) and on the relative values of the indices of refraction. If the first medium is air ($n_1 = 1$) and we consider normal incidence, a film with optical thickness H, of $\lambda_0/4$, $3\lambda_0/4$, $5\lambda_0/4, \ldots$ has a maximum reflectivity if $n_2 > n_3$ and a minimum if $n_2 < n_3$. A common example of the latter case is an antireflection coating consisting of magnesium fluoride ($n_2 \sim 1.38$) on glass ($n_3 \sim 1.5$). For normal incidence and m odd the reflectivity is given by

$$R = \left(\frac{n_1 n_3 - n_2^2}{n_1 n_3 + n_2^2}\right)^2 \tag{8.31}$$

and would be exactly zero if $n_2 = \sqrt{n_1 n_3}$. Thus, the possibility of tailoring n has some exciting possibilities. For a GaAs laser the important optical interface is the GaAs/air interface with a reflectivity of 0.32. When m is even and the optical thickness has values $\lambda_0/2$, $2\,\lambda_0/2$, $3\,\lambda_0/2 \ldots$ then

$$R = \left(\frac{n_1 - n_3}{n_1 + n_3}\right)^2$$

for normal incidence.

In a superlattice of period h the dielectric constant ϵ is also a periodic function given by

$$\epsilon(z + jh) = \epsilon(z) \qquad j = 1, 2, \ldots, N \tag{8.32}$$

The stack may be designated as in Fig. 8.11 and we consider the special case when the two basic layers are of the same optical thickness $\lambda_0/4$, so that

$$n_2 h_2 = n_3 h_3 \tag{8.33}$$

Then the total reflectivity is given by

$$R = \left[\frac{1 - \dfrac{n_l}{n_1}\left(\dfrac{n_2}{n_3}\right)^{2N}}{1 + \dfrac{n_l}{n_1}\left(\dfrac{n_2}{n_3}\right)^{2N}}\right]^2 \tag{8.34}$$

where N is the total number of double layers. If $n_2 < n_3$, the reflectivity can approach unity for sufficiently large N.

For example, for AlAs $n_2 = 3.18$ and for GaAs $n_3 = 3.62$, assuming $n_l = n_1$, $R = 0.86$ for $N = 10$ and $R = 0.99$ for $N = 20$. It is precisely these high reflec-

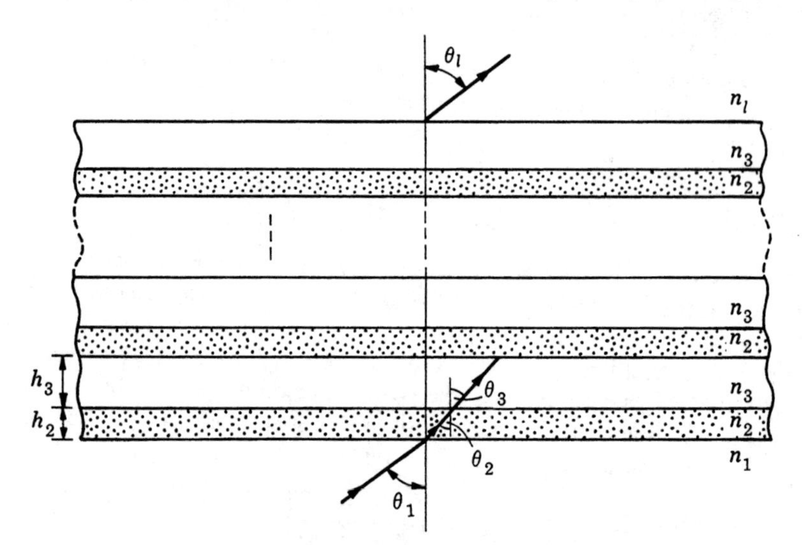

FIGURE 8.11 A periodic multilayer.

tivities that are needed for laser structures where superior reflectivity may be crucial to lasing action.

8.4.4 Edge and Surface Lasers

The more common configuration of laser diodes is for light emission from the edge of the diode (Fig. 8.12) where the light is parallel to the epitaxial layers (perpendicular to the growth direction). In these horizontal edge-emitting lasers, the light and current vectors are orthogonal.

Another configuration, for surface-emitting lasers, uses the basic laser concept in conjunction with a Bragg grating constructed of alternating layers of semiconducting material. Thus the "optical stack" consists of layered semiconductor structures optimized to control the optical path as well as the electron flow. The light and current vectors are parallel. One example of such a system might consist of 50 semiconductor interfaces comprising top and bottom quarter-wave reflectors, sandwiching a quantum-well laser (Fig. 8.13). This structure is called a vertical-cavity surface-emitting laser.

The optical axis of vertical-cavity surface-emitting lasers is along the epitaxial growth direction of the semiconductor crystal. Due to the short gain path of vertical-cavity lasers, it is required that the reflectivity of the mirrors defining the Fabry–Perot étalon be close to unity. As shown in Fig. 8.13 a high reflectivity is achieved through use of metallic reflectors and quarter-wave ($\lambda/4$) semiconductor multilayer Bragg reflectors. The metallic reflector is of crucial importance. It serves two pur-

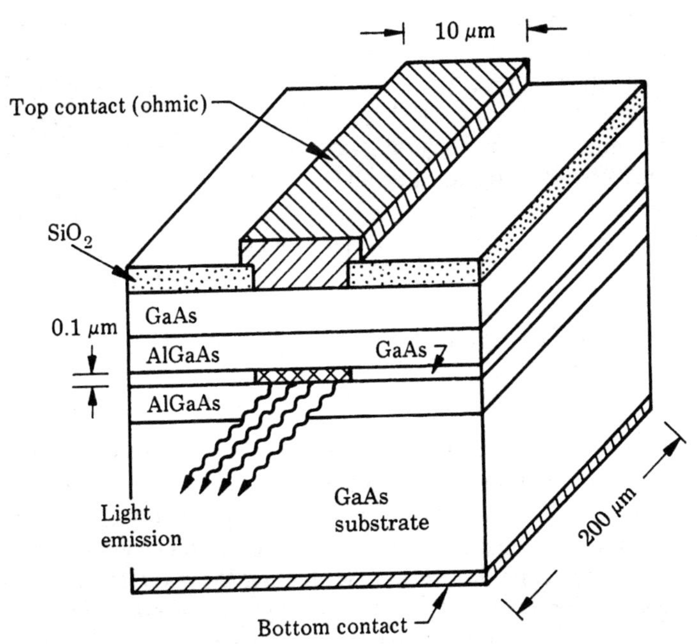

FIGURE 8.12 Schematic of edge-emitting laser.

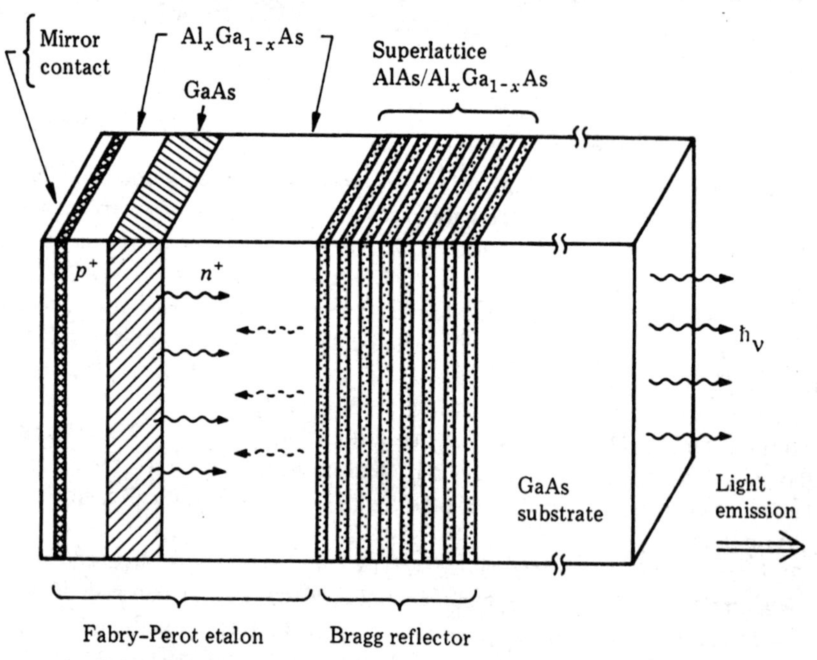

FIGURE 8.13 Surface-emitting laser using Bragg reflector.

poses, functioning as both mirror and ohmic electrical contact. The parallel direction of optical propagation and current density vector in vertical-cavity lasers requires the simultaneous realization of reflector and contact. The metallic mirror–contact concept has a number of advantages including low series resistance, low thermal resistance, and small current spreading due to the proximity of active layer and contact.

A surface emitting laser, consisting of a superlattice mirror, active layer and metallic contact can be grown using an all heteroepitaxy configuration. Light emerging perpendicular to the top surface has distinct advantages over the in-plane emission of an "edge emitter" for practical applications.

8.5 Electron Transport

8.5.1 Heterojunction Bipolar Transistors (HBT)

The bipolar transistor operates with the emitter current overwhelmingly carried by the injection of carriers from the emitter into the base. For an n–p–n transistor, this condition means that electrons are injected into the base with negligible hole injection into the emitter from the base. In conventional homojunction silicon transistors this is achieved by making the donor concentration N_D in the emitter much greater than the acceptor concentration N_A in the base. In heterojunction bipolar transistors, it is achieved by growing a wider-band-gap emitter on a narrower-band-gap base (Kroemer, 1982). In silicon structures, for example, the base can be formed by a

GeSi alloy, (E_G(GeSi) < E_G(Si)), and the emitter by an epitaxial Si layer (Bean, 1988).

The ratio of injected electrons n_p to injected holes p_n for an abrupt homojunction is proportional to the ratio of dopant concentrations in the emitter and base:

$$\frac{n_p}{p_n} = \frac{N_D}{N_A} \tag{8.35}$$

where we assume that the majority carrier concentrations are equal to the dopant concentration (i.e., $n_n = N_D$ and $p_p = N_A$). This quantity must be considerably greater than unity for effective transistor action.

The difficulty with a lightly doped base is that the series resistance of the base is high, and this factor limits high-current and high-frequency applications. It would be desirable to increase the dopant concentration in the base without degrading the injection criterion.

The dopant concentration in the base can be increased if the emitter has a wider band gap E_G than the base. Figure 8.14 shows an energy-band diagram for a wide-band material, such as AlGaAs ($Al_xGa_{1-x}As$), grown epitaxially on GaAs, in (a) thermal equilibrium and (b) with forward bias voltage V_{BE} applied between base and emitter and reverse bias voltage V_{CB} applied between collector and base. The valence band discontinuity ΔE_V adds directly to the barrier for hole injection. In analogy to the homojunction bipolar transistor, the ratio of injected electrons to that of injected holes for the heterojunction bipolar transistor is

$$\frac{n_p}{p_n} = \frac{N_D}{N_A} \exp(\Delta E_G/kT) \tag{8.53}$$

where $\Delta E_G = \Delta E_C + \Delta E_V$. At room temperature, $kT = 0.025$ eV, so that even a small change of 0.15 eV in ΔE_G leads to a value of exp $(\Delta E_G/kT) = 320$. Thus the dopant concentration in the base of a heterojunction can exceed that in the emitter by orders of magnitude without degrading the injection ratio. The possibility of a substantial increase in N_A (base doping) without reducing the injection ratio implies a lower resistivity base. This results in higher operating speeds.

8.5.2 Doping: Control and Modulation

The ultimate speed of electronic devices is limited by the fundamental transport properties of the charge carriers (electrons and holes) in the semiconductor system. Such electrons and holes originate from the introduction of dopants into the semiconductor. With thin film growth techniques, one can incorporate specialized spatial configurations of dopants to maximize carrier mobility and device speed.

In semiconductor systems the speed of carriers is characterized by the mobility of the charge carriers in the system. For simple systems the mobility is given by:

$$\mu = e\tau/m^*$$

where m^* is the effective mass and τ represents the mean free path between scattering events which can change the direction (elastic scattering) and/or energy (inelastic

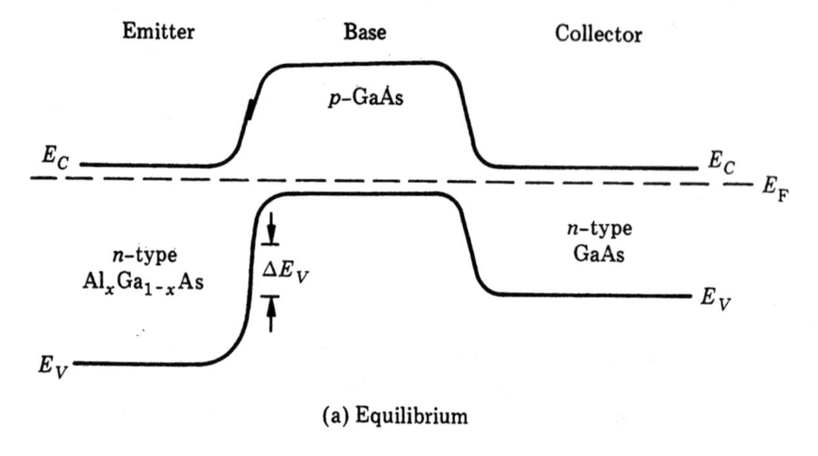

(a) Equilibrium

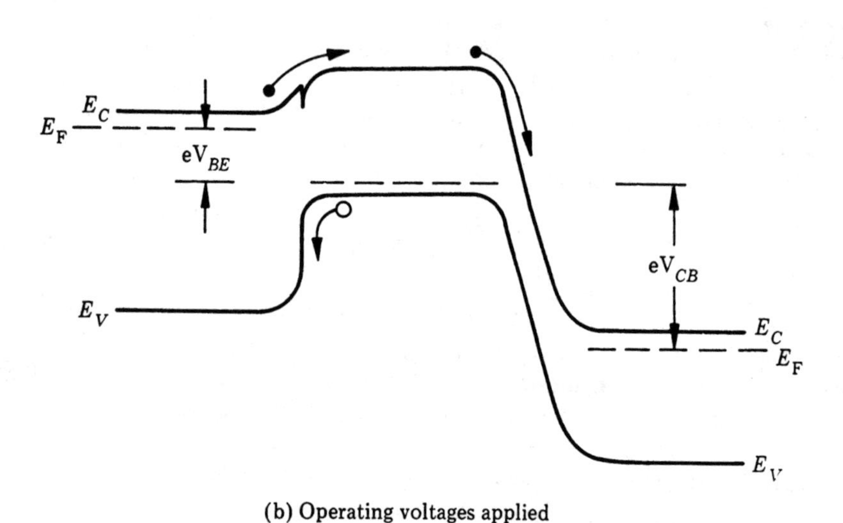

(b) Operating voltages applied

FIGURE 8.14 Energy-band diagrams of an n–p–n heterojunction bipolar transistor (HBT) (a) in thermal equilibrium and (b) under active mode of operation.

scattering) of the carrier. It is this scattering rate ($1/\tau$) that represents the fundamental limit to the mobility. For elastic scattering the principal contributions to the scattering are (1) electron–phonon scattering, (2) random alloy scattering, and (3) ionized impurity scattering. Electron–phonon scattering is that process in which the electron senses the imperfections in the crystal lattice due to the thermal vibrations of the atoms in the crystal. As verified by experiment the phonon contribution to scattering is a strong function of temperature.

Random alloy scattering arises from the fluctuations in the potential experienced by the charge carriers. Thus pure Si and GaAs have no random alloy contribution to their scattering, but AlGaAs is expected to have a substantial contribution as long as the Ga and Al atoms are truly random. (It is interesting to note that monolayer growth could, in principle, simulate a system where the Al and Ga are not random. Nevertheless most AlGaAs growth does result in a random distribution of Ga and Al and contains a random alloy scattering component.) Again it is basically the fluctuations in potential due to the alloy constituents which cause the scattering.

Ionized impurity scattering is the dominant scattering mechanism for highly doped material. The basic interaction is that the screened coulomb field, originating from the ionized impurity atom, scatters the electron. The theoretical treatment of ionized impurity scattering starts with the basic scattering concepts of simple Rutherford scattering.

Clever use of thin film growth has been used to minimize this dominant component to the scattering. The idea is illustrated in Fig. 8.15. A quantum well is formed that will become the conducting channel for the electrons. Dopants are introduced in the AlGaAs at a distance, a "setback," such that there is a reasonable probability for electrons to fall into the well. Once in the well the electrons are kept at a large distance from the ionized impurities, thus minimizing and almost turning off ionized impurity scattering. An example of the large increases in mobility achieved through this scheme over the years is illustrated in Fig. 8.16. The largest gains are at low temperature; at room temperature, the total improvement in mobility is not as great because the phonon contribution is significant.

8.5.3 Electron Confinement in a Heterostructure

One can take advantage of the band-gap discontinuity in a heterojunction and accumulate electrons in the "notch" in the conduction band. For example, a heavily doped, wide-band-gap semiconductor, AlGaAs, can be grown on an epitaxial layer of high-purity, high-mobility GaAs as shown in Fig. 8.17. The electrons associated with the donors in the AlGaAs transfer to the GaAs by surmounting the barriers and accumulating in the notch region in the conduction band. The electrons in the undoped GaAs channel are separated from the donors in the heavily doped wide-band

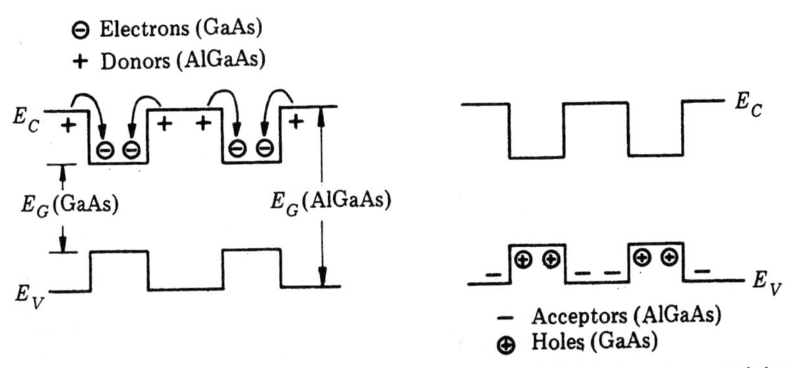

FIGURE 8.15 Undoped GaAs quantum well between AlGaAs layers containing (a) donors or (b) acceptors. The arrows show the transition of electrons and holes from the AlGaAs to the GaAs wells.

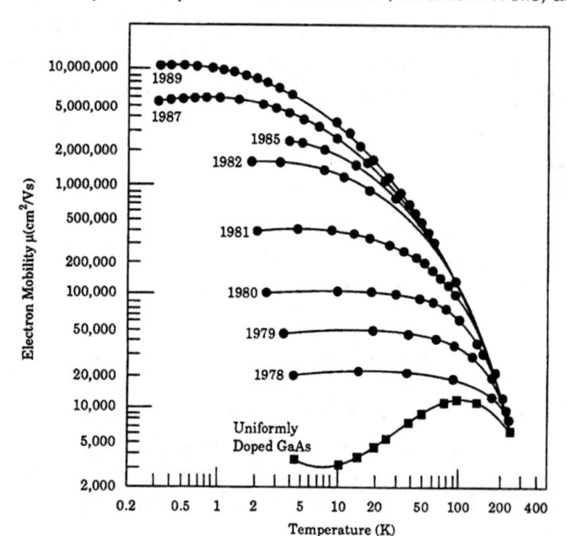

FIGURE 8.16 The mobility versus temperature of electrons in undoped GaAs wells showing the improvement in growth and mobility with time (from Pfeiffer et al., 1989).

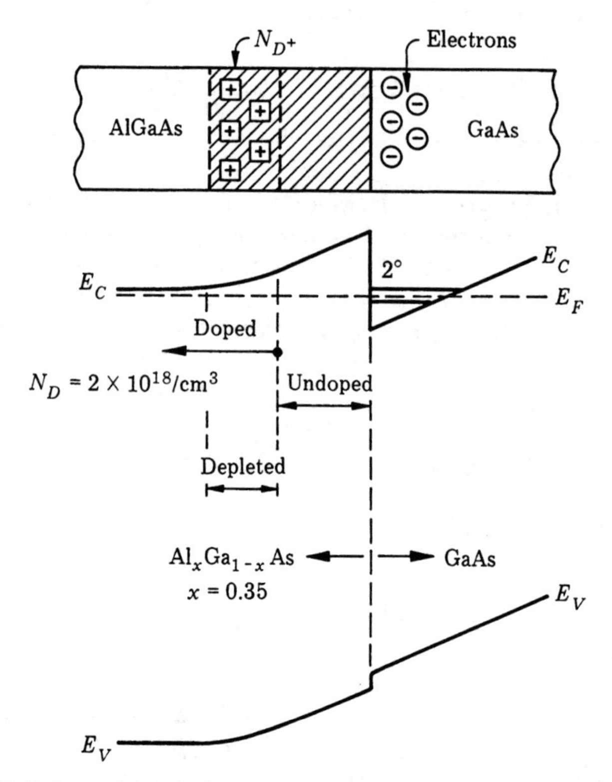

FIGURE 8.17 Schematic of structure and energy-band diagram showing the formation of an electron gas at an interface. The doped layer of AlGaAs is separated from the GaAs by an undoped AlGaAs spacer layer.

AlGaAs by a thin (≃10 nm) undoped AlGaAs spacer region. Since there are no ionized donors to act as scattering centers in the high-purity GaAs, the electrons have a mobility that is much higher than in GaAs with a donor concentration equivalent to the concentration of electrons in the notch.

The energy-band discontinuity ΔE_C is the key to spatial separation of electrons and donors. The structure, shown in Fig. 8.17, is formed by $Al_xGa_{1-x}As$ (E_G = 1.80 eV at x = 0.3) in contact with GaAs (E_G = 1.4 eV). A thin (10 nm) layer of undoped AlGaAs is grown on the undoped GaAs followed by a heavily doped ($N_D = 2 \times 10^{18}/cm^3$) layer of AlGaAs. Electrons depleted from the AlGaAs accumulate in the GaAs in the thin layer defined by the potential "notch." This establishes a plane or two-dimensional electron gas of carriers. These structures are used in field-effect transistors (FET) for high-frequency performance; such devices are described by acronyms such as HEMT (high-electron-mobility transistors) or MODFET (modulation doped).

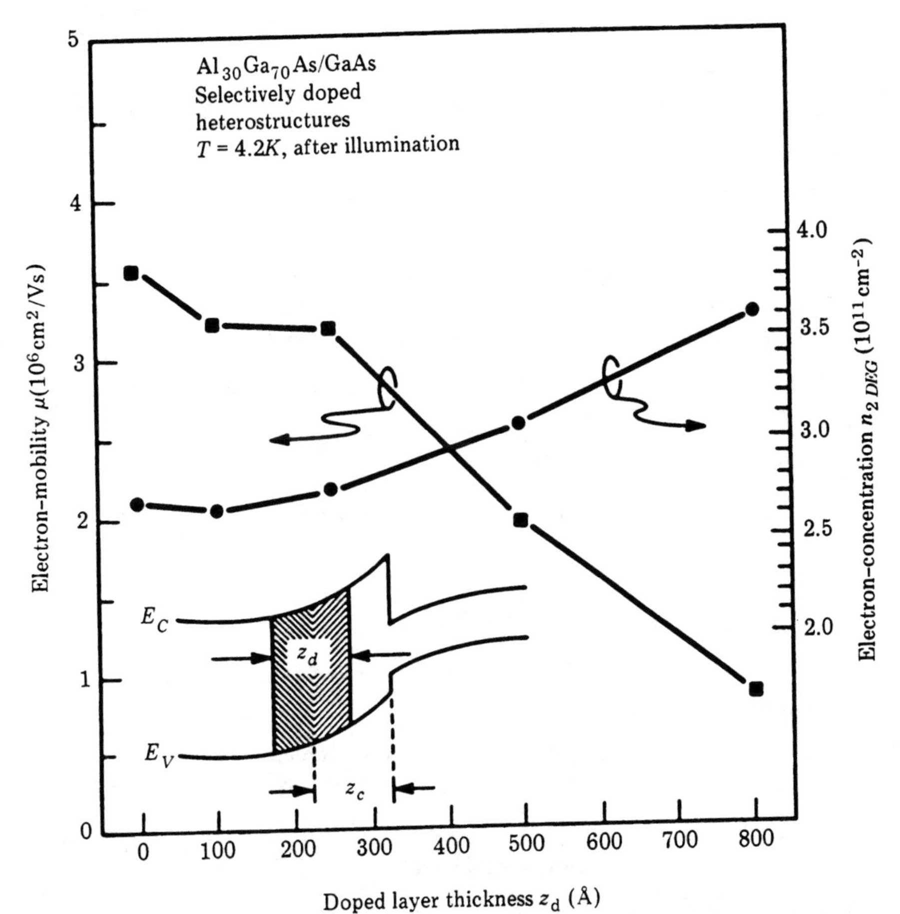

FIGURE 8.18 Electron mobility and concentration versus doped-layer thickness in a selectively doped AlGaAs layer on undoped GaAs (from Schubert et al., 1989).

Even in the modulated doping scheme the electron still senses the ionized impurities. For the same number of dopants/cm^2 one can show that this residual coulomb scattering effect can be minimized through *delta doping*, the introduction of dopants into a very thin layer—ultimately one monolayer. It is left as a problem to prove this for the simple case of an electron passing two ionized charge centers. The effect of delta doping is shown in Fig. 8.18 which illustrates the change in mobility, measured at low temperature, for the modulated doping scheme with the dopant layer approaching a delta structure, that is, $z_d = 0$.

References

1. E. G. Bauer et al., Department of Energy report, ''Fundamental Issues in Heteroepitaxy,'' *J. Mater. Res.* 5, 852 (1990).
2. J. C. Bean, ''Silicon-Based Semiconductor Heterostructures,'' Chapter 11 in *Silicon Molecular Beam Epitaxy*, edited by E. Kasper and J. C. Bean, CRC Press, Boca Raton FL (1988).
3. M. Born and E. Wolf, *Principles of Optics*, 6th edition, Pergamon Press, Oxford, (1980).
4. H. C. Casey, Jr., and M. B. Panish, *Heterostructure Lasers*, Academic Press, Orlando (1978).
5. R. Dingle, editor, *Applications of Multiquantum Wells, Selective Doping, and Superlattices*, Vol. 24 in *Semiconductors and Semimetals*, edited by R. K. Willardson and A. C. Beer, Academic Press, New York (1987).
6. L. Esaki, ''Compositional Superlattices,'' Chapter 6 in Parker (1985).
7. M. Jaros, *Physics and Applications of Semiconductor Microstructures*, Oxford University Press, New York (1989).
8. H. Kroemer, ''Heterojunction Bipolar Transistors and Integrated Circuits,'' *Proc. IEEE* 70, 13 (1982).
9. S. Luryi, Chapter 2 in *High Speed Semiconductor Devices*, edited by S. M. Sze, Wiley–Interscience, New York (1990).
10. J. W. Mayer and S. S. Lau, *Electronic Materials Science for Integrated Circuits in Si and GaAs*, Macmillan, New York (1990).
11. M. A. Morrison, T. L. Estle, and N. F. Lane, *Quantum States of Atoms, Molecules and Solids* (Prentice-Hall, Englewood Cliffs, New Jersey, 1976).
12. G. C. Osbourne, ''Electronic Properties of Strained-Layer Superlattices,'' *J. Vac. Sci. Technol.* B1, 379 (1983).
13. E. H. C. Parker, editor *The Technology and Physics of Molecular Beam Epitaxy*, Plenum Press, New York (1985).

14. T. P. Pearsall, editor, *Strained Layer Superlattices: Science and Technology*, Vol. 33 in *Semiconductors and Semimetals*, edited by R. K. Willardson and J. C. Beer, Academic Press, New York (1990).

15. L. Pfeiffer, K. W. West, H. L. Stormer, and K. W. Baldwin, "Electron Mobilities Exceeding 10^7 cm^2/VS in Modulation Doped GaAs", *Appl. Phys. Lett.* 55, 1888 (1989).

16. S. T. Picraux, B. L. Doyle, and J. Y. Tsao, "Structure and Characterization of Strained Layer Superlattices" in Pearsall (1990).

17. E. F. Schubert, L. Pfeiffer, K. W. West, and A. Izabelle, "Dopant distribution for maximum carrier mobility in selectively doped $Al_{0.30}Ga_{0.70}As/GaAs$ heterostructures" *Appl. Phys. Lett.* 54, 1350 (1989).

18. M. Shinada and S. Sugano, "Interband Optical Transitions in Extremely Anisotropic Semiconductors," *J. Phys. Soc. Jap.* 21, 1936–1946 (1966).

19. W. T. Tsang and E. F. Schubert, "Extremely High-quality $Ga_{0.47}In_{0.53}As/InP$ Quantum Wells Grown by Chemical Beam Epitaxy," *Appl. Phys. Lett.* 49, 229 (1986).

20. W. T. Tsang, "Quantum Confinement Heterostructure Semiconductor Lasers," Chapter 7 in Dingle (1987).

21. C. Weisbuch, "Fundamental Properties of III–V Semiconductor Two-Dimensional Quantized Structures: The Basis for Optical and Electronic Device Applications," Chapter 1 in Dingle (1987).

Problems

8.1 For a 2 nm InSb quantum well sandwiched between CdTe layers, calculate the $n = 1$ and $n = 2$ energy levels for electrons ($m^* = 0.015m_e$) and holes ($m^* = 0.3m_e$). Assume the potential well is infinite.

8.2 Compare the energy dependence of the one, two and three dimensional density of states of a free electron gas where the energy dispersion relation is isotropic. Indicate the units.

8.3 For silicon where the relative dielectric constant $\epsilon_r = 11.9$, assume that electron and hole effective masses $= m_e$ and calculate:
(a) The three-dimensional binding energy of the exciton for $n = 1, 2$.
(b) The Bohr radius a_B.
(c) The two-dimensional exciton energy levels for $n = 1$ and $n = 2$.

8.4 For light with energy of 1.24 eV, in a medium 1 with refractive index 3.5, incident at the interface with medium 2 (air) having refractive index 1.0, calculate the reflection (r_{12}) and transmission (t_{12}) coefficients for

(a) Normal incidence.
(b) Incidence at angle $\theta = 15°$.

8.5 Compare the total reflectivity of superlattices having 5, 10 and 20 layers, where the two individual layers are of the same optical thickness $\lambda_0/4$ and the refractive indices are $n_2 = 2.28$ (ZnS) and $n_3 = 3.33$ (GaP).

8.6 Consider an electron passing two ionized charges as shown in the sketch provided. The sketch simulates an electron in a low-band-gap layer guided past ionized dopants. Show that the scattering potential is a minimum when $\Delta = 0$. For $\Delta = 0$, the dopant configuration is a δ-doped structure and illustrates that delta doping is the configuration with the least scattering for a given number of dopant atoms/cm^2. (Hint: Expand the coulomb potential using the setback r_0 and Δ.)

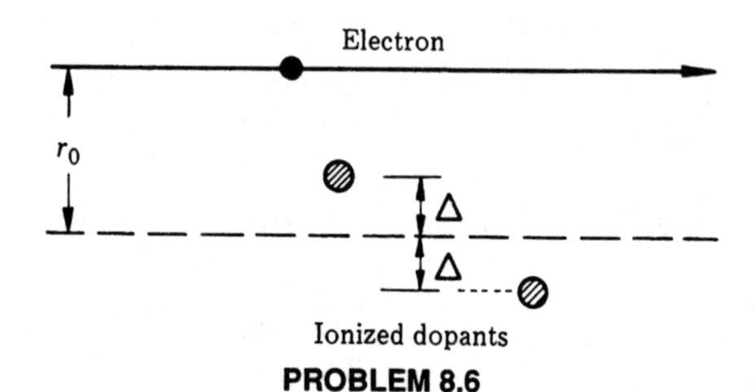

PROBLEM 8.6

8.7 For a one-dimensional quantum well 20 nm wide in GaAs with $m^* = 0.067m_e$, bounded by infinite potential barriers, calculate
(a) The energy E_1 of the first level.
(b) The density of states in that level.
(c) The energy difference between levels E_1 and E_2 and the wavelength of light equal to this difference.

8.8 Accompanying this problem is photoluminescence data from a set of GaAs–Al$_x$Ga$_{1-x}$As quantum well samples with varying thickness. The data corresponds to measurements at room temperature ($E_G = 1.420$ eV) and low temperature ($E_G = 1.512$ eV). The optical transition corresponds to the $n = 1$ electron state and to the $n = 1$ heavy hole state. Using $(m^*/m)_e = .065$ and $(m^*/m)_h = 0.47$ calculate the transition energy as a function of well thickness for 50Å, 100Å, 200Å, and 400Å wells using the "infinite well" approximation. Compare explicitly to the data.

8.9 The fundamental equation which defines the solid state laser considers the attenuation of light intensity associated with a "round trip" in the laser, $e^{-2\alpha L}R^2$ and the gain due to stimulated emission, e^{2gL}, where g is the gain/length, R the reflectivity and α the light absorption coefficient. The threshold gain, g_{th}, is given by

$$R^2 e^{-2\alpha L} \cdot e^{2g_{th}L} = 1.$$

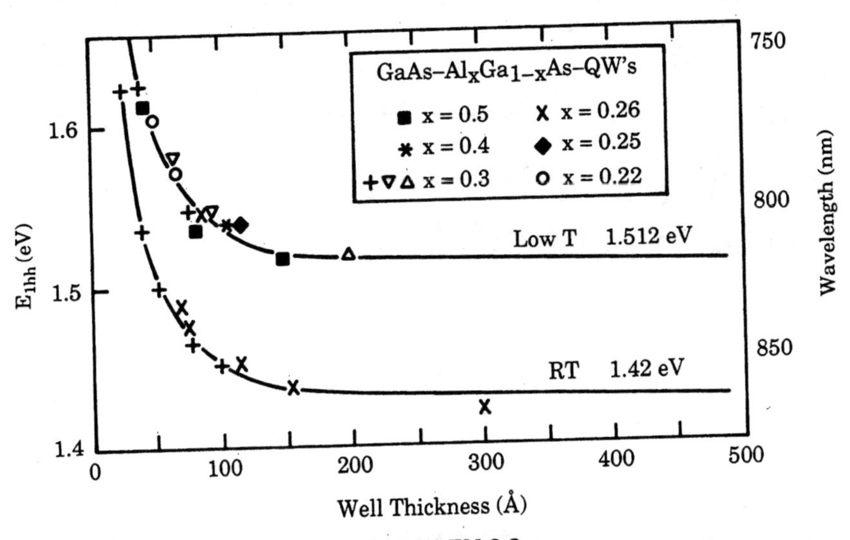

PROBLEM 8.8

From G. Livesco, M. Angell, J. Filipe and W. H. Knox, *J. of Elec. Materials, 19,* 937 (1990).

The gain is a function of the applied current density, J, through the relation:

$$g = (g_0/J_0)\left(\frac{J}{d} - J_0\right)$$

where d is the thickness of the active layer. For a typical Al-GaAs/GaAs/AlGaAs laser $g_0/J_0 = 5 \times 10^{-2}$ cm · μ/A and $J_0 = 4.5 \times 10^3$ A/cm² · μ. Using the dimensions given in Fig. 8.12 and the following data (n (GaAs) $= 3.5$; $\alpha = 10^2$/cm), calculate (a) g_{th}, (b) J, at $g = 10\, g_{th}$, (c) I, the current applied.

8.10 The purpose of this problem is to compare the energy levels of a finite well to an infinite quantum well (Eq. 8.5). Standard quantum mechanics texts show that the energy levels for a finite well of depth V_0 are given by the transcendental equation:

$$\alpha = \cos\left[\frac{\gamma\alpha - (n - 1)\pi}{2}\right]$$

for $n = 1, 2, \cdots$. In this equation $\alpha = (1 + E/V_0)^{1/2}$ and $\gamma = \sqrt{\frac{2m^* V_0 L^2}{\hbar^2}}$.

Consider a well formed by $L = 10$ nm GaAs layer surrounded by AlAs(AlAs/GaAs (10 nm)/AlAs) and assume that the GaAs band is equidistant from the AlAs conduction and valence bands. Calculate the energy level of the first excited state of the finite well and compare to the infinite well solution. Take $m^*/m = 0.06$. (Be careful to note the different zero energy in the two problems; for the finite well $V = -V_0$, $0 < Z < L$, and for the infinite well $V = 0$, $0 < Z < L$.)

CHAPTER 9

Schottky Barriers and Interface Potentials

9.0 Introduction

In previous chapters we have been concerned with the formation of epitaxial semiconductor layers on semiconductors and with the electrical characteristics of heterostructures, quantum wells, and superlattices. At this point, we consider the electrical contact to these structures. One of the last steps in the fabrication of any electronic device, whether it is a simple p–n junction diode or a sophisticated field-effect transistor (FET), is the deposition of a metal layer for contact formation. In integrated circuits, the deposition of metal layers, is called metallization and can require several deposition steps. Figure 9.1 shows a cross section of a metal–oxide–silicon FET, a MOSFET, with silicide contacts to silicon, tungsten contacts to the silicide in holes (vias) in the insulating layers of SiO_2, and finally two metal layers that provide interconnects between transistors (metal 1) as well as connections (metal 2) to contact pads at the periphery of the integrated circuit.

In Fig. 9.1, the electrical contact to the single-crystal and polycrystalline silicon (poly Si) is achieved with a silicide, which is a metal–silicon compound such as PtSi or $TiSi_2$. The silicides are low resistivity and hence other metals such as tungsten form low-resistance contacts to silicides. The prime reason for the use of silicides in Si integrated circuits is that the silicide/silicon interface is planar and stable to temperatures well above 800°C. The most silicon-rich compound—often the disilicide, MSi_2, with M = refractory metal—does not react with Si unless processed at temperatures so high that the silicide decomposes.

Pure metals in contact with semiconductors such as gallium arsenide or silicon are not stable under thermal processing. Under typical integrated-circuit processing conditions, temperature cycles of 400°C to 450°C are often encountered after the metallization steps. For aluminum contacts to silicon, these processing temperatures lead to dissolution of silicon into aluminum to satisfy the solubility requirements. In spite of the low solubility of silicon in aluminum (≃0.5 atomic percent at 400°C), pronounced pits or "spikes" are formed because the aluminum is in contact with silicon only in localized areas defined by windows in the oxide layer and also because silicon can migrate for substantial distances in aluminum (Fig. 9.2). Partial solutions to this pitting problem are provided by designing the circuit layout to restrict the length of aluminum lines adjacent to contact windows (design rules) and by co-

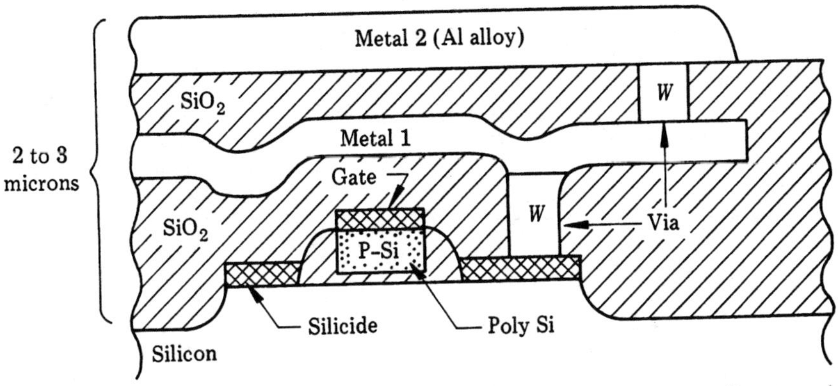

FIGURE 9.1 Cross-sectional view of multilayer metallization to a field effect transistor, a component of an integrated circuit. The silicide layers make contact to silicon. The W plugs in the oxide openings (vias) carry electrical signals from the silicide to the metal interconnects.

depositing 0.5 atomic percent of silicon in the aluminum. These solutions are not adequate in tightly packed large-scale integrated circuits where the minimum feature size may be 0.5 microns or below and where *p–n* junction depths may be 0.2 to 0.3 microns below the surface. The pits may be shallow, but the penetration of aluminum into even shallow pits will short out the p–n junction current–voltage characteristics. The silicide layer is therefore introduced as a diffusion barrier between the Al overlayer and the silicon substrate. It not only prevents Al penetration but also provides a smooth, uniform, planar contact to the silicon.

The emphasis of this chapter is on the electrical evaluation of silicide contacts to silicon. Silicides are formed either by codepositing metal and Si atoms (Fig. 9.3) or by depositing a metal layer and then forming the silicide layer in a subsequent heat treatment. In this chapter we are only interested in the existence of a well-defined crystalline silicide layer in contact with silicon. The methods by which the silicide is formed are discussed in subsequent chapters. Codeposition of metal and silicon results in an amorphous layer whose crystallization kinetics are described in Chapter 10; the formation of silicides by the reaction of metal and silicon is treated in Chapter 12.

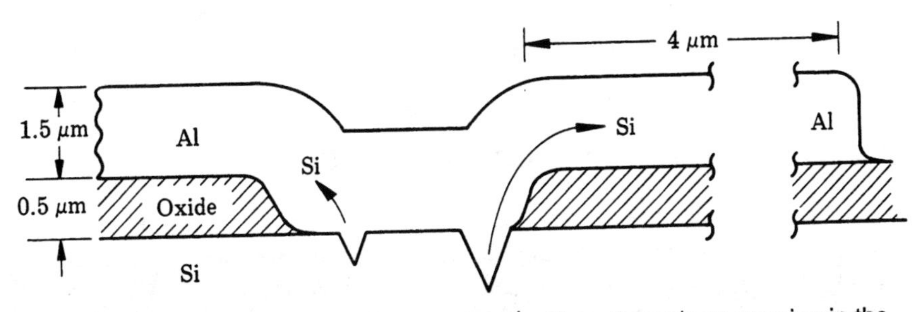

FIGURE 9.2 An aluminum layer in contact with silicon through an opening in the oxide layer. Heat treated at temperatures of 400°C to 450°C, the silicon dissolves into the aluminum and the aluminum penetrates into the pits forming spikes.

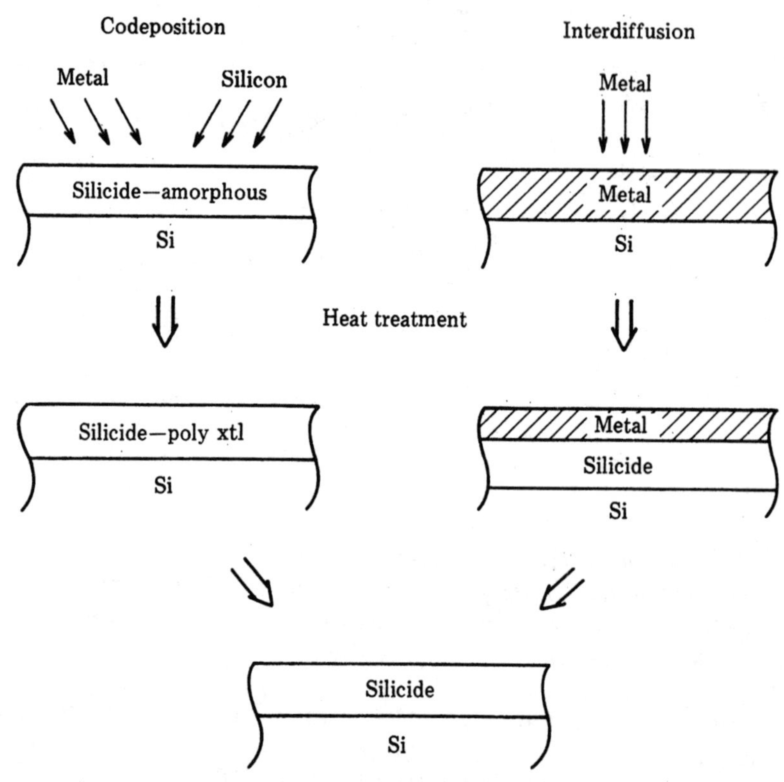

FIGURE 9.3 Silicide layers on silicon are formed either by codeposition of metal and silicon atoms or by interdiffusion of a deposited metal layer. Heat treatment is required to convert the amorphous codeposited layer to a polycrystalline material, or to permit diffusion of metal or silicon across the growing silicide layer.

The electrical evaluation of a silicide/silicon or a metal/semiconductor structure leads to a determination of the potential barrier at the interface. These structures are called *Schottky barrier diodes* and their current–voltage ($I-V$) and capacitance–voltage ($C-V$) characteristics are determined to a large extent by the height of the potential barrier. These barriers are common in semiconductor heterostructures and superlattices as we have shown in the previous chapter.

9.1 Metal–Semiconductor Contacts

The metal film in contact with the surface region of a semiconductor device provides the interface between the outside electrical circuitry and the internal current carriers, the electrons and holes. It is by means of this metal–semiconductor contact that voltages are applied to the devices and currents are transported. The current–voltage ($I-V$) characteristics of a metal contact on a *lightly doped* semiconductor are nearly identical to the $I-V$ characteristics of a p–n junction diode. A metal contact to a *heavily doped* semiconductor generally offers no impedence to current flow and acts as an ohmic contact with a linear $I-V$ characteristic.

The metal–semiconductor contact and its rectifying current–voltage ($I-V$) behavior were extensively studied in the 1930s. The name "Schottky barrier diode" for the metal–semiconductor contact is in honor of Werner Schottky, one of the early researchers in the field. Schottky barrier or surface barrier are synonymous names for the metal contact to lightly doped semiconductors.

9.1.1 Work Function and Electron Affinity

To develop the systematics of the metal–semiconductor barrier behavior we first consider the energy required to remove an electron from a solid to vacuum. We define the vacuum level to indicate an electron with zero kinetic energy. For a metal, the potential difference φ_w or energy difference $e\varphi_w$ between the Fermi level E_F and the vacuum level is called the work function. Figure 9.4a shows the energy-band diagram for a metal and semiconductor separated by vacuum. The energy difference between the vacuum level and the bottom of the conduction band (E_C) in the semiconductor is defined as the electron affinity $e\chi$. The electron affinities of Ge, Si, and GaAs are close to 4.0 eV (Sze, 1981).

When the metal is in contact with an n-type semiconductor (Fig. 9.4b) in thermal equilibrium, Fermi levels line up and a potential barrier $e\varphi_B$ is established, which electrons in the metal must surmount in going from metal to semiconductor. The equalization of levels produces a space-charge region (described in Section 9.1.2) in the semiconductor and a potential barrier $e\varphi_S$ in the n-type semiconductor near the metal–semiconductor interface. Electrons in the bulk of the semiconductor must traverse the barrier $e\varphi_S$ to go into the metal.

The barriers in the simplest case are

$$e\varphi_B = e\varphi_W - e\chi \tag{9.1}$$

$$e\varphi_S = e\varphi_B - (E_C - E_F) \tag{9.2}$$

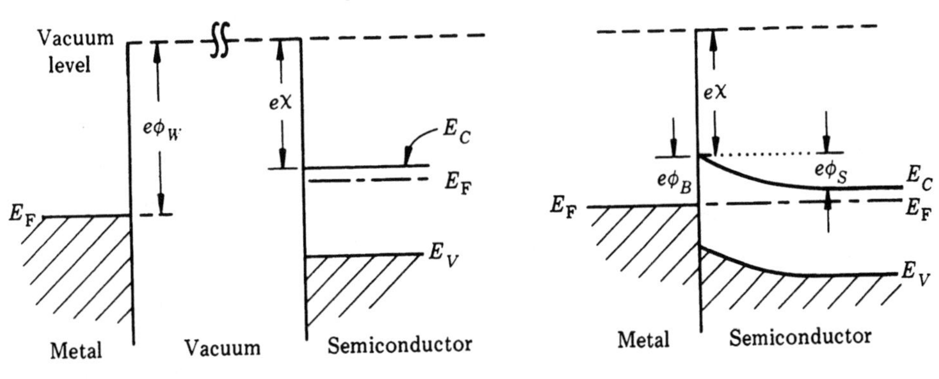

(a) Separation by vacuum (b) Metal/semiconductor in contact

FIGURE 9.4 Energy-band diagrams for a metal and an n-type semiconductor that are (a) separated by vacuum and (b) formed into a metal–semiconductor contact.

where $(E_C - E_F) = eV_n$ is determined by the concentration of dopant atoms in the semiconductor. Methods to measure the barrier height $e\varphi_B$ are given in Section 9.2.

9.1.2 Band-bending and Depletion Region

The presence of a space-charge region in the n-type semiconductor near the metal interface requires an internal electric field. This is established by a sheet of charge per unit area Q^- at the interface that is balanced by the fixed positive charge Q^+ in the semiconductor (Fig. 9.5). The mobile charge carriers are swept out of the space-charge region by the electric field so that the positive charge is

$$Q^+ = eN_D^+ d_n \tag{9.3}$$

where N_D^+ is the concentration of ionized donors (fixed in the lattice and thus immobile) and d_n is the width of the space-charge region in the semiconductor. Since

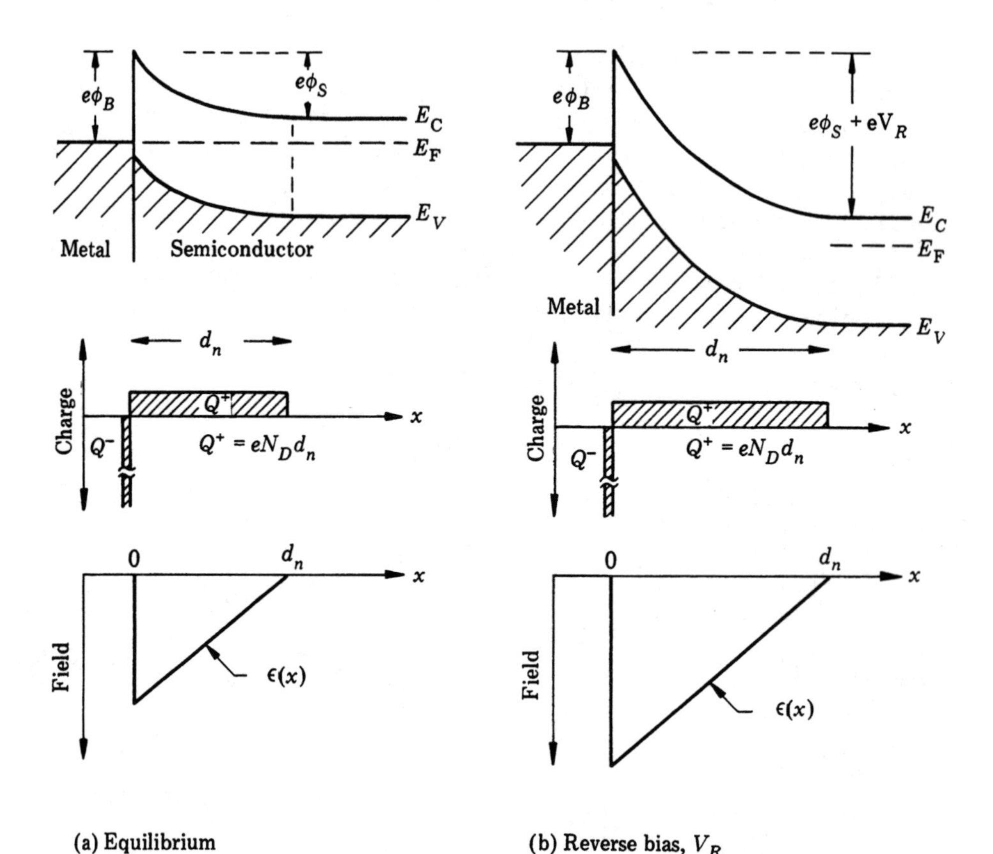

(a) Equilibrium (b) Reverse bias, V_R

FIGURE 9.5 Energy-level diagram, charge distribution, and electric field for a metal–semiconductor contact (a) in equilibrium and (b) under reverse bias with positive voltage V_R applied to the n-type semiconductor.

the carriers (electrons) are swept out or depleted from the space-charge region, it is called the *depletion region.*

The electric field distribution $\mathscr{E}$ in the semiconductor is found from Poisson's equation:

$$\frac{d^2V}{d^2x} = -\frac{d\mathscr{E}}{dx} \tag{9.4}$$

We employ the permittivity $\epsilon = \epsilon_r\epsilon_0$ with ϵ_0 the permittivity of vacuum ($\epsilon_0 = 8.85 \times 10^{-14}$ F/cm) and ϵ_r the relative dielectric constant (ϵ_r(Si) = 11.9, ϵ_r(GaAs) = 13.1, and ϵ_r(Ge) = 16.0). For silicon, $\epsilon = 1.05 \times 10^{-12}$ F/cm. Treating only the lightly doped n-type layer

$$\frac{d\mathscr{E}}{dx} = -\frac{eN_D^+}{\epsilon} \tag{9.5}$$

Integration of Eq. (9.5) with the boundary condition $\epsilon(d_n) = 0$ gives

$$\epsilon(x) = -\frac{eN_D^+}{\epsilon}(d_n - x) \tag{9.6}$$

which leads to the linear variation of the electric field (Fig. 9.5). The negative sign indicates that the electric field is directed toward the metal–semiconductor interface. Integration of Eq. (9.6) gives the potential distribution,

$$V(x) = -\frac{eN_D^+}{\epsilon}\left(d_n x - \frac{x^2}{2}\right) \tag{9.7}$$

The potential decreases from the interface into the semiconductor as shown in the top part of Fig. 9.5.

For a potential φ_S, the width of the depletion region d_n is given from Eq. (9.7) with $x = d_n$ as

$$d_n = \left(\frac{2\epsilon\varphi_S}{eN_D}\right)^{1/2} \tag{9.8}$$

where we have assumed that all the donors are ionized so that $N_D = N_D^+$.

Under a positive voltage V_R applied to the n-type semiconductor (referred to as *reverse bias* as this is the blocking or rectifying current of the diode), all the voltage will appear across the semiconductor in the space-charge region (Fig. 9.5b). The width of the depletion region is increased,

$$d_n = \left[\frac{2\epsilon(\varphi_S + V_R)}{eN_D}\right]^{1/2} \tag{9.9}$$

The width of the depletion region can be determined from the small-signal capacitance C of a metal–semiconductor junction of area A,

$$C = \frac{A\epsilon}{d_n} \tag{9.10}$$

9.2 Schottky Barrier Characteristics

The metal–semiconductor or Schottky barrier diode exhibits rectifying behavior. Figure 9.6 shows energy levels versus distance for a Schottky barrier with negative voltage V_F applied to the n-type semiconductor (this is *forward bias*). The potential barrier $e\varphi_B$ on the metal side remains unchanged under the applied voltage. The potential barrier the electrons must surmount in going from semiconductor to metal is changed by the applied voltage. The barrier is φ_s for V = 0; it decreases for forward-bias voltage V_F to $\varphi_S - V_F$ and increases to $\varphi_S + V_R$ under reverse-bias voltage V_R.

The change in the potential determines the electron flux (direction and magnitude) in the metal–semiconductor barrier. Without applied voltage, the electron current flux from metal to semiconductor $J_{\mathrm{M \to S}}$ is equal to that from semiconductor to metal $J_{\mathrm{S \to M}}$. That is,

$$J_{\mathrm{M \to S}} = J_{\mathrm{S \to M}} \tag{9.11}$$

and the net current flux $J = J_{\mathrm{M \to S}} - J_{\mathrm{S \to M}}$ is zero.

To determine the I–V characteristics, we will first find the number of electrons that have energies greater than the potential barrier. For the metal, we can use Maxwell–Boltzmann statistics for electrons at energies many kT above the Fermi energy E_{F}. The number of electrons in the metal that have energies greater than the barrier height $e\varphi_B$ will be proportional to $\exp(-e\varphi_B/kT)$. This electron emission current is in the flux direction of reverse bias (positive voltage applied to the n-type semiconductor) and is independent of applied voltage as we have assumed that φ_B is fixed. In the semiconductor, the number of electrons able to surmount the potential barrier

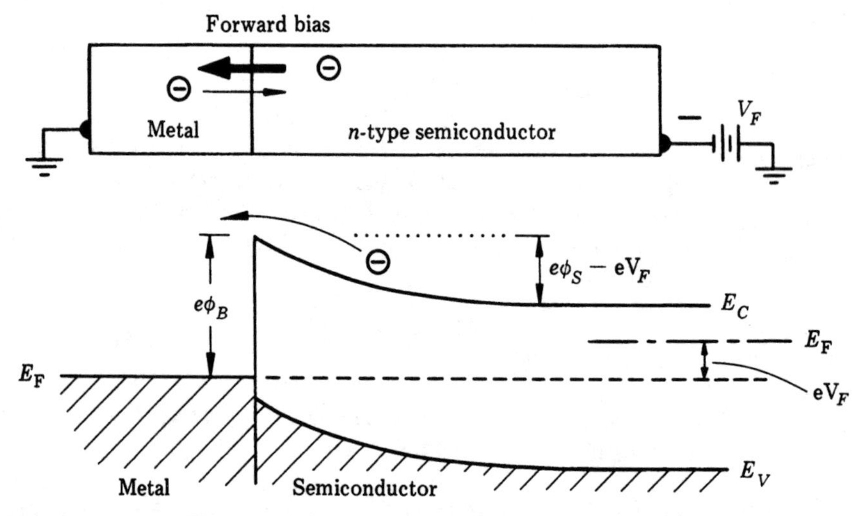

FIGURE 9.6 Schematic of a metal–semiconductor contact under forward bias showing electrons injected into the metal and an energy-level diagram for a negative voltage applied to the n-type semiconductor.

$e\varphi_S$ is proportional to $\exp(-e\varphi_S/kT)$. Under zero applied voltage the two electron currents are equal, as Eq. (9.11) states.

Under forward bias (negative voltage on an n-type semiconductor) all the voltage V_F appears in the barrier region and the potential barrier in the semiconductor is reduced from $e\varphi_S$ to $(e\varphi_S - eV_F)$. Under forward bias the number of electrons that can flow into the metal is now increased by a factor $\exp(eV_F/kT)$. Then $J_{S \to M} >> J_{M \to S}$ and the forward-bias current density J_F is

$$J_F = J_{ST} \exp(eV_F/kT) \tag{9.12}$$

where J_{ST} is a constant, independent of voltage. Under reverse-bias voltage V_R the barrier in the semiconductor is increased to $\varphi_S + V_R$ and $J_{S \to M} << J_{M \to S}$. The current density under reverse bias is

$$J_R = J_{ST} = J_{M \to S} \tag{9.13}$$

where $J_{M \to S}$ is independent of voltage as we neglect the change in φ_B due to image forces. The metal–semiconductor current density versus voltage characteristics can be written

$$J = J_{ST}[\exp(eV/kT) - 1] \tag{9.14}$$

The constant J_{ST} is derived in the next section (9.3) to be

$$J_{ST} = AT^2 \exp(-e\varphi_B/kT) \tag{9.15}$$

where A is Richardson's constant ($A \simeq 100\ \text{Amp/cm}^2\text{K}^2$).

9.3 Schottky Barrier Measurement

There are three techniques commonly used for measuring Schottky barrier heights of metal–semiconductor contacts; they are current–voltage, capacitance–voltage, and photoresponse techniques. The current–voltage (I–V) technique is based on electrical transport of majority carriers across the contact. We shall discuss it first.

9.3.1 Thermionic Emission and Current Voltage Technique

We consider the contact on an n-type semiconductor, so the majority carrier is the electron. The theory of electrical transport of carriers across the contact is based on *thermionic emission* (i.e., the flux of electrons emitting from the free surface of a metal or a semiconductor). In emission from a free surface, the electrons must have sufficient kinetic energy to overcome the work function of the metal or the electron affinity of the semiconductor. In a metal–semiconductor contact the potential barrier that the carriers have to overcome is much lower; we have shown in Eq. (9.1) that the Schottky barrier potential (or height) is

$$\varphi_B = \varphi_W - \chi \tag{9.16}$$

where φ_w and χ are the work function of the metal and the electron affinity of the semiconductor, respectively. In thermionic emission theory we can use the condition of thermodynamic equilibrium to assume that the flux of electrons emitting from the surface must equal to the flux of electrons impinging on the surface. It is easy to formulate the latter.

Using the same approach as in Chapter 1 for vacuum deposition of atoms on a surface, we have the flux of electrons impinging on a surface (or current density, which is defined as the number of electrons passing through a unit area in a unit time) given by

$$J = env \tag{9.17}$$

where e is the electrical charge of an electron, n is the electron concentration or the number of electrons per unit volume, and v is the mean velocity of electrons.

For the mean thermal velocity of electrons v we use Eq. (A.13) in Appendix A and assume the direction of transport is in the x-direction.

$$\begin{aligned} J &= en \int_{v_{0x}}^{\infty} \sqrt{\frac{m_e^*}{2\pi kT}}\, v_x \exp\left(-\frac{m_e^* v_x^2}{2kT}\right) dv_x \\ &= en \left(\frac{kT}{2\pi m_e^*}\right)^{1/2} \exp\left(-\frac{m_e^* v_{0x}^2}{2kT}\right) \end{aligned} \tag{9.18}$$

where v_{0x} is the minimum velocity required for an electron to go over the barrier and m_e^* is the effective mass of electron. Electrons with a velocity less than v_{0x} do not contribute to J. At thermal equilibrium, we have

$$\tfrac{1}{2} m_e^* v_{0x}^2 = e\varphi_S \tag{9.19}$$

where φ_S is the built-in voltage in the semiconductor. Under forward or reverse bias,

$$\tfrac{1}{2} m_e^* v_{0x}^2 = e(\varphi_S \mp V) \tag{9.20}$$

respectively, where V is the applied voltage. Taking the forward-biased case specifically,

$$J_{\mathrm{S \to M}} = en \left(\frac{kT}{2\pi m_e^*}\right)^{1/2} \exp\left[-\frac{e\varphi_S - eV}{kT}\right] \tag{9.21}$$

where the subscript S→M indicates the current goes from the semiconductor to the metal. For the electron concentration n of an n-type semiconductor, we use Maxwell–Boltzmann statistics,

$$n = N_C \exp\left(-\frac{E_C - E_F}{kT}\right) \tag{9.22a}$$

where N_C is the density of states in the conduction band. E_C is the bottom of the conduction band and $E_C - E_F = eV_n$. Then

$$n = 2\left(\frac{2\pi m_e^* kT}{h^2}\right)^{3/2} \exp\left(-\frac{eV_n}{kT}\right) \tag{9.22b}$$

Substituting the last equation into Eq. (9.21) we obtain

$$J_{S\to M} = A^*T^2 \exp\left(-\frac{e\varphi_B}{kT}\right)\exp\left(\frac{eV}{kT}\right) \tag{9.23}$$

where $\varphi_B = \varphi_S + V_n$ is the Schottky barrier height, and

$$A^* = \frac{4\pi e m_e^* k^2}{h^3} \tag{9.24}$$

is the Richardson constant for thermionic emission of electrons from semiconductor to metal. We use the asterisk to indicate the use of an electron effective mass.

Since the number of electrons flowing from the metal into the semiconductor remains unchanged with the applied voltage, the flux must be equal to that at thermal equilibrium, hence

$$J_{M\to S} = -A^*T^2 \exp\left(-\frac{e\varphi_B}{kT}\right) \tag{9.25}$$

Summing Eqs. (9.23) and (9.25), we have

$$\begin{aligned} J_n &= \left[A^*T^2 \exp\left(-\frac{e\varphi_B}{kT}\right)\right]\left[\exp\left(\frac{eV}{kT}\right) - 1\right] \\ &= J_{ST}\left[\exp\left(\frac{eV}{kT}\right) - 1\right] \end{aligned} \tag{9.26}$$

where the saturation current density,

$$J_{ST} = A^*T^2 \exp\left(-\frac{e\varphi_B}{kT}\right) \tag{9.27}$$

Thermionic emission is further modified by taking into account the quantum-mechanical transmission effect and electric-field-enhanced emission; the modification can be included in the Richardson's constant and the new constant is called the effective Richardson's constant, A^{**}. For n-Si and p-Si, $A^{**} \cong 110$ and 30 amp/cm^{2}K^2, respectively. For n-GaAs and p-GaAs, $A^{**} \cong 8.4$ and 120 amp/cm^{2}K^2, respectively.

There is a lowering of the Schottky barrier height under an applied voltage, due to an image force; this is defined as *Schottky barrier lowering* $\Delta\varphi$ (see the book by Sze for details of this lowering). If the zero-field barrier height is φ_{B0} (i.e., $\varphi_{B0} = \varphi_B + \Delta\varphi$), we rewrite Eq. (9.26) as

$$J_n = A^{**}T^2 \exp\left(-\frac{e\varphi_{B0}}{kT}\right)\exp\left[\frac{e(\Delta\varphi + V)}{kT}\right] \tag{9.28}$$

when $V > 3kT/e$ and $\Delta\varphi$ is a function of the applied voltage. Then the relationship between J_n and V is not

$$J_n \sim \exp\left(\frac{eV}{kT}\right)$$

but rather

$$J_n \sim \exp\left(\frac{eV}{nkT}\right) \tag{9.29}$$

where n is the ideality factor and

$$n = \frac{e}{kT}\frac{\partial V}{\partial(\ln J_n)}. \tag{9.30}$$

A typical value of n, obtained from the slope of $\ln J_n$ against V is about 1.01 to 1.05. A larger value of n is often due to a large recombination current near the contact region. The barrier height φ_{B0} is calculated from J_{ST} which is the extrapolated value of J_n at zero voltage,

$$\varphi_{B0} = \frac{kT}{e}\ln\left(\frac{A^{**}T^2}{J_{ST}}\right) \tag{9.31}$$

Fig. 9.7 shows a plot of $\ln J_n$ versus V for an Ir/n-Si contact after annealing at 300°C for 1 hour. We find that $n = 1.04$ and $\varphi_{B0} = 0.93$ eV.

For a high-Schottky-barrier contact on n-Si as shown in Fig. 9.7, room temper-

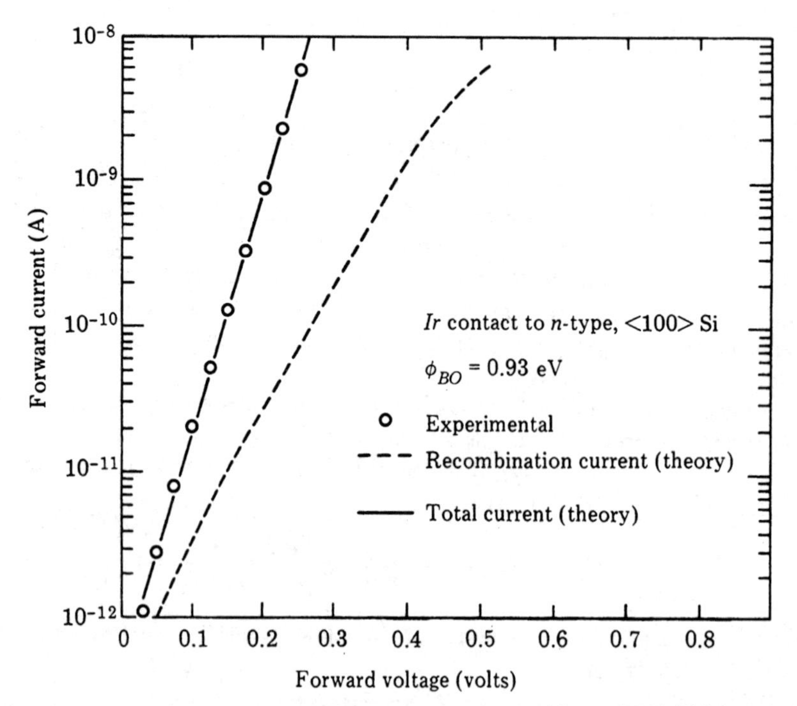

FIGURE 9.7 Forward current–voltage measurement of Ir on ⟨100⟩ Si heated at 300°C for 1 hour. Computer fitting of the total current across the Schottky contact is shown to equal to the sum of the diffusion and recombination currents (taken from Ohdomari, Kuan, and Tu, 1979).

ature measurements give a sufficient linear region in the ln J_n versus V plot for slope and extrapolation determinations, so that the obtained values of n and φ_{BO} are reliable. For a low-Schottky-barrier contact, the thermionic emission is high and saturation can occur at low enough voltages that the linear region is not well defined for a reliable measurement of n and φ_{BO}. To overcome this problem, the $I-V$ measurement must be carried out at lower temperatures; this temperature dependence exemplifies the nature of thermionic emission in Schottky behavior.

Figure 9.8 shows a set of ln J_n versus V curves of $TiSi_2$ contacts on n-Si measured in the temperature range from room temperature to -90°C. An extended linear region is obtained at the lower temperatures with $n \cong 1.01$ and $\varphi_{BO} = 0.6$ eV. Similar measurements on p-Si showed a barrier height of 0.52 eV, indicating that the sum of Schottky barrier heights on n-type and p-type Si equals the band gap of Si, $E_G = 1.12$ eV, which is expected.

Schottky barriers of silicides on Si have been studied extensively for device applications and for understanding metal–semiconductor interfaces. From the point of view of interfacial chemistry (i.e., to vary the metal in contact with Si), many transition metal disilicides can be formed at thermal equilibrium on n-Si and p-Si. A systematic comparison can be made among them, since errors due to sample preparation and measurement technique ($I-V$) can be minimized by keeping them the same for these disilicides. Figure 9.9 shows a plot of φ_{BO} against T_{eu} (the eutectic temperature) between Si and disilicide. The reason of plotting φ_{BO} and T_{eu} together

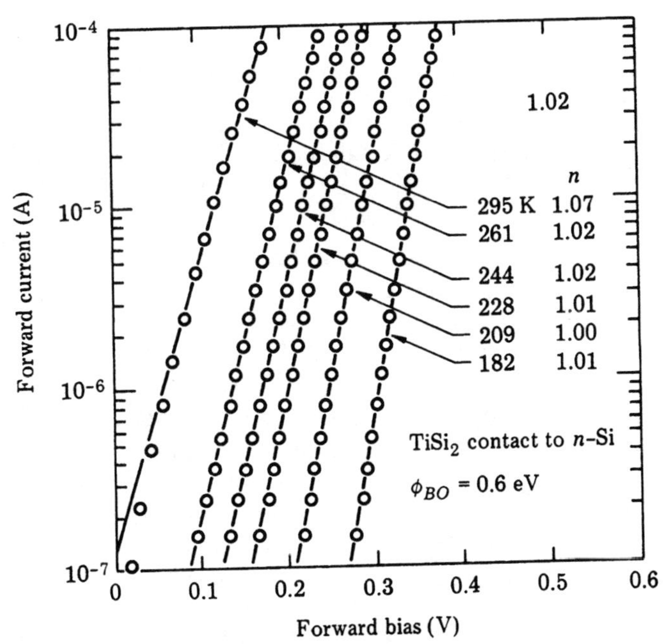

FIGURE 9.8 Forward current–voltage characteristics of Ti on n-type Si(100) as a function of measurement temperature for samples annealed at 1023 K for 1 hour. Diode diameter is 129μm (taken from Aboelfatoh and Tu, 1986).

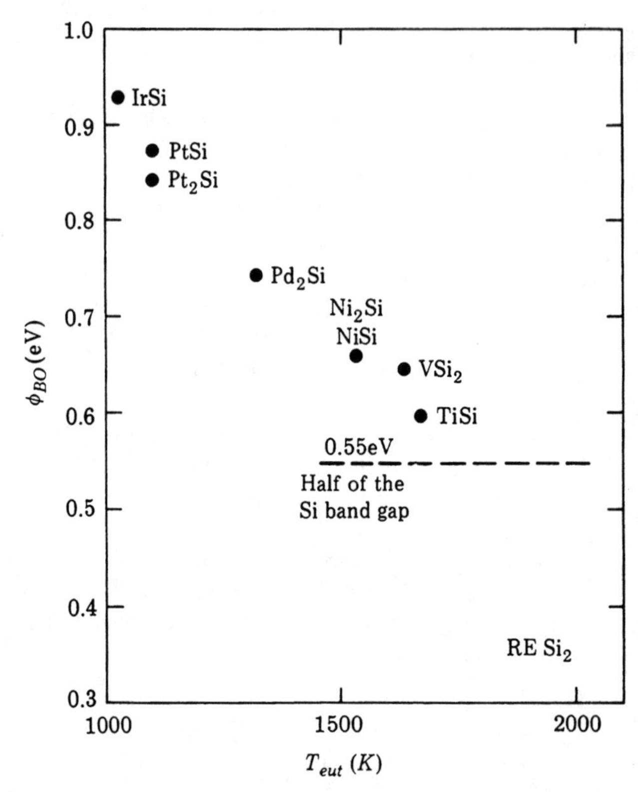

FIGURE 9.9 Plot of Schottky barrier height against eutectic temperature of near-noble, refractory, and rare-earth metal silicides. The barrier height values (except that of TiSi) were measured by *I–V* technique at room temperature using 1/16 mil diameter diodes on n-Si of ~ 10 Ω-cm (taken from Tu, Thompson, and Tsaur, 1981).

is that both are interfacial properties. Note that the rare-earth metal disilicides have the lowest Schottky barrier height of 0.4 eV on n-Si.

We can also vary interfacial structure without changing the chemistry at the contact. For example, epitaxial and non-epitaxial $NiSi_2$ can be grown on ⟨111⟩ Si, and their Schottky barrier heights are found to be different. The Schottky barrier height of the B-type $NiSi_2$ (where the silicide has a twinning orientation relationship with the Si) is found to be 0.8 eV, and that for non-epitaxial $NiSi_2$ is about 0.66 eV. In the case of A-type $NiSi_2$ (where the silicide has the same orientation as the Si), the Schottky barrier height varies between the 0.8 eV and 0.66 eV values. The finding that Schottky barrier height depends on interfacial structure and chemistry is not a surprise. These are intrinsic properties of an interface, similar to an interfacial energy.

9.3.2 Capacitance–Voltage and Photoresponse Techniques

The capacitance–voltage technique of measuring Schottky barrier height relies on the formation of a depletion layer in the semiconductor in contact with a metal.

Assuming an abrupt junction between the metal and the semiconductor, and that the ions are uniformly distributed in the depletion layer, the relation between the capacitance/area and applied voltage is given from Eqs. (9.9) and (9.10),

$$\frac{1}{C^2} = \frac{2\left(\varphi_s + V - \dfrac{kT}{e}\right)}{e\epsilon N_D} \tag{9.32}$$

where ϵ is the permittivity, N_D the ionized donor concentration, and kT/e is a correction for mobile carriers. A plot of $1/C^2$ against the reverse-bias V gives φ_S at $1/C^2 = 0$. The term kT/e represents mobile carriers and it is negligible at low voltages. Since $\varphi_{BO} = -\varphi_S + V_n$, we obtain the Schottky barrier height when V_n is calculated from the doping level, which is obtained from N_D in the slope of the plot. In Fig. 9.10, the $C-V$ plots of 1 mm diameter diodes of Ir/Si annealed at 600°C for 1 hour are shown, and the measured Schottky barrier heights are given.

Photoresponse is a direct measure of the photoyield (photocurrent density per incident photon flux) across the Schottky barrier. When the photon energy ($h\nu$) is greater than the barrier height by a factor of several kT, the photoyield (Y) is given approximately by the Fowler formula,

$$Y = B(h\nu - e\varphi_{BO})^2 \tag{9.33}$$

where B is an efficiency factor, being almost independent of $h\nu$ for most metal–semiconductor interfaces. The Fowler plot, which plots the square root of

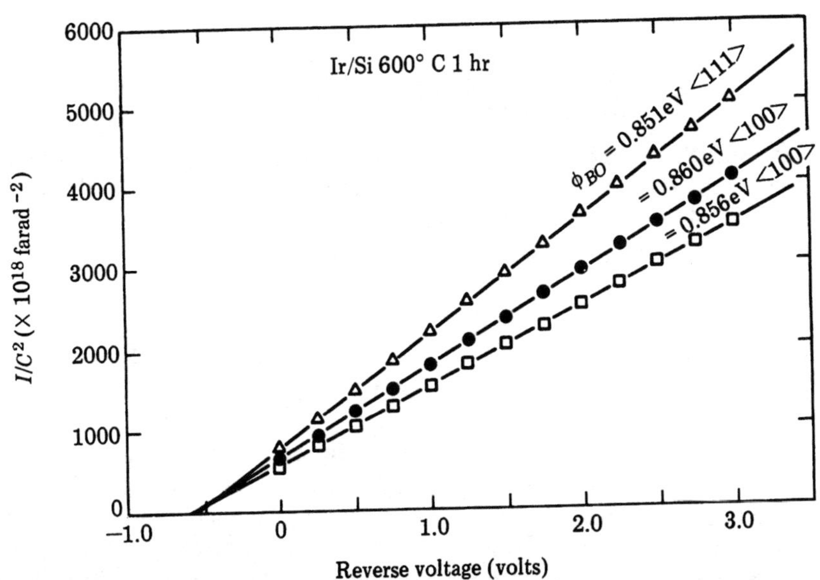

FIGURE 9.10 $1/C^2$ versus reverse voltage plot for 1 mm diameter diodes on ⟨111⟩ and ⟨100⟩ Si annealed at 600°C for 1 hour (taken from Ohdomari, Kuan, and Tu, 1979).

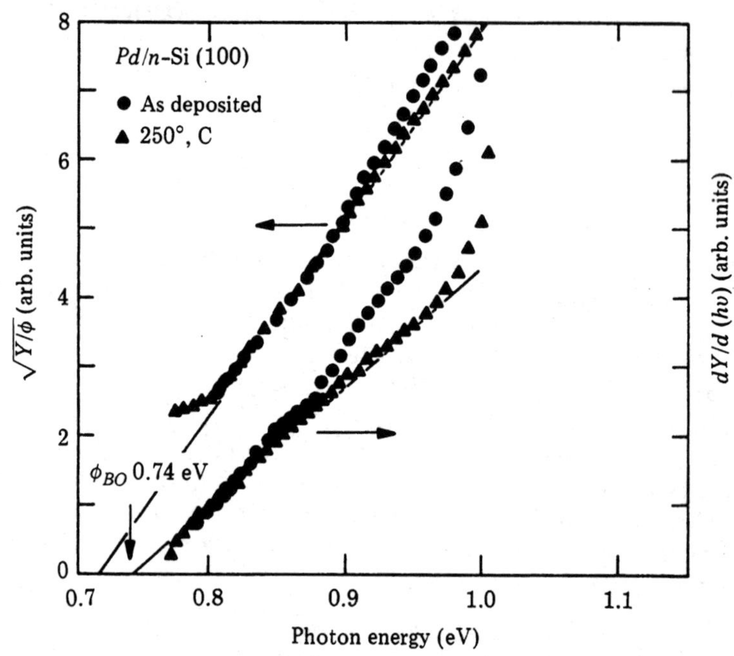

FIGURE 9.11 Square-root plot and first-derivative plot for Pd on n-Si diodes before and after annealing at 250°C for 30 minutes (taken from Okumura and Tu, 1983).

photoyield as a function of $h\nu$, gives the barrier height at $h\nu = 0$. By differentiating Y with respect to $h\nu$, we have

$$\frac{dY}{dh\nu} = 2B(h\nu - e\varphi_{BO}) \tag{9.34}$$

It shows that by plotting the first derivative $dY/dh\nu$ against $h\nu$, we again obtain the barrier height at $h\nu = 0$. In Fig. 9.11, a square-root plot and a first-derivative plot of Pd/n-Si diodes before and after an annealing at 250°C for 30 minutes are shown. The measured Schottky barrier height is about 0.72 to 0.74 eV.

9.4 Effects of Surface States, Damage, and Parallel Contacts on Schottky Barrier

If the semiconductor surface has a high density of surface states, electrons will occupy these surface states up to the energy which is equal to the Fermi level of the semiconductor. The ions will form a space-charge region and cause band bending such that a potential barrier is formed. When the semiconductor is in contact with a metal, the metal will be able to reach equilibrium by charge redistribution with the surface states, and will hardly affect the space-charge region (depletion layer) in the semiconductor. For electrons to go from the semiconductor to the metal, they have to overcome the barrier height established by the surface states (not by the metal). The Schottky barrier is thus independent of the metal or the metal work function. In such a case, we regard the Fermi level at the interface as "pinned" by the surface states, resulting in a fixed Schottky barrier height which is determined by the doping and surface properties of the semiconductor.

Typically, Schottky contacts to n-type Ge and to n-type GaAs are pinned at about 0.5 eV and 0.9 eV below E_C, respectively, independent of metal. On the other hand, we have shown in the last section that silicide contacts to n-Si are unpinned.

Lattice defects due to damage in the semiconductor in the vicinity of the interface (caused by surface preparation and metal deposition) will serve as carrier recombination centers, adding a recombination current to the thermionic emission current in $I-V$ measurement. This results in a higher ideality factor and a lower Schottky barrier. Similarly, in a photoresponse measurement, it will decrease the collection efficiency of the photoemission current. In $C-V$ measurements, an effective deep-level recombination center shifts the Fermi level toward itself and compensates for the shallow dopant levels, resulting in a highly resistive region between the metal and the space-charge region. This is equivalent to adding a capacitance in series to the depletion layer capacitance. The highly resistive region will decrease the measured capacitance and raise the C^{-2} versus V curve. Consequently, the estimated barrier height is higher than the real one. The discrepancy between barrier height values measured by $I-V$ and $C-V$ techniques is generally caused by defects in the semiconductor. However, if the metal–semiconductor interface has parallel diodes, the difference is more serious.

In a real metal–semiconductor contact, a nonuniform interface forms a parallel contact to the semiconductor. For example, silicide formation on Si may not be complete and uniform due to a native oxide on the Si surface, and the metal may react with compound semiconductors to form a mixture of two intermetallic compounds. Different intermetallic compounds or phases will have different Schottky barrier heights in the contact.

Figure 9.12 shows schematically a two-phase parallel contact in $I-V$ measurement. The low-barrier phase will dominate the current transport across the contact. In a $C-V$ measurement, since capacitance is directly proportional to area, the major phase which occupies the larger area of the contact will dominate the measured barrier height. Clearly the results of $I-V$ and $C-V$ measurements of the same parallel contact can be very different. In analyzing a simple parallel contact of two phases, there are three variables: the potential heights of the high and low barriers, φ_{h0} and φ_{l0}, and the ratio of their areas S_l/S_h (or S_l/S where the total contact area $S = S_h + S_l$).

In the case of current–voltage measurement, we have the total current

$$I = I_h + I_l \tag{9.35}$$

where

$$I = SJ = SA^{**}T^2 \exp\left(-\frac{e\varphi_{ap}}{kT}\right)\left[\exp\left(\frac{eV}{kT}\right) - 1\right] \tag{9.36}$$

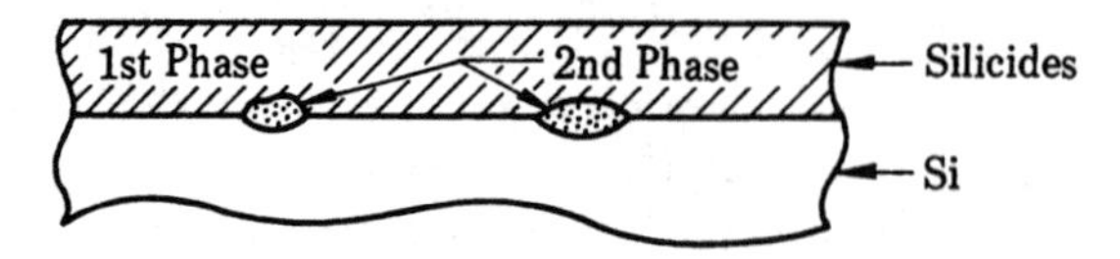

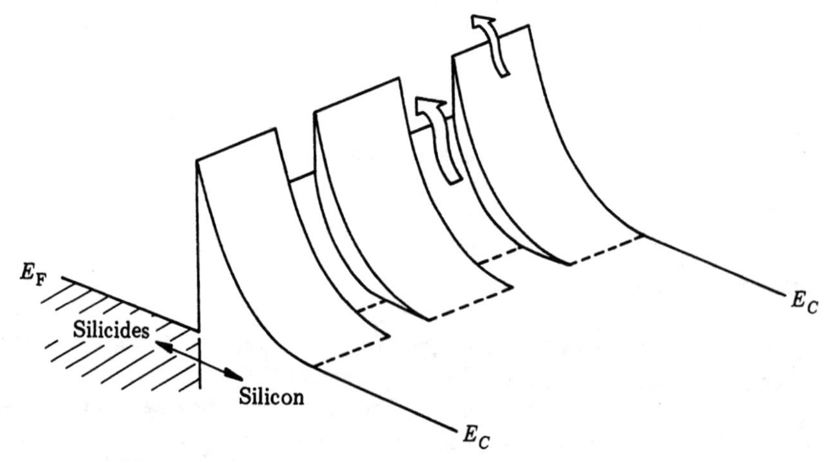

FIGURE 9.12 Schematic cross section of a parallel contact on Si and the corresponding energy-band diagrams (taken from Ohdomari and Tu, 1980).

where φ_{ap} is the apparent barrier height of the parallel contact, and the expressions for I_h and I_l are similar. By cancelling the common terms, we have

$$S \exp\left(-\frac{e\varphi_{ap}}{kT}\right) = S_h \exp\left(-\frac{e\varphi_{h0}}{kT}\right) + S_l \exp\left(-\frac{e\varphi_{l0}}{kT}\right) \tag{9.37}$$

and by rearranging terms, we have

$$\varphi_{ap} = -\frac{kT}{e} \ln\left\{\frac{S_l}{S}\left[\exp\left(-\frac{e\varphi_{l0}}{kT}\right) - \exp\left(-\frac{e\varphi_{h0}}{kT}\right)\right] + \exp\left(-\frac{e\varphi_{h0}}{kT}\right)\right\} \tag{9.38}$$

To see the dominant effect of the low-barrier phase on the apparent barrier height as given by Eq. (9.38), we take $\varphi_{h0} = 0.87$ eV (PtSi on n-Si) and $\varphi_{l0} = 0.66$ eV ($NiSi_2$ on n-Si) and we plot in Fig. 9.13 the apparent barrier height φ_{ap} against S_l/S, the fraction of area of the low-barrier phase. This shows that even when $S_l/S = 0.1$, φ_{ap} is dominated by φ_{l0}. So I–V measurements are sensitive to a low-barrier phase in the contact. In the figure, the crosses are experimental data and they agree well with the calculated curve given by Eq. (9.38). Other combinations of φ_{h0} and φ_{l0} and S_l/S can be calculated using Eq. (9.38).

The dominating effect of the low-barrier phase in I–V measurement of a parallel contact leads to an interesting prediction. Since the opposite phase will become the low-barrier phase when the semiconductor type is changed (e.g., from n-type to p-type), one expects that the sum of the apparent barriers of a set of parallel diodes

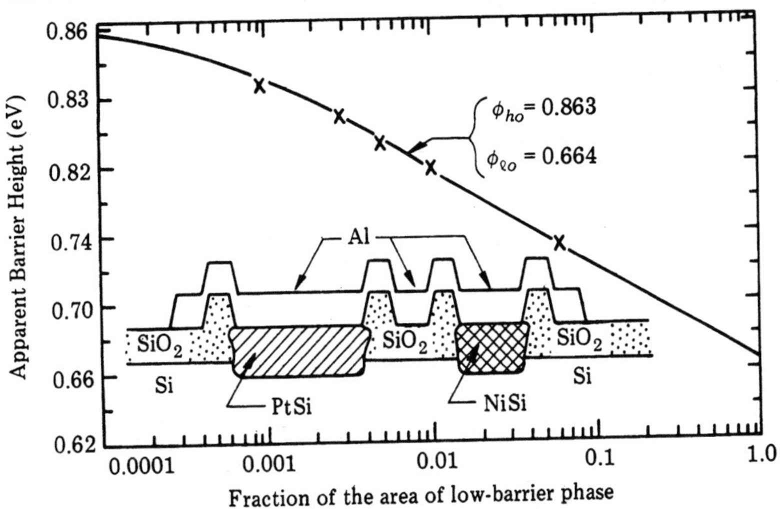

FIGURE 9.13 Measured (I–V) and computed apparent Schottky barrier heights of parallel contacts of PtSi and NiSi plotted against the fraction of the contact area occupied by NiSi (taken from Ohdomari and Tu, 1980).

on n-type and p-type semiconductors will always be less than the band gap. By defining $\Delta\varphi_B = \varphi_{h0}^{n} - \varphi_{l0}^{n}$ (i.e., on an n-type semiconductor) and neglecting the difference in the mass of an electron and a hole, we have

$$\varphi_{ap}^{n} + \varphi_{ap}^{p} = E_G - \frac{kT}{e}\ln\left\{\frac{2S_l}{S}\left(1 - \frac{S_l}{S}\right) \times \left[\cosh\left(\frac{e\Delta\varphi_B}{kT}\right) - 1\right] + 1\right\} \tag{9.39}$$

where φ_{ap}^{n} and φ_{ap}^{p} are the apparent Schottky barrier heights of a parallel contact measured on n-type and p-type semiconductors, respectively. Equation (9.39) shows that at $\Delta\varphi_B = 0$, the sum equals the band gap, which is expected for a uniform contact. When $\Delta\varphi_B \neq 0$, (i.e., when the contact is non-uniform), we plot the sum against $\Delta\varphi_B$ for two values of S_l/S (0.5 and 0.05 or 0.95) in Fig. 9.14. This shows that the sum is much more sensitive to $\Delta\varphi_B$ than to S_l/S and is less than the band gap except at $\Delta\varphi_B = 0$. Experimental data of a Ti–Pd_2Si pair with area ratio of 60 to 40 showed $\varphi_{ap}^{n} = 0.54$ eV and $\varphi_{ap}^{p} = 0.44$ eV. Another pair of $GdSi_2$– PtSi with the same area ratio showed $\varphi_{ap}^{n} = 0.46$ eV and $\varphi_{ap}^{p} = 0.30$ eV. Their sums are indicated by the bars in Fig. 9.14, in good agreement with the calculated curves.

In capacitance–voltage measurement of a parallel contact, the apparent junction capacitance C_{ap} can be given by

$$C_{ap} = C_h + C_l \tag{9.40}$$

where C_h and C_l are the junction capacitances (i.e., junction area times the depletion layer capacitance per unit area) of the high barrier and low barrier, respectively. Writing

$$C_{ap} = S\left[\frac{e\epsilon N_D}{2\left(V_{ap} - V - \frac{kT}{e}\right)}\right]^{1/2} \tag{9.41}$$

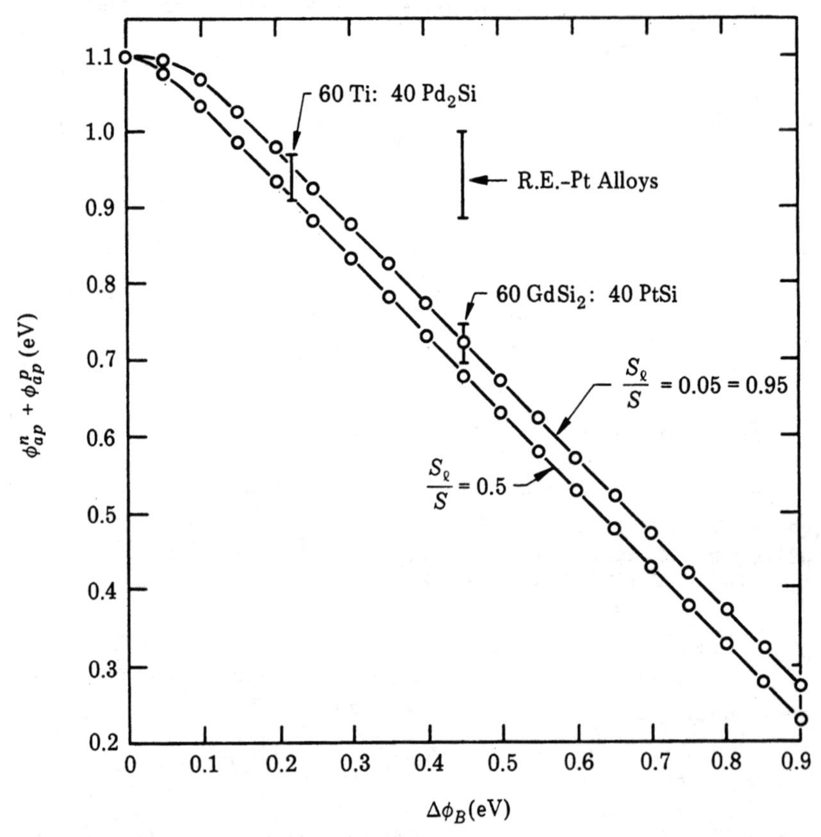

FIGURE 9.14 The sum of the apparent barrier heights of parallel diodes on n-type and p-type Si plotted versus the difference in the barrier heights of the individual phases on n-Si (taken from Thompson and Tu, 1982).

and using similar equations for C_h and C_l, where for simplicity we set

$$V + \frac{kT}{e} = 0 \tag{9.42}$$

which corresponds to applying a very small negative bias ($= -kT/e$) to the junction), we obtain

$$\left(\frac{1}{V_{\text{ap}}}\right)^{1/2} = \left(1 - \frac{S_l}{S}\right)\left(\frac{1}{V_{\text{bh}}}\right)^{1/2} + \frac{S_l}{S}\left(\frac{1}{V_{\text{bl}}}\right)^{1/2} \tag{9.43}$$

Hence, by measuring V_{bh} and V_{bl} which are the built-in voltages of the high barrier and the low barrier, respectively, we can plot V_{ap} against S_l/S. In Fig. 9.15, the solid curve is plotted by taking $V_{\text{bh}} = 0.57$ eV (PtSi on n-Si) and $V_{\text{bl}} = 0.37$ eV (NiSi on n-Si). The dots are experimental data of parallel contacts consisting of PtSi and $NiSi_2$. Although the scattering of the data points are large, the agreement is still good. It is clear from the curve that the apparent barrier height measured by the C–V technique is dominated by the phase which occupies the larger area.

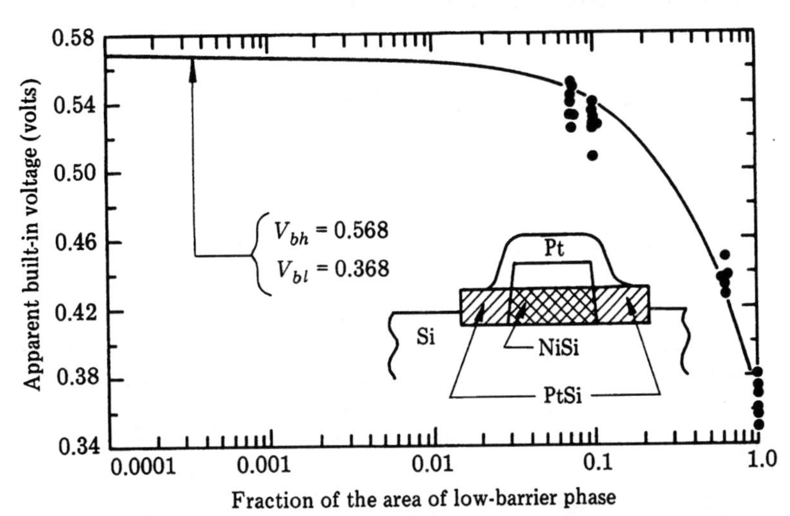

FIGURE 9.15 Measured ($C-V$) and computed built-in voltage of parallel contacts of PtSi and NiSi plotted against the fraction of the contact area occupied by NiSi (taken from Ohdomari and Tu, 1980).

In the above calculation, the high barrier and the low barrier have been assumed to be isolated or noninteractive. The boundary between them is assumed to be abrupt, yet in reality the difference between V_{bh} and V_{bl} will generate an electric field which redistributes space charges until a local equilibrium is reached. This means that the boundary region is expected to be smooth, which, in turn, modifies the apparent Schottky barrier to a higher value. When an alloy of Pt and rare-earth metal is made to contact n-type and p-type silicon, the sum of the apparent barrier heights is greater than the calculated value, as shown in Fig. 9.14, by about 0.1 eV.

9.5 Ohmic Contacts

Certain contacts to a semiconductor device in an integrated circuit should impose no restriction to the flow of charge carriers. Such contacts are called *ohmic contacts* because the current between contact and semiconductor should be linear with voltage depending only on the resistivity of the semiconductor and with no resistance attributed to the contact itself. The contact is typically a metal film deposited on the semiconductor. The contact resistance R_C defined at zero applied voltage is

$$R_C = \left(\frac{dJ}{dV}\right)^{-1} \tag{9.44}$$

with units for R_C of $\Omega\text{-cm}^2$. A usable contact should have a resistance of 10^{-7} ohm-cm^2 or lower. For the metal–semiconductor diode of Section 9.3, the current density is

$$J = J_{\text{ST}}[\exp(eV/kT) - 1]$$

so that

$$R_C = \frac{kT}{eJ_{ST}} \tag{9.45}$$

with $J_{ST} \simeq 120$ amp/cm$^2K^2$ $\times$ T^2 $\exp(-e\varphi_B/kT)$. For a barrier height of 0.6 eV at 300 K, the contact resistance will be about 10 Ω-cm^2, a value orders of magnitude too high for applications in integrated circuits.

The relation for contact resistance given in Eq. (9.45) was based on thermionic emission where the carriers must surmount the potential barrier. There is another mechanism for current transport across a contact: tunneling. In this case transmission J_T of the electrons through the barrier by tunneling occurs more efficiently than thermionic emission over the barrier. Metal–semiconductor ohmic contacts are made in heavily doped semiconductors where the Fermi level lies above the conduction band edge (Fig. 9.16a). Consequently with even a small forward bias, the electrons in the conduction band are lined up in energy with a large density of unoccupied states (Fig. 9.16b). The transmission probability for a tunneling transition is high if

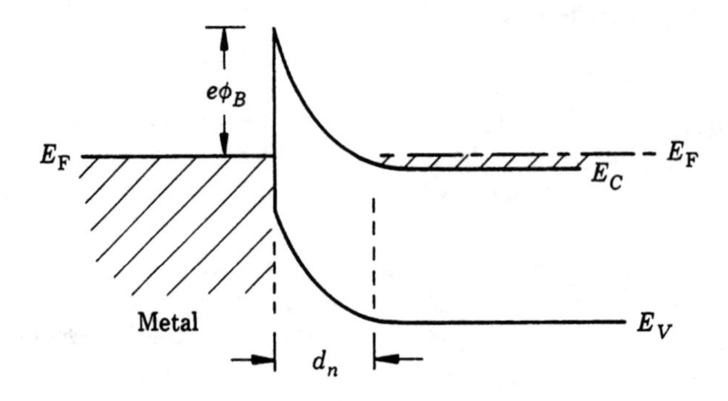

(a) Thermal equilibrium

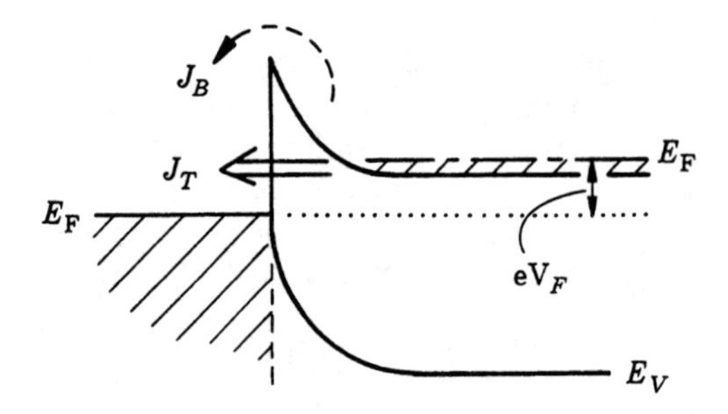

(b) Forward

FIGURE 9.16 Energy-level diagrams for a metal–semiconductor contact on a heavily doped, degenerate semiconductor in (a) equilibrium and (b) forward bias.

the barrier width is small, typically 3 to 5 nm. As a first approximation we can use the width of the depletion region at zero applied bias as an estimate of the width of the tunnel barrier. From Eq. (9.8), $d_n \simeq (2\epsilon\varphi_B/eN_D)^{1/2} = 2.5$ nm for $\epsilon = 10^{-12}$ F/cm, $\varphi_B = 0.6$ eV, and $N_D = 10^{20}/\text{cm}^3$. At this high doping concentration of 10^{20} ionized donors/cm^3, thin tunnel barriers and ohmic contacts can be achieved.

For Si structures, doping concentrations of $10^{20}/\text{cm}^3$ can be achieved with ion implantation. Ohmic contacts are made with either metal or silicide layers in contact with the heavily doped material. For GaAs structures, the formation of ohmic contacts is much more difficult. The barrier heights on n-type GaAs are high, typically 0.8 eV, and it is difficult to achieve electron concentrations greater than about $5 \times 10^{18}/\text{cm}^3$. Low-resistance ohmic contacts utilizing tunneling through the barrier are not feasible on material doped to $10^{19}/\text{cm}^3$ or below.

In order to achieve ohmic contacts to GaAs, the general approach is to form or grow a semiconductor layer on top of the GaAs, where it is easy to make low-resistance contacts to the layers. For example, Woodall et al. (1981) formed ohmic contacts on an epitaxial layer of $Ga_{1-x}In_xAs$ graded in composition from essentially InAs at the surface to GaAs at the interface. It is also possible to use the dissolution and subsequent regrowth of GaAs in contact with metal layers to form a heavily doped layer containing Ge or Sn (Lau and Wang, 1989). Thermally stable and low-resistance ohmic contacts to n-type GaAs have also been made by using In-based alloys such as NiInW (Murakami and Price, 1987).

References

1. J. C. Bean, ''Silicon-Based Semiconductor Heterostructures,'' Chapter 11 in *Silicon Molecular Beam Epitaxy*, edited by E. Kasper and J. C. Bean, CRC Press, Boca Raton FL (1988).
2. H. C. Casey, Jr. and M. B. Panish, *Heterostructure Lasers* (Part A: ''Fundamental Principles'' and Part B: ''Materials and Operating Characteristics'', Academic Press, Orlando FL (1978).
3. N. G. Einspruch and W. R. Wisseman, *GaAs Microelectronics*, Vol. II in *VLSI Electronics for Microstructure Science*, edited by N. G. Einspruch, Academic Press, Orlando FL (1985).
4. D. A. Fraser, *The Physics of Semiconductor Devices*, Oxford University Press, Oxford (1977).
5. H. Kroemer, ''Heterojunction Bipolar Transistors and Integrated Circuits,'' *Proc. IEEE* 70, 13 (1982).
6. J. W. Mayer and S. S. Lau, *Electronic Materials Science for Integrated Circuits in Si and GaAs*, Macmillan, New York (1990).
7. E. H. C. Parker editor, *The Technology and Physics of Molecular Beam Epitaxy*, Plenum Press, New York (1985).

8. N. J. Shah and S. S. Pei, *AT&T Technical Journal* 68 (No. 1), 19–28(1989).

9. S. M. Sze, *Physics of Semiconductor Devices*, Wiley, New York (1969).

10. S. Wang, *Fundamentals of Semiconductor Theory and Device Physics*,'' Prentice–Hall, Englewood Cliffs NJ (1989).

11. C. A. Wert and R. M. Thomson, *Physics of Solids*, McGraw–Hill, New York (1964).

12. P. K. L. Yu and P. C. Chen, ''GaAs Optoelectronic Device Technology,'' in *Introduction to GaAs Technology*, edited by C. Wang, Wiley, New York (1988).

Specific References

1. M. O. Aboelfatoh and K. N. Tu, ''Schottky Barrier Heights of Ti and $TiSi_2$ on n-type and p-type Si,'' *Phys. Rev. B* 34, 2311–2318 (1986).

2. M. Murakami and W. H. Price, ''Thermally stable, low resistance NiInW ohmic contacts to n-type GaAs,'' *Appl. Phys. Lett.* 51, 664 (1987).

3. I. Ohdomari, T. S. Kuan, and K. N. Tu, ''Microstructure and Schottky Barrier Height of Iridium Silicide Formed on Silicon,'' *J. Appl. Phys*, 50, 7020 (1979).

4. I. Ohdomari and K. N. Tu, ''Parallel Silicide Contacts,'' *J. Appl. Phys.* 51, 3735, 205 (1980).

5. T. Okumura and K. N. Tu, ''Analysis of Parallel Schottky Contacts by Differential Internal Photoemission Spectroscopy,'' *J. Appl. Phys*, 59, 922 (1983).

6. G. Ottaviani, K. N. Tu, and J. W. Mayer, ''Interfacial Reaction and Schottky Barrier in Metal–Silicon Systems,'' *Phys. Rev. Lett.* 44, 284 (1980).

7. R. D. Thompson and K. N. Tu, ''Schottky Barrier of Nonuniform Contacts to n-type and p-type Silicon,'' *J. Appl. Phys.* 53, 4285 (1982).

8. K. N. Tu, R. D. Thompson, and B. Y. Tsaur, ''Low Schottky Barrier of Rare Earth Silicides on n-Si,'' *Appl. Phys. Lett*, 38, 626 (1981).

9. R. T. Tung, *Phys. Rev. Lett.* 52, 461 (1984).

10. L. C. Wang and S. S. Lau, ''Low Resistance Nonspiking Ohmic Contact for AlGaAs/GaAs High Electron Mobility Transistors Using the Ge/Pd scheme,'' *Appl. Phys. Lett.* 54, 2678 (1989).

11. M. Wittmer, P. Oelhafen, and K. N. Tu, ''Schottky Barrier Formation at Ir/Si Interfaces,'' *Phys. Rev. B* 35, 9073–9084 (1987).

12. J. M. Woodall, J. L. Freeouf, G. D. Pettit, T. N. Jackson, and P. Kirchner, *J. Vac Sci. Technol* 19, 626 (1981).

Problems

9.1 The effective densities of states in the conduction band N_C and valence band N_V for GaAs are $4.7 \times 10^{17}/\text{cm}^3$ and $7.0 \times 10^{18}/\text{cm}^3$, respectively at 300 K. What are the ratios of effective to free electron masses, m^*/m_0, for electrons and holes?

9.2 For effective masses m^* equal to 1.0 and 0.1 of the free electron mass, calculate the densities of states in the conduction band N_C and the value of $E_C - E_F$ for $n = 10^{16}$ electrons/cm^3 at 300 K (use Maxwell–Boltzman statistics).

9.3 For a semiconductor with an effective density of states in the conduction band of $10^{19}/\text{cm}^3$ at 300 K, what would be the position of the Fermi level relative to E_C if the concentrations of electrons were 10^{15} and $10^{17}/\text{cm}^3$? What would be the values at 400 K?

9.4 The density of states is proportional to $E^{1/2}$ in a three-dimensional sysem. What is the energy dependence in two-dimensional and one-dimensional systems?

9.5 **(a)** An electron is confined in a one-dimensional potential well where $L = 2 \times 10^{-10}$ m. Calculate the energies E_1 for $n = 1$ and E_2 for $n = 2$ in eV.
(b) For free electrons in a cube of dimension $L = 10^{-2}$ m where $n_x^2 = n_y^2 = n_z^2$, calculate the value of n_x for electrons with energy $= 1.0$ eV.

9.6 You place a Schottky barrier with area of 10^{-4} cm^2 on n-type Si where the barrier height $\varphi_B = 0.7$ volts and there are 10^{16} ionized donors/cm^3 in the n-type region at 300 K where $N_C = 2.8 \times 10^{19}/\text{cm}^3$. Calculate
(a) The potential $e\varphi_S$ in the n-type region.
(b) The width of the depletion region and the maximum field strength at the interface for a 5-volt reverse bias.
(c) Compare the number/cm^2 of ionized donors in the depletion layer with the number of atoms in a monolayer ($6.8 \times 10^{14}/\text{cm}^3$).
(d) What is the value of the capacitance at 5-volt reverse bias?

9.7 For a Schottky barrier on n-type Si where $\varphi_S = 0.5$ volts and $V_n = 0.2$ volts, calculate the saturation current density J_{ST} at 300 K and at 400 K for
(a) The free electron mass.
(b) An effective electron mass ratio $m^*/m_0 = 0.1$.

9.8 For a Schottky barrier on n-type GaAs at 300 K where $\varphi_B = 0.9$ eV and there are 10^{17} electrons/cm^3, what is the value of φ_S, A^*, and the saturation current density using an electron effective mass ratio $m^*/m_0 = 0.1$?

9.9 In a two-silicide system making contact with n-type Si ($N_D = 10^{16}/\text{cm}^3$), the high-barrier ($\varphi = 0.8$ eV) silicide covers $0.9S$. The low-barrier silicide has $\varphi = 0.4$ eV with $S_l = 0.1S$ at 300 K. Calculate
(a) The apparent barrier height φ_{ap}.
(b) The ratio of currents for $S_l = 0$ to $S_l = 0.1$.
(c) The ratio of capacitances for $S_l = 0$ to $S_l = 0.1$.

CHAPTER 10

Solid Phase Amorphization, Crystallization and Epitaxy

10.0 Introduction

In Chapter 5, we discussed thin film growth modes and emphasized the effects of surface and interface energies on those modes. In Chapter 7, the epitaxial growth of a semiconductor on another semiconductor and the growth of superlattices were treated without considering the effect of surface energy. We could do this because we dealt with semiconductors having roughly the same surface energy (as in the case of homoepitaxy). However, the influence of surface energies cannot be ignored in growing a superlattice of dissimilar materials—e.g., Si and $CoSi_2$. The latter is a metallic compound, and has a cubic CaF_2 crystal structure and a lattice parameter with a very good match to Si.

To grow a superlattice of dissimilar materials, we encounter the difficulty of asymmetrical growth. The problem is depicted in Fig. 10.1 by a trilayer of Si/silicide/Si. If the silicide is assumed to have a lower surface energy than that of Si, the growth of the silicide on Si obeys the Frank–Van der Merwe mode of layered growth. Then, in the growth of Si on the silicide, the Si should follow the Volmer–Weber mode of island growth and ball up. The larger the difference in their surface energies, the greater the wetting angle between them. Another way to state the asymmetrical growth problem is that when material A wets material B, B will not wet A. For example, gasoline spreads out over the surface of water, but water does not spread out on gasoline; it forms drops. Therefore, it is intrinsically difficult to grow a superlattice of two materials of very different surface energies.

In selecting a pair of materials for superlattice or three-dimensional structure growth, surface energies as well as the interfacial energy between the materials must be considered. It is easy to check the interfacial misfit since the lattice parameters are generally known, and even the thermal strain can be calculated by knowing thermal expansion coefficients. Yet surface energies are generally not available.

In this chapter, an indirect method of measuring surface energy by the kinetics of nucleation is described. The crystallization of amorphous-$CoSi_2$ and amorphous-Si thin films are taken as examples. They serve as the link between treatment of epitaxial growth and consideration of other kinetic processes in thin films. These other processes will be the subject of the remainder of this book.

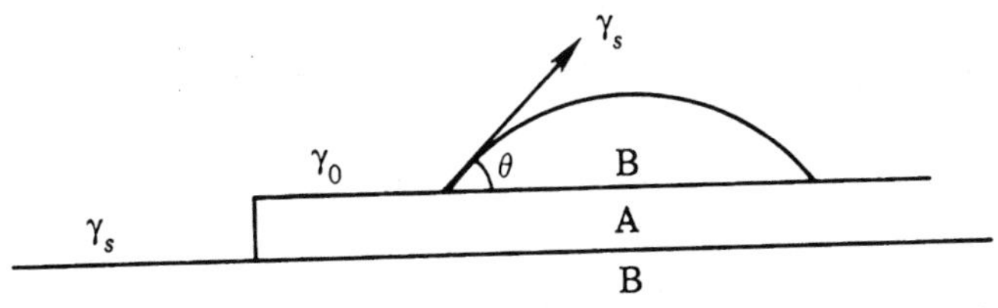

FIGURE 10.1 Asymmetrical growth morphology where A wets B, and therefore B will not wet A but balls up on A.

Since amorphous thin films are metastable, we first address the subject of metastable states and consider the formation of metastable phases by thin film reactions. It is intriguing that we can achieve solid phase amorphization (e.g., to form amorphous alloys) by slow thin film reactions. The basic nature of these reactions will be discussed. Then we shall cover amorphous-to-crystalline transformations and the measurement of surface energy by nucleation. Finally we shall present solid phase epitaxy as a combined process of crystallization and epitaxial growth. Solid phase epitaxy is another mode of crystallization that occurs when atoms reorder on the underlying single-crystal substrate.

10.1 The Metastable State

The thermodynamic reference of a metastable state is the corresponding stable state. Metastable states occur because of kinetic limitations. A stable state has the lowest free energy at a given set of thermodynamic variables, and is defined by equilibrium phase diagrams. In Fig. 10.2a, a schematic pressure–temperature phase diagram of a pure substance (such as H_2O) is shown. The equilibrium or stable state is defined when the two variables (p and T) are given. While the phase boundaries indicate the reversible transition between the phases, it is possible to cross these phase boundaries without a transition, to obtain metastable phases. Examples are an undercooled liquid without solidification or a supersaturated gas without condensation, as shown by the arrows A and B, respectively, in Fig. 10.2a. In formation of these metastable phases, there is no configurational change and at the same time the nucleation of the corresponding stable phase is suppressed. The barrier to nucleation can be better explained by Fig. 10.2b, where an isotherm in the pressure–volume phase diagram is shown by the solid line. The horizontal portion of the isotherm depicts the coexistence of the liquid and gas phases in equilibrium. The isotherm can be extended into the metastable regions, as shown by the two short broken lines. The pressure of the supersaturated gas is greater than the equilibrium pressure (i.e., $p/p_0 > 1$). As discussed in Section 5.8, nucleation requires $p/p_0 > 1$. Nucleation, however, constitutes a barrier to stable phase formation and enables the metastable phase to exist. For example, a slightly superheated liquid can exist without gas bubbles being formed inside the liquid because the nucleation of a gas bubble has to overcome a nucleation barrier.

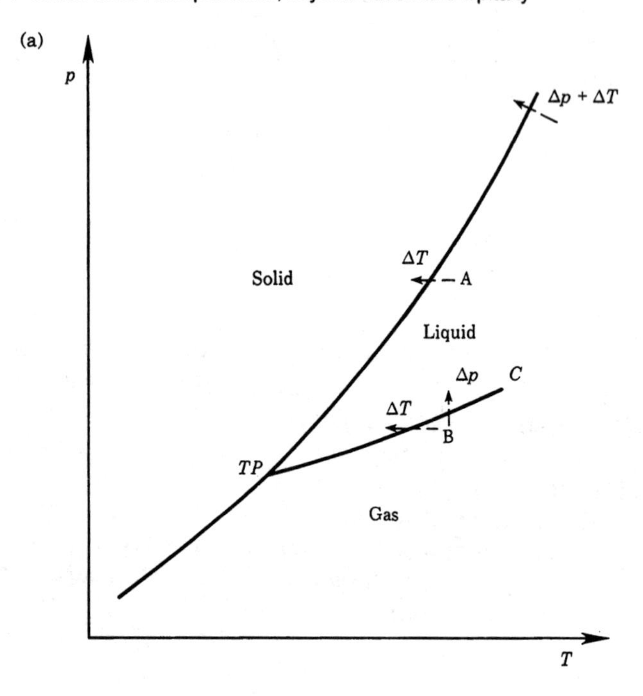

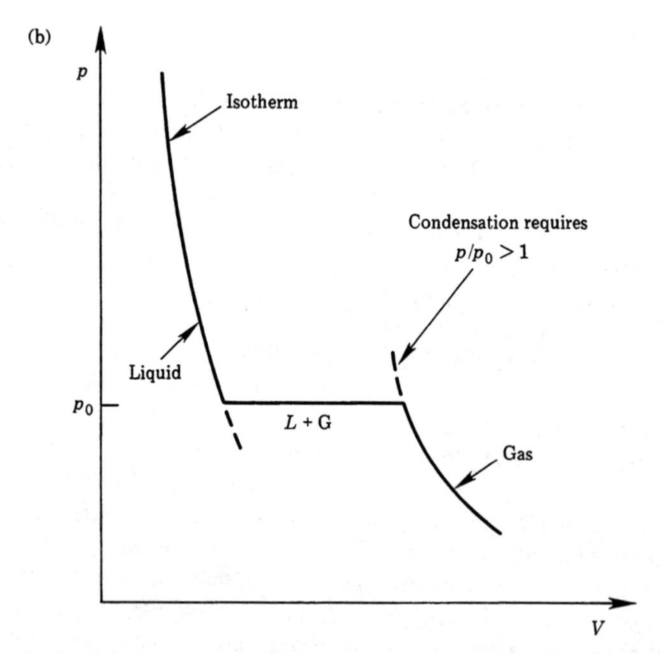

FIGURE 10.2 (a) A schematic equilibrium pressure–temperature phase diagram. The arrows indicate kinetic processes which can lead to metastable phase formation. (b) A schematic equilibrium pressure–volume phase diagram. The broken curves indicate the metastable states.

Metastable phases can also be formed by cooling under a high pressure—e.g., to form diamonds. In such a case, nucleation and growth do occur and there is a configurational change. But only nucleation and growth of the metastable diamond phase occur, not nucleation and growth of the stable graphite phase.

The above examples show that the processes which lead to metastable phase formation all involve a change in temperature and/or pressure. The change can be very fast. Explosion and quenching are processes of rapid change in pressure and temperature which produce metastable phases. In contrast, metastable phases can be formed in thin film reactions at constant temperature and constant pressure. For example, a bilayer thin film of Ni and Zr reacts at 250°C and at one atmospheric pressure of He ambient to grow amorphous Ni–Zr alloys. The question here is, if temperature and pressure are kept constant, what do we vary to produce the metastable amorphous phase?

For a thermodynamic process between equilibrium states having temperature and pressure as the variables, we showed in Chapter 3 that the energy change is described by Gibbs function. If there is more than one component, we have

$$dG = -S\,dT + V\,dp + \sum_i \mu_i\,dN_i \tag{3.10}$$

where μ is the chemical potential. For a binary system in reaction at constant p and T, we have

$$dG = (\mu_2 - \mu_1)\,dN_1$$

using $(N_1 + N_2)$ as a constant. Hence, the thermodynamic variable here is N, the number of particles per unit volume, or the composition changes during the reaction. The reaction must always occur in the direction towards equilibrium (i.e., to lower its free energy). We have

$$\frac{dG}{dN_1} = 0 \tag{10.1}$$

which means that the reaction seeks a minimum in the free energy versus composition curve, namely the curve of $G = G(C_1)$ where $C_1 = N_1/(N_1 + N_2)$. In a typical binary system which forms intermetallic compounds and metastable phases, there can be several minima in the $G(C_1)$ curve. The question is, which one will emerge from the reaction? To answer this fundamental question, we examine Fig. 10.3.

A schematic free energy versus composition diagram for a binary system of A and B is shown in Fig. 10.3, where the free energy curves of an amorphous alloy, α-AB, and an intermetallic compound, $A_\gamma B$, are shown. The reaction between A and B can form either one. The composition of the product phase is defined by the tangents extending from a point of the free energy curve (not necessarily the minimum) to the chemical potential of the element A or B at the vertical axis on either side of the diagram. It is not possible to go from the elements to the amorphous

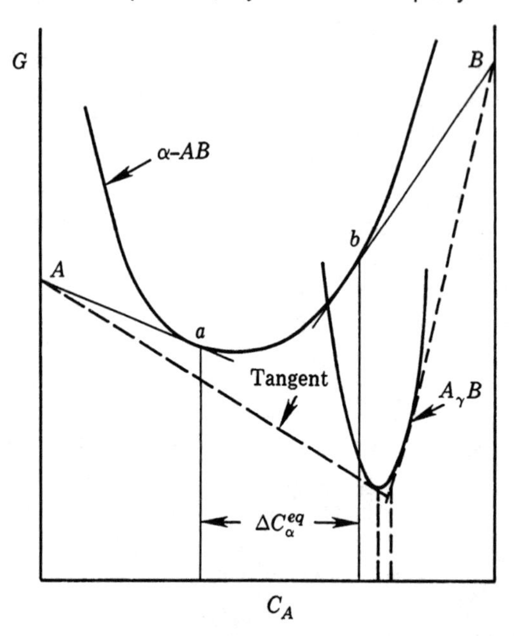

FIGURE 10.3 Gibbs free energy G curves as a function of composition C in a binary system of A and B forming an amorphous alloy of α-AB and an intermetallic compound of $A_\gamma B$. The broken tangent line which connects the compound phase to the terminal phase A lies below the G-curve of the amorphous phase.

alloy or to the intermetallic compound by a continuous change in composition. That process requires the free energy to decrease all the time, but part of the curve goes up in energy as the composition changes. This is also true for the transition between the amorphous alloy and the compound. Production of the alloy or compound can only occur by a jump from one composition to another, which means that it has to occur by nucleation. Because the nucleation barrier can be different for different phases, there is no assurance that the stable phase or the phase which provides the largest free energy change is the one which has the lowest nucleation barrier. In such a situation, if the amorphous alloy can nucleate preferentially and grow rapidly, we have solid phase amorphization. In Table 10.1 we list the kinetic processes which form metastable phases. These phases can be formed by fast processes such as quenching from the liquid state or by slow processes such as solid-state interdiffusion. In the next section, we discuss formation of amorphous allows by solid-state reactions in thin film systems.

10.2 Solid Phase Amorphization

Amorphous thin solid films are metastable phases which can transform irreversibly into stable crystalline phases. We define an amorphous solid as one without long range order. The nearest-neighbor distance may be preserved as in amorophous Si

TABLE 10.1 Kinetic Processes to Form Metastable Phases

Variable	Slow Process	Fast Process
Temperature	Slow cooling to form supercooled liquid	Quenching, Flash heating
Pressure	Slow build-up of the pressure in a supersaturated gas	Explosion, mechanical mixing
Concentration	Slow reaction by solid-state amorphization, oxidation of Si to form amorphous SiO_2	Ion implantation

where the first-neighbor covalent bond lengths are virtually unchanged (see Fig. 10.4). However, the bond angles are changed so that the second-nearest-neighbor position correlation is lost. Amorphous thin films can be made by ultrafast processes which require a rapid change in temperature, such as quenching of a liquid alloy or vapor deposition of a binary thin film on a cold substrate. Amorphous layers can also be made by ion implantation which can be viewed as a rapid change in composition. Accompanying an ultrafast process in which there is a rapid change in temperature, pressure, or composition (or in a combination of them), there is a jump (or a drop) in energy. In other words, there is a high rate of energy change. In this section, we will consider the formation of amorphous alloys by a "slow" reaction of bilayer or multilayer thin films. We shall also consider the rate of energy change in the slow reaction.

Amorphous layers can be formed in the reactions of crystalline films of two materials A and B in a number of systems such as Ni–Zr, Ce–Co, Rh–Si and Ti–Si. To show the amorphization, we present in Fig. 10.5a and 10.5b a pair of bright-field transmission electron micrographs of an amorphous Rh–Si alloy layer between a polycrystalline Rh thin film and (100)-oriented single-crystal Si. Figure 10.5a shows the cross-section image of the sample annealed at 200°C, and Fig. 10.5b shows the image of the same area after a consecutive anneal at 260°C. It is evident from

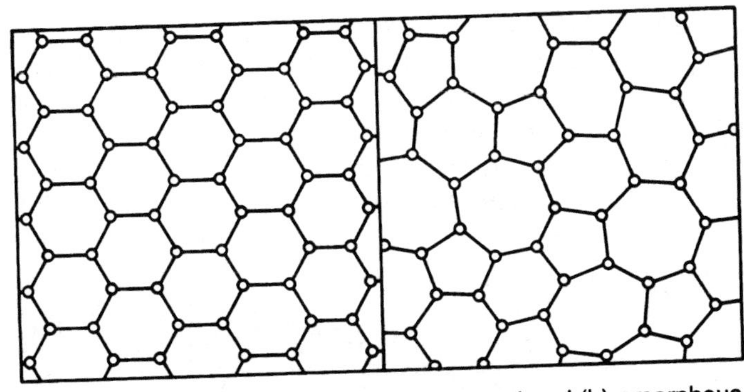

FIGURE 10.4 Schematic diagram of (a) single-crystal and (b) amorphous two-dimensional network.

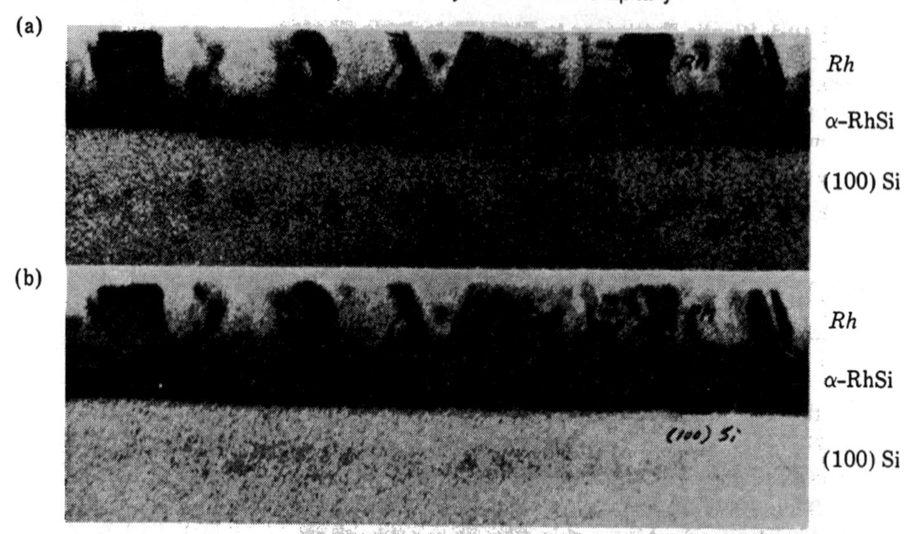

FIGURE 10.5 Bright-field transmission electron microscopic images of a cross-sectional area of a polycrystalline Rh thin film on (100)-oriented single-crystal Si (a) after annealed at 200°C, and (b) after a consecutive anneal at 260°C. The top layer is Rh, the bottom layer is Si, and the middle layer is the amorphous Rh–Si alloy. (Courtesy of S. R. Herd, IBM Research Division)

comparing the images of the same grains in the Rh thin film that the amorphous alloy layer has grown thicker. No Rh–Si intermetallic compound was observed in the sample during these anneals.

To explain this reaction between crystalline films to form an amorphous layer the first condition is that the free energy change ΔG of the reactions is negative. This is shown in Fig. 10.6 where we consider the reaction paths and energy changes of a bilayer thin film of A and B reacting to form a metastable phase of α-AB against the formation of an equilibrium intermetallic compound $A_\gamma B$. The broken line represents the reaction path leading to the metastable phase with a kinetic barrier of ΔH_1 (per atom) and an energy change of ΔG_1 (per atom). The solid line represents the reaction path leading to the stable (equilibrium) phase with a kinetic barrier of ΔH_0 and an energy change of ΔG_0. The magnitude of ΔG_0 is greater than ΔG_1 by definition, and ΔH_0 is greater than ΔH_1 by assumption.

To analyze the reaction, we take ΔH_0 and ΔH_1 to be the activation energy barriers of nucleating the critical nuclei of the stable phase and of the amorphous phase at the interface, respectively. The ratio of their nucleation numbers following classical nucleation theory (see Turnbull, 1956) is

$$\frac{N_0}{N_1} = \exp\left[-\frac{(\Delta H_0 - \Delta H_1)}{kT}\right] \tag{10.2}$$

where $\Delta H_0 = A_0(\gamma_0)^3/(\Delta G_0)^2$ and $\Delta H_1 = A_1(\gamma_1)^3/(\Delta G_1)^2$, and A_0 and A_2 are geometrical shape factors and γ_0 and γ_1 are the average surface energy per atom of

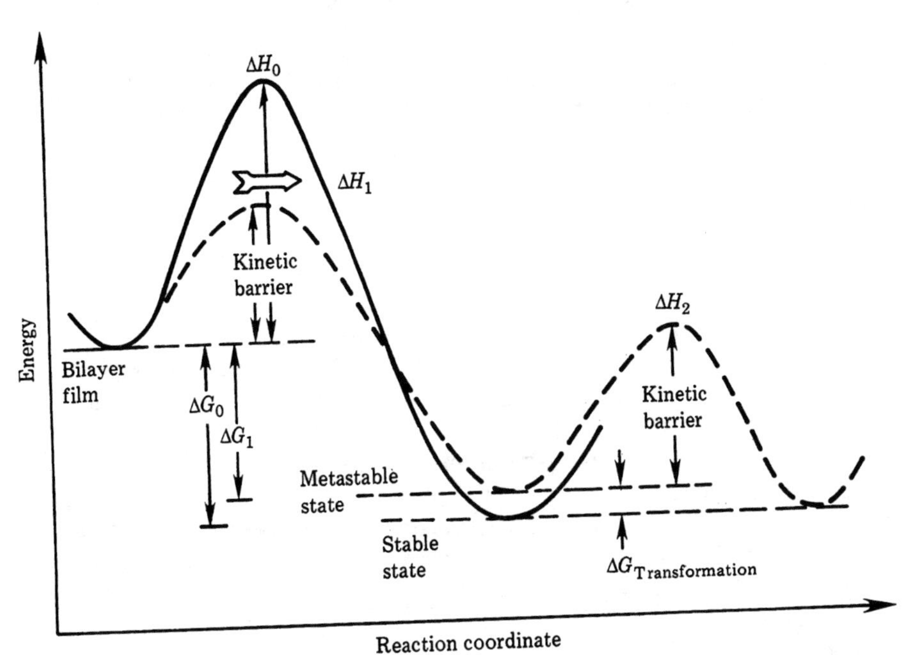

FIGURE 10.6 A schematic diagram of two kinetic paths of the reaction of a bilayer thin film A/B leading respectively to the formation of the stable ($A_\gamma B$) and the metastable (α-AB) phases. The solid curve leads directly to the stable phase. The broken curve goes first to the metastable phase before reaching the stable phase.

the critical nuclei of the stable and the amorphous phases, respectively. The ratio can be written as

$$\frac{N_0}{N_1} = \exp - \left\{ \frac{A_0 \gamma_0^3}{\Delta G_0^2 kT} \left[1 - \frac{A_1}{A_0} \left(\frac{\gamma_1}{\gamma_0} \right)^3 \left(\frac{\Delta G_0}{\Delta G_1} \right)^2 \right] \right\} \tag{10.3}$$

Since the formation energies ΔG_0 and ΔG_1 are actually the driving forces of the reactions, the second condition of metastable phase formation is that the magnitude of ΔG_1 is close to that of ΔG_0, so that kinetics rather than driving force dominates the reaction. Since we have assumed $\Delta H_0 > \Delta H_1$, it follows that $\gamma_0 > \gamma_1$, and we have $N_1 > N_0$ from Eq. (10.3) when the ratio of the shape factors is ignored. While the above consideration explains the greater nucleation rate of the metastable phase, we still need to know how to prevent the reaction product from going to the equilibrium state. Consider the broken curve in Fig. 10.6, extended to show a barrier of ΔH_2. We assume that this kinetic barrier (ΔH_2) which prevents a further transformation from occurring must be as high as or preferably higher than ΔH_1. Otherwise the reaction will proceed to the equilibrium state or to another metastable state.

To present this picture in a quantitative way, we shall follow the reaction of a bilayer of Ni/Zr film to form the amorphous NiZr alloy and later to crystallize the alloy. The heat of formation of the amorphous alloy by solid-state reaction (i.e., ΔG_1) has been measured to be about 0.4 eV/atom. The heat of crystallization of amorphous NiZr alloys (i.e., $\Delta G_0 - \Delta G_1$), is about 0.05 eV/atom, over a wide alloy-composition range. The large difference between 0.4 and 0.05 supports the assumption that the heat of formation (the driving force of reaction) of the metastable phase is comparable to that of the stable phase. When the heat of transformation (or heat of crystallization) from the metastable to the stable state is small as compared to their heat of formation, the effect of the driving force is less important and the reaction is controlled by kinetics. So, the reaction favors the metastable state if its activation barrier of reaction is lower.

The activation energy of growth of the amorphous alloy by solid-state reaction was measured to be about 1.0 to 1.4 eV/atom, and the initial growth rate of the alloy layer at 300°C is about 1 nm/min. The activation energy of crystallization of amorphous NiZr alloys (i.e., ΔH_2), over a wide alloy-composition range, was found to vary from 2.5 to 5 eV/atom. These values suggest that the amorphous alloy is confined in the metastable state after formation. Also, this is because the heat of crystallization is small. On the other hand, the assumption that ΔH_0 is greater than ΔH_1 can be justified by the fact that the temperature necessary to form a crystalline compound is higher than that to form the corresponding amorphous alloy. To cite an example from another thin-film reaction, in this case the reaction of Ti with Si, the formation of crystalline TiSi occurs above 550°C and the formation of amorphous TiSi occurs around 450°C.

The amorphous phase nucleates because its barrier is lower than that of the crystalline phase. This has been proposed as a key condition for the occurrence of solid phase amorphization. According to Eq. (10.3), if the difference between ΔG_0 and ΔG_1 is about 10% as in the Ni–Zr case, a 10% decrease in surface energy will make the nucleation of the amorphous phase overwhelming. Clearly, the assurance of metastability of the amorphous phase is by assuming no nucleation of the crystalline phases, but the crystalline phases must have a finite chance to nucleate, otherwise our consideration is incomplete. Indeed, the incubation time of the crystalline nucleus has been defined as part of the metastable period of the amorphous phase. In addition, the effect of growth of the amorphous phase on the nucleation of the crystalline phase has to be considered.

In the inset of Fig. 10.7, we sketch between the two terminal phases A and B an amorphous layer of α-AB which has grown to a thickness of X and a compound $A_\gamma B$ of thickness W. We define W to be the initial thickness for the compound to form a continuous layer to prevent the amorphous layer from further growth; the grains of the compound have nucleated and have grown to join each other. In Fig. 10.7, the tangent which connects the compound and the terminal phase A lies below the G-curve of the amorphous phase; the amorphous phase is unstable relative to them, so it will disappear. We define the period of metastability of the amorphous phase to be the period before the crystalline phase has reached the thickness W.

In the competing growth between two phases, the change of free energy is rate dependent. Time-dependent (kinetics) rather than time-independent (end states) free energy changes are considered. The free energy change in a given time t_0 is

$$-\Delta G = \int_0^{t_0} - \left(\frac{d\Delta G}{dt}\right) dt \tag{10.4}$$

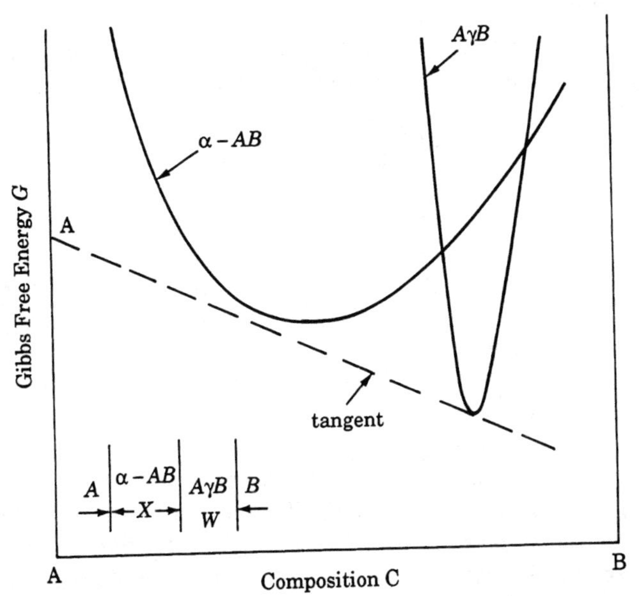

FIGURE 10.7 Gibbs free-energy *G* curves as a function of composition *C* in a binary system of *A* and *B* forming an amorphous alloy of α-*AB* and an intermetallic compound of $A_\gamma B$. The broken tangent line which connects the compound phase to the terminal phase *A* lies below the *G* curve of the amorphous phase.

and the rate of energy change is

$$-\frac{d\Delta G}{dt} = -\left(\frac{d\Delta G}{dx}\right)\left(\frac{dx}{dt}\right) = Fv \tag{10.5}$$

where $F = -d\Delta G/dx$ is the driving force of the reaction or chemical affinity, and $v = dx/dt$ is its growth or reaction rate. The concept of rapid change in energy (e.g., a high rate of energy change in quenching or implantation) has been used by Turnbull (1981) to describe the formation of metastable structures. As long as the metastable phase can nucleate and can grow fast, it will lead to a large free energy change and remain metastable. When the driving forces for forming the crystalline phase and the amorphous phase are comparable, the latter can quickly lead to a large free energy change in the initial period if it can nucleate and grow fast. So after nucleation the fast growth of the amorphous phase is favored. Nevertheless, the growth will not go far. As the amorphous phase gets thicker, it becomes diffusion-controlled and must slow down. Then, the characteristics of fast growth disappear.

The effect of the initial rapid growth is that the fast-moving front will not allow enough time for the crystalline phase to nucleate. When the growth rate v slows down to the point where the time required for the amorphous layer to advance a distance $W(t = W/v)$ is comparable to the time needed for forming the continuous crystalline layer, the crystalline phase begins to dominate the reaction kinetics. The time the crystalline layer needs is the sum of its incubation time to nucleate plus the grain growth time for the grains to join each other. The formation of the stable phase is delayed due to the rapid growth of the metastable phase.

To formulate this picture, we consider the situation at the amorphous-crystalline interface when the amorphous phase reaches a critical thickness. We have

$$-\left(\frac{d\Delta G}{dt}\right)_{\alpha} = -\left(\frac{d\Delta G}{dt}\right)_{c} \tag{10.6}$$

where the left-hand and the right-hand terms represent the amorphous and the crystalline sides of the interface, respectively. This defines a ''critical thickness'' of the amorphous phase beyond which it is unstable. The critical thickness is a measure of the metastability of the amorphous phase. Therefore, an amorphous phase cannot grow to arbitrary thickness by interdiffusion. If multilayered thin films are used where the layer thickness is less than the critical thickness, a rather thick amorphous layer can be obtained.

The metastability of phases produced by slow amorphization at constant temperature and pressure is a kinetic issue in thin film reactions. We assume that the free energy change of the reaction is negative. When the driving forces of forming the amorphous phase and its competing crystalline phase are comparable, kinetics dominates the reaction. The amorphous phase formed is stable if no nucleation of the crystalline phases can occur. If the nucleation rate of the crystalline phases is finite, the growth of the amorphous phase begins with a high rate of free energy change so that its fast-moving front prevents the nucleation of the crystalline phase. Its metastability prevails until the growth finally slows down to the point where crystallization occurs.

10.3 Solid Phase Crystallization and Avrami's Equation

The phase transformation from an amorphous state to a crystalline state is an important subject when we use alloyed thin films and semiconducting films. Upon codeposition or sputtering of alloy or semiconductor films at room temperature, the films are often amorphous (e.g., a Si film deposited onto a room temperature substrate is amorphous). The metastability of an amorphous film depends on its behavior of crystallization.

In crystallization, the product can have a different composition from the amorphous phase by phase separation into a crystalline and an amorphous phase, or into two crystalline phases; in either case a certain amount of long-range diffusion is involved. On the other hand, a simpler case is where the crystalline phase has the same composition as the amorphous phase; no long-range diffusion is required in the process. For this reason, it is often preferable to process an amorphous deposited compound film by crystallization, because it takes less time than producing the same compound film by interdiffusion of a multilayer film with the same overall composition. For example, the refractory metal silicide films WSi_2 used as gates in metal–oxide–silicon (MOS) devices have been processed by crystallizing amorphous silicide films rather than by interdiffusing metal and silicon. Another example is the codeposited amorphous film of $CoSi_2$ which crystallizes without a change in composition. Figure 10.8 shows the morphology of partially crystallized $CoSi_2$ circular grains from an amorphous $CoSi_2$ thin film, observed by means of transmission electron microscopy (TEM).

In considering the kinetics of crystallization, we shall limit ourselves to the processes of nucleation and growth of crystalline grains which have the same composition as the amorphous film.

Figure 10.9 shows a series of bright-field transmission electron microscopic images of nucleation and growth of circular grains of $CoSi_2$ in the amorphous film. In the initial state, there is a small number of grains nucleated randomly. They are random in space; however, they are not random in time, and are assumed to have a steady-state nucleation rate. The size and the number of grains increase with time. Impingement of grains occurs and a grain boundary is formed between grains. The process continues and leads to the formation of a polycrystalline microstructure. If we can control the nucleation and growth processes for the evolving microstructure of a thin film, we can in turn control the extrinsic film properties. Therefore we wish to know how a microstructure develops and how to modify it by tailoring nucleation and growth.

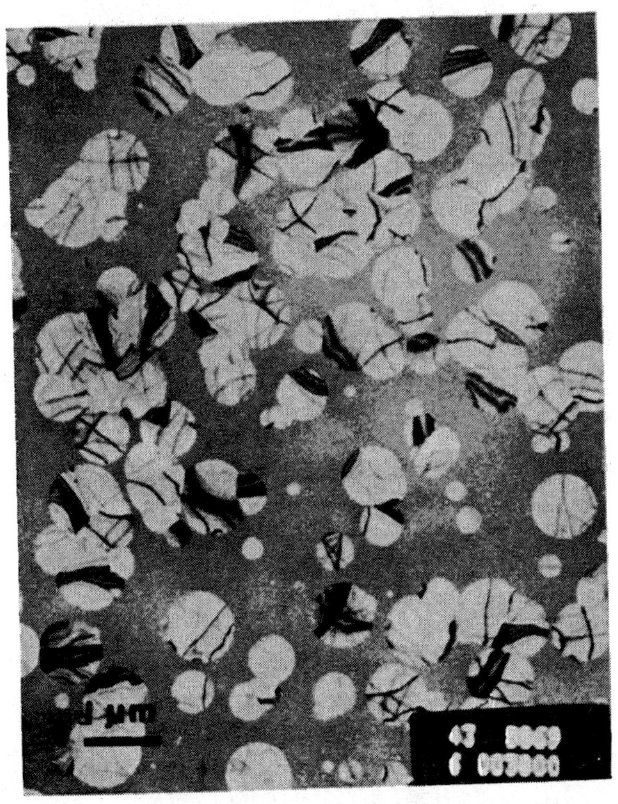

FIGURE 10.8 A bright-field transmission electron micrography image of a partially transformed amorphous $CoSi_2$ film at 150°C. The largest circular grains are about 1 micron in diameter (taken from Tu et al., 1987).

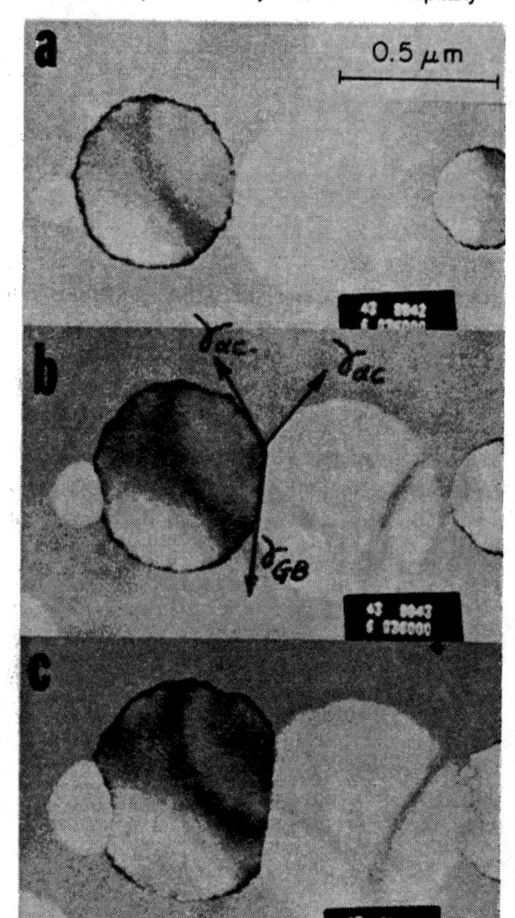

FIGURE 10.9 A sequential growth of two circular grains and the formation of a grain boundary between them (taken from Tu et al., 1988).

To describe the crystallization process as nucleation and growth, we employ the classical analysis of phase transformation by Johnson and Mehl, (1939), and Avrami (1941). They present the following equation,

$$X_T = 1 - \exp(-X_{\text{ext}}) \tag{10.7}$$

to describe the overall behavior, where X_T is defined as the fraction of volume transformed and X_{ext} as the fraction of the extended volume. To understand the equation, we consider V to be the total volume involved in the transformation, in which V_U is the untransformed volume and V_T is the transformed volume,

$$V = V_U + V_T \tag{10.8}$$

or

$$1 = X_U + X_T \tag{10.9}$$

where $X_U = V_U/V$ and $X_T = V_T/V$ are the fraction of volume untransformed and transformed, respectively. If we assume that the transformation occurs by nucleation and growth and that the nucleation is random in time and in space, and if we take R_N to be the nucleation rate (i.e., the number of nuclei per unit time per unit volume), we then have, in the interval between t and $t + d\tau$, a number of new transformed regions nucleated in V_U,

$$N = R_N V_U \, d\tau \tag{10.10}$$

If we further assume that each of the transformed regions has the same isotropic growth rate of R_G, the volume of a transformed region which originated at τ is

$$\begin{aligned} V_\tau &= \frac{4\pi}{3} R_G^3 (t - \tau)^3 \qquad &\text{for } t > \tau \\ V_\tau &= 0 &\text{for } t \leq \tau \end{aligned} \tag{10.11}$$

Hence, in the period $d\tau$, the differential change of the untransformed volume is

$$-dV_U = NV_\tau = V_\tau R_N V_U \, d\tau \tag{10.12}$$

The negative sign is due to the reduction of the untransformed volume. From Eq. (10.8), we have $-dV_U = dV_T$, so for the change of the transformed volume, we have

$$dV_T = V_\tau R_N V_U \, d\tau \tag{10.13}$$

By integration,

$$V_T = \frac{4\pi}{3} \int_{\tau=0}^{\tau=t} V_U R_N R_G^3 (t - \tau)^3 \, d\tau \tag{10.14}$$

In the initial stage of the transformation, we can assume that $V_T << V_U$. Therefore we take $V_U = V$ and if we assume that the nucleation rate is constant in time, we have from the last equation,

$$X_T = \frac{V_T}{V} = \frac{\pi}{3} R_N R_G^3 t^4 \tag{10.15}$$

The expression of Eq. (10.15) is correct for the initial stage of transformation where the transformed regions do not interfere with each other (i.e., no overlapping and no impingement occurs). But the expression becomes incorrect for later stages of transformation when overlapping and impingement do occur as shown in Fig. 10.9. To treat the interference, we introduce the concept of the "extended volume" V_{ext} in the period between t and $t + d\tau$ such that

$$dV_{ext} = V_\tau R_N (V_U + V_T) \, d\tau \tag{10.16}$$

where dV_{ext} is the differential change in volume in transformed and untransformed regions. This equation is similar to Eq. (10.13), except that Eq. (10.16) includes the random nucleation and isotropic growth in both the transformed and the untransformed regions. The nuclei in the transformed region are the "phantom nuclei" of

Avrami since their growth does not affect the untransformed region. In other words, we do not gain any more transformation by their growth.

To see the origin and physical meaning of the "phantom nuclei," we shall take a two-dimensional example. We imagine rain falling onto a pond as shown in Fig. 10.10. As raindrops hit the surface of the pond, they produce circular ripples which spread out. When the raindrops are few and far apart from each other, the ripples can grow for a while without interference. When it rains harder, interference or impingement of ripples occurs. Moreover, some of the raindrops will fall inside the existing ripples and produce smaller ripples within larger ones. We can regard these later raindrops as "phantom nuclei" since they originate changes within the transformed region but are ineffective in transforming the untransformed region. Only the transformation in the untransformed region is of interest and importance.

To consider random nucleation in time and in space we must, as in the case of raindrops, include the phantom nuclei. From Eq. (10.8), we have

$$\left(1 - \frac{V_T}{V}\right)(V_U + V_T) = V_U \tag{10.17}$$

By substituting this into Eq. (10.13), we obtain

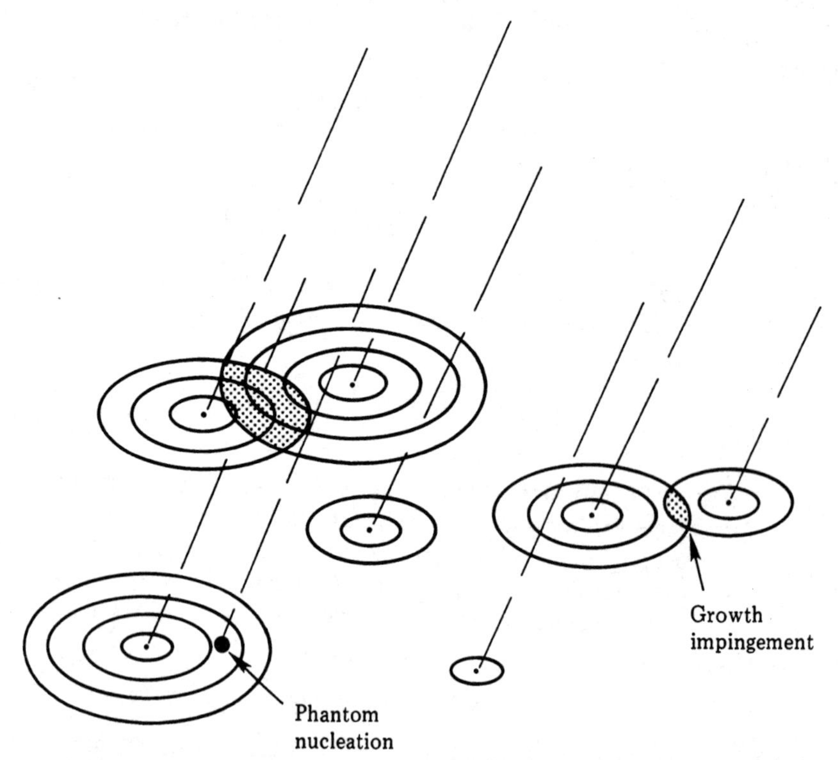

FIGURE 10.10 A schematic picture depicting the falling of raindrops on the surface of a pond. The overlapping of ripples (growth impingement) and a later raindrop fallings within a ripple (phantom nucleation) are indicated.

$$dV_T = \left(1 - \frac{V_T}{V}\right) V_\tau R_N (V_U + V_T)\, d\tau \tag{10.18}$$
$$= \left(1 - \frac{V_T}{V}\right) dV_{ext}$$

The last substitution is by using Eq. (10.16) to introduce dV_{ext} in Eq. 10.18. The fraction that affects the untransformed region is $(1 - V_T/V)$. Clearly, this is the fraction of interest and Eq. (10.18) is the essence of the Johnson, Mehl, and Avrami analyses of phase transformation. By rearrangement of Eq. (10.18), we obtain

$$dV_{ext} = \frac{-V\, d\left(1 - \frac{V_T}{V}\right)}{\left(1 - \frac{V_T}{V}\right)} \tag{10.19}$$

and by integration we have

$$V_{ext} = -V \ln\left(1 - \frac{V_T}{V}\right) + \text{constant} \tag{10.20}$$

The constant is zero since $V_{ext} = 0$ when $V_T = 0$. Now we obtain Avrami's equation,

$$X_T = 1 - \exp(-X_{ext}) \tag{10.7}$$

where the fractions $X_T = V_T/V$ and $X_{ext} = V_{ext}/V$ have been defined in Eq. (10.9). Specifically, the latter is given by

$$X_{ext} = \int_{\tau=0}^{\tau=t} \frac{4\pi}{3} R_N R_G^3 (t - \tau)^3\, d\tau \tag{10.21}$$

where R_N is the nucleation rate and R_G is the growth rate, and t is the time of transformation. Equation (10.21) shows that knowledge of the nucleation and growth rates determines X_{ext}, and in turn X_T from Eq. (10.7). Experimentally, X_T could be measured directly in a thin film by imaging techniques, or indirectly by changes in resistivity, x-ray diffraction intensity, or heat produced during the transformation.

If the nucleation is random and continuous at a constant rate, and the growth is isotropic and linear with time, we can express the integration in Eq. (10.21) as

$$X_{ext} = \frac{\pi}{3} R_N R_G^3 t^4 \tag{10.22}$$

and

$$X_T = 1 - \exp(-Kt^4) \tag{10.23}$$

where

$$K = \frac{\pi}{3} R_N R_G^3 = \frac{\pi}{3} R_{N_0} R_{G_0}^3 \exp\left(-\frac{\Delta H_N + 3\Delta H_G}{kT}\right) \tag{10.24}$$

and both R_N and R_G have been assumed to follow Boltzmann's distribution, so that

$$R_N = R_{N_0} \exp\left(\frac{-\Delta H_N}{kT}\right) \tag{10.25}$$

and

$$R_G = R_{G_0} \exp\left(\frac{-\Delta H_G}{kT}\right) \tag{10.26}$$

and ΔH_N and ΔH_G are the activation enthalpies of nucleation and of growth, respectively. Experimentally X_T is measured as a function of time and temperature, and a constant value is chosen (e.g., $X_T = 0.5$). Then we have from Eq. (10.23),

$$Kt^4 = \text{constant} \tag{10.27}$$

and by taking the logarithm of Eq. (10.27),

$$-\frac{\Delta H_N + 3\Delta H_G}{kT} + 4 \ln t = \text{constant} \tag{10.28a}$$

This shows that if we plot $\ln t$ versus $1/kT$, the slope gives us the activation enthalpy of transformation

$$\Delta H_T = \frac{(\Delta H_N + 3\Delta H_G)}{4} \tag{10.28b}$$

It means that if we can measure ΔH_N or ΔH_G separately, we know the other.

In the above analysis we have assumed a t^4 dependence in nucleation and growth. In general the time dependence is not necessarily to the power of 4. It may depend not only on the dimension of the transformation (whether it is one-dimensional, two-dimensional, or three-dimensional) but also on the mode of nucleation and growth. For example, the nucleation can be instantaneous, meaning that there is no time dependence and that all the nuclei are present in the very beginning of the transformation, yet it can also have a constant rate or a varying rate which depends on time. The growth can be isotropic or anisotropic, and it can be linear with time or it can vary with the square root of time when it is diffusion-limited. To include all these possibilities, we rewrite Eq. (10.23) as another form of Avrami's equation,

$$X_T = 1 - \exp(-Kt^n) \tag{10.29}$$

or

$$X_T = 1 - \exp\left[-\left(\frac{t}{\tau_0}\right)^n\right] \tag{10.30}$$

where n is defined as the mode parameter of transformation and $K = (\tau_0)^{-n}$. At $t = \tau_0$, we have $X_T = 1 - 1/e = 0.63$. By twice taking the logarithm of Eq. (10.29), we have

$$\ln[-\ln(1 - X_T)] = n \ln t + \text{constant} \tag{10.31}$$

Thus, the slope of a plot of $\ln[-\ln(1 - X_T)]$ versus $\ln t$ gives the value of n.

For example, in the case of a solid-state precipitation, if the nucleation is instantaneous and homogeneous and the growth is diffusion-controlled and isotropic in three-dimensions, the mode parameter n is $3/2$. In textbooks on phase transformation, a table of n and the corresponding modes of transformation can be found. The measurement of n alone does not determine the mode of transformation uniquely since this is a combination of nucleation and growth. To determine the mode, we must study the morphology of the transformation and measure the processes of nucleation and growth separately.

10.4 Measurement of Crystallization of Amorphous Thin Films

To apply Eq. (10.23) to a phase change, we measure X_T as a function of time and temperature. Its spatial distribution is determined by the morphology. We can measure X_T directly or indirectly. For example, in a bulk sample, we can use sectioning techniques to measure the transformed regions directly, yet it is tedious. Indirect methods such as integrated intensity of x-ray diffraction, resistivity change, and differential scanning calorimetry have been used and they all rely on assuming that the measured quantity is proportional to the volume of phase change. This assumption must be checked carefully.

We shall consider the thin film example of the crystallization of amorphous $CoSi_2$ to crystalline $CoSi_2$ of the CaF_2 structure. Transmission electron microscopy was used to measure directly the nucleation rate in the untransformed region and the growth rate of each region, which permits the determination of X_{ext} if the nucleation is random and if its rate and the growth rate are constant. Image analysis of the micrographs determines X_T directly, assuming a two-dimensional growth. This assumption is reasonable provided that the width of the transformed region is much greater than film thickness. The resistivity change was used to measure X_T indirectly as a function of time and temperature. The data can be compared to those from image analysis and used to determine the activation enthalpy of transformation and the mode parameter.

Figure 10.8 shows the bright-field transmission electron micrograph image of a partially transformed amorphous $CoSi_2$ film. The circular images are those of crystalline $CoSi_2$. The distribution of these images is random and the bend contour within the images shows that each is a single crystal and that there is no preferred orientation. The nucleations of these grains are independent events and they occur randomly in time and space. Figure 10.9 shows the sequential growth of a set of these circular grains and the formation of a grain boundary between impinging neighbors. The growth is isotropic in a macroscopic sense, for the shape remains circular during growth. By measuring the diameters of the grains as a function of time, the growth rate can be determined and is found to be constant (see Fig. 10.11). Since we can count the number of the circular grains in a given area as a function of time and temperature, we can measure the nucleation rate.

Since the nucleation and growth behavior of $CoSi_2$ satisfies the condition demanded by the Johnson, Mehl, and Avrami mode of phase transformation, we can

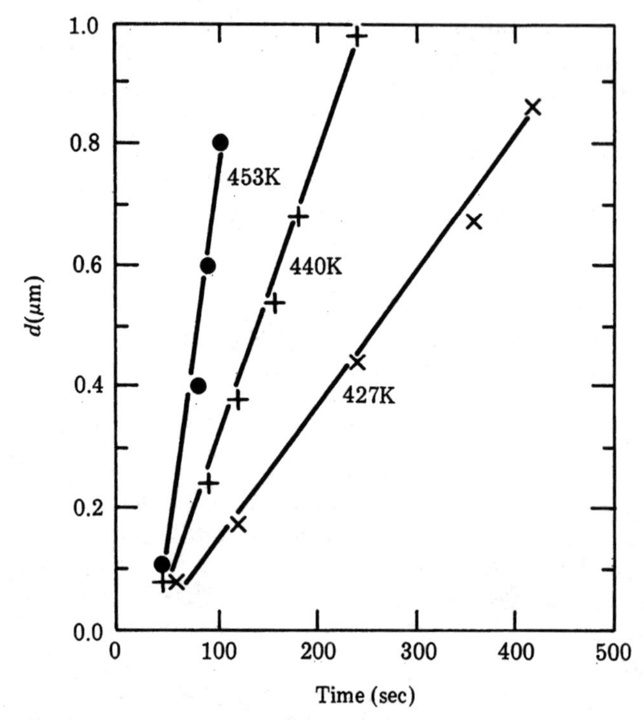

FIGURE 10.11 The measured diameters of crystalline $CoSi_2$ grains annealed at 427,444, and 453 K as a function of time.

apply Eq. (10.23) to the amorphous-to-crystalline transformation of $CoSi_2$. Figure 10.12 shows a plot of two resistivity changes of an amorphous $CoSi_2$ film in van der Pauw's geometry deposited on an oxidated Si substrate and ramp-annealed at 0.5°C/min and 3°C/min in situ in a He furnace from room temperature to 500°C. The resistivity changes show that the amorphous-to-crystalline transformation occurs around 150°C. Figure 10.13 shows the change of X_T at four isothermal annealings at temperatures around 150°C. The changes are expressed in terms of resistivities where ρ_α, ρ_c, and ρ_t are the resistivity of the amorphous and the crystalline $CoSi_2$ and the resistivity of the film at time t, respectively. The plot assumes a linear relation of

$$X_T = \frac{\rho_t - \rho_\alpha}{\rho_c - \rho_\alpha} \tag{10.32}$$

The linear relation has been found to exist in the range of 0 to 0.4 of X_T, by a comparison of results of resistivity changes with values of X_T measured directly by image analysis. In Fig. 10.13, we draw a horizontal line at $X_T = 0.4$ to intercept the four annealing curves so that the corresponding annealing times for them can be decided from the horizontal axis. The mode parameter n is found to be 4 in the initial stage of the reaction, while it decreases to 3 in the later stage of the reaction as shown in Fig. 10.14. The amorphous $CoSi_2$ thin films used for electron microscopy and for resistivity measurements were deposited together in the same run, so that

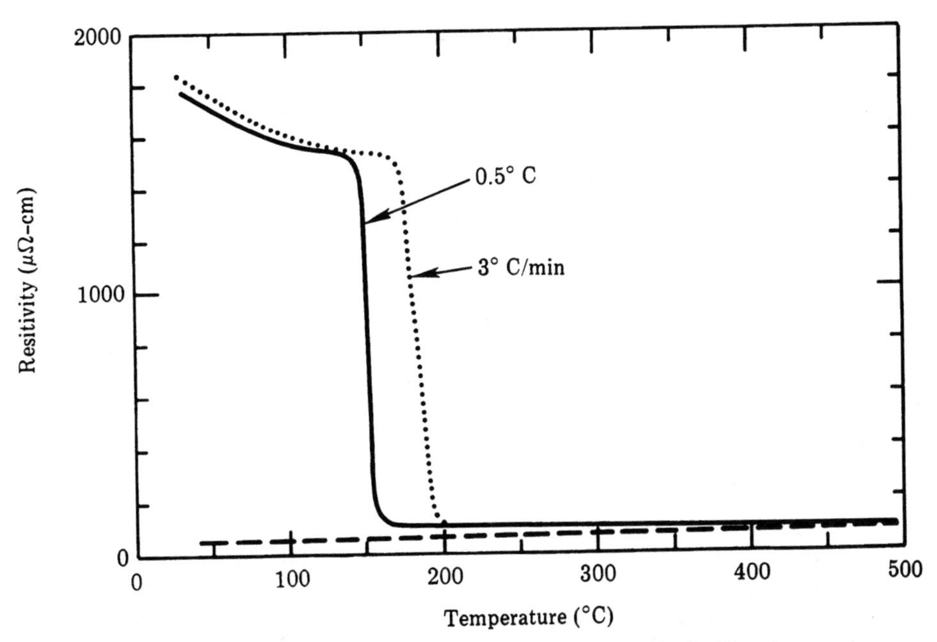

FIGURE 10.12 A plot of resistivity changes of amorphous $CoSi_2$ films in van der Pauw's geometry deposited on oxidated Si wafer and ramp-annealed from room temperature to 500°C at 0.5°C/min and 3°C/min in situ in a He furnance (taken from Cros et al., 1987).

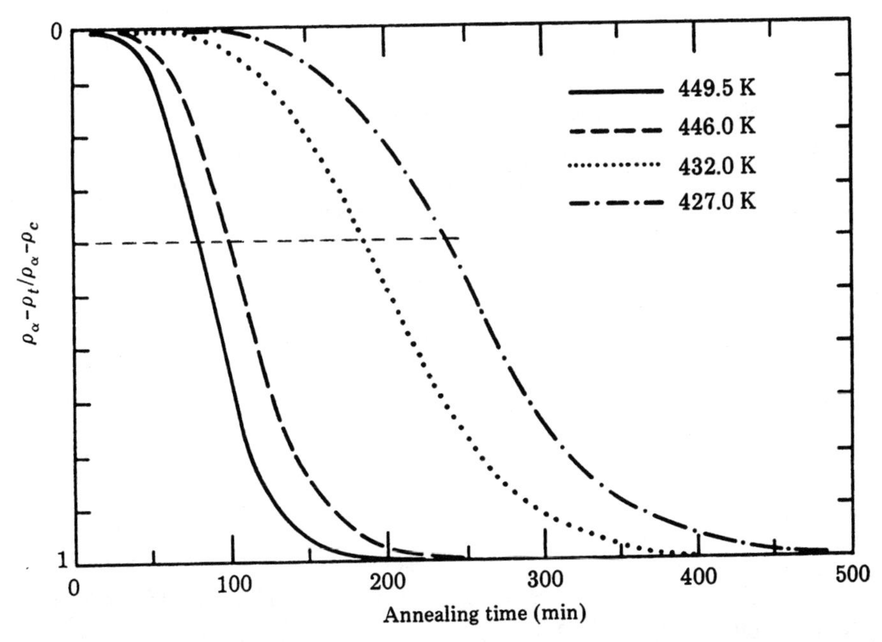

FIGURE 10.13 The fractional volume change X_T expressed in terms of ρ_α, ρ_C, and ρ_t; they are the resistivity of the amorphous and the crystalline $CoSi_2$ films and the resistivity of the partially transformed film at time *t*. The horizontal line is drawn at X_T = 0.4 (taken from Cros et al., 1987).

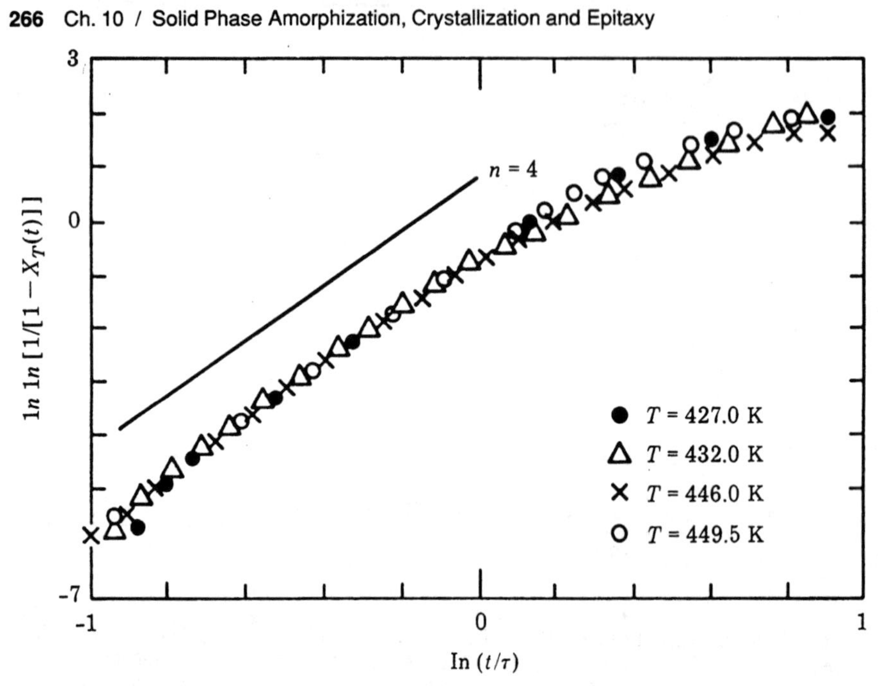

FIGURE 10.14 Plot of $\ln[-\ln(1 - X_T)]$ versus ln t for determining the mode parameter in crystallization of amorphous $CoSi_2$ thin films.

results can be compared directly. On the basis of these data and the use of Eq. (10.28), the activation enthalpy of the amorphous-to-crystalline transformation of $CoSi_2$ is found to be

$$\Delta H_T = \frac{\Delta H_N + 3\Delta H_G}{4} = 0.9 \text{ eV/atom} \tag{10.33}$$

As Fig. 10.8 shows, we can count the number of grains and also the increase of the number of grains in a given area, especially in the untransformed area, as a function of time and temperature. We can measure the nucleation rate and determine the activation enthalpy of nucleation. Similarly we can measure the growth rate (as shown in Fig. 10.11) and hence the activation enthalpy of growth. Therefore in this case we can measure ΔH_T, ΔH_N, and ΔH_G independently and we can use Eq. (10.33) to check them. However, here we deduce ΔH_N by measuring ΔH_T and ΔH_G independently. This is because in Section 10.6, where we shall compare the crystallization behavior of amorphous $CoSi_2$ to that of amorphous Si, we note that it is not yet possible to measure ΔH_N for Si independently. So for a direct comparison between $CoSi_2$ and Si, we choose to measure ΔH_T and ΔH_G separately and deduce ΔH_N from them.

For the growth of crystalline $CoSi_2$, the value of ΔH_G has been reported to be 1.09 eV/atom and it then gives $\Delta H_N = 0.3$ eV/atom. The heat of crystallization of amorphous $CoSi_2$ has been measured to be $\Delta H_{AC} = 0.05$ eV/atom. In Table 10.2, we summarize these measured values.

TABLE 10.2 A Comparison of Kinetic Data of $CoSi_2$ and Si

Properties	$CoSi_2$	Si
Overall reaction, ΔH_T	0.9 eV/atom	3.4 eV/atom
Growth, ΔH_G	1.09 eV/atom	2.7 eV/atom
Nucleation, ΔH_N	0.3 eV/atom	5.5 eV/atom
Heat of crystallization, ΔH_{AC}	0.05 eV/atom	0.1 eV/atom
Interfacial Energy, γ_{AC}	0.11 eV/atom (or 180 erg/cm^2)	0.3 eV/atom (or 480 erg/cm^2)
Critical Nucleus	about 12 atoms	about 110 atoms

10.5 Calculation of Critical Nucleus and Interfacial Energy from Nucleation

Figure 5.17 shows schematic curves depicting the nucleation behavior in which an activation energy is required to form the critical nucleus. To form a nucleus of N atoms, the energy change ΔH_N is

$$\Delta H_N = -N\Delta H_{AC} + bN^{2/3}\gamma_{AC} \tag{10.34}$$

where ΔH_{AC} and γ_{AC} are heat (per atom) of amorphous-to-crystalline transformation and interfacial energy (per atom) between the crystalline and the amorphous phases, respectively, and b is a geometrical constant depending on the shape of the nucleus. For forming the critical nucleus,

$$\frac{\partial \Delta H_N}{\partial N} = 0$$

We obtain the activation enthalpy of nucleation

$$\Delta H_N = \frac{4b^3}{27}\frac{\gamma_{AC}^3}{\Delta H_{AC}^2} \tag{10.35}$$

and the number of atoms in the critical nucleus (see Problem 10.9)

$$N_{\text{crit}} = \frac{8b^3}{27}\left(\frac{\gamma_{AC}}{\Delta H_{AC}}\right)^3 = \frac{2\Delta H_N}{\Delta H_{AC}} \tag{10.36}$$

We note that N_{crit} does not depend explicitly on b. By rearranging Eq. (10.35), we have

$$b\gamma_{AC} = \left(\frac{27\Delta H_N \Delta H_{AC}^2}{4}\right)^{1/3} \tag{10.37}$$

Both Eq. (10.36) and Eq. (10.37) show that we determine N_{crit} and $b\gamma_{AC}$ by knowing ΔH_N and ΔH_{AC}. Taking their values from Table 10.2 for $CoSi_2$, we obtain

$$N_{\text{crit}} = \frac{2 \times 0.3}{0.05} = 12 \text{ atoms}$$

$$b\gamma_{AC} = \left(\frac{27 \times 0.3 \times 0.05^2}{4}\right)^{1/3} = 0.17 \text{ eV/atom}$$

The number of atoms in the critical nucleus of $CoSi_2$ is equal to the number of atoms in a unit cell of $CoSi_2$. To calculate the interfacial energy γ_{AC}, we need to determine the geometrical factor b. We shall consider two extreme cases. First, in the limit of N approaching unity, b also approaches unity. Secondly, when N is large, if we assume that the nucleus has a cubic shape with six square surfaces, $b = 6$. Then if we minimize the surface area by assuming a spherical shape, $b = 4.8$. Therefore in general b is greater than unity, but it is only slightly greater than unity for a critical nucleus which contains a small number of atoms. This is because b is limited by the relation that

$$bN^{2/3} \leq N \tag{10.38}$$

This means that the number of atoms on the surface of the nucleus cannot be greater than the number of atoms in the nucleus. For the present case of $N_{crit} = 12$, we have $1 < b \leq 12^{1/3}$ ($= 2.3$). Using the relation $b\gamma_{AC} = 0.17$, we find that γ_{AC} varies from 0.17 to 0.08 eV/atom when b varies from 1 to 2.3. Taking the average value, we have $\gamma_{AC} = 0.12 \pm 0.05$ eV/atom.

We can make a refinement of the value of γ_{AC} as follows. In the crystallization of amorphous thin films, heterogeneous nucleation is assumed where the nucleus starts from the free surface of the amorphous thin films (Fig. 10.15). We also assume that the atoms forming the free surface of the nucleus have the same energy as those forming the free surface of the amorphous film, so that they make no contribution to the barrier of nucleation. The nucleation barrier comes only from the atoms forming the interface between the nucleus and the amorphous matrix. For this reason, we take $bN^{2/3} < N$ or $b < N^{1/3}$. Among the 12 atoms in the critical nucleus of $CoSi_2$, if we assume that only 7 to 9 of them are interfacial atoms, it means that b takes the values of 1.34 to 1.7 and $\gamma_{AC} = 0.11 \pm 0.01$ eV/atom.

In Table 10.2, we list the calculated values of N_{crit} and γ_{AC} as well as the data of crystallization of amorphous Si for comparison. Analysis of the kinetics of crystallization of amorphous Si is presented in the next section.

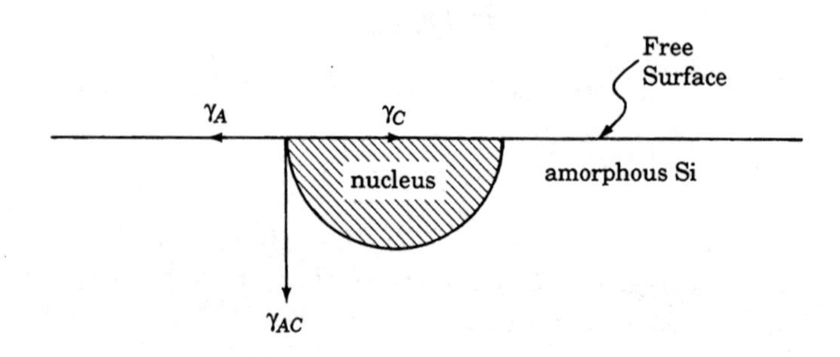

FIGURE 10.15 Schematic diagram of heterogeneous nucleation of a nucleus on the free surface of an amorphous film.

10.6 Solid Phase Epitaxy (Without a Medium)

Solid phase epitaxy is the conversion of a nonepitaxial film to an epitaxial film on a single-crystal substrate. The transformation occurs in the solid state, unlike liquid phase epitaxy or vapor phase epitaxial deposition. A typical example is the growth of an epitaxial Si film by crystallization on a Si single-crystal substrate. It can be done directly or with a transport medium. For example, we can use ion implantation to convert a surface layer of single-crystal Si into amorphous Si, then we can crystallize the amorphous Si by converting it back to epitaxial single-crystal Si, as shown in Fig. 10.16a. Or we can deposit amorphous Si on Pd_2Si on (100) Si and achieve epitaxial growth of Si on the substrate via the Pd_2Si layer, as shown in Fig. 10.16b. Solid phase epitaxy has received wide attention because it is both a routine processing step in integrated circuit technology and also a way to study the interface between amorphous Si and crystalline Si. We can use slow annealing in a furnace or rapid annealing by laser heating. A further example of solid phase epitaxy, important from the viewpoint of device application, is SOI (silicon on insulator), which can be achieved by combining solid phase epitaxy and lateral growth of Si over SiO_2 surfaces.

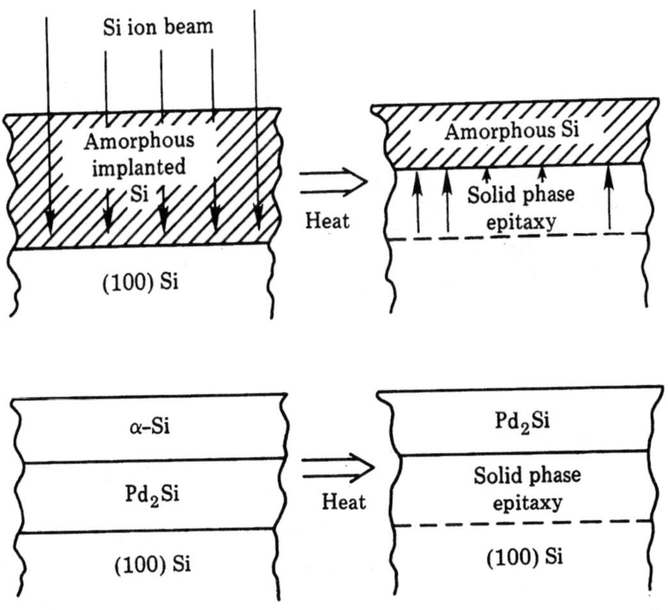

FIGURE 10.16 (a) The formation of an amorphous layer of Si in (100) single crystals by ion implantation of Si ions to a dose of about 10^{15} ions/cm^2 and the subsequent epitaxial recrystallization at temperature of 500°C with a growth rate of about 1 Å/sec. (b) Schematic diagram of solid phase epitaxial growth using a transport medium.

We first discuss the measurement of solid phase epitaxial growth of Si on Si by ion channeling and ion backscattering spectrometry. It is a very unique measurement of thin film homoepitaxial growth. In the experiment, an ion beam is directed along a major crystallographic direction of the single-crystal wafer, so that channeling occurs in the substrate but not in the amorphous Si film. The amorphous Si backscatters the ion beam more than the Si substrate; thus, the channeling effect differentiates the two regions of the same element because of their structural difference. Figure 10.17 shows a set of Rutherford backscattering spectrometry (RBS) data for the regrowth of amorphous Si on a (100) Si surface upon annealing at 550°C. The width of the backscattered signal from the amorphous Si can be used to calculate the thickness of the amorphous Si layer with an in-depth resolution better than 20 nm. By measuring the thickness changes as a function of time and temperature, the regrowth rate can be determined.

Another technique which has been used to study solid phase epitaxial growth is the time-resolved reflectivity (TRR) measurement. The index of refraction of amor-

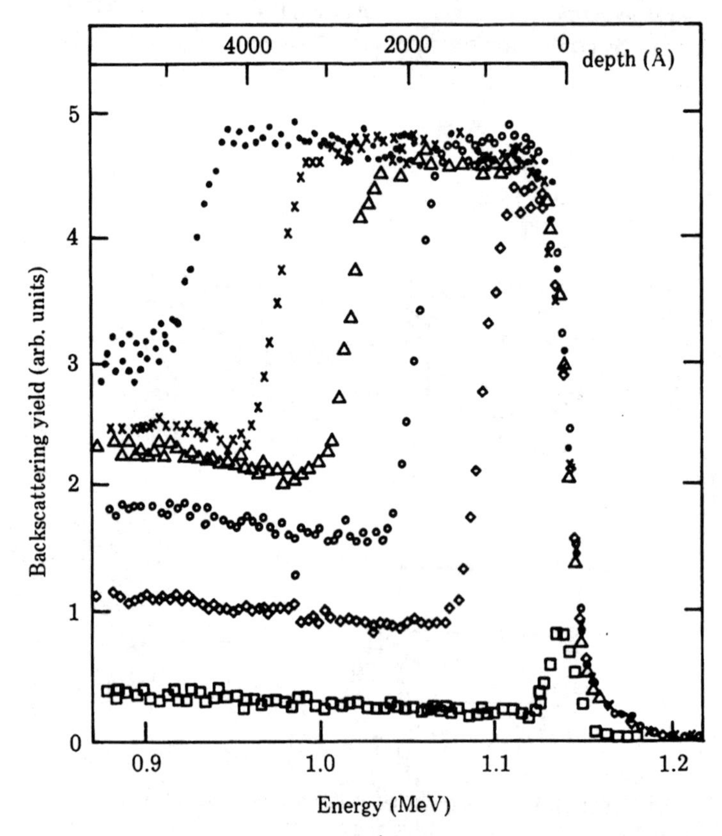

FIGURE 10.17 Aligned spectra for 2 MeV^{4}He ions incident on silicon samples implanted at LN_2 temperature ($\simeq$ 80K), preannealed at 400°C for 60 min and annealed at 550°C, showing ⟨100⟩ spectra on a sample with the surface normal 0.3° off the ⟨100⟩ axis. The depth scale is calculated assuming the bulk density of Si (taken from Lau and VanderWeg, 1978).

phous Si in the visible spectrum is substantially greater than that of crystalline Si. The index difference can produce a reflection from the amorphous-to-crystalline interface which interferes with the surface reflection. The interference allows the thickness of the amorphous film to be determined with an in-depth resolution of about 3 nm. Figure 10.18 shows the calculated reflectivity at λ = 632.8 nm (He–Ne laser beam) against film thickness for an amorphous Si film on (100) Si substrate at 650°C. The incremental thickness between successive interference maximum and minimum is 32.6 nm. As shown in Fig. 10.18, the technique is quite sensitive for amorphous Si film of thicknesses up to 300 nm. By measuring the time rate change of the reflectivity as a function of temperature, the regrowth velocity of amorphous Si on crystalline Si surfaces can be determined and expressed as

$$v = v_0 \exp\left(-\frac{\Delta H_G}{kT}\right)$$

For (100) Si, the data from RBS and TRR methods are shown in Fig. 10.19. The best fit to the data yields ΔH_G = 2.7 eV and $v_0 = 3 \times 10^{-8}$ cm/sec. A typical growth rate, say at 624°C, is about 2.3 nm/sec and the growth rate is constant over a thickness of 250 nm of amorphous Si.

Growth rates have been observed to depend on substrate orientation, as shown in Fig. 10.20. The atomistic picture of the interfacial process is of interest. Models have been proposed which can reasonably explain the orientation dependence. They assume crystallization occurs by rearrangement of bonds propagated along (111) planes. These planes are normal to the <111> direction and hence the growth in this direction is slow compared to that in the <100> direction.

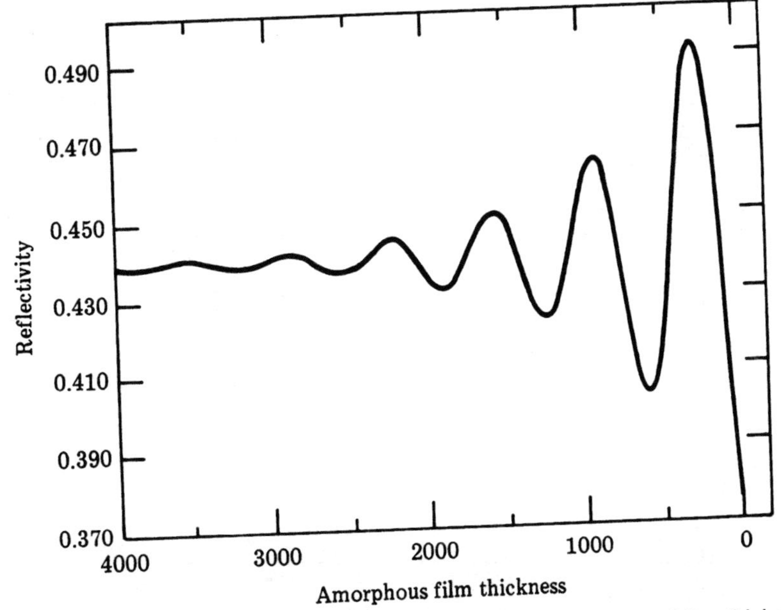

FIGURE 10.18 Calculated reflectivity at λ = 6328Å as a function of film thickness for α-Si on c-Si substrate at 650°C (taken from Olson and Roth, 1988).

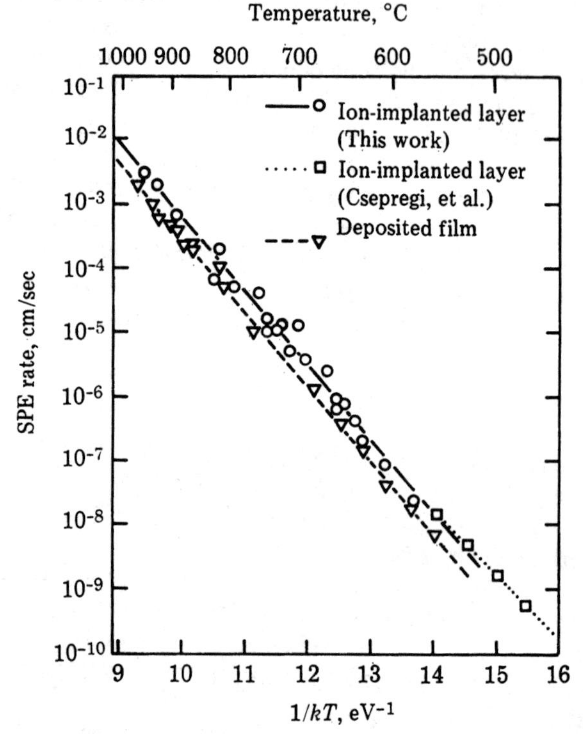

FIGURE 10.19 Temperature dependence of intrinsic solid phase growth velocity in Si^+-implanted and e-beam evaporated (deposited) α-Si. Low temperature implanted film data of Csepregi, Mayer, and Sigmon (1976) are also shown (taken from Olson and Roth, 1988).

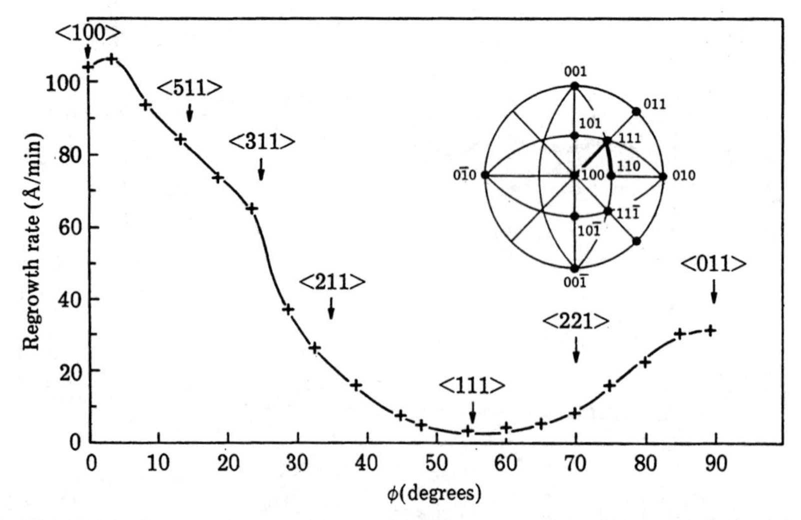

FIGURE 10.20 Regrowth rate of Si samples as a function of substrate orientation (taken from Lau and VanderWeg, 1978).

Knowing ΔH_G, we can separate out ΔH_N, according to Eq. (10.28), if ΔH_T (which is the overall activation enthalpy of crystallization) is determined. The technique of time-resolved reflectivity has been applied to measure the kinetics of crystallization of amorphous Si deposited on SiO_2, an inert substrate. In this case, random nucleation and growth replace solid phase epitaxial growth.

For the crystallization of 200 nm of amorphous Si on an oxidized Si substrate in ultra high vacuum (UHV), the activation enthalpy of transformation is $\Delta H_T = 3.4$ eV per atom (Olson and Roth, 1988). On the basis of three-dimensional growth, we obtain $\Delta H_N = 5.5$ eV by taking $\Delta H_G = 2.7$ eV. The value of ΔH_T measured in air is higher and is about 4.0 eV, indicating that the nucleation is sensitive to film surfaces and is heterogeneous. Using calorimetry measurement, the heat of crystallization of amorphous Si has been determined to be 0.1 eV/atom, and we shall take this value for ΔH_{AC}.

On the basis of Eq. (10.36) and Eq. (10.37), we calculate

$$N_{\text{crit}} = \frac{2 \times 5.5}{0.1} = 110 \text{ atoms}$$

$$b\gamma_{AC} = \left(\frac{27 \times 5.5 \times 0.1^2}{4}\right)^{1/3} = 0.72 \text{ eV/atom}$$

To evaluate γ_{AC}, we estimate the magnitude of b by following the reasoning given in the last section that $1 < b < 110^{1/3}$ ($= 4.8$). We obtain $0.72 > \gamma_{AC} > 0.15$ eV/atom.

To refine γ_{AC} by considering heterogeneous nucleation, we show in Fig. 10.15 a hemispherical nucleus on the film surface where we have assumed that the surface energy per atom of the amorphous film γ_A is the same as that of nucleus γ_C. The volume of the hemispheric nucleus is $N_{\text{crit}}\Omega = 110 \times 0.02 = 2.2$ nm^3, where $\Omega = 0.02$ nm^3 is the atomic volume of Si based on Eq. (1.16). The radius of the nucleus is about 1 nm. The number of atoms on the shell of the hemisphere is about 65, in which about 20 of them are on the rim of the shell. The latter can be regarded as atoms on both the free surface and the interface, hence their contribution to the nucleation barrier is not well defined. For simplicity we take 55 ± 10 to be the interfacial atoms. Then we have $b = 2.4 \pm 0.4$ and $\gamma_{AC} = 0.30 \pm 0.04$ eV/atom.

In Fig. 10.15, the wetting angle (i.e., the angle between γ_C and γ_{AC}) is 90° when $\gamma_A = \gamma_C$. If the actual wetting angle can be measured or if both γ_A and γ_C can be measured independently, the geometrical shape of the nucleus is given and we should have a better estimate of the number of interfacial atoms and of γ_{AC}.

For comparing the crystallization of amorphous Si to that of amorphous $CoSi_2$, we list the measured and calculated values of Si in Table 10.2 We see that the barrier of nucleation (i.e., the interface energy) for Si is much greater than that for $CoSi_2$. Also to nucleate a cluster of crystalline Si requires a larger number of atoms than to do the same with $CoSi_2$.

We must emphasize that the calculated values of γ_{AC} and N_{crit} depend on the accuracy of measured energy values and on the geometrical shape factor. The preceding analysis is reasonable for the purpose of comparison between $CoSi_2$ and Si, but the absolute values of γ_{AC} and N_{crit} must be taken with caution. The γ_{AC} of the critical nucleus may not be the same as that of a bulk material in view of the fact that the former is a very small cluster of atoms. Nevertheless the concept of a nucleation barrier is correct.

If ΔH_N is not much greater than ΔH_{AC} in Eq. (10.36), we have a small critical nucleus which contains only a few atoms. Since the number of possible arrangements of a few atoms is finite, we might be able to consider atomistic nucleation. An atomistic model of heterogeneous nucleation for the condensation of vapor on solid surfaces, such as Ag on NaCl and Au on MgO, has been developed by Rhodin and Walton (1963). In the model the low energy states of 1 to 4 atoms with an oriented configuration on the single crystal substrate surfaces have been investigated. When N_{crit} is small, it seems that we might be able to link the classical model with the atomistic model of nucleation. Hence, the accurate measurements of ΔH_N and ΔH_{AC} for the determination of N_{crit} are of interest.

The finding that γ_{AC} of Si is much higher than that of $CoSi_2$ has a very important implication for the growth of three-dimensional structures or superlattices of silicide and Si. This is because we can argue that the surface energy of Si is much higher than that of $CoSi_2$. On the basis of the analysis of wetting of a surface given in Chapter 2, we see that $CoSi_2$ can grow epitaxially on Si due to their extremely good match in lattice parameter, but Si tends to grow epitaxially on $CoSi_2$ with pinhole formation. Pinhole formation is due to the fact that Si will not wet the surface of $CoSi_2$. This is a key issue of solid phase epitaxial growth of heterophase structures. We must try to reduce the surface energy of Si or to increase that of $CoSi_2$ during the growth process.

We have measured the interface energy between the amorphous and the crystalline phases of $CoSi_2$ and between the same phases of Si, but we have not measured their surface energies. Nevertheless, knowing the interface energy, we are able to determine the grain boundary energy using the equilibrium condition—i.e., the sine law obeyed by the three surface tensions at the triple point, as shown in Fig. 10.9. In turn, by extending the grain boundary to the free surface of the sample, we can determine the energy of the surface that is at equilibrium with the grain boundary.

10.7 Solid Phase Epitaxy (With a Medium)

We can achieve solid phase epitaxy (SPE) with a medium for mass transport. For example, we can interpose a layer of Pd_2Si or Au between amorphous Si and crystalline Si to achieve SPE, as shown in Fig. 10.16b. The advantage of the medium is that it can reduce the growth temperature. For that reason, the medium chosen should contain an element which reacts easily with the growth materials (one of the noble or near-noble metals is best for Si). We recall that solid phase epitaxy without a medium is an interfacial reaction process; there is no long-range atomic diffusion. When a transport medium is introduced, the process becomes a combination of long-range diffusion and interfacial reaction. This becomes clear in the following. In the case c-Si/Au/α-Si, we can imagine that the Si atoms in the α-Si dissolve into Au and deposit on the c-Si to achieve solid phase epitaxial growth. In the case of c-Si/

Pd_2Si/α-Si, we can again imagine that the Pd_2Si serves as a transport medium for Si atoms to go from α-Si to c-Si. However, there is another possibility. We can imagine that Pd_2Si decomposes into Pd and Si at the interface with c-Si. The Si from the decomposition will grow epitaxially on the c-Si and the Pd diffuses across the remaining Pd_2Si to the interface with α-Si and reacts with α-Si to form new layers of Pd_2Si. In other words, the Pd_2Si decomposes at one side and regrows at the other side. Therefore, we can have two very different processes, with Si the diffusing species in one case and Pd in the other. Obviously, the question then is, which one of these two takes place in solid phase epitaxial growth? The difference, we note, is in diffusing species. In Chapter 12 we shall discuss in detail diffusion-controlled and interfacial-reaction-controlled processes. In Chapter 11 we shall apply marker analysis to determine the dominant diffusing species in interdiffusion. Here we study the kinetic process in solid phase epitaxy with a medium, and introduce the method of performing marker analysis.

In Fig. 10.21, we sketch a layered structure of $A(c)/A_\beta B/A(\alpha)$, where $A_\beta B$ serves as the transport medium of A atoms to go from A(α) to A(c). We assume that

1. J_A is the flux of direct transport of A atoms from the amorphous layer to the crystalline layer,
2. J_B is the flux of B atoms due to decomposition of $A_\beta B$ at the $A_\beta B/A(c)$ interface, and
3. J_A^1 is the release of A atoms due to the decomposition of $A_\beta B$ at the $A_\beta B/A(c)$ interface; these grow on A(c).

We note that $J_A^1 = -\beta J_B$ by definition. Within the compound $A_\beta B$,

$$J_A = -D_\beta^A \frac{\partial C_\beta^A}{\partial x} \tag{10.39}$$

$$J_B = -D_\beta^B \frac{\partial C_\beta^B}{\partial x} \tag{10.40}$$

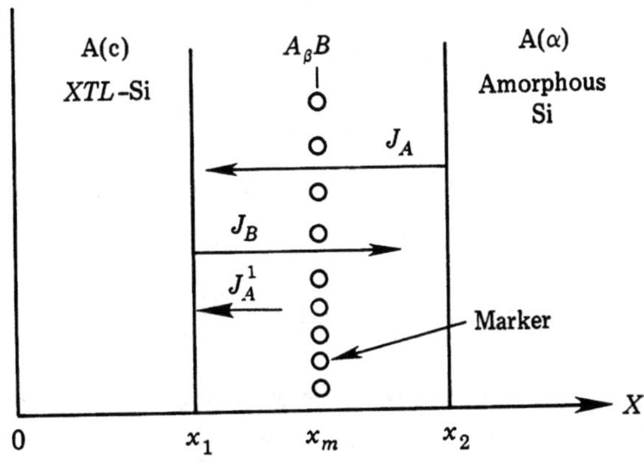

FIGURE 10.21 Diagram of solid phase expitaxial growth by transport of A(α) through a transport medium $A_\beta B$ to crystallize on the crystal surface of A(c).

At position x_1, the growth of the crystalline phase A(c) is

$$C_C^A \frac{dx_1}{dt} = -J_A - J_A^1 \tag{10.41}$$

where C_C^A is the concentration of A atoms in the crystalline phase A(c). If we let

$$\Omega_C^A = \frac{1}{C_C^A} \tag{10.42}$$

which is the volume per A atoms in A(c), we have

$$\begin{aligned} \frac{dx_1}{dt} &= -J_A\Omega_C^A - J_A^1\Omega_C^A \\ &= -J_A\Omega_C^A + \beta J_B\Omega_C^A \end{aligned} \tag{10.43}$$

At the position x_2, if we assume the layer thickness of $A_\beta B$ is invariant during SPE (the layer moves as a unity), then

$$\frac{dx_2}{dt} = \frac{dx_1}{dt} \tag{10.44}$$

Concerning the marker position x_m, we can regard its motion as the sum of dx_1/dt and the relative motion of the marker with respect to x_1,

$$\frac{dx_m}{dt} = \frac{dx_1}{dt} - J_B\Omega_\beta \tag{10.45}$$

Equation (10.45) is obtained from considering the fact that the marker motion with respect to x_1 is due to the flux of J_B only. For each B atom, a molecule of $A_\beta B$ is decomposed from the layer, so that the volume of a molecule is removed from between x_1 and x_m. Therefore, if we take Ω_β as the molecular volume of an $A_\beta B$ molecule in the compound, we have $J_B\Omega_\beta$ as the relative motion of the marker with respect to x_1; it is negative because it represents reduction. Also

$$\Omega_\beta = \Omega_\beta^B + \beta\Omega_\beta^A \tag{10.46}$$

where Ω_β^B and Ω_β^A are the atomic volumes of B and A atoms in the compound $A_\beta B$. However, we can assume that the atomic volume of A in the amorphous, in the crystalline, and in the compound are the same,

$$\Omega_\alpha^A = \Omega_C^A = \Omega_\beta^A = \Omega_A \tag{10.47}$$

Now, if we substitute J_B from Eq. (10.45) into Eq. (10.43) we have

$$\frac{dx_1}{dt} = -J_A\Omega_A + \beta\Omega_A\left(\frac{-d(x_m - x_1)}{dt}\right)\frac{1}{\Omega_\beta} \tag{10.48}$$

By rearranging terms, we obtain from the preceding equation,

$$J_A = -\frac{1}{\Omega_\beta}\frac{d}{dt}\left[\beta x_m + \left(\frac{\Omega_\beta}{\Omega_A} - \beta\right)x_1\right] \tag{10.49}$$

Also rearranging Eq. (10.45), we obtain

$$J_B = -\frac{1}{\Omega_\beta}\frac{d}{dt}(x_m - x_1) \tag{10.50}$$

We see from Eq. (10.50) that if $x_m - x_1 =$ constant, which means there is no marker motion, $J_B = 0$ and $J_A = -C_A dx_1/dt$. In this case the epitaxial growth is due to J_A alone, or to the direct transport of A from A(α) to A(c).

If the marker migrates, however, there is some decomposition at the interface. To estimate the amount of decomposition, we evaluate the flux ratio,

$$R = \frac{\int_0^t |J_A| dt}{\int_0^t |J_B| dt} \tag{10.51}$$

and we have from Eq. (10.39) and Eq. (10.40)

$$R = \frac{D_\beta^A}{D_\beta^B} \qquad (\text{since } C_\beta^A + C_\beta^B = 1) \tag{10.52}$$

Then from Eq. (10.49) and Eq. (10.50), we have

$$\frac{D_\beta^A}{D_\beta^B} = \frac{\beta x_m + \left(\frac{\Omega_B}{\Omega_A} - \beta\right) x_1}{x_m - x_1} \tag{10.53}$$

This shows that by measuring x_m we can determine the ratio of the fluxes, and in turn decide whether the solid phase epitaxy is dominated by interfacial decomposition or not. There is a question about the initial position of the marker. Unlike the reaction between A and B to form a compound $A_\beta B$, where we place the marker at the original interface, in SPE we can place a marker at either the A(c)/$A_\beta B$ interface or at the $A_\beta B$/A(α) interface. So we must define x_m and x_1 at the beginning of the reaction and modify Eq. (10.53) accordingly.

In solid phase epitaxy using Pd_2Si as a medium to transport Si, marker experiments showed that interfacial decomposition dominated the reaction. We use the example of solid phase epitaxy with a transport medium to introduce the concept of interdiffusion in a layered thin film structure. In the next chapter, as in the present example, the use of markers and the measurement of their motion during interdiffusion are key features in the analysis.

References

1. M. Avrami, *J. Chem. Phys.* 7, 1103 (1937); 8, 212 (1940); 9 177 (1941).

2. J. W. Christian, *The Theory of Transformation in Metals and Alloys*, Pergamon Press, New York (1965).

3. A. Cros, K. N. Tu, D. A. Smith, and B. Z. Weiss, *Appl. Phys. Lett.* 52, 1311 (1988).

4. L. Csepregi, J. W. Mayer and T. W. Sigmon, *Appl. Phys. Lett.* 29, 92 (1976).

5. U. Gösele and K. N. Tu, *J. Appl. Phys.* 66, 2619 (1989).

6. U. Gösele and K. N. Tu, *J. Appl. Phys.* 53, 3252 (1982).

7. M. H. Grabow and G. H. Gilmer, *Surf. Sci.* 194, 333 (1988).

8. J. C. Hensel, A. F. J. Levi, R. T. Tung, and J. M. Gibson, *Appl. Phys. Lett.* 47, 151 (1985).

9. S. R. Herd, K. N. Tu, and K. Y. Ahn, *Appl. Phys. Lett.* 42, 597 (1983).

10. W. L. Johnson, *Progress in Materials Science* 30, 80 (1986).

11. W. A. Johnson and R. F. Mehl, *Trans. AIME* 135, 416 (1939).

12. S. S. Lau and W. F. Van der Weg, Chapter 12 in *Thin Films: Interdiffusion and Reactions*, edited by J. M. Poate, K. N. Tu, and J. W. Mayer, Wiley–Interscience, New York (1978).

13. S. B. Newcomb and K. N. Tu, *Appl. Phys. Lett.* 48, 1436 (1986).

14. G. L. Olson and J. A. Roth, "Kinetics of Solid Phase Crystallization in Amorphous Silicon," *Materials Science Reports* 3, No. 1 (1988).

15. T. N. Rhodin and D. Walton, in *Metal Surfaces*, edited by W. B. Robertson and N. Gjostein, American Society for Metals, Chapter 8, Metals Park, Ohio (1963).

16. R. B. Schwarz and W. L. Johnson, *Phys. Rev. Lett.* 51, 415 (1983).

17. I. S. Servi and D. Turnbull, *Acta Met.* 1, 161 (1966).

18. D. A. Smith, K. N. Tu, and B. Z. Weiss, *Ultramicroscopy* 30, 90 (1989).

19. F. Spaepen in *Phase Transitions in Condensed Systems—Experiments and Theory* (p. 161), edited by G. S. Cargill III, F. Spaepen, and K. N. Tu, MRS Symposium Proceedings No. 57, Materials Research Society, Pittsburgh (1987).

20. W. A. Tiller, *The Science of Crystallization*, Cambridge University Press (1991).

21. K. N. Tu, S. R. Herd, and U. Gösele, *Phys. Rev. B.* 43, 1198 (1991).

22. K. N. Tu and J. W. Mayer, in *Thin Films: Interdiffusion and Reactions* (p. 359), edited by J. M. Poate, K. N. Tu, and J. W. Mayer, Wiley–Interscience, New York, (1978).

23. K. N. Tu, D. A. Smith, and B. Z. Weiss, *Phys. Rev. B.* 36, 8948 (1987).

24. D. Turnbull, "Phase Changes", *Solid State Physics* 3, 226 (1956).

25. D. Turnbull, "Metastable Structures," *Metallurgical Trans. A* 12A, 695 (1981).

Problems

10.1 In a semiconductor process cobalt and silicon have been deposited as an amorphous film with a 1 : 2 ratio. Given the nucleation rates per sec $R_{N0} = 0.8 \times 10^{+5}$ and $R_N = 16.7$ at 410 K, and the growth rates $R_G = 7.2 \times 10^{-7}$ and 3×10^{-6} at 410 and 430 K respectively, how long does it take for 95% of the film to crystallize at 435 K? Also, what is the activation enthalpy of the amorphous-to-crystalline transition? Do not use the value in Table 10.2. Assume that the growth rate is isotropic and linear with time, and that nucleation is random and continuous.

10.2 In a semiconductor process $CoSi_2$ is used in a superlattice structure. Given the nucleation rate $R_{N0} = 0.8 \times 10^5$, the growth rate $R_{G0} = 1.7 \times 10^7$, and the activation enthalpy of the amorphous-to-crystalline transformation is 0.9 eV, derive an equation for the time required for a resistivity of 1000 μΩ/□ as a function of temperature. Also, use the derived equation to find time needed at 167°C. The amorphous and crystalline resistivities are 1500 μΩ/□ and 75 μΩ/□ respectively. Assume that the growth rate is isotropic and linear with time, and that nucleation is random and continuous.

10.3 **(a)** The formation of amorphous TiSi takes place at 450°C (723 K). Assuming that $\Delta H_0 - \Delta H_1 = 0.5$ eV($\Delta H_0 = 2$ eV and $\Delta H_1 = 1.5$ eV), and that N_0, the initial rate of nucleation, is 1 nm/minute, find $N_,$ the nucleation rate of amorphous TiSi.

(b) The formation of crystalline TiSi takes place at 750°C (1023 K). Given that $\Delta H_2 - \Delta H_1 = 1.0$ eV ($\Delta H_2 = 2.5$ eV), find N_2, the rate of nucleation for crystalline TiSi.

10.4 **(a)** Integrate Eq. (10.20) to obtain Eq. (10.21) as a result.

(b) If the nucleation rate is 10^{-6}/sec and the growth rate is 10^{-4}/sec, what are the times of reaction for $X_T = 0.6$ and $X_T = 0.47$.

10.5 For a bilayer thin film, we are given the activation energy barrier of nucleating critical nuclei of stable phase and of amorphous phase to be 0.6 eV/atom and 0.4 eV/atom, respectively.

(a) Find ratio of nucleation rates at 200°C. What phase will appear initially?

(b) For transition from amorphous to crystalline phase, what is the critical size for nuclei? Assume that $a = b = 1$, $\gamma_{AC} = 0.5$ eV/atom, and $\Delta H_{AC} = 0.1$ eV/atom.

10.6 An experiment was performed in which SPE was used to grow crystalline Si from an amorphous layer with a transport layer of WSi_2 in between. A marker layer of atoms was placed in the silicide to monitor the growth. The rate of change of the position of the marker layer was found to be 10 Å/sec. The atomic concentrations of Si and W are .05 atom/Å^3 and .06 atom/Å^3 and do not vary between crystalline and compound. Determine whether the growth in this experiment was limited by direct transport or by interfacial decomposition. Assume flux rates are constant and the crystal Si growth rate $= 15\text{Å}/\text{sec}$.

10.7 In this chapter the Avrami equation was developed for spherical nuclei that nucleate at random in time and space, and grow at a constant rate. This derived equation was of the form.

$$X_T = 1 - \exp(-Kt^4)$$

Now in a thin planar amorphous film a spherical nucleus will cease growing normal to the plane of the film once its diameter is the height of the film. At this point it grows as a circular disk in the plane of the film—that is, in a two-dimensional fashion. Derive the Avrami equation for nucleating circular discs of thickness Y. Neglect the time it takes a spherical nucleus to become a disk.

10.8 Given that the nucleation rate of the stable phase is 0.04% of the metastable nucleation rate, calculate the minimum barrier height necessary to prevent the reaction from continuing to a stable state once an amorphous state is reached. The activation energy for the stable phase is 1.69 eV/atom and $T = 300°\text{C}$.

10.9 In the solid state phase transformation of α to β, consider the homogeneous nucleation of a spherical β nucleus in the matrix of α. Let $\gamma_{\alpha\beta}$ be the interfacial energy per unit area and $\Delta H_{\alpha\beta}$ be the heat of transformation per unit volume of the nucleus. Determine the radius and activation energy of nucleation of the critical nucleus. What is the volume of the critical nucleus and what is the number of atoms in the critical nucleus? Compare the number with Eq. (10.36).

CHAPTER 11

Interdiffusion

11.0 Introduction

If two solids are put together, they generally interdiffuse. Interdiffusion is treated here because integrated circuits are made of layered thin film structures that can interdiffuse during the thermal processing steps in fabrication. This is one of the generic topics in thin film science. To discuss it, we shall start from what we already know about bulk diffusion couples (see Fig. 11.1).

Two infinite rods A and B are joined together and heated so that interdiffusion occurs. They are assumed to have the same crystal structure (e.g., fcc, as silver and gold). When they interdiffuse they form a continuous solid solution that has the same crystal structure. The complication caused by a change in crystal structure is thereby eliminated. Interdiffusion with only compositional change covers, however, only a minor set of cases, because when many solids interdiffuse they form compounds. Most of the compounds have a different crystal structure from A or B. Compound formation during interdiffusion is presented later in this chapter as an introduction to the general topic of thin film reactions covered in Chapter 12.

11.1 Interdiffusion to Form Solid Solutions

When A and B interdiffuse, the concentration profile broadens as shown in Fig. 11.2. The mathematical form of this profile is determined by considering the very general

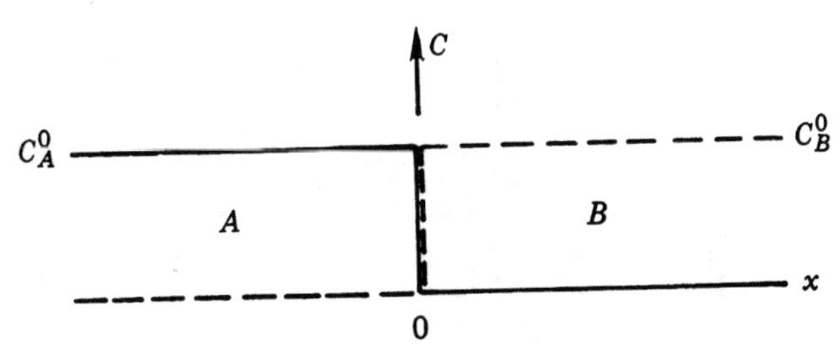

FIGURE 11.1 A sketch of a bulk diffusion couple.

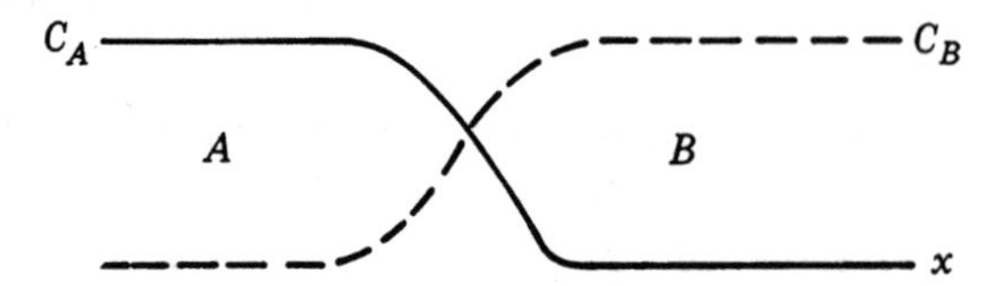

FIGURE 11.2 A sketch of the compositional profile of the diffused couple.

diffusion equation assumed in the one-dimensional case,

$$\frac{\partial C_A}{\partial t} = D \frac{\partial^2 C_A}{\partial x^2} \tag{11.1}$$

The solution in the case of a thin layer of A diffusing into a semi-infinite bar of B with a unit cross-sectional area, as given in Chapter 3, is

$$C_A = \frac{bC_A^0}{\sqrt{4\pi Dt}} \exp\left(-\frac{x^2}{4Dt}\right) \tag{11.2}$$

where C_A is the concentration of A atoms/cm^3 and C_A^0 is the initial concentration of A atoms/cm^3 in the thin layer of thickness b before interdiffusion.

The thin layer solution of Fig. 11.2 is adapted to semi-infinite diffusion couples by dividing A into many vertical slices as shown in Fig. 11.3. Each, of thickness $\Delta\alpha$ and unit cross-sectional area, will have a solution that is similar to Eq. (11.2), and the individual contributions are summed. A solution is

$$C_A(x,t) \cong \frac{C_A^0}{\sqrt{4\pi Dt}} \sum_{i=1}^{n} \exp\left[-\frac{(x+\alpha_i)^2}{4Dt}\right]\Delta\alpha_i \tag{11.3}$$

where α_i is the distance of ith slice to the origin.

The summation is expressed by an integral,

$$C_A(x,t) = \frac{C_A^0}{\sqrt{4\pi Dt}} \int_{-\infty}^{0} \exp\left[-\frac{(x+\alpha)^2}{4Dt}\right] d\alpha \tag{11.4}$$

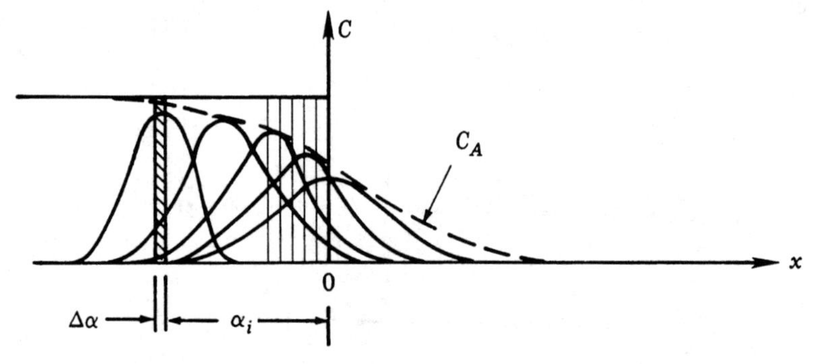

FIGURE 11.3 The compositional profile of C_A is the sum of all the exponential curves diffusing out from each slab of thickness $\Delta\alpha$.

A new dimensionless variable η is introduced,

$$\eta = \frac{x + \alpha}{2\sqrt{Dt}}$$

where

$$\eta = \frac{x}{2\sqrt{Dt}} \qquad \text{when } \alpha = 0$$

$$\eta = -\infty \qquad \text{when } \alpha = -\infty$$

$$d\eta = \frac{d\alpha}{2\sqrt{Dt}}$$

Then Eq. (11.4) can be rewritten as:

$$\begin{aligned} C_A(x,t) &= \frac{C_A^0}{\sqrt{\pi}} \int_{-\infty}^{x/2\sqrt{Dt}} \exp(-\eta^2)\, d\eta \\ &= \frac{C_A^0}{\sqrt{\pi}} \left[\int_{-\infty}^{0} \exp(-\eta^2)\, d\eta + \int_{0}^{x/2\sqrt{Dt}} \exp(-\eta^2)\, d\eta \right] \qquad (11.5) \\ &= \frac{C_A^0}{2} \left[1 + \text{erf}\left(\frac{x}{2\sqrt{Dt}} \right) \right] \end{aligned}$$

Now the first term by definition is an error function which equals $\sqrt{\pi}/2$, and the second is the standard form of the error function, and

$$\begin{aligned} \text{erf}(z) &= \frac{2}{\sqrt{\pi}} \int_0^z \exp(-\eta^2)\, d\eta \\ \text{erf}(0) &= 0 \qquad (11.6) \\ \text{erf}(\infty) &= 1 \\ \text{erf}(-z) &= -\text{erf}(z) \end{aligned}$$

From the definition of the error function, it is very easy to see that

$$\int_{-\infty}^{0} \exp(-\eta^2)\, d\eta = -\int_0^{-\infty} \exp(-\eta^2)\, d\eta = \int_0^{\infty} \exp(-y^2)\, dy$$

where $y = -\eta$.

An important physical meaning of this solution, Eq. (11.5), is that at $x = 0$, $C_A = C_A^0/2$ (see Fig. 11.4).

As a function of time and temperature, interdiffusion causes a change in composition profile. The midpoint of composition is always at $x = 0$ from the mathematical solution, as shown in Fig. 11.4. But this is wrong from an experimental viewpoint, since the solution does not take into account the mechanism of diffusion. It has always been found that the concentration at the midpoint is not one-half the original composition; rather, it is either a little bit lower or a little bit higher. The

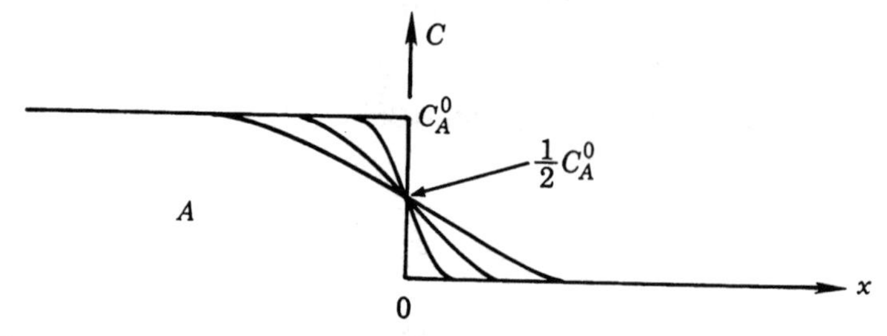

FIGURE 11.4 A sketch showing that the midpoint of the compositional curve is always at the original position of the interface between A and B according to Eq. (11.5).

derivation does not contain the physical picture of diffusion in solids because A diffusing into B is not equal to B diffusing into A. Mathematically we assumed that the two fluxes were the same. In reality they are not. Diffusion requires exchange between an atom and a vacancy, and some atoms exchange with vacancies faster, while some exchange slower.

Another way of looking at the discrepancy between the mathematical solution and experimental observation is that the measured compositional profile of C_A is not due to the diffusion of A atoms alone. The profile is caused by the combined diffusion of both A and B atoms in opposite directions. There is a relative motion between them. The interdiffusing fluxes of A and B are generally unequal; one is faster, and the fluxes must be balanced by a vacancy flux. This flux has led to the observation of the Kirkendall effect and void formation. Also the diffusivity in Eq. (11.4) is assumed to be independent of composition, which again is untrue.

11.2 Kirkendall Effect

Kirkendall, in the early 1940s, did a simple experiment. He and his associates took a piece of brass and put two pieces of copper on each side of the brass. Brass is 70 atomic percent copper and 30 atomic percent zinc. They put some molybdenum (moly) wires at the original interface. At the interdiffusion temperatures the moly wires do not react with the brass or the copper; they serve as markers to indicate any displacement from the original position. Kirkendall used two arrays of markers as shown in Fig. 11.5 and measured the spacing between the two. After interdiffusion the markers were found to have moved closer together. There is zinc in brass and no zinc in the copper, so during interdiffusion copper will diffuse in and zinc will diffuse out. If the markers move closer together, it must mean that the flux of zinc diffusing out is greater than the flux of copper into the brass. There is a loss of material from the inside to the outside, therefore the marker is displaced inward.

Figure 11.6 shows that A and B interdiffuse, the flux of A is greater than that of B. In this case A is zinc and B is copper. If more A (zinc) diffuses into the B (copper) with the marker position fixed, more material is on the right. Since this is an internal reaction, the center of gravity has to remain the same. Therefore, if the sample is

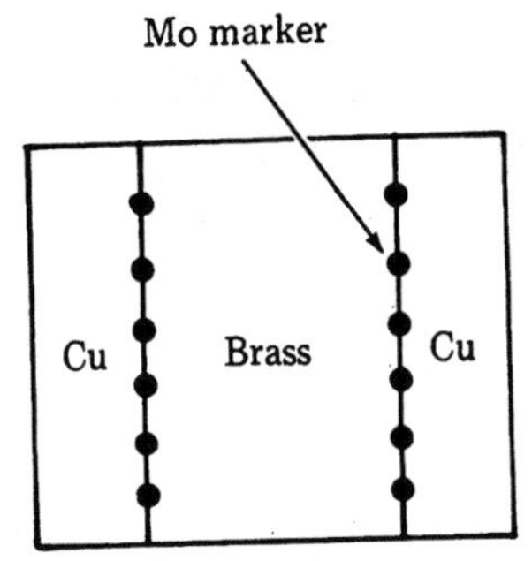

FIGURE 11.5 A sketch of a bulk diffusion couple of copper and brass with Mo markers at their interfaces.

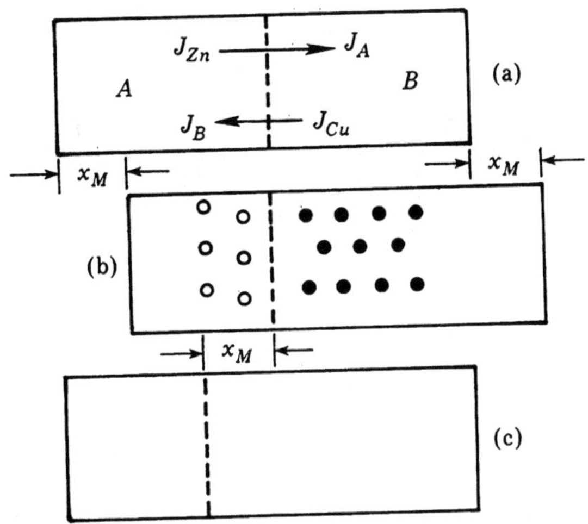

FIGURE 11.6 A sketch showing the marker displacement due to uneven interdiffusional fluxes.

put back to the original alignment, the marker has to move to the left. From this study, it is very clear that the diffusion coefficients of A and B are not the same. If the diffusion coefficients are not equal, the fluxes are not equal. In this kind of interdiffusion, the diffusivities of Cu and Zn are unequal, so that there are two unknowns in this problem. The solution requires two independent equations.

11.3 Boltzmann–Matano Analysis

The first equation discussed below is from the Boltzmann–Matano analysis. It solves the diffusion equation when $D = D(C)$ (i.e., the diffusivity is a function of composition), and we obtain from it a solution of $D(C)$. Section 3.5 described Fick's Second Law expressed mathematically as:

$$\frac{\partial C}{\partial t} = \frac{\partial}{\partial x}\left(D\frac{\partial C}{\partial x}\right) = \frac{\partial D}{\partial x}\frac{\partial C}{\partial x} + D\frac{\partial^2 C}{\partial x^2} \tag{11.7}$$

First assume that

$$C(x, t) = C(\eta)$$

and

$$\eta = \frac{x}{t^{1/2}}$$

Then, by differentiation

$$\frac{\partial C}{\partial t} = \frac{dC}{d\eta}\frac{\partial \eta}{\partial t} = -\frac{1}{2}\frac{x}{t^{3/2}}\frac{dC}{d\eta} = -\frac{\eta}{2t}\frac{dC}{d\eta}$$

$$\frac{\partial C}{\partial x} = \frac{dC}{d\eta}\frac{\partial \eta}{\partial x} = \frac{1}{t^{1/2}}\frac{dC}{d\eta}$$

$$\frac{\partial^2 C}{\partial x^2} = \frac{\partial}{\partial x}\left(\frac{\partial C}{\partial x}\right) = \frac{\partial}{\partial \eta}\frac{\partial \eta}{\partial x}\left(\frac{\partial C}{\partial x}\right) = \frac{1}{t}\frac{d^2 C}{d\eta^2}$$

$$\frac{\partial D}{\partial x} = \frac{\partial D}{\partial \eta}\frac{\partial \eta}{\partial x} = \frac{1}{t^{1/2}}\frac{dD}{d\eta}$$

Equation (11.7) becomes

$$-\frac{\eta}{2t}\frac{dC}{d\eta} = \frac{1}{t^{1/2}}\frac{dD}{d\eta}\cdot\frac{1}{t^{1/2}}\frac{dC}{d\eta} + \frac{D}{t}\frac{d^2 C}{d\eta^2}$$

Drop t:

$$-\frac{\eta}{2}\frac{dC}{d\eta} = \frac{dD}{d\eta}\frac{dC}{d\eta} + D\frac{d^2 C}{d\eta^2}$$

$$= \frac{d}{d\eta}\left(D\frac{dC}{d\eta}\right)$$

Because these are total differentials, we drop $1/d\eta$ and integrate both sides:

$$-\frac{1}{2}\int_0^{C'} \eta\, dC = \int_0^{C'} d\left(D\frac{dC}{d\eta}\right) = \left[D\frac{dC}{d\eta}\right]\Bigg|_0^{C'} \tag{11.8}$$

where C' is an arbitrary concentration, $0 < C' < C_0$, and C_0 is the concentration of A at $x = -\infty$. Now consider the physical picture of the interdiffusion case. Consider, at a given time (i.e., t is fixed, so $d\eta \propto t^{-1/2}\,dx$), both ends of the diffusion couple (Fig. 11.7); $dC/d\eta = 0$ at $C = 0$ and at $C = C_0$. Therefore, in Eq. (11.8), integration from $C = 0$ to $C = C_0$ along the "vertical" axis gives

$$-\frac{1}{2}\int_0^{C_0} \eta\, dC = D\frac{dC}{d\eta}\bigg|_{C=C_0} - D\frac{dC}{d\eta}\bigg|_{C=0}$$

$$= 0 - 0 = 0$$

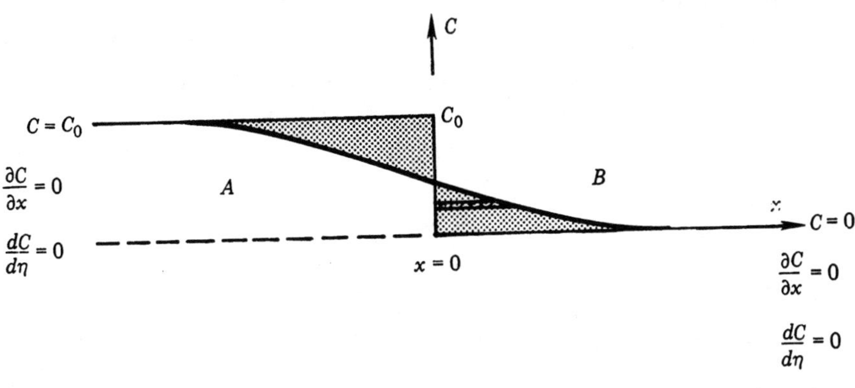

FIGURE 11.7 A sketch of the Matano interface to show that the integration of Eq. (11.8) is along the vertical axis. The position of the Matano interface is chosen to make the two hatched areas equal.

This means that

$$\int_0^{C_0} x\,dC = 0 \tag{11.9}$$

because the interdiffusion is considered at a fixed time, so that the variable η is the same as x. This relationship, Eq. (11.9), defines the *Matano interface,* where the quantity of A atoms $(1 - C_A)$ that have been removed from the left of the interface is equal to the quantity C_A that has been added to the right. In the sketch in Fig. 11.7, the shaded area in A (left of interface) is equal to the shaded area in B. The reference interface is defined as Matano's interface. It is not at the same location as the original interface.

The Matano interface defines the location of the origin of the x-axis (i.e., $x = 0$) for the integration in Eq. (11.8). Otherwise, the integral of $-\frac{1}{2}\int_0^{C'} \eta\,dC$ is "undetermined" because it is integrated over C (the vertical axis) and x is arbitrary—until the location of $x = 0$ is defined with Matano's interface,

$$-\frac{1}{2}\int_0^{C'} \eta\,dC = D\left(\frac{dC}{d\eta}\right)\bigg|_{C'}$$

Convert η to x and define D to be the interdiffusion coefficient $\tilde{D}$, so that

$$\tilde{D}(C') = -\frac{1}{2t}\left(\frac{dx}{dC}\right)\bigg|_{C'} \int_0^{C'} x\,dC \tag{11.10}$$

This equation indicates that measurement of C as a function of x (i.e., the concentration profile of A) gives $\tilde{D}$ by the graphical method (t is given), using the slope and the shaded area as shown in Fig. 11.8. Now, a very important point is that $\tilde{D}$ is defined as *interdiffusion coefficient.* It is an averaged sum of the diffusion of the two components and is measured on the basis of Boltzmann–Matano analysis.

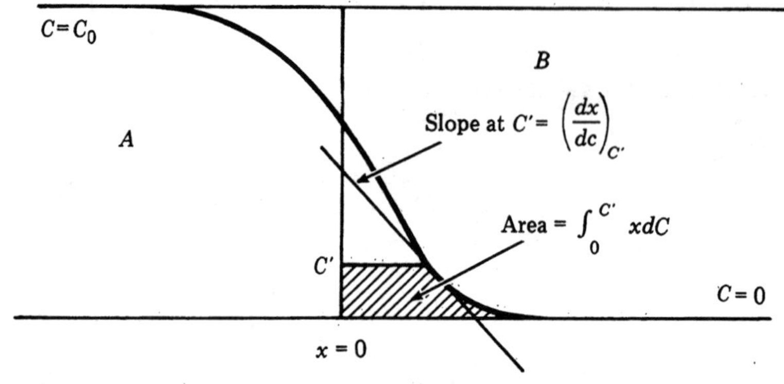

FIGURE 11.8 A sketch to show the geometrical application of the Boltzmann–Matano solution to obtain the interdiffusion coefficient.

11.4 Interdiffusion Coefficient

There are three kinds of diffusivities. One is called the chemical interdiffusion coefficient, or interdiffusion coefficient, which is measured by the Boltzmann–Matano analysis. This coefficient is expressed by

$$\tilde{D} = C_B D_A + C_A D_B \tag{11.11}$$

The *intrinsic diffusion coefficient* of species A (or B) in the AB alloy is given by D_A (or D_B). For example, D_A and D_B characterize the diffusion of gold and silver, respectively in gold–silver alloys. These are also called chemical diffusion coefficients.

The third kind of diffusivity is called the *tracer diffusion coefficient.* Tracer diffusion, with no chemical force, is a random walk, such as the diffusion of radioactive Au isotope in pure Au. Tracer diffusivity is related to chemical (intrinsic) diffusivity by

$$D_A = D_A^*\left(1 + \frac{\partial \ln \gamma_A}{\partial \ln C_A}\right) \tag{11.12}$$

where γ_A is the activity coefficient as discussed in Chapter 3, Section 3.2. A different situation exists if we have a homogeneous alloy of Au and Ag and put a trace of radioactive gold (Au*) on it and diffuse. The diffusivity that we measure, in this case, is the tracer diffusion coefficient of gold in this homogeneous alloy. There is no chemical driving force because there is no concentration gradient. We note that the tracer diffusivity of Au* in the alloy is different from the tracer diffusivity of Au* in pure Au. The two can be related by the factor $(1 + bx)$ where b is the solute–solvent interaction parameter and x is the solute concentration in the dilute solution.

For the interdiffusion of a piece of Au and a piece of Ag, the interdiffusion coefficient is measured by the Matano process. It determines how fast they mix together: how fast gold diffuses in the concentration gradient or how fast silver does so. There is a chemical force because of homogenization. The chemical (intrinsic) diffusion coefficients of Au and Ag are contained in the interdiffusion coefficient. To determine these coefficients, Darken's analysis is used.

11.5 Darken's Analysis

In the case of Boltzmann–Matano analysis, the object is to define the reference plane of $x = 0$, which permits calculation of the interdiffusion coefficient. The concept of reference plane or coordinate is most important in considering interdiffusion phenomena. This is also true in Darken's analysis of marker motion.

In an interdiffusion experiment, both A atoms and B atoms are moving: they move opposite and relative to each other. It is the relative motion which requires a clear definition of the reference frame for analysis. The concentration profile of either A or B used in the Boltzmann–Matano analysis is not due to the diffusion of either A or B alone, but rather it is due to both. The motion of a B atom as seen from the outside consists of the motion of the B atom plus that of all the A atoms. In other words, the B atom diffuses in a moving framework of A. Some examples of relative motion are helpful here. If we drop a tennis ball on a flowing river, the ball produces ripples on the surface of the river. Standing on the riverbank, we see the combined movement of the flow of the river plus the radial spreading out of the ripples. On the other hand, if we could sit on the tennis ball, we would only see the radial motion of the ripples. Another example is that of sitting in a theater in front of a rotating stage. An actor drops a ball vertically downward. The ball bounces up and down. From the actor's viewpoint the ball does not change position except up and down, but we, the audience, see that the ball moves in arc-like motion due to the rotating motion of the stage. To decouple the two kinds of motion, we must measure the motion from the standpoint of the tennis ball or of the actor. In interdiffusion, markers are required to determine the flux relative to the moving framework. This is where Darken's analysis begins.

The marker moves during interdiffusion and the marker displacement or its velocity can be measured. The laboratory frame is taken as the reference, as shown in Fig. 11.9. The marker is placed at the interface and is moving with a velocity v. For component A, we have

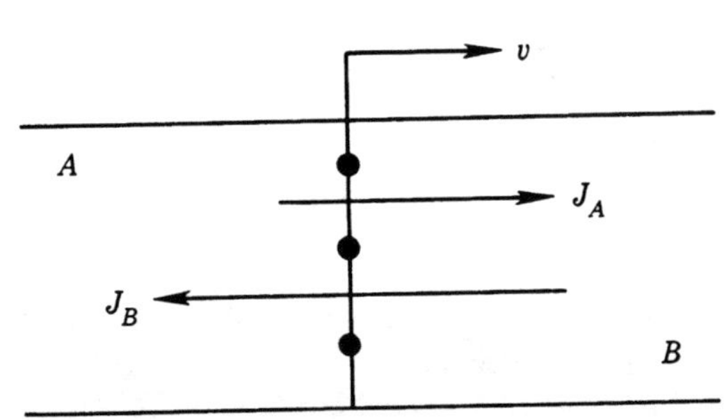

FIGURE 11.9 A sketch of marker motion with a velocity v due to the difference in opposite diffusional fluxes.

$$J_A = -D_A \frac{\partial C_A}{\partial x} + C_A v \tag{11.13}$$

where C_A is number of A atoms per unit volume and J_A is the flux of A. Similarly,

$$J_B = -D_B \frac{\partial C_B}{\partial x} + C_B v \tag{11.14}$$

We have

$$C = C_A + C_B$$

which is constant (material conserved).

It implies

$$\frac{\partial C_A}{\partial x} = -\frac{\partial C_B}{\partial x} \tag{11.15}$$

The net flux is

$$J = J_A + J_B$$

and

$$\frac{\partial C}{\partial t} = -\nabla \cdot \mathbf{J}$$

or

$$\frac{\partial C}{\partial t} = \frac{\partial}{\partial x}\left[D_A \frac{\partial C_A}{\partial x} + D_B \frac{\partial C_B}{\partial x} - Cv\right] \tag{11.16}$$

Since the total concentration C is constant, $\partial C/\partial t = 0$,

$$D_A \frac{\partial C_A}{\partial x} + D_B \frac{\partial C_B}{dx} - Cv = \text{constant} = K$$

Consider the point at the end of the sample, $x = 0$, where no interdiffusion occurs.

$$\frac{\partial C_A}{\partial x} = \frac{\partial C_B}{\partial x} = 0$$

Also, since there is no marker motion at $x = 0$, $v = 0$. Therefore the constant K is zero, and

$$v = \frac{1}{C}\left[D_A \frac{\partial C_A}{\partial x} + D_B \frac{\partial C_B}{\partial x}\right] \tag{11.17}$$

This is the marker velocity measured from the laboratory framework. The reference coordinate (the coordinate where the $x = 0$ origin is) is set at the end of the diffusion couple where no diffusion occurs. Now, consider the following equation

$$\frac{\partial C_A}{\partial t} = -\nabla \cdot \mathbf{J_A}$$

$$= \frac{\partial}{\partial x}\left[D_A \frac{\partial C_A}{\partial x} - C_A v\right]$$

Substitute into it the expression for v from Eq. (11.17) and insert $(C_A + C_B)/C$ (equals unity) in the first term. Then

$$\frac{\partial C_A}{\partial t} = \frac{\partial}{\partial x}\left\{\frac{C_A + C_B}{C} D_A \frac{\partial C_A}{\partial x} - \frac{C_A}{C}\left[D_A \frac{\partial C_A}{\partial x} + D_B \frac{\partial C_B}{\partial x}\right]\right\}$$

$$= \frac{\partial}{\partial x}\left[\frac{C_B}{C} D_A \frac{\partial C_A}{\partial x} - \frac{C_A}{C} D_B \frac{\partial C_B}{\partial x}\right]$$

From Eq. (11.15), $\partial C_A/\partial x = -\partial C_B/\partial x$, so that

$$\frac{\partial C_A}{\partial t} = \frac{\partial}{\partial x}\left[\frac{C_B}{C} D_A \frac{\partial C_A}{\partial x} + \frac{C_A}{C} D_B \frac{\partial C_A}{\partial x}\right]$$

$$= \frac{\partial}{\partial x}\left[(N_B D_A + N_A D_B)\, \partial C_A/\partial x\right] \tag{11.18}$$

where $N_A = C_A/C$ and $N_B = C_B/C$ are the fraction of A and B atoms in the unit volume, respectively.

Equation (11.18) is Fick's second law of diffusion of A and, as in Eq. (11.11), we can take

$$\tilde{D} = N_B D_A + N_A D_B \tag{11.19}$$

where $\tilde{D}$ is defined as the interdiffusion coefficient.
Substituting $\partial C_B/\partial x = -\partial C_A/\partial x$ into equation 11.17, the velocity is

$$v = \frac{1}{C}\left[D_A \frac{\partial C_A}{\partial x} - D_B \frac{\partial C_A}{\partial x}\right]$$

$$= (D_A - D_B)\left[\frac{\partial}{\partial x}(C_A/C)\right]$$

Therefore

$$v = (D_A - D_B)\, \partial N_A/\partial x \tag{11.20}$$

From Eq. (11.19) and Eq. (11.20), D_A and D_B can be determined, by obtaining $\tilde{D}$ from the Boltzmann–Matano analysis and v from the marker displacement measurement. An example of marker analysis of interdiffusion in a bulk couple of Pb and PbIn alloy is given by Campbell et. al. (1976).

The terms D_A and D_B are the chemical diffusion coefficients. If the solution is an ideal solution (i.e., ΔH mixing (enthalpy of mixing) is zero), then D_A and D_B are the same as the tracer diffusion coefficients, D_A^* and D_B^*, in the homogeneous solution of the same composition. On the other hand, if the solution is nonideal and letting γ_A be the activity coefficient (Eq. 3.18), then

$$D_A = D_A^*\left(1 + \frac{\partial \ln \gamma_A}{\partial \ln N_A}\right)$$

and

$$D_B = D_B^*\left(1 + \frac{\partial \ln \gamma_B}{\partial \ln N_B}\right)$$

Furthermore,

$$\tilde{D} = (N_B D_A^* + N_A D_B^*)\left(1 + \frac{\partial \ln \gamma_A}{\partial \ln N_A}\right)$$

11.6 Interdiffusion to Form Intermetallic Compounds

The above treatment considers the interdiffusion between two pieces of solids which form continuous solid solutions, such as silver and gold or copper and brass. These are binary systems in which interdiffusion leads to compositional change but no structural change. For example, if the constituents are face-centered cubic structures, their solid solution also has face-centered cubic structure. Structural continuity is a prerequisite of compositional continuity; the latter is required in the Matano analysis. However, more often than not, the binary systems of interest to us form intermetallic compounds.

These compounds tend to have a narrow composition range and crystal structure different from that of their constituents or of the end phases. The difficulty in dealing with interdiffusion in such cases is that more than one intermetallic compound may be formed (see Fig. 11.10). If there are at least two unknowns for each of the compounds (i.e., the chemical diffusion coefficients of the two constituents), there are $2N$ unknowns for N compounds. The problem is intractable due to the number of unknowns. Besides, the compound interfaces are discontinuities (i.e., nondifferentiable), so Matano's analysis fails at these points.

However, we emphasize that in the cases where the binary systems can form only one intermetallic compound, Matano's analysis can in principle be applied. We shall illustrate it. In Fig. 11.11, a single intermetallic compound $A_\beta B$ forms between A and B. The Matano interface is defined such that the two shaded areas are equal.

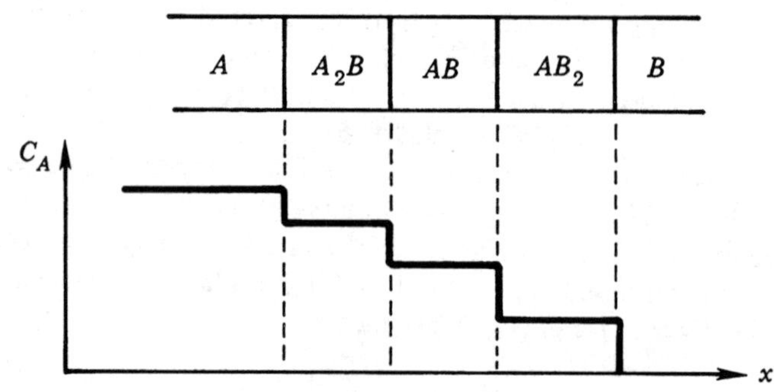

FIGURE 11.10 A sketch of compositional changes in multiple compound formation in a bulk diffusion couple.

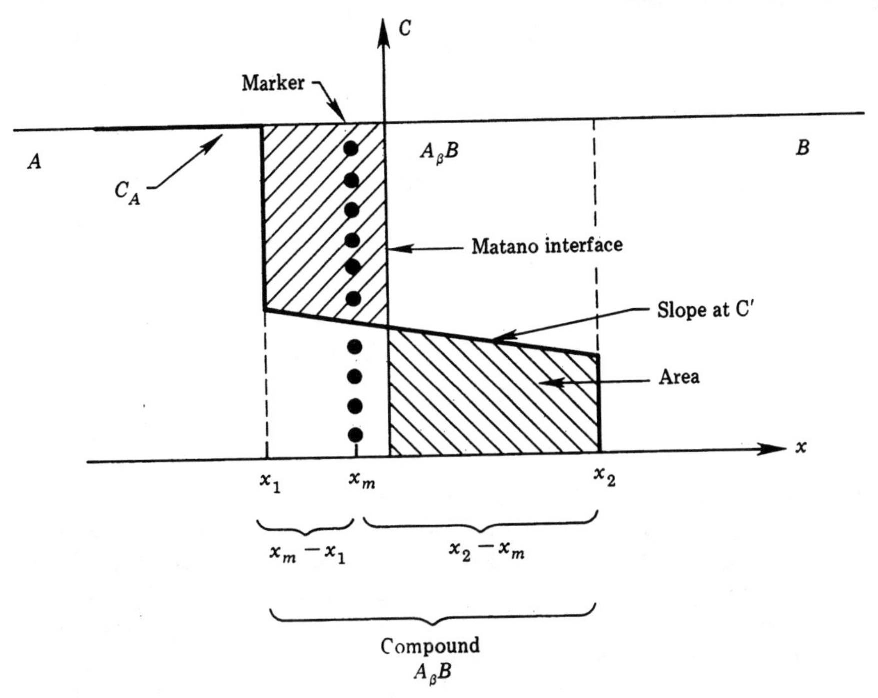

FIGURE 11.11 A diagram of single compound formation in a diffusion couple, showing the Matano interface and marker displacement.

The chemical interdiffusion coefficient, at the point C' in the compound, can be calculated from the slope at C' and the area of the shaded area between $C = 0$ and $C = C'$. Because $A_\beta B$ is a compound, the concentration gradient across it is small. If C' is moved a bit, the slope and the shaded area do not change much. In this case, the interdiffusion coefficient in the compound has a negligible dependence on concentration. Physically, this is usually correct! Nevertheless, for a single-layer compound formation, it is much easier to determine its chemical interdiffusion coefficient by measuring its thickness as a function of time and temperature.

If it is known that the thickness or the position of interfaces of a compound layer is proportional to $(\text{time})^{1/2}$, the growth of the layer is diffusion-controlled. Thus, it is possible to measure the interdiffusion coefficient of each compound layer by simply assuming that

$$x_i^2 = 4\tilde{D}_i t \tag{11.21}$$

Yet the chemical (intrinsic) diffusion coefficient of each constituent cannot be measured except by using tracer techniques or by marker analysis. If the diffusing species has no long-lifetime tracer, not much can be done with tracer techniques.

For marker analysis of a single compound layer, assume that the growth of the layer is diffusion-controlled. If the marker is placed at the original interface between A and B, each B atom passing the marker will form a molecule of $A_\beta B$, and each time βA atoms pass the marker, they will form a molecule of $A_\beta B$ on the other side.

If we assume atomic volumes to be the same,

$$\frac{J_B}{\beta J_A} = \frac{x_m - x_1}{x_2 - x_m}$$

From the flux equations

$$J_A = -D_\beta^A \frac{\partial C_A}{\partial x} \quad \text{and} \quad J_B = -D_\beta^B \frac{\partial C_B}{\partial x}$$

we have

$$\frac{D_\beta^B}{D_\beta^A} = \frac{\beta(x_m - x_1)}{x_2 - x_m} \tag{11.22}$$

$$(x_2 - x_1)^2 = 4\tilde{D}_\beta t \tag{11.23}$$

where we have used β as a subscript to indicate diffusion in the $A_\beta B$ layer. Therefore from measurement of the marker displacement x_m and knowledge of the composition $A_\beta B$, we can determine the ratio of the diffusion coefficients from Eq. (11.22). From knowledge of $(x_2 - x_1)$ we can find the interdiffusion coefficient using Eq. (11.23). An example of marker analysis of the formation of Al_3Ti intermetallic compound in a bimettalic thin film couple of Al and Ti is given by Tardy et al. (1985).

11.7 Analysis of Growth of Layered Compounds

Rather than addressing the complicated problem of the simultaneous growth of several intermetallic compounds in a bulk diffusion couple, we shall discuss the issue of "missing intermetallic compounds." It is sometimes observed that in a bulk diffusion couple, not all the equilibrium intermetallic compounds will grow in the couple; that is, one or two are missing. Kidson was among the first to address this issue in the analysis of layered growth of compounds.

When we discuss thin film reactions, we will show that the phenomenon of "single-phase" formation is very common. Single-phase formation means that except for one particular phase, the equilibrium compounds do not grow in thin film reactions. Thus the concept of missing phases and why they are missing becomes important. Kidson's analysis gives a general expression of Eq. (11.21), which will be used later in thin film reactions analysis. It shows that all phases should grow if they are diffusion controlled.

Kidson first considers the motion of a single interface with a discontinuous change of composition across the interface. Such an interface can be regarded as one of the two sides (interfaces) of a compound layer. Let the phases α and β be separated by the interface as shown in Fig. 11.12. The position of the interface is represented by $\xi_{\alpha\beta}$ and its velocity relative to the Matano interface can be defined as $d\xi/dt$. Taking the relationship that $J = Cv$ (where C and v are concentration and velocity, respectively), at the interface the fluxes are

$$(C_{\alpha\beta} - C_{\beta\alpha}) \frac{d\xi_{\alpha\beta}}{dt} = J_{\alpha\beta} - J_{\beta\alpha} \tag{11.24}$$

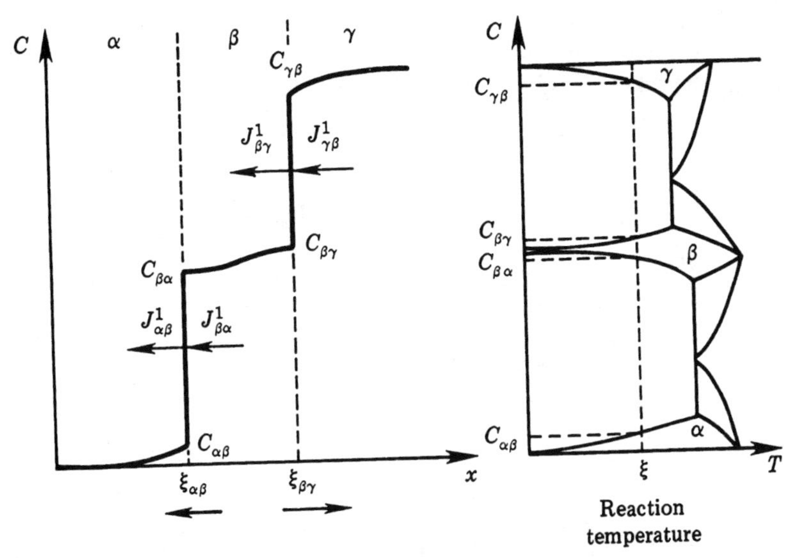

FIGURE 11.12 A schematic binary phase diagram for a single intermetallic compound, and a sketch of the compositional profile during compound formation.

Then

$$\frac{d\xi_{\alpha\beta}}{dt} = \frac{1}{C_{\alpha\beta} - C_{\beta\alpha}}\left[\left(-D\frac{\partial C}{\partial x}\right)_{\alpha\beta} - \left(-D\frac{\partial C}{\partial x}\right)_{\beta\alpha}\right] \tag{11.25}$$

where

$$J_{\alpha\beta} = -D\left.\frac{\partial C}{\partial x}\right|_{\alpha\beta}$$

and

$$J_{\beta\alpha} = -D\left.\frac{\partial C}{\partial x}\right|_{\beta\alpha}$$

as given by Fick's first law.

To evaluate $(\partial C/\partial x)_{\alpha\beta}$ and $(\partial C/\partial x)_{\beta\alpha}$, Boltzmann–Matano's analysis is used with the assumption that

$$C(x,t) = C(\eta)$$

where, as in Section 11.3,

$$\eta = \frac{x}{\sqrt{t}}$$

and

$$\frac{\partial C}{\partial x} = \frac{1}{t^{1/2}}\frac{dC}{d\eta}$$

Next, the concentrations at the interface (i.e., $C_{\alpha\beta}$ and $C_{\beta\alpha}$) are assumed to remain constant (equilibrium values) with respect to time and position. In turn

$$\frac{dC(\eta)}{d\eta} = f(\eta)$$

where $f(\eta)$ = constant if η is constant. Therefore, Eq. (11.25) can be rewritten as

$$\begin{aligned} \frac{d\xi_{\alpha\beta}}{dt} &= \frac{1}{C_{\alpha\beta} - C_{\beta\alpha}} \left[-D \left.\frac{\partial C}{\partial \eta}\right|_{\alpha\beta} + D \left.\frac{\partial C}{\partial \eta}\right|_{\beta\alpha} \right] \frac{1}{t^{1/2}} \\ &= \frac{1}{C_{\alpha\beta} - C_{\beta\alpha}} [(K_{ji}D)_{\beta\alpha} - (K_{ij}D)_{\alpha\beta}] \frac{1}{t^{1/2}} \end{aligned} \tag{11.26}$$

where

$$K_{ij} = \left(\frac{dC}{d\eta}\right)_{ij} = \sqrt{t}\left(\frac{\partial C}{\partial x}\right)_{ij}$$

Integration of Eq. (11.26) gives

$$\begin{aligned} \xi_{\alpha\beta} &= 2\left[\frac{(DK)_{\beta\alpha} - (DK)_{\alpha\beta}}{C_{\alpha\beta} - C_{\beta\alpha}}\right] t^{1/2} \\ &= A_{\alpha\beta} t^{1/2} \end{aligned} \tag{11.27}$$

This shows the parabolic time dependence of the interface movement. The parameter $A_{\alpha\beta}$ will be positive, zero, or negative depending on the relative magnitudes of the terms $(DK)_{\alpha\beta}$ and $(DK)_{\beta\alpha}$. Correspondingly the interface $\xi_{\alpha\beta}$ will move to the right, remain stationary, or move to the left.

Now we consider the growth of an intermetallic compound phase by interdiffusion and represent the case by the following expression:

$$\xi_{\alpha\beta} = 2\left[\frac{(DK)_{\beta\alpha} - (DK)_{\alpha\beta}}{C_{\alpha\beta} - C_{\beta\alpha}}\right] t^{1/2}$$

$$\xi_{\beta\gamma} = 2\left[\frac{(DK)_{\gamma\beta} - (DK)_{\beta\gamma}}{C_{\beta\gamma} - C_{\gamma\beta}}\right] t^{1/2}$$

The width of the β phase is

$$W_\beta = \xi_{\beta\gamma} - \xi_{\alpha\beta} = B_\beta t^{1/2} \tag{11.28}$$

Again, a parabolic time dependence of growth of the compound layer thickness is shown. This approach can be extended to the growth of N intermetallic compounds in interdiffusion.

Concerning the absence of certain phases, it is tempting to argue that in the kinetic expression of $A_{\alpha\beta}$, B_β and so on, the phase widths can be zero or negative which means no growth. However, such a consideration must be rejected. This is due to the fact that in Kidson's analysis, the growth is diffusion-controlled, that is

$$x^2 = At$$

so

$$\frac{dx}{dt} \propto \frac{A}{x}$$

This means that when a layer thickness x shrinks to zero ($x \to 0$), the growth velocity dx/dt goes to infinity. The layer cannot shrink away in the diffusion-controlled mode. In other words, the chemical potential gradient becomes infinite as $x \to 0$. Therefore, while Kidson's analysis shows parabolic time dependence for growth of compound layers, it fails to explain the missing phase phenomenon. The absence of certain phases is addressed in Chapter 12.

11.8 The Prediction of First Phase Formation

In this section we shall discuss two key questions about single phase formation: which one forms first and how can we predict its formation.

In Chapter 10 we discussed solid phase reactions at constant temperature and pressure and showed that the thermodynamic criterion of these reactions is:

$$\frac{dG}{dN_1} = 0 \tag{10.2}$$

While this equation demands a minimum in the free energy vs. composition curve, it does not specify any particular minimum. We have an undetermined case when there is more than one minimum in the curve, or more than one intermetallic compound in the phase diagram. Obviously, we need one more criteria in order to choose which compound is the first to form. There have been proposals to select the one compound in the binary phase diagram which is closest to the 50-50 composition, or the one next to the lowest eutectic point, or the one having the highest melting point, etc. While all of them seem reasonable and show a certain degree of agreement with the observed first phases in thin film reactions, they nevertheless are empirical criteria and they all fail to predict metastable phases, such as an amorphous alloy, to be the first phase of formation. The failure is because they all rely on equilibrium phase diagrams and they largely ignore the kinetic consideration. We know that amorphous alloys grow in thin film reactions and in essence it is a kinetic consideration.

In a first order phase transition, the compositional change is *discontinuous*, which means that the formation of a new phase must occur by nucleation. Since nucleation cannot occur without surmounting an energy barrier, the barrier imposes a selection of phase formation because the barrier will be different for different phases. The kinetic consideration leads to the criteria of first phase formation that has the lowest nucleation barrier. The nucleation barrier is given by the activation energy,

$$\Delta H_N = \frac{4b^3}{27} \frac{\gamma_{AC}^3}{\Delta H_{AC}^2} \tag{10.34}$$

Although equation (10.34) was derived for an amorphous-to-crystalline transformation, it does contain the key elements of the nucleation barrier in general, that a low value of ΔH_N is obtained by having a low interfacial energy and/or a high heat of transformation of the nucleus. Since the nucleus may interface with both A and B in the reaction between A and B, it is the sum of the interfacial energies which is minimized. For this reason, the nucleus tends to favor a simple crystal structure, an epitaxial interface with one of the reacting phases, and to take place at a high energy site such as the triple junction of grains. Recall that in Chapter 10, we explained solid phase amorphization by assuming that when the heat of transformation of an amorphous alloy and a crystalline compound are comparable, the amorphous alloy can become the first phase to form if it has a lower nucleation barrier or a lower interfacial energy.

To determine the nucleation barrier in Eq. (10.34), we need to know the geometrical constant, interfacial energy, and heat of transformation. Among them the heat of transformation can be measured readily, the geometrical constant can be taken to be the same for comparison, so it is the interfacial energy term which is unavailable for prediction. This points to the need of basic understanding and measurement of interfacial energies.

After nucleation the phase must grow. This is not an issue if the phase is the only one that can nucleate. When there are two or more phases whose nucleation barriers are nearly equal, the phase selection is then by competitive growth. Competition in growth between two layered compounds will be treated in detail in the next chapter.

References

1. D. R. Campbell, K. N. Tu, and R. E. Robinson, *Acta Met.* 24, 609 (1976).
2. E. G. Colgan, ''A Review of Thin Film Aluminide Formation,'' *Materials Science Reports* 5, 1–44 (1990).
3. L. Darken, *Trans. AIME* 174, 184 (1948).
4. U. Gösele, K. N. Tu, and R. D. Thompson, *J. Appl. Phys.* 53, 8759 (1982).
5. G. V. Kidson, *J. Nuclear Materials* 3, 21 (1961).
6. S. R. Shatynski, J. P. Hirth, and R. A. Rapp, *Acta Met.* 24, 1071 (1976).
7. P. G. Shewmon, *Diffusion in Solids*, 2nd edition, The Minerals, Metals, and Materials Society, Warrendale PA (1989).
8. J. Tardy and K. N. Tu, *Phys. Rev. B.*, 32, 2070 (1985).
9. K. N. Tu, ''Metal–Silicon Reaction,'' Chapter 7 in *Advances in Electronic Materials* edited by B. W. Wessels and G. Y. Chin, American Society for Metals, Metals Park OH (1985).

Problems

11.1 Material A reacts with material B to form the compound A_2B. Markers were placed at the A/B interface before reaction. After anneal at 250°C for 15 min, the sample was examined; the marker positions are shown in the figure accompanying this problem. Calculate the diffusion coefficient of each constituent, A and B, in A_2B.

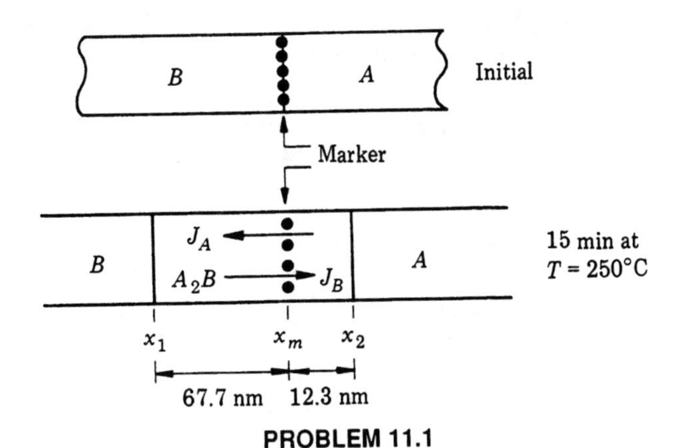

PROBLEM 11.1

11.2 Consider a couple made of elements A and B with wire markers placed at the metallurgical boundary. The couple is annealed at 150°C for two hours and a single compound A_2B is formed at the interface. You determine the kinetic data shown in the figure for this problem. Determine the diffusion coefficients: $\tilde{D}$, D_A and D_B.

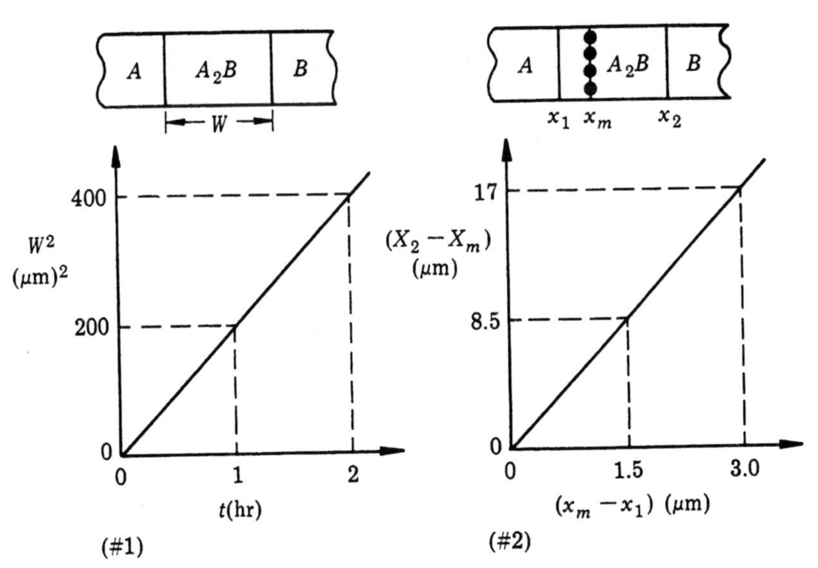

PROBLEM 11.2

11.3 In Section 11.3, the interdiffusion coefficient $\tilde{D}$ was derived for the diffusion-limited case by substituting $\eta = x/t^{1/2}$. For a reaction-limited case, $x \propto t$. Use the substitution $\gamma = x/t$.

(a) Is the Matano interface at a different location?

(b) Give the new form of $\tilde{D}(C')$ in the format of Eq. (11.10).

11.4 A diffusion couple is formed by joining Alloy 1 ($A_{.50}B_{.50}$) and Alloy 2 (B) and heating for 40 hours. Assume that $D_A = D_B = \tilde{D} = 3.04 \times 10^{-7}$ cm²/sec (constant and independent of composition).

(a) What is the concentration of A at a distance $x = 0.2$ cm and $x = -0.2$ cm?

(b) For a longer anneal time t_2, the same concentration of A (as in (a) above at $x = 0.2$ cm) was found at $x = 0.4$ cm. What is t_2?

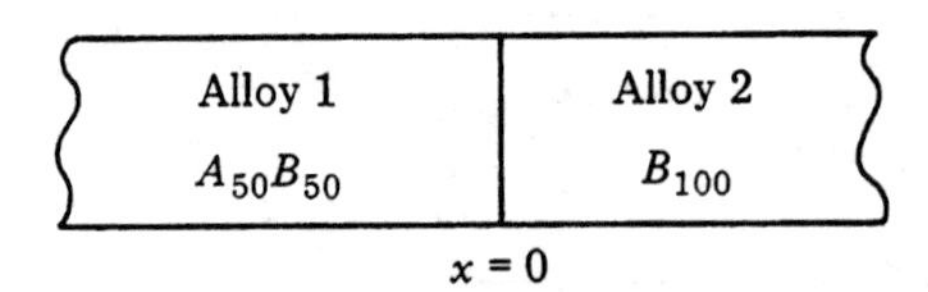

PROBLEM 11.4

11.5 **(a)** Estimate the interdiffusion coefficient $\tilde{D}$ from Fig. 11.11 if $x_1 - x_2 = 6$ microns, $C_A = 4 \times 10^{22}/\text{cm}^3$, $C_A(x_1) = 1.8 \times 10^{22}/\text{cm}^3$, and $C_A(x_2) = 1.2 \times 10^{22}/\text{cm}^3$ for an interdiffusion time of 4 hours. Use Eq. (11.10).

(b) If $\tilde{D} = 2 \times 10^{-13}$ cm²/sec and $v = 4 \times 10^{-11}$ cm/sec, calculate D_A and D_B.

11.6 A compound A_2B is formed by heating for 10^4 seconds. Measured from x_1 ($x_1 = 0$), $x_m = 40$ nm and $x_2 = 200$ nm. Calculate $\tilde{D}$ and D_A/D_B.

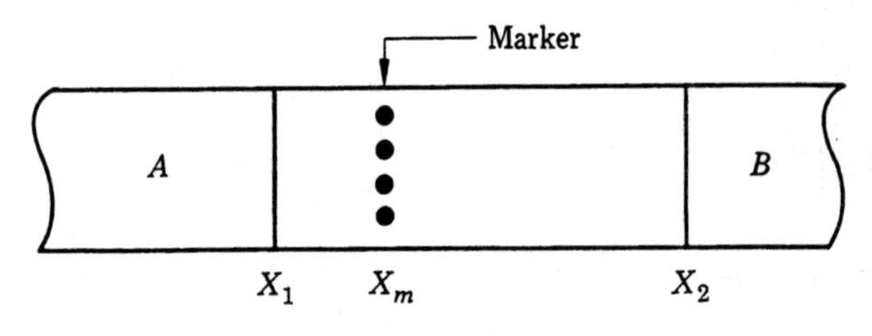

PROBLEM 11.6

11.7 Referring to Fig. 11.12, calculate the position of the $\alpha\beta$ interface given the following data.

$$\text{diffusion time} = 30 \text{ min}$$

$$D_{\beta\alpha} = 5 \times 10^{-14} \text{ cm}^2\text{sec}^{-1}$$

$$D_{\alpha\beta} = 3 \times 10^{-14}\ \text{cm}^2\text{sec}^{-1}$$

$$\frac{\partial C_{\beta\alpha}}{\partial x} = 1.79 \times 10^{26}\ \text{atoms cm}^{-1}$$

$$\frac{\partial C_{\alpha\beta}}{\partial x} = 1.43 \times 10^{25}\ \text{atoms cm}^{-1}$$

$$C_{\alpha\beta} = 2 \times 10^{22}\ \text{atoms cm}^{-3}$$

$$C_{\beta\alpha} = 1 \times 10^{22}\ \text{atoms cm}^{-3}$$

(In this problem x is measured in atoms/cm^2)

11.8 A Pb–Pb(50 atomic percent)In bulk diffusion couple is produced with Al marker wires. The Pb side is a single crystal and the PbIn side is a large-grained polycrystal (2 mm avg. grain size); thus the effects of grain boundary diffusion will be negligible. After a 118 hour anneal at 173°C, the concentration profile A2 of In (see figure) was obtained using an energy dispersive x-ray analysis unit on a scanning electron microscope. Use the Matano–Boltzmann analysis to determine the interdiffusion coefficient $\tilde{D}$ at the Matano interface. Use Darken's relations to obtain the intrinsic diffusivities of In and Pb. You may use the empirical relationship that the velocity of the Al markers = $x_m/2t$. The Al markers have moved 67 μm into the alloy. (Note: The Matano interface has an In concentration of 0.41).

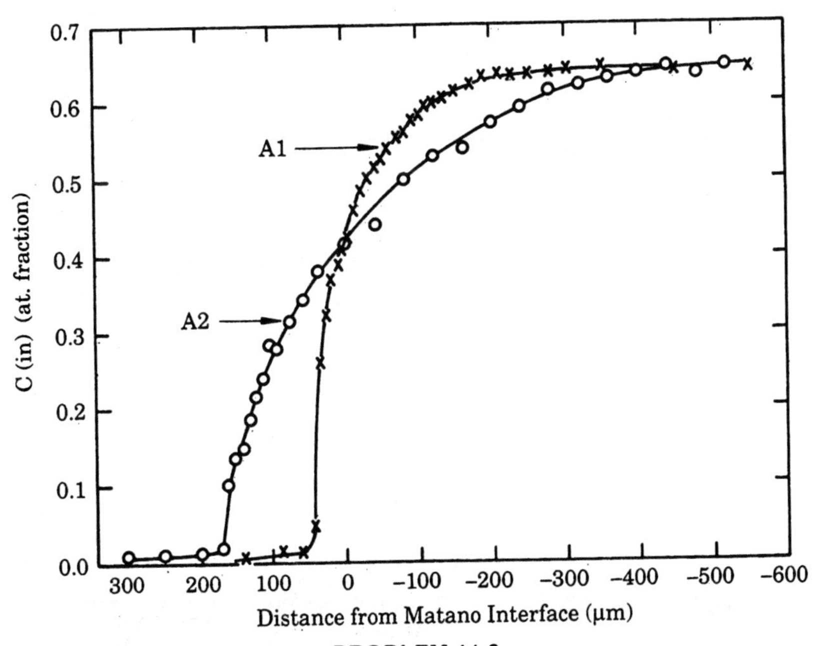

PROBLEM 11.8

CHAPTER 12

Thin Film Reactions

12.0 Compound Formation in Bulk Couples and Thin Films

In Chapter 11, we mentioned that in bulk interdiffusion couples, often one or two of the compounds are missing. This has become a dominant issue in the study of thin film reactions. Our first example is a gold–aluminum bulk couple (Philofsky,

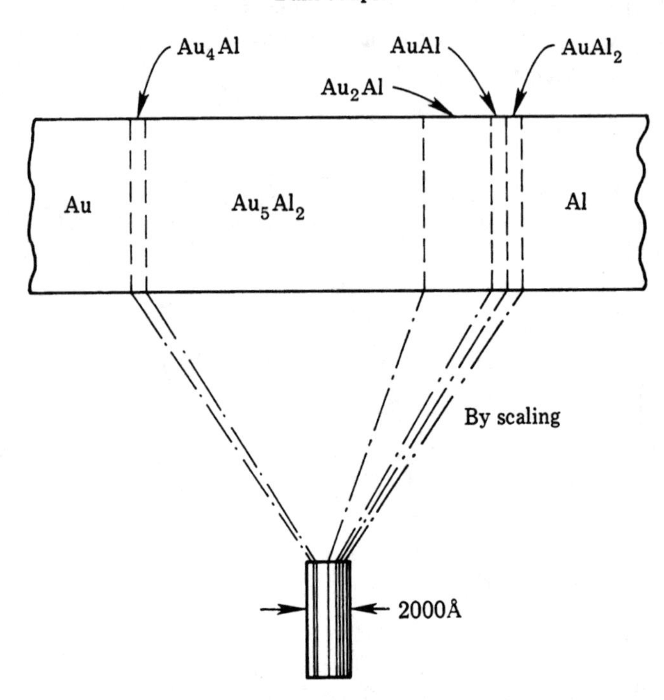

FIGURE 12.1 A sketch of intermetallic compound formation in an Au–Al bulk diffusion couple annealed at 460°C for 100 minutes (Philofsky, 1970). When scaling down to the thickness of thin films, it may be expected that all five compounds can still coexist.

1970) interdiffused at 460°C for 100 minutes; five compounds were found, as sketched in Fig 12.1, and they are formed in the correct order as expected from the binary Au–Al phase diagram. However if this couple is annealed at a different temperature such as 200°C, $AuAl_2$ and AuAl are absent. If we instead interdiffuse a bimetallic thin film, gold and aluminum, at 200°C, there is only one dominant phase formed with a composition of Au_2Al (Campisano et. al., 1975). The other phases do not form simultaneously; rather they form one by one sequentially as shown in Fig. 12.2. We cannot scale down the dimension from a bulk couple to a thin film couple and expect to find the same reaction product.

For the next example, we choose silicon and nickel. An optical cross-sectional picture is shown in Fig. 12.3. Nickel has been bonded to silicon, pressed, and then heated to 850°C for eight hours. The figure shows that several compounds form simultaneously.

On the other hand, in a thin film case, if we deposit a nickel film on a silicon wafer and anneal at 250°C for one hour it will form Ni_2Si. Figure 12.4 shows the backscattering spectrum with a step in the signal heights indicating formation of

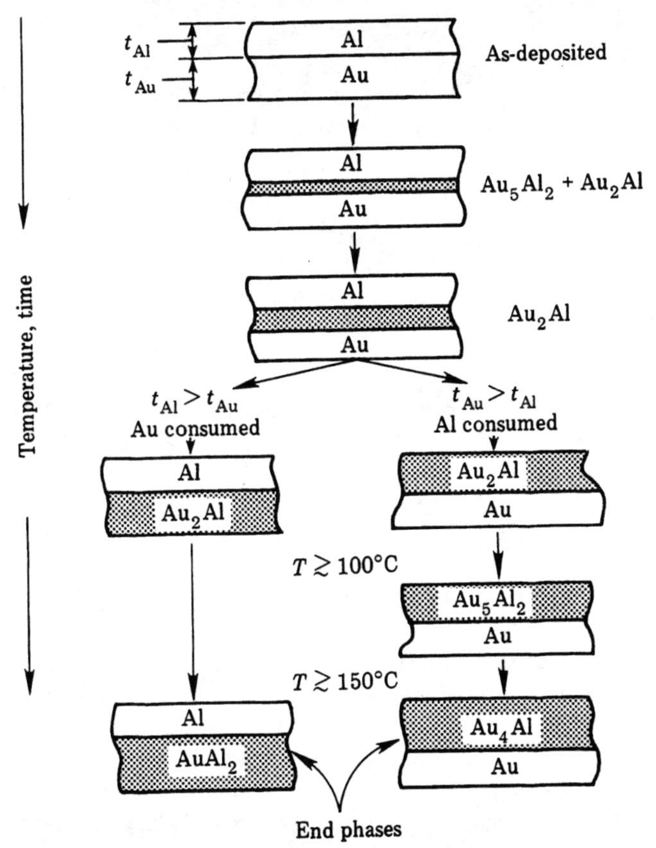

FIGURE 12.2 Compound formation in Au–Al thin film couples annealed at 250°C for 8 hours (Campisano et al., 1975).

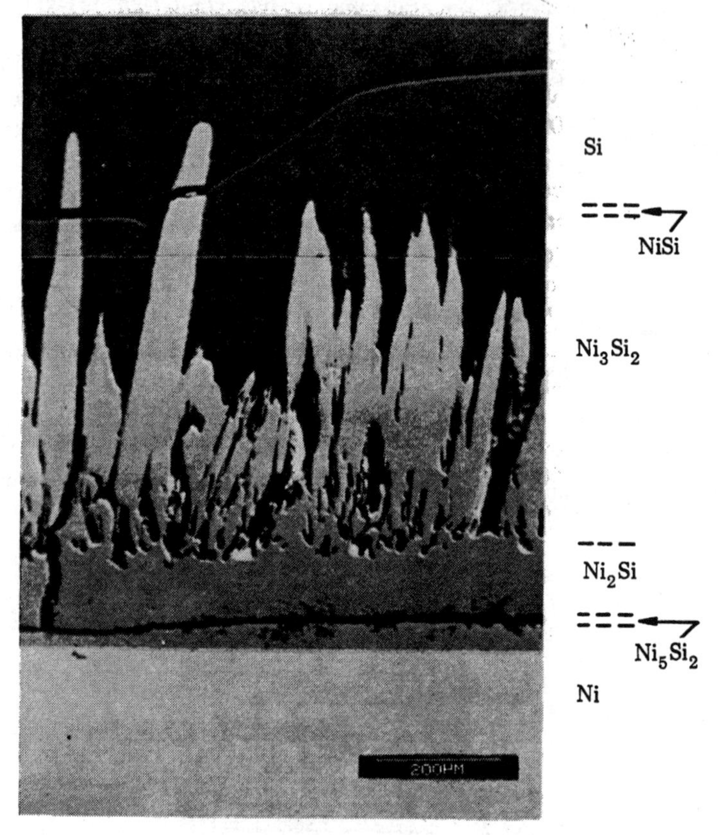

FIGURE 12.3 Cross section of Ni/Si bulk couple annealed at 850°C for 8 hours (Tu et al., 1983).

nickel silicide (Ni_2Si). If we anneal for four hours, the steps thicken. If we take an x-ray of that sample we see many reflections that can be identified with Ni_2Si, as shown in Fig. 12.5.

It shows that in reacting a nickel film with a silicon wafer, the first phase formed is Ni_2Si. When all the nickel is consumed, then NiSi is formed and after that $NiSi_2$, as shown in Fig. 12.6. If we put a thin silicon film on a thick nickel substrate, again the first phase formed is Ni_2Si, but the subsequent phases are different and are rich in nickel.

With a nickel film on a silicon wafer, the composition of the whole sample is very Si rich. According to thermodynamics, the phase in equilibrium with silicon should be $NiSi_2$. Yet the first phase formed is Ni_2Si and the second is NiSi. We cannot predict the phase sequence just by looking at the phase diagram. Phase formation is a kinetic phenomenon. At first glance one would doubt the "single phase phenomenon" because thermodynamics predicts that all phases should be there. One would suspect that multiple phases were present but that the spatial resolution was not adequate to detect the missing phases.

The missing phase question can be settled by the techniques of atomic imaging. In the cross-section transmission electron microscope imaging of nickel silicide on

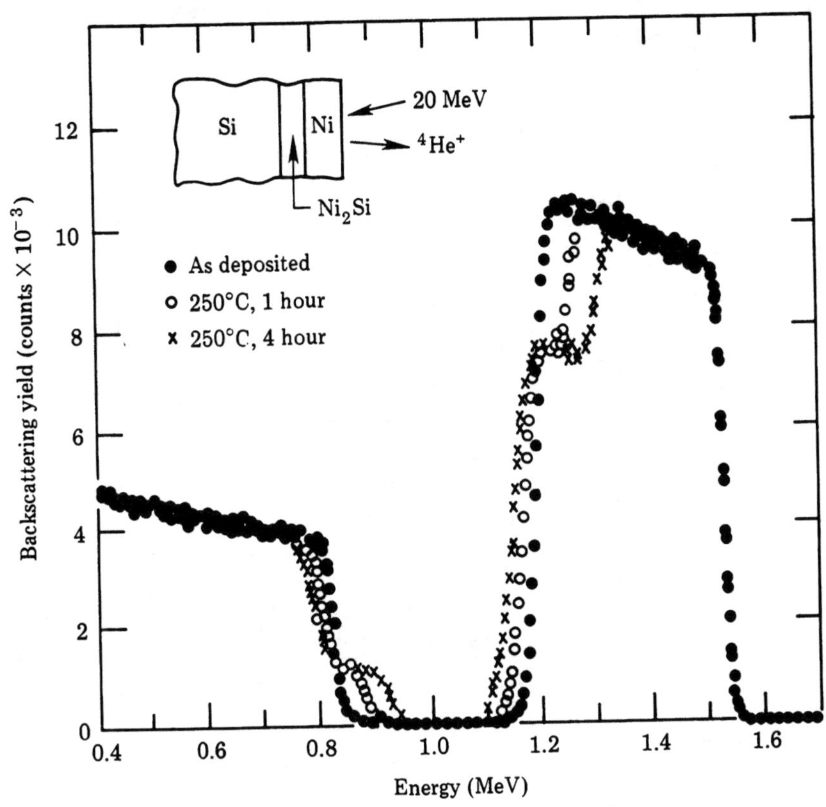

FIGURE 12.4 Rutherford backscattering spectra of a 2000Å Ni film on Si before and after annealing at 250°C for 1 and 4 hours.

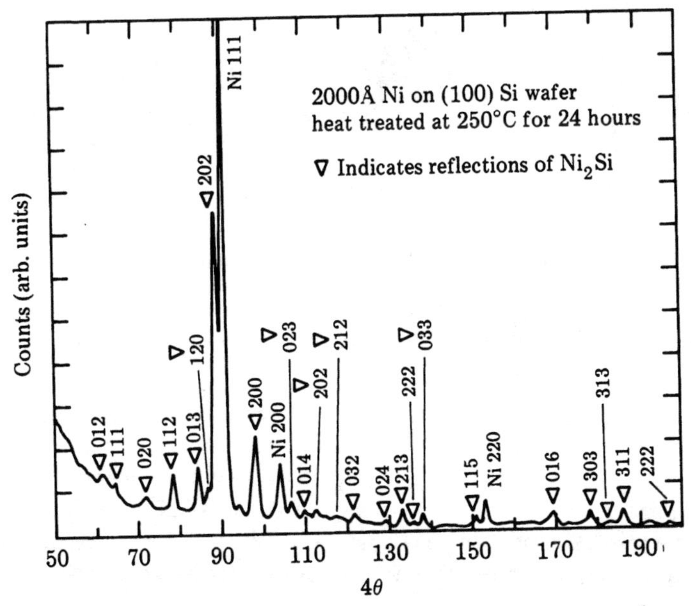

FIGURE 12.5 Grazing incidence x-ray diffraction spectrum of a 2000Å Ni film on Si after annealing at 250°C for 24 hours.

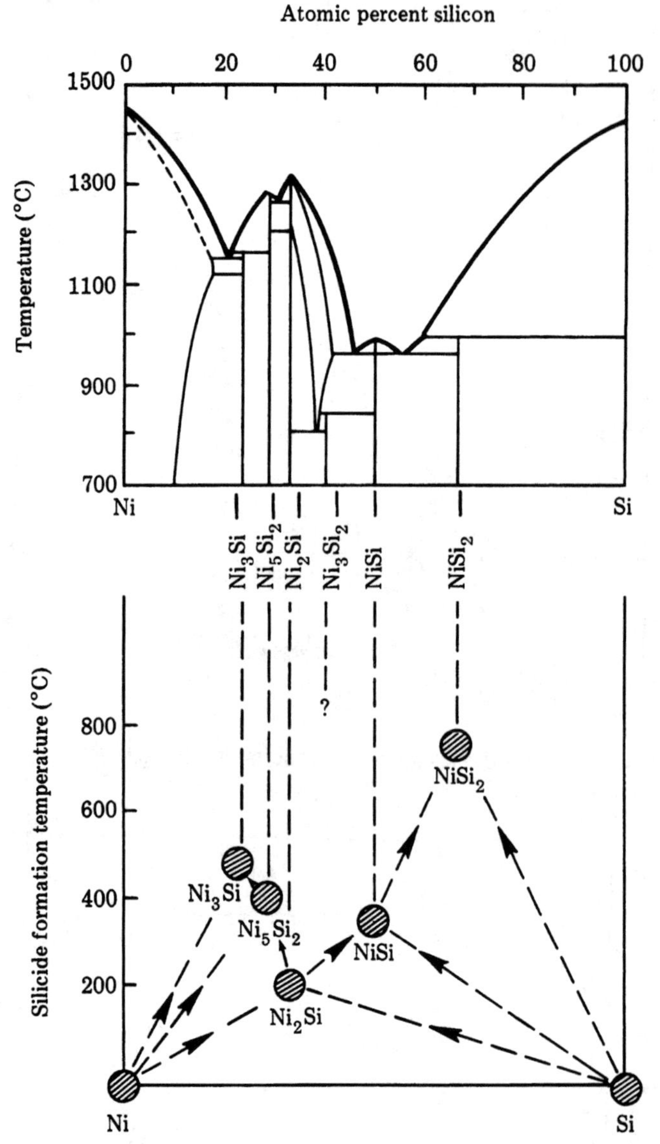

FIGURE 12.6 Phase sequence of formation in Ni/Si system.

silicon (see Fig. 12.7), the spacing between dots is about 3 angstroms. With this binary system, we should be able to tell whether there is any layer of compound missing. If the compound is there, it must have at least the dimension of a unit cell and it can be resolved easily. The image of cobalt silicide on silicon (Fig. 12.8) indicates a resolution of a single atomic layer. Whether there is a compound or not is no longer the question because we can resolve single layers. The missing phases are truly missing.

Furthermore, let us take an example of the decomposition of nickel silicide, as shown in Fig. 12.9. If we put nickel on NiSi on silicon, the nickel and NiSi react to

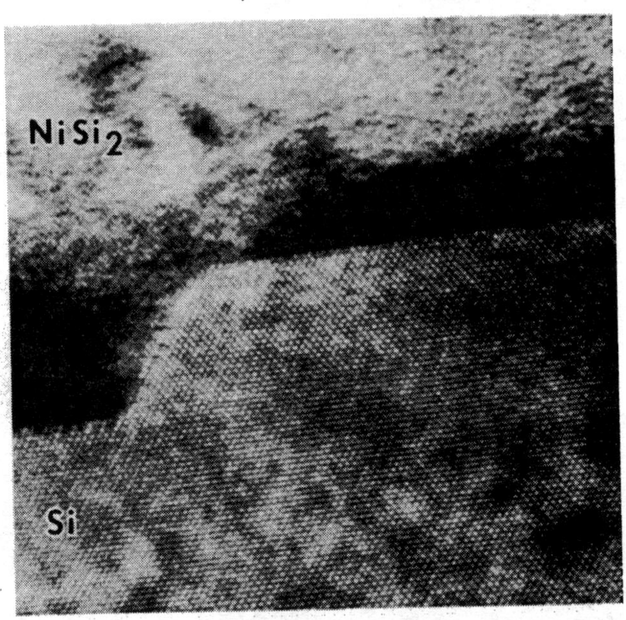

FIGURE 12.7 High-resolution image of $NiSi_2$ on (111) Si (Foell et al., 1982).

form Ni_2Si. The latter will then grow and consume NiSi. So if NiSi is the second phase, it cannot coexist with the first phase, which will appear and become dominant. Similarly, if we have CoSi formed and we deposit Co on top, CoSi will be converted into Co_2Si. Because CoSi has a higher free energy of formation than Co_2Si, formation energy is not controlling the selection of phase; rather, kinetic behavior is responsible.

In the next section, we shall introduce the concept of reaction processes controlled by kinetics. In Table 12.1, we list the characteristics of the first silicide phase formed

TABLE 12.1 Comparison of the Three Transition Metal Silicide Classes*

Characteristics	Near Noble Metal (Ni, Pd, Pt, Co., . . .)	Refractory Metal (W, Mo, V, Ta, . . .)	Rare Earth Metal (Eu, Gd, Dy, Er, . . .)
1st Phase formed	M_2Si	MSi_2	MSi_2
Formation Temperature	~200°C	~600°C	~350°C
Growth Rate	$x^2 \propto t$	$x \propto t$	?
Activiation Energy of Growth	1.1–1.5 eV	>2.5 eV	?
Dominant Diffusion Species	Metal	Si	Si
Barrier Height to n-Si	0.66–0.93 eV	0.52–0.68 eV	~0.40 eV
Resistivity	20–100μΩ-cm	13–1000μΩ-cm	100–300μΩ-cm

*[illegible] 4372 (1982).

FIGURE 12.8 High-resolution image of epitaxial Si grown on a 7 nm $CoSi_2$ film on (111) Si (courtesy of J. L. Batstone, IBM Research Division).

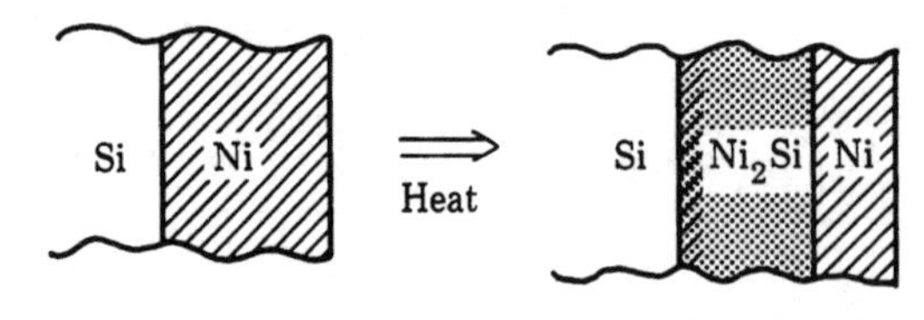

(a) Ni on Si; Ni_2Si formed by reaction of Ni with Si

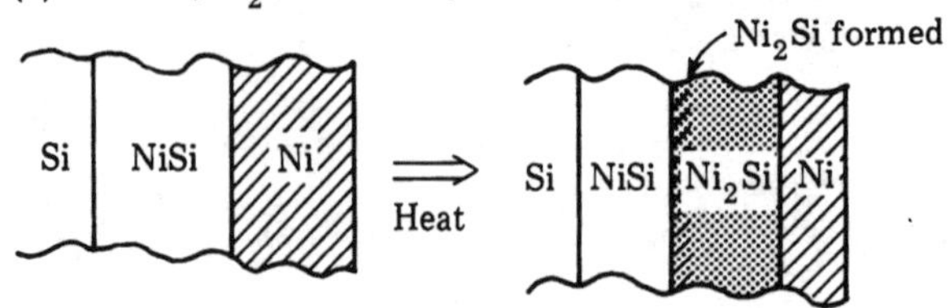

(b) Ni on NiSi; Ni_2Si formed by reaction of Ni with NiSi

FIGURE 12.9 Formation of Ni_2Si (Hung and Mayer, 1983).

between silicon and three classes of transition metals: near-noble metals, refractory metals and rare-earth metals. This table shows that there is a pattern to the formation of the first phase. Specifically, with near-noble metals (M) on Si, M_2Si grows with a square root of time dependence by metal transport, yet with refractory metals (M) on Si, MSi_2 grows with linear time dependence by Si transport. These different time behaviors are due to diffusion and interfacial-reaction-controlled growth kinetics.

12.1 Thin Film Reactions: Diffusion and Reaction Control

In order to explain the "missing compound" phenomenon and in turn the single-phase growth in thin film reactions, we shall treat the layered growth of a compound in a more general manner. Specifically, we shall consider diffusion-controlled and interfacial-reaction-controlled kinetic processes together. We first define that in a growth kinetic process if

$$\frac{J}{-\dfrac{\partial C}{\partial x}} = \text{constant} \tag{12.1}$$

we are in a diffusion-controlled regime. On the other hand, if

$$\frac{J}{C} = \text{constant} \tag{12.2}$$

we are in a reaction-controlled regime. Here J is the flux and C is the concentration. A schematic atomistic picture of the reaction-controlled process with a driving force F' is shown in Fig. 12.10. If $\lambda F'/kT << 1$ for this interface-controlled reaction, the net frequency of forward jump $\nu_n = \nu\lambda F'/kT$ (Chapter 3). The interfacial velocity v is given by

$$v = \lambda\nu_n = \nu\lambda^2 F'/kT$$

and the flux J is

$$J = Cv = C\frac{\nu\lambda^2}{kT}F'$$

We have shown in Chapter 3 that if we take the driving force

$$F = -\partial\mu/\partial x$$

and assume a dilute ideal solution, we obtain

$$\frac{J}{-(\partial C/\partial x)} = \nu\lambda^2 = D$$

It is a diffusion-controlled process (recall Eq. (12.1)). On the other hand, if we take the driving force to be constant and ignore the concentration gradient, we obtain

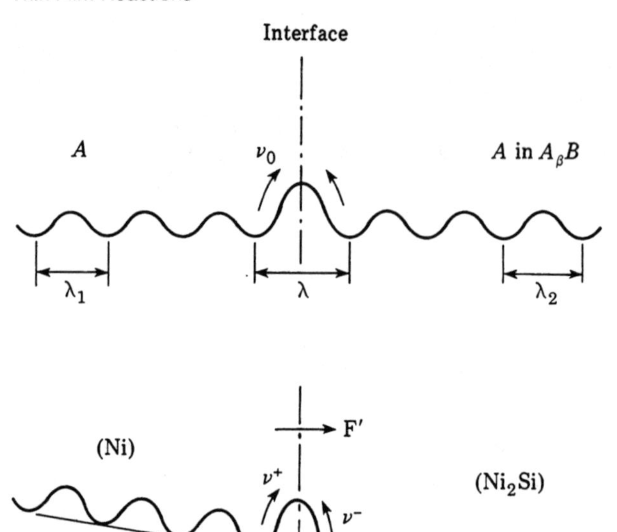

FIGURE 12.10 Schematic of energy barriers and jump frequency near the interface between A and $A_\beta B$ (Ni and Ni_2Si) for (a) equilibrium and (b) during reaction.

$$\frac{J}{C} = \frac{\nu\lambda^2}{kT} F' = \frac{D}{kT} F' = MF' = K$$

$$dx/dt = \text{constant}$$

where K is defined as a reaction-controlled interfacial constant. It has the dimension of velocity and is a measure of the "mobility" of the atoms or of an interface. A diffusion-controlled layer growth will show a parabolic relationship as given by Kidson's analysis. An interfacial-reaction-controlled growth will show an $x \propto t$ linear relationship (since velocity $= dx/dt =$ constant).

Since the concept of an interfacial-reaction-controlled process is less common than the diffusion-controlled, let us consider two examples. If one drives to New York from New Jersey over the George Washington Bridge, one must stop at the toll station to pay a fee. On holidays, cars jam up in front of the booth, and the slow rate of passing through the toll station (interface control) is independent of the jam (supersaturation). After passing the toll, one can drive away at the speed limit (diffusion control).

Another example is from crystal growth. It has been known for a long time that the growth rate on (100) or (111) surfaces of a crystal is much (many orders of magnitude) slower than what is predicted from supersaturation. The growth depends on the availability of surface steps of a screw dislocation on the growth surface (interface control). Growth can only occur on the steps and one rotation of the screw dislocation leads to the growth of one atomic layer, so it is very slow. The process does not depend on diffusion nor on supersaturation. The growth rate is constant with time.

In Eq. (3.5) we derived the relation between velocity and its driving force:

$$v = MF$$

It is worth pointing out that this is a powerful relation in kinetic analysis. From it we have derived the diffusion equation as well as the interfacial reaction equation. It can be used to describe grain growth. It also appears in other branches of science. For example, in electrical conductivity, if we let $\langle v \rangle$ be the drift velocity of electrons, we have

$$v = \mu e \mathscr{E}$$

where μ, e, and $\mathscr{E}$ are mobility, charge of the electron, and the applied field, respectively, and $e\mathscr{E}$ equals the force acting on the electron. Now if we compare atomic flux (J) to electrical current density (j), we have respectively $J = Cv$ and $j = en\langle v \rangle$, where n is the concentration or number of electrons per unit volume. Electrical current density is thus equal to charge e times flux of electrons $n\langle v \rangle$, or number of charged particles passing a unit area in a unit time.

12.2 Growth of a Layered Compound

We shall consider the growth of a layered compound controlled by both diffusion and interfacial reaction. The growth occurs because of a combination of atomic fluxes which diffuse to the interfaces and of the reactions which take place at the interfaces. We shall analyze these two processes, diffusion and reaction, together. We shall consider the growth of a single layer of a compound first, then the competing growth of two layered compounds. We represent a compound by $A_\alpha B$, where α means there are α A atoms for one B atom in the compound. For Ni_2Si, $\alpha = 2$; for $NiSi_2$, $\alpha = 1/2$. For pure Ni, $\alpha = \infty$, which means that there are an infinite number of Ni atoms for each Si atom. For pure Si, then, $\alpha = 0$.

In the case of Ni_2Si the concentration of Ni in the compound is

$$\frac{\alpha}{\alpha + 1} = \frac{2}{2 + 1} = \frac{2}{3}$$

and the concentration of Si in the compound is

$$\frac{1}{\alpha + 1} = \frac{1}{2 + 1} = \frac{1}{3}$$

We shall use this fractional representation for concentration, assuming that the atomic volumes Ω of the different elements are identical in the compound, so that they cancel out.

We shall consider the growth of a layered compound of $A_\beta B$ between $A_\alpha B$ and $A_\gamma B$ as shown in Fig. 12.11a. For simplicity we assume that the concentration of A

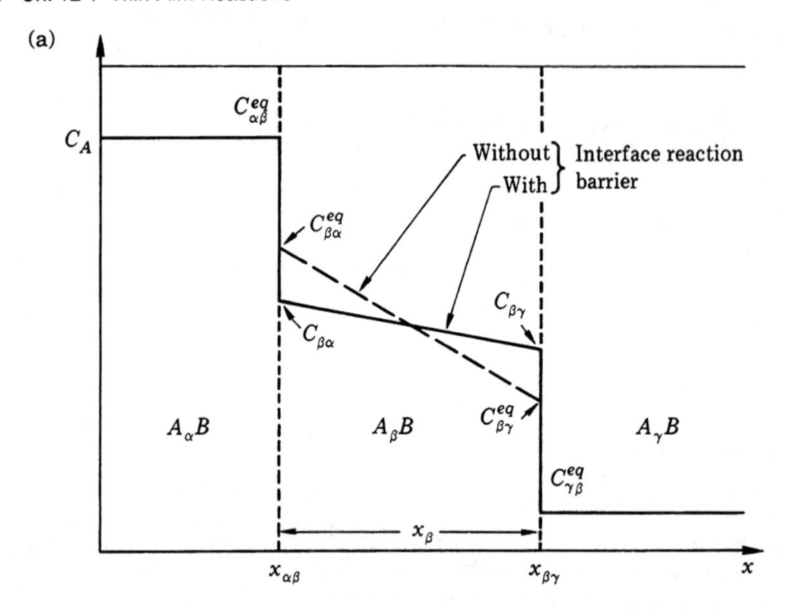

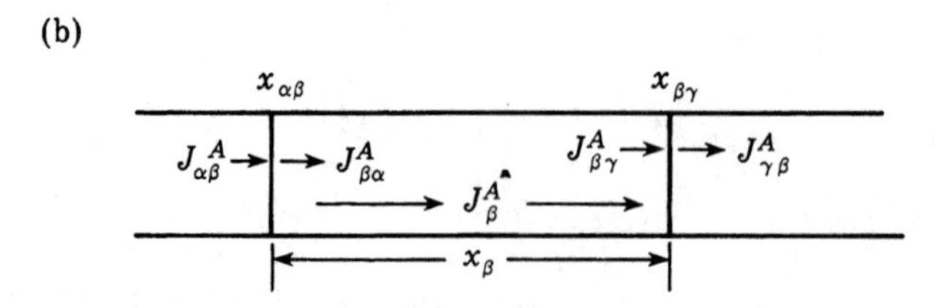

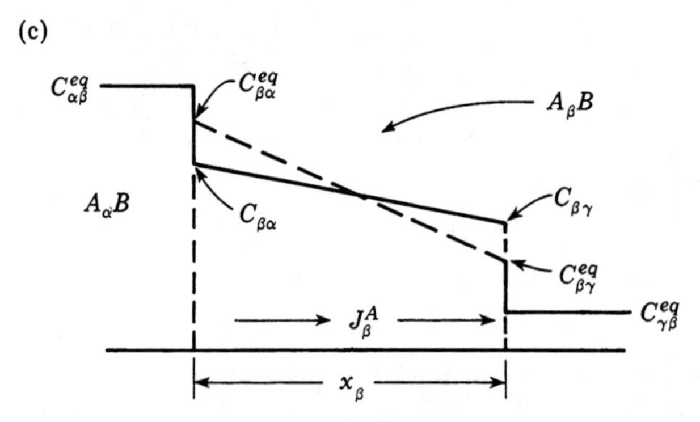

FIGURE 12.11 (a) Schematic of concentration profile (Gösele and Tu, 1982). (b) Flux diagram. (c) Schematic of the flux and the interfacial concentrations.

in $A_\alpha B$ is flat ($A_\alpha B$ is pure A so $\alpha = \infty$). Also the concentration of B in $A_\gamma B$ is flat ($A_\gamma B$ is pure B so $\gamma = 0$).

In Kidson's analysis (Section 11.7) of a diffusion-controlled growth, the concentrations at the interfaces of the compound are assumed to be constants and equal to their equilibrium values. Here, we assume that they are not the equilibrium values ($C_{\beta\alpha} < C_{\beta\alpha}^{eq}$ and $C_{\beta\gamma} > C_{\beta\gamma}^{eq}$). These inequalities specify that $C_{\beta\alpha}$ is less than and that $C_{\beta\gamma}$ is greater than equilibrium value. Consider for the moment that $A_\beta B$ is a liquid solution and is dissolving A atoms from $A_\alpha B$, which is pure A. If the dissolution rate is extremely high and is only limited by how fast the A atoms can diffuse away, then the liquid will be able to maintain the equilibrium concentration of A near the interface even though A atoms are being drained away to the other end through the $A_\beta B$ phase. On the other hand, if the process of breaking A atoms from the $A_\alpha B$ surface is slow, and any A atom which becomes free from the $A_\alpha B$ surface can quickly diffuse away, then the liquid will not be able to maintain the equilibrium concentration of A at the interface. This *reaction-controlled* process is the slower one, and $C_{\beta\alpha} < C_{\beta\alpha}^{eq}$.

At the other end of the $A_\beta B$ phase, A atoms are incorporated into the $A_\gamma B$ surface for the growth of the latter. If the incorporation can take place as soon as the atoms arrive at the surface, the equilibrium concentration will be maintained. If the surface process is slow, A atoms will accumulate in front of the surface. It becomes supersaturated and greater than the equilibrium value, and the process is therefore reaction controlled.

Now, let us consider the interface which is moving with a velocity $v = dx_{\alpha\beta}/dt$,

$$
\begin{aligned}
(C_{\alpha\beta}^{eq} - C_{\beta\alpha})\frac{dx_{\alpha\beta}}{dt} &= J_{\alpha\beta}^{A} - J_{\beta\alpha}^{A} \\
&= \left(-\tilde{D}_\alpha \frac{dC_\alpha^A}{dx}\right) - \left(-\tilde{D}_\beta \frac{dC_\beta^A}{dx}\right) \\
&= \tilde{D}_\beta \frac{dC_\beta^A}{dx} = -J_\beta^A
\end{aligned}
\tag{12.3}
$$

The term $J_{\alpha\beta}^{A}$ goes to zero because we have assumed that $A_\alpha B$ is pure A; the concentration of A is flat and its gradient $= 0$. The last equality in Eq. (12.3) is the definition of a flux equation (see Fig. 12.11b for the flux diagram) and it shows that $J_{\beta\alpha}^{A} = J_\beta^A$. In the compound $A_\beta B$ we can assume a linear concentration gradient, so

$$\frac{dC_\beta^A}{dx} = \frac{C_{\beta\alpha} - C_{\beta\gamma}}{x_\beta} \tag{12.4}$$

Therefore, we have

$$(C_{\alpha\beta}^{eq} - C_{\beta\alpha})\frac{dx_{\alpha\beta}}{dt} = \tilde{D}_\beta \frac{C_{\beta\alpha} - C_{\beta\gamma}}{x_\beta} = -J_\beta^A \tag{12.5}$$

If we consider the flux from the viewpoint of a reaction-controlled process we have

$$J_\beta^A = (C_{\beta\alpha}^{eq} - C_{\beta\alpha})K_{\beta\alpha} \tag{12.6}$$

where $K_{\beta\alpha}$ is defined as the interfacial reaction constant at the $x_{\alpha\beta}$ interface. It has units of velocity and indicates the rate of removing A from the $A_\alpha B$ surface. It is defined as

$$\frac{J_\beta^A}{C_{\beta\alpha}^{eq} - C_{\beta\alpha}} = K_{\beta\alpha} \tag{12.7}$$

If there is no surface sluggishness, $C_{\beta\alpha}$ will approach $C_{\beta\alpha}^{eq}$. Because of sluggishness, however, the actual concentration at the interface is lower than the equilibrium value, so $K_{\beta\alpha}$ is a measure of the actual flux J_β^A leaving the interface with respect to the concentration change at the interface. The sluggish reaction is the cause of the reduction of the flux. We note that if $C_{\beta\alpha}$ approaches $C_{\beta\alpha}^{eq}$, the concentration difference across the interface is smaller, yet the interface velocity is greater. When the process is reaction controlled, the concentration difference across the interface is greater, yet the interface velocity is slower and the actual flux is smaller. In this case, the mobility of the interface is lower.

We then have similarly at the $x_{\beta\gamma}$ interface,

$$\begin{aligned}(C_{\beta\gamma} - C_{\gamma\beta}^{eq})\frac{dx_{\beta\gamma}}{dt} &= -\tilde{D}_\beta\left(\frac{dC_\beta^A}{dx}\right)\\ &= -\tilde{D}_\beta\frac{C_{\beta\alpha} - C_{\beta\gamma}}{x_\beta} \qquad (12.8)\\ &= J_\beta^A\end{aligned}$$

$$J_\beta^A = (C_{\beta\gamma} - C_{\beta\gamma}^{eq})K_{\beta\gamma} \tag{12.9}$$

From Eq. (12.6) and Eq. (12.9) we have, respectively

$$\frac{J_\beta^A}{K_{\beta\alpha}} = C_{\beta\alpha}^{eq} - C_{\beta\alpha} \tag{12.10}$$

$$\frac{J_\beta^A}{K_{\beta\gamma}} = C_{\beta\gamma} - C_{\beta\gamma}^{eq} \tag{12.11}$$

By adding Eq. (12.10) and Eq. (12.11)

$$J_\beta^A\left(\frac{1}{K_{\beta\alpha}} + \frac{1}{K_{\beta\gamma}}\right) = (C_{\beta\gamma} - C_{\beta\alpha}) + (C_{\beta\alpha}^{eq} - C_{\beta\gamma}^{eq}) \tag{12.12}$$

Let

$$\frac{1}{K_\beta^{eff}} = \frac{1}{K_{\beta\alpha}} + \frac{1}{K_{\beta\gamma}} \tag{12.13}$$

and from Eq. (12.5),

$$C_{\beta\gamma} - C_{\beta\alpha} = -\frac{J_\beta^A x_\beta}{\tilde{D}_\beta}$$

Also let (see Fig. 12.11c),

$$\Delta C_\beta^{eq} = C_{\beta\alpha}^{eq} - C_{\beta\gamma}^{eq} \tag{12.14}$$

and we obtain

$$J_\beta^A\left(\frac{1}{K_\beta^{eff}} + \frac{x_\beta}{\tilde{D}_\beta}\right) = \Delta C_\beta^{eq} \tag{12.15a}$$

or

$$J_\beta^A = \frac{\Delta C_\beta^{eq} K_\beta^{eff}}{\left(1 + \frac{x_\beta K_\beta^{eff}}{\tilde{D}_\beta}\right)}. \tag{12.15b}$$

Now to calculate the thickening rate of $A_\beta B$, we obtain from Eq. (12.5) and Eq. (12.8) respectively

$$\frac{dx_{\alpha\beta}}{dt} = \frac{-1}{C_{\alpha\beta}^{eq} - C_{\beta\alpha}} J_\beta^A$$

$$\frac{dx_{\beta\gamma}}{dt} = \frac{1}{C_{\beta\gamma} - C_{\gamma\beta}^{eq}} J_\beta^A$$

$$\begin{aligned} \frac{dx_\beta}{dt} &= \frac{d}{dt}(x_{\beta\gamma} - x_{\alpha\beta}) \\ &= \left[\frac{1}{C_{\beta\gamma} - C_{\gamma\beta}^{eq}} + \frac{1}{C_{\alpha\beta}^{eq} - C_{\beta\alpha}}\right] J_\beta^A \\ &= G_\beta J_\beta^A \end{aligned} \tag{12.16}$$

where G_β is an inverse concentration parameter for $A_\beta B$ (see Fig. 12.11c for the interfacial concentrations). By substituting J_β^A from Eq. (12.15b)

$$\frac{dx_\beta}{dt} = \frac{G_\beta \Delta C_\beta^{eq} K_\beta^{eff}}{1 + x_\beta \frac{K_\beta^{eff}}{\tilde{D}_\beta}}. \tag{12.17}$$

If we define a "changeover" thickness x_β^* for the change from reaction-controlled to diffusion-controlled, where $x_\beta^* = \tilde{D}_\beta / K_\beta^{eff}$

$$\frac{dx_\beta}{dt} = \frac{G_\beta \Delta C_\beta^{eq} K_\beta^{eff}}{1 + \frac{x_\beta}{x_\beta^*}} \tag{12.18}$$

For a large x_β^* (i.e., $x_\beta / x_\beta^* \to 0$), $\tilde{D}_\beta >> K_\beta^{eff}$ and the process is reaction controlled. We then have

$$\frac{dx_\beta}{dt} \simeq G_\beta \Delta C_\beta^{eq} K_\beta^{eff} \quad \text{or} \quad x_\beta \propto t$$

For a small x_β^* (i.e., $x_\beta / x_\beta^* >> 1$), $\tilde{D}_\beta << K_\beta^{eff}$ and the process is diffusion controlled.

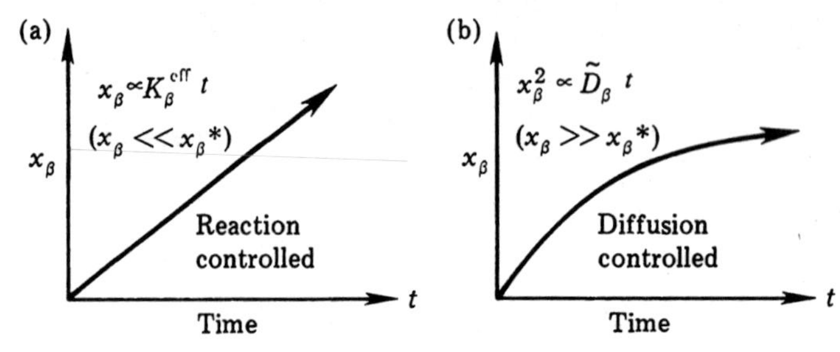

FIGURE 12.12 The layer width x_β versus time t for (a) reaction-controlled and (b) diffusion-controlled compound growth.

We can drop the ''1'' in the denominator, and then

$$\frac{dx_\beta}{dt} \simeq G_\beta \Delta C_\beta^{eq} \frac{\tilde{D}_\beta}{x_\beta} \quad \text{or} \quad x_\beta^2 \propto t$$

This demonstrates the well-known relationships that in an interface-reaction-controlled growth, $x_\beta \propto t$(see Fig. 12.12a), but in a diffusion-controlled growth, $x_\beta \propto t^{1/2}$ (see Fig. 12.12b). Furthermore, a reaction-controlled growth will always change over to a diffusion-controlled growth if the layer thickenss x_β has grown sufficiently large that $x_\beta >> x_\beta^*$.

12.3 Growth of Two Layered Compounds

In Section 12.2, we discussed the growth of a single compound layer where both diffusion-controlled and interfacial-reaction-controlled processes are operative simultaneously. We shall next consider the growth of two layered compounds. This will allow us to analyze their competition in growth, which eventually leads to the vanishing of one of them. Thus we will determine the kinetic criteria of ''missing compounds.''

In Fig. 12.13a we have the two compounds $A_\beta B$ and $A_\gamma B$ growing between $A_\alpha B$ and $A_\delta B$. Again we assume that $A_\alpha B$ is pure A and that $A_\delta B$ is pure B. The equilibrium concentrations C_{ij}^{eq} and nonequilibrium concentrations C_{ij} at the interfaces x_{ij} are not shown, but they follow from those in Fig. 12.11a. At the $x_{\alpha\beta}$ interface, we have as before in Eq. (12.3)

$$(C_{\alpha\beta}^{eq} - C_{\beta\alpha}) \frac{dx_{\alpha\beta}}{dt} = -J_\beta^A \tag{12.19}$$

At the $x_{\beta\gamma}$ interface, we have in Fig. 12.13b,

$$\begin{aligned}(C_{\beta\gamma} - C_{\gamma\beta}) \frac{dx_{\beta\gamma}}{dt} &= J_{\beta\gamma}^A - J_{\gamma\beta}^A \\ &= \left[-\tilde{D}_\beta \frac{(C_{\beta\alpha} - C_{\beta\gamma})}{x_\beta} \right] - \left[-\tilde{D}_\gamma \frac{(C_{\gamma\beta} - C_{\gamma\delta})}{x_\gamma} \right] \end{aligned} \tag{12.20}$$

(a)

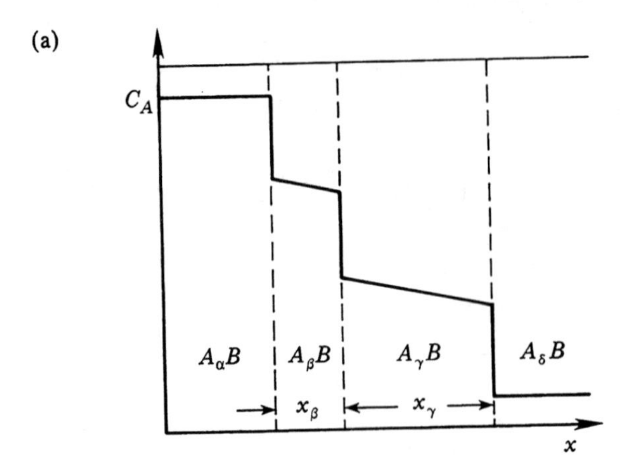

(b)

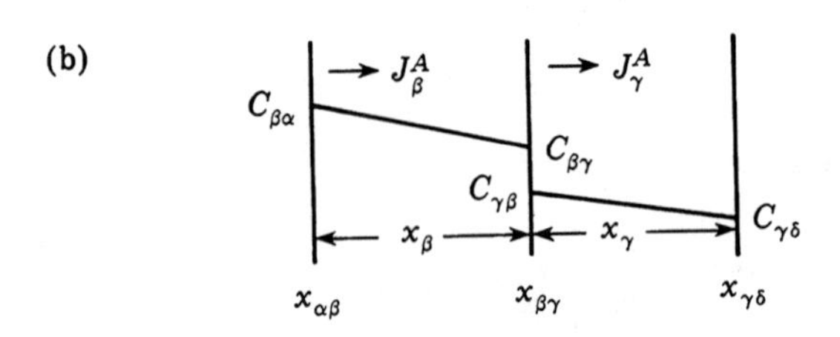

(c)

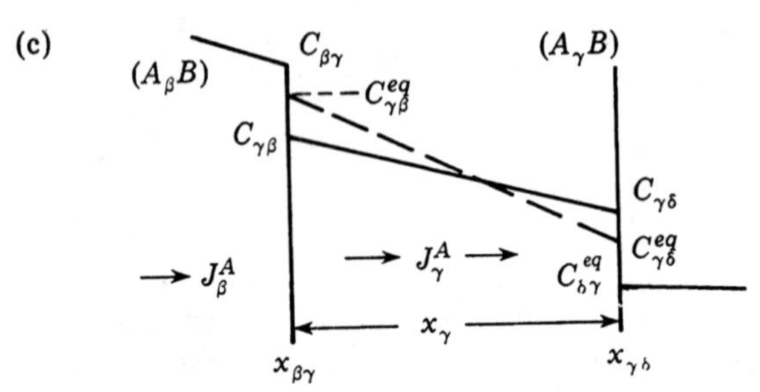

FIGURE 12.13 (a) Concentration profile of A atoms in the simultaneous growth of two layered compounds. (b) Schematic of the flux and interfacial concentrations of the two growing layers. (c) Schematic of the interfacial concentrations of the $A_\gamma B$ phase.

$$= J^A_\beta - J^A_\gamma$$

At the $x_{\gamma\delta}$ interface (see Fig. 12.13c),

$$(C_{\gamma\delta} - C^{eq}_{\delta\gamma}) \frac{dx_{\gamma\delta}}{dt} = J^A_{\gamma\delta} - J^A_{\delta\gamma}$$

$$= -\tilde{D}_\gamma \frac{dC_\gamma^A}{dx} \tag{12.21}$$

$$= J_\gamma^A$$

where $J_{\delta\gamma}^A = 0$ because $A_\delta B$ is pure B. The thickening of $A_\beta B$

$$\frac{dx_\beta}{dt} = \frac{d}{dt}(x_{\beta\gamma} - x_{\alpha\beta})$$

$$= \left[\frac{1}{C_{\beta\gamma} - C_{\gamma\beta}} + \frac{1}{C_{\alpha\beta}^{eq} - C_{\beta\alpha}}\right] J_\beta^A - \frac{1}{C_{\beta\gamma} - C_{\gamma\beta}} J_\gamma^A$$

$$\frac{dx_\beta}{dt} = G_\beta J_\beta^A - G_{\beta\gamma} J_\gamma^A \tag{12.22}$$

where

$$G_{\beta\gamma} = \frac{1}{C_{\beta\gamma} - C_{\gamma\beta}}$$

Similarly, the thickening of $A_\gamma B$

$$\frac{dx_\gamma}{dt} = \frac{d}{dt}(x_{\gamma\delta} - x_{\beta\gamma})$$

$$= \left[\frac{1}{C_{\gamma\delta} - C_{\delta\gamma}^{eq}} + \frac{1}{C_{\beta\gamma} - C_{\gamma\beta}}\right] J_\gamma^A$$

$$- \left[\frac{1}{C_{\beta\gamma} - C_{\gamma\beta}}\right] J_\beta^A$$

$$\frac{dx_\gamma}{dt} = G_\gamma J_\gamma^A - G_{\gamma\beta} J_\beta^A \tag{12.23}$$

Equations (12.22) and (12.23) are the pair of equations describing the simultaneous growth of x_β and x_γ, the thicknesses of the layered compounds $A_\beta B$ and $A_\gamma B$. To consider whether one of them may vanish due to competition, we include the effects of diffusion-controlled and reaction-controlled kinetics in the following discussion.

For the two interfaces of $A_\beta B$, we have

$$J_{\beta\alpha}^A = (C_{\beta\alpha}^{eq} - C_{\beta\alpha})K_{\beta\alpha}$$

$$J_{\beta\gamma}^A = (C_{\beta\gamma} - C_{\beta\gamma}^{eq})K_{\beta\gamma}$$

For the two interfaces of $A_\gamma B$, we have

$$J_{\gamma\beta}^A = (C_{\gamma\beta}^{eq} - C_{\gamma\beta})K_{\gamma\beta}$$

$$J_{\gamma\delta}^A = (C_{\gamma\delta} - C_{\gamma\delta}^{eq})K_{\gamma\delta}$$

Following the same procedure as we did in the last section, we can derive

$$J_\beta^A = \frac{\Delta C_\beta^{eq} K_\beta^{eff}}{1 + x_\beta \dfrac{K_\beta^{eff}}{\tilde{D}_\beta}} \tag{12.24}$$

$$J_\gamma^A = \frac{\Delta C_\gamma^{eq} K_\gamma^{eff}}{1 + x_\gamma \dfrac{K_\gamma^{eff}}{\tilde{D}_\gamma}} \tag{12.25}$$

where

$$\Delta C_\beta^{eq} = C_{\beta\alpha}^{eq} - C_{\beta\gamma}^{eq}$$

$$\Delta C_\gamma^{eq} = C_{\gamma\beta}^{eq} - C_{\gamma\delta}^{eq}$$

$$\frac{1}{K_\beta^{eff}} = \frac{1}{K_{\beta\alpha}} + \frac{1}{K_{\beta\gamma}}$$

$$\frac{1}{K_\gamma^{eff}} = \frac{1}{K_{\gamma\beta}} + \frac{1}{K_{\gamma\delta}}$$

Now if we substitute Eq. (12.24) and Eq. (12.25) into Eq. (12.22) and Eq. (12.23), we have the two relations for discussion of the competing growth of the two layered compounds (i.e., the criteria for their growth or shrinkage). It is easy to see that there is a positive and a negative term in both Eq. (12.22) and Eq. (12.23), so the layers can either grow or shrink.

For $A_\beta B$, we know that

$$\frac{dx_\beta}{dt} > 0 \qquad \Rightarrow \text{it grows}$$

$$\frac{dx_\beta}{dt} < 0 \qquad \Rightarrow \text{it shrinks}$$

From eq. (12.22) this means

$$G_\beta J_\beta^A - G_{\beta\gamma} J_\gamma^A > 0 \quad \text{for growth}$$

$$< 0 \quad \text{for shrinkage}$$

or

$$\frac{J_\beta^A}{J_\gamma^A} > \frac{G_{\beta\gamma}}{G_\beta} \quad \text{for growth of } A_\beta B$$

$$\frac{G_{\beta\gamma}}{G_\beta} = r_1$$

and r_1 is a ratio of composition, to be derived in the following.

For the concentration parameter G_β,

$$G_\beta = \frac{1}{C_{\beta\gamma} - C^{eq}_{\gamma\beta}} + \frac{1}{C^{eq}_{\alpha\beta} - C_{\beta\alpha}}$$

Recall that the concentration of A in $A_\beta B$ is $\beta/(1 + \beta)$. If we take

$$C_{\beta\alpha} \simeq C_{\beta\gamma} = C^{eq}_{\beta} = \frac{\beta}{1 + \beta}$$

(in most cases, $C_{\beta\alpha} - C_{\beta\gamma}$ is less than a few percent)

$$C^{eq}_{\alpha\beta} = \frac{\alpha}{1 + \alpha} \quad \text{and} \quad C^{eq}_{\gamma\beta} = \frac{\gamma}{1 + \gamma}$$

$$G_\beta = \frac{1}{\dfrac{\alpha}{1 + \alpha} - \dfrac{\beta}{1 + \beta}} + \frac{1}{\dfrac{\beta}{1 + \beta} - \dfrac{\gamma}{1 + \gamma}}$$

$$= (1 + \beta)^2 \left(\frac{1}{\alpha - \beta} + \frac{1}{\beta - \gamma} \right)$$

$$= \frac{(1 + \beta)^2(\alpha - \gamma)}{(\alpha - \beta)(\beta - \gamma)}$$

$$G_{\beta\gamma} = \frac{1}{\dfrac{\beta}{\beta + 1} - \dfrac{\gamma}{\gamma + 1}} = \frac{(1 + \beta)(1 + \gamma)}{\beta - \gamma}$$

$$r_1 = \frac{G_{\beta\gamma}}{G_\beta} = \frac{(1 + \beta)(1 + \gamma)}{(\beta - \gamma)} \times \frac{(\alpha - \beta)(\beta - \gamma)}{(1 + \beta)^2(\alpha - \gamma)}$$

$$= \frac{(1 + \gamma)(\alpha - \beta)}{(1 + \beta)(\alpha - \gamma)}$$

We repeat, if

$$\frac{J^A_\beta}{J^A_\gamma} > r_1, \quad A_\beta B \text{ grows}$$
$$\frac{J^A_\beta}{J^A_\gamma} < r_1, \quad A_\beta B \text{ shrinks} \tag{12.26}$$

Similarly, the criteria for $A_\gamma B$ to grow or to shrink are

$$G_\gamma J^A_\gamma - G_{\gamma\beta} J^A_\beta > 0 \quad \text{for growth}$$
$$< 0 \quad \text{for shrinkage}$$

or

$$\frac{J^A_\beta}{J^A_\gamma} < \frac{G_\gamma}{G_{\gamma\beta}} \quad \text{for } A_\gamma B \text{ to grow}$$

$$\frac{G_\gamma}{G_{\gamma\beta}} = \frac{(1 + \gamma)(\beta - \delta)}{(1 + \beta)(\gamma - \delta)} = r_2$$

where again r_2 is a constant of composition as derived previously for r_1 with

$$G_{\gamma\beta} = G_{\beta\gamma}$$

$$G_\gamma = \frac{(1 + \gamma)^2(\beta - \delta)}{(\beta - \gamma)(\gamma - \delta)}$$

See Eq. (12.22) and Eq. (12.23) for comparison. So

$$\frac{J_\beta^A}{J_\gamma^A} < r_2 \quad \text{for } A_\gamma B \text{ growth}$$
$$> r_2 \quad \text{for } A_\gamma B \text{ shrinkage} \tag{12.27}$$

Note that there is a change of the inequality sign for the growth and shrinkage of $A_\beta B$ and $A_\gamma B$, respectively. We see that

$$\frac{r_1}{r_2} = \frac{(\alpha - \beta)(\gamma - \delta)}{(\alpha - \gamma)(\beta - \delta)} < 1$$

This follows from the relations

$$(\alpha - \gamma) > (\alpha - \beta)$$
$$(\beta - \delta) > (\gamma - \delta)$$

(see the compositional profile shown in Fig. 12.13). Therefore

$$r_2 > r_1$$

Now if

$$r_1 < \frac{J_\beta^A}{J_\gamma^A} < r_2$$

both $A_\beta B$ and $A_\gamma B$ grow together. If the ratio J_β^A/J_γ^A is outside the range of r_1 to r_2, then only one of them will grow and the other will shrink. Specifically, if

$$\frac{J_\beta^A}{J_\gamma^A} < r_1$$

we see from Eq. (12.26) that $A_\beta B$ shrinks. Since $r_2 > r_1$, this also means

$$\frac{J_\beta^A}{J_\gamma^A} < r_1 < r_2$$

and then we see from Eq. (12.27) that $A_\gamma B$ grows. On the other hand, if

$$\frac{J_\beta^A}{J_\gamma^A} > r_2$$

we find that $A_\gamma B$ shrinks while $A_\beta B$ grows. We illustrate the above behaviors in

(a)

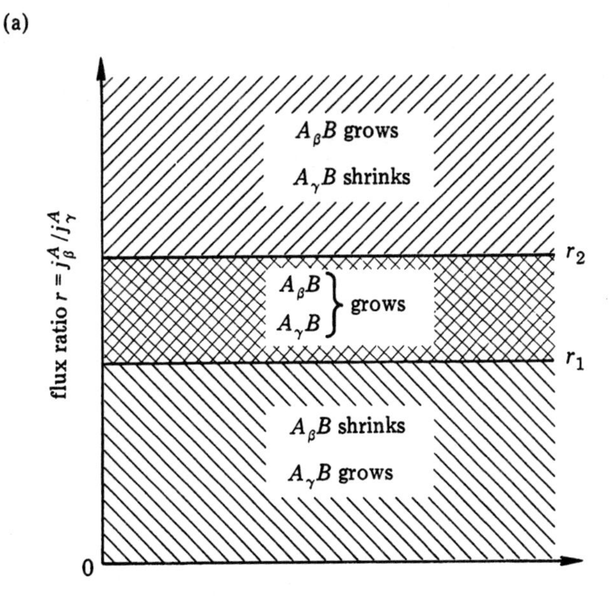

(b)

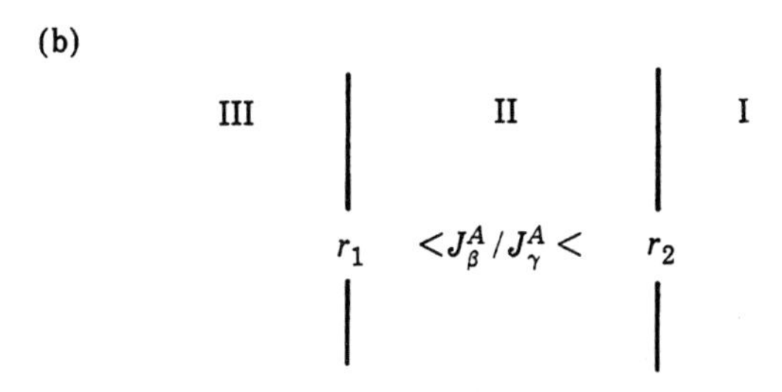

FIGURE 12.14 (a) Growth and shrinkage regimes for the couple shown in Figure 12.13 (Gösele and Tu, 1982). (b) The three regions of the ratio J_β^A / J_γ^A.

Fig. 12.14a. The question whether the shrinking compound will shrink away completely or not depends on the diffusion-controlled or interface-controlled behavior to be discussed.

Let us consider a specific case where one layer shrinks away completely. We assume that each layer can grow either diffusion controlled or interface controlled. Then we have three combinations

1. Both are diffusion controlled.
2. Both are interface controlled.
3. One is diffusion controlled and the other is interface controlled.

Let us take the last case and assume that

1. $A_\beta B$ is *interface* controlled (i.e., $x_\beta << x_\beta^*$)

$$J_{\beta}^{A} = \frac{C_{\beta}^{eq}K_{\beta}^{eff}}{1 + x_{\beta}\dfrac{K_{\beta}^{eff}}{\tilde{D}_{\beta}}} = \frac{\Delta C_{\beta}^{eq}K_{\beta}^{eff}}{1 + \dfrac{x_{\beta}}{x_{\beta}^{*}}}$$

$$\simeq \Delta C_{\beta}^{eq}K_{\beta}^{eff}$$

2. $A_{\gamma}B$ is *diffusion* controlled (i.e., $x_{\gamma} >> x_{\gamma}^{*}$)

$$J_{\gamma}^{A} = \frac{\Delta C_{\gamma}^{eq}K_{\gamma}^{eff}}{1 + x_{\gamma}\dfrac{K_{\gamma}^{eff}}{\tilde{D}_{\gamma}}} = \frac{\Delta C_{\gamma}^{eq}K_{\gamma}^{eff}}{1 + \dfrac{x_{\gamma}}{x_{\gamma}^{*}}}$$

$$= \frac{\Delta C_{\gamma}^{eq}\tilde{D}_{\gamma}}{x_{\gamma}}$$

The flux ratio is

$$\frac{J_{\beta}^{A}}{J_{\gamma}^{A}} = \frac{\Delta C_{\beta}^{eq}K_{\beta}^{eff}}{\Delta C_{\gamma}^{eq}\tilde{D}_{\gamma}} \cdot x_{\gamma} \tag{12.28}$$

Now we consider the three regions of the ratio as shown in Fig. 12.14b. In region I,

$$\frac{J_{\beta}^{A}}{J_{\gamma}^{A}} > r_2 > r_1$$

so that $A_{\gamma}B$ shrinks and $A_{\beta}B$ grows according to the criteria of Eqs. (12.26) and (12.27). The ratio will decrease because x_{γ} decreases (see Eq. (12.28) until it becomes equal to r_2, then $A_{\gamma}B$ will stop shrinking, and both compounds will grow together thereafter. This later situation is represented by region 2 where both grow simultaneously.

In region 3 (this is the most interesting case),

$$\frac{J_{\beta}^{A}}{J_{\gamma}^{A}} < r_1 < r_2$$

$A_{\beta}B$ will shrink and $A_{\gamma}B$ will grow and we see from Eq. (12.28) that the ratio does not depend on x_{β},the thickness of the shrinking $A_{\beta}B$, but only on x_{γ}, the thickness of the growing $A_{\gamma}B$. Thus as long as x_{γ} is below the critical thickness x_{γ}^{crit}

$$x_{\gamma}^{crit} = \frac{\Delta C_{\gamma}^{eq}\tilde{D}_{\gamma}}{\Delta C_{\beta}^{eq}K_{\beta}^{eff}} \cdot \frac{J_{\beta}^{A}}{J_{\gamma}^{A}} \tag{12.28a}$$

the layer $A_{\beta}B$ can shrink away entirely. In the case that layer $A_{\gamma}B$ has grown to the critical thickness before $A_{\beta}B$ shrinks away, then they can coexist and grow together. This can be seen in Eq. (12.28): as x_{γ} increases, the ratio increases, and it soon approaches r_1. We also employ Eq. (12.28) to obtain (by rearranging) the expression given for the critical thickness.

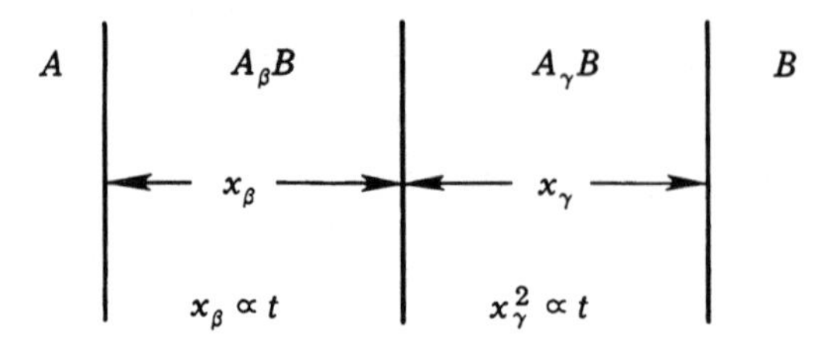

$\dot{x}_\beta = dx_\beta/dt = \text{CONST}$ $\qquad dx_\gamma/dt \propto 1/x_\gamma$

FIGURE 12.15 Schematic diagram showing the competing growth between two layered compounds.

We can interpret the situation in region 3 simply as shown in Fig. 12.15. While $A_\beta B$ grows interface-controlled at a constant velocity (since $x \propto t$), $A_\gamma B$ grows diffusion-controlled with a velocity inversely proportional to thickness. When x_γ is small (below the critical thickness), we have $\dot{x}_\gamma > \dot{x}_\beta$, so $\dot{x}_\gamma$ takes over as long as $\dot{x}_\beta$ is constant. Beyond the critical thickness, $\dot{x}_\gamma$ slows down because $\dot{x}_\gamma \propto 1/x_\gamma$. Then $A_\beta B$ has an opportunity to grow together with $A_\gamma B$.

The critical thickness x_γ^{crit} is estimated to be of the order of magnitude of microns, and is measured by lateral growth in films, which will be discussed in the next section. Since most thin film diffusion couples have thicknesses in the hundreds of nm, we see only a single compound layer growth, if the above criteria are satisfied. The conclusion is that we can explain single phase growth.

In the case of a metal film deposited on Si wafer, such as Ni on Si wafer, we need to discuss the second phase growth from the viewpoint of "supply limitation" or "source exhaustion", which means that the Ni is completely reacted in forming Ni_2Si. Assume that $A_\gamma B$ is the first phase formed, as just discussed (x_γ is below the critical thickness, x_γ^{crit}) and also assume that the end element B has been consumed. We now examine Eq. (12.22) and Eq. (12.23) and let J_γ^A go to zero. Then,

$$\frac{dx_\beta}{dt} = G_\beta J_\beta^A$$

$$\frac{dx_\gamma}{dt} = -G_{\gamma\beta} J_\gamma^A$$

This means x_γ will shrink and now x_β will grow. So the second phase (say NiSi) starts to grow (see Fig. 12.16).

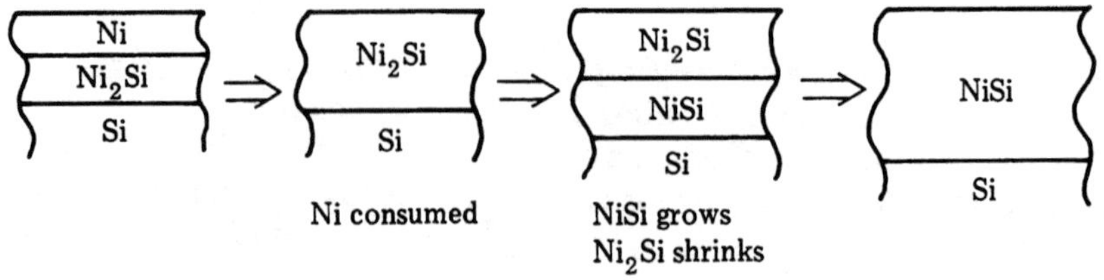

Ni / Ni_2Si / Si $\Rightarrow$ Ni_2Si / Si $\Rightarrow$ Ni_2Si / NiSi / Si $\Rightarrow$ NiSi / Si

Ni consumed

NiSi grows
Ni_2Si shrinks

FIGURE 12.16 Sequential growth of Ni_2Si and NiSi in a source-limited case.

12.4 Lateral Diffusion Couples

We have derived an expression for the critical thickness of a single compound layer. The critical thickness defines the thickness beyond which the second compound layer can coexist with the first one. In most thin film reactions, we cannot observe the critical thickness because the total film thickness is much thinner than the critical thickness. The sequential second compound formation in a thin film is not due to the fact that the first one has reached its critical thickness; rather, its source of supply is exhausted.

The existence of a critical thickness can be verified from study of lateral diffusion couples of thin films (Zheng et al., 1982 and Blanpain et al., 1990). In Fig. 12.17, a schematic diagram of a thin film laterial diffusion couple is shown. The advantages of using such samples are the long diffusion distances (same as bulk samples) and the ease of performing compositional and structural measurements (using thin film techniques such as Rutherford backscattering spectroscopy and transmission electron microscopy).

Figure 12.18 shows a bright-field TEM image of a Ni–Si couple annealed at 750°C, forming five intermetallic compounds of Ni_3Si, Ni_5Si_2, Ni_2Si, Ni_3Si_2, and NiSi. Four of the compounds are the same as those observed in the bulk couple shown in Fig. 12.3. By measuring the compound formation at short intervals of time, the sequence of formation can be followed. In Fig. 12.19, a plot of width (thickness) of compounds against annealing time is shown. The first compound is Ni_2Si, and it reaches a thickness of 20μm before the second compound NiSi appears. The lateral diffusion couple and the conventional bulk diffusion couple behave alike. In the Al/Pd system, the first phase Al_3Pd_2 reached a critical length of about 6 μm before the second phase Al_3Pd began to grow (Blanpain et al., 1990).

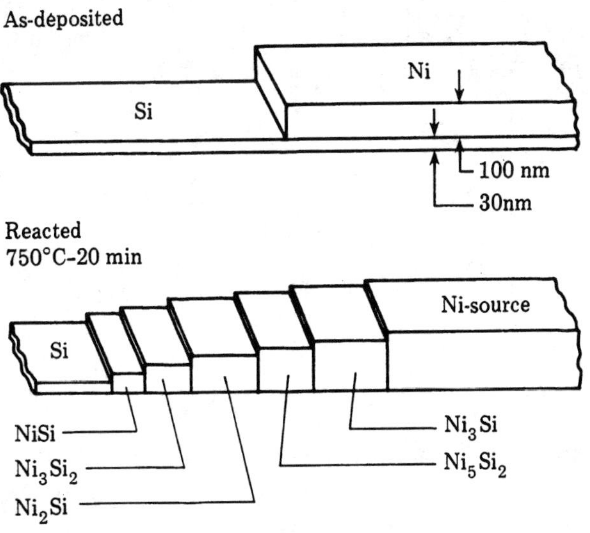

FIGURE 12.17 Schematic diagram showing a Ni/Si lateral diffusion couple before and after annealing at 750°C for 20 minutes (S. H. Chen et al., 1985b).

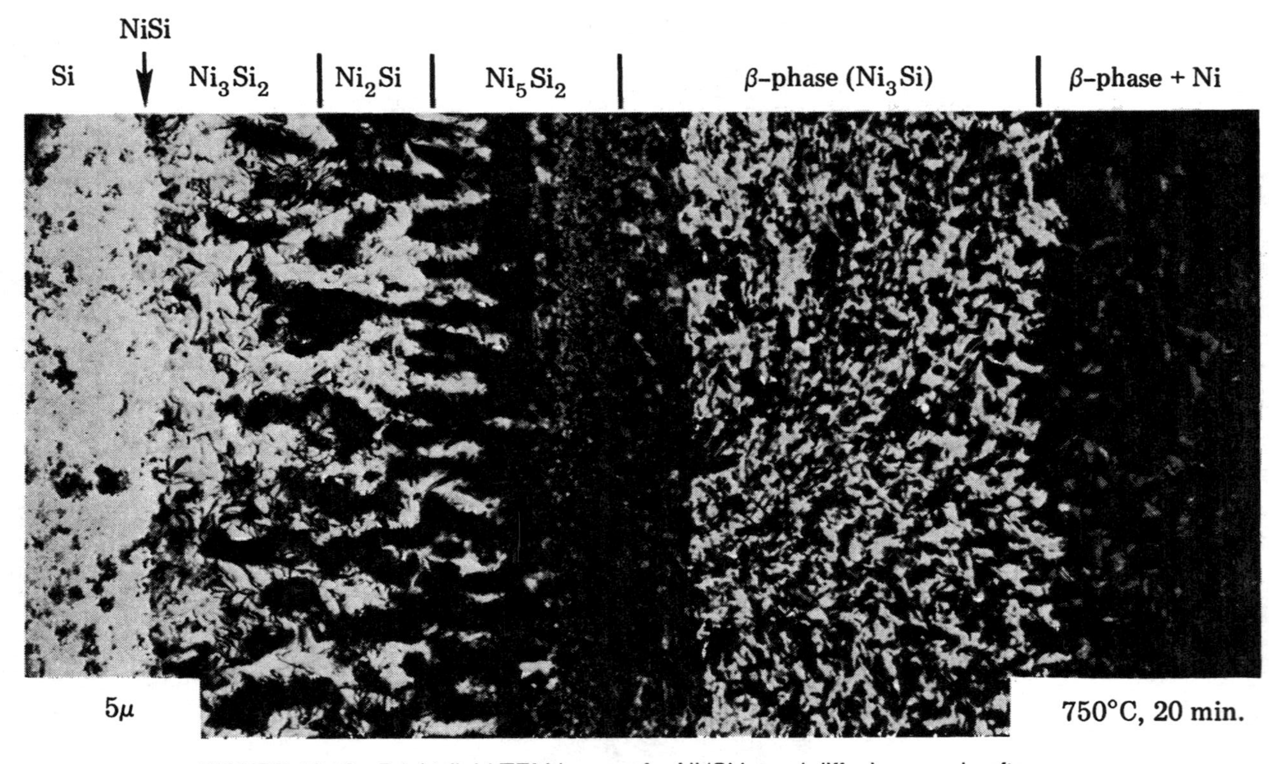

FIGURE 12.18 Bright-field TEM image of a Ni/Si lateral diffusion couple after annealing at 750°C for 20 minutes (S. H. Chen et al., 1985a).

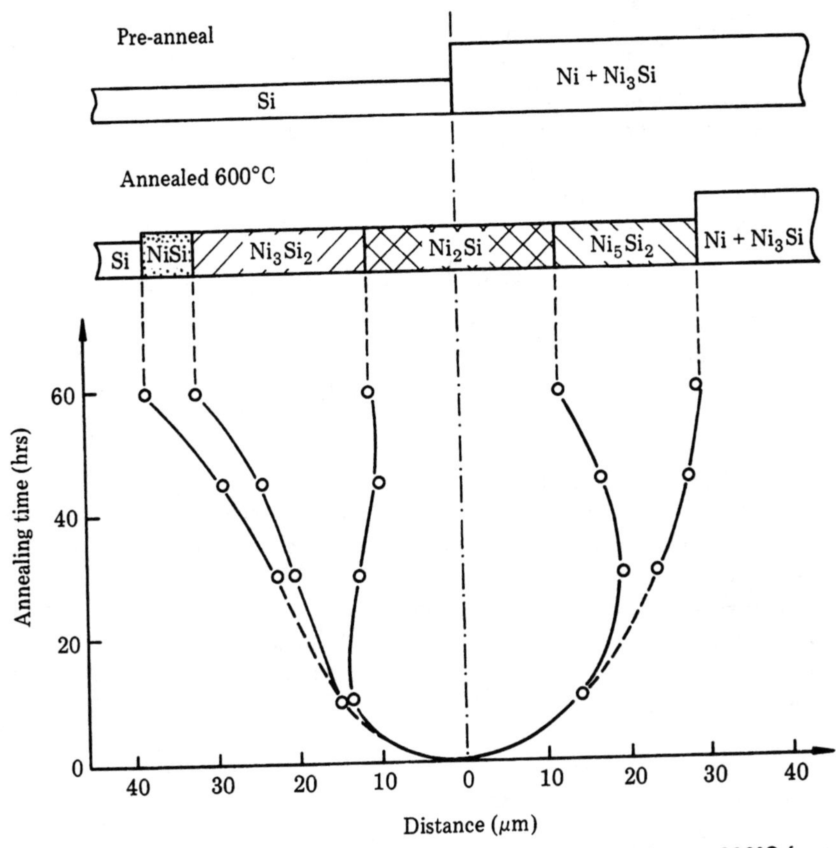

FIGURE 12.19 Length of individual phases versus annealing time at 600°C for a Ni/Si lateral diffusion couple (Zheng, 1985).

12.5 Kinetic Parameters and Measurements

In the beginning of this chapter, we discussed the experimental observation of compound formation in thin film reactions by optical and electron transmission microscopy, Rutherford backscattering, and x-ray diffraction. We emphasized the combined use of several experimental techniques to unravel the reaction kinetics. We correlated morphological and structural change with kinetic behavior. We also emphasized atomic resolution observation in order to understand atomic processes.

In measuring reaction kinetics, the procedure is to determine the transition of events as a function of time and temperature and to develop a method of observing this in a controlled manner. The goal is to analyze or understand the kinetic processes and parameters which govern the reaction. In Chapter 2 we discussed the thermodynamic variables: p and V, T and S, μ and N. In kinetics, the basic parameters of interest are: atomic diffusivities, surface and interface energies, interfacial reaction constants, and enthalpy (heat) of formation.

To relate these parameters to a reaction, we recall Fig. 10.4, where the activation barrier to a reaction consisted of surface energies spent in nucleation and of the activation energy of atomic diffusion. To overcome the barrier, heat is supplied. The reaction produces a product because of the gain in formation energy of the product, which is the driving force of the reaction. Under a driving force, the reaction rate can be diffusion controlled or interfacial reaction controlled.

Note that the formation energy gained (or released) from the reaction is generally dissipated because we use external heat to maintain the temperature of the reaction and keep it going. On the other hand, we could burn a piece of wood or a rod of magnesium and the reaction would be self-sustained. It is sustained by the energy released from the burning. We can employ a similiar process in thin films. Using an alternating layered structure of, for example, twenty layers of Ni and amorphous Si, each layer about 5 nm in thickness, we can achieve a self-sustained explosive reaction. The very thin layer thickness allows a high rate of reaction between layers, and the heat released simultaneously from all the layers can provide a sufficient amount of heat to keep the reaction going once it is ignited. These self-propagating explosive reactions have also been found in metal–metal multilayered thin films (Ma et al., 1990).

To measure the heat of compound formation, the technique used is differential scanning calorimetry. The heating is typically performed by ramping the temperature up at a constant rate until the reaction occurs. Table 12.2 lists the heat of formation of most silicides. As we have discussed in Chapter 2, the heat of compound formation is of the order of 1 eV/atom (or 23 Kcal/mole). When the heat in Kcal/mole is divided by the number of atoms in the molecule, the unit is changed to Kcal/g-atom. In Table 12.2, take Mg_2Si for example; its heat of formation is 6.2 Kcal/g-atom or 18.6 Kcal/mole.

The measurement of surface and interface energies has been explained in Chapters 2 and 10; we will not repeat it here. The kinetic parameters left for our consideration are diffusivities and interfacial reaction constants. While the subject of diffusion was discussed in Chapter 3, we did not address the problem of how to deduce the diffusivity in a thin film reaction. Because most solid-state reactions are found to be diffusion controlled, it is not surprising that most kinetic studies in thin films have been performed to determine diffusivities (mainly the activation enthalpy) rather than interfacial reaction constants; see Table 12.1.

A reaction by interdiffusion in a bilayered thin film structure can produce a solid solution, a sequence of intermetallic compounds, or an eutectic structure. We have discussed the first two in our treatment of bulk diffusion couples in Chapter 11. The last one occurs for example, in Al/Si and Au/Si binary systems.

To analyze a thin film reaction, we must first find out what is the product. This information can be obtained by x-ray or electron diffraction. Then we need to know the morphological and compositional changes with time and temperature during the reaction. For example, in Fig. 12.4, the RBS spectra show that Ni_2Si grows in a layered mode (there are steps in the compositional profile) and that the growth is diffusion controlled (the width of the step changes with the square root of annealing time). Although RBS has in-depth resolution of 20 nm, it lacks lateral resolution.

TABLE 12.2 Free Energy of Formation ΔH of Silicides*

Silicide	ΔH kcal g-atom^{-1}	Silicide	ΔH kcal g-atom^{-1}	Silicide	ΔH kcal g-atom^{-1}
Mg_2Si	6.2	Ti_5Si_3	17.3	V_3Si	6.5
		$TiSi$	15.5	V_5Si_3	11.8
$FeSi$	8.8	$TiSi_2$	10.7	VSi_2	24.3
$FeSi_2$	6.2				
		Zr_2Si	16.7	Nb_5Si_3	10.9
Co_2Si	9.2	Zr_5Si_3	18.3	$NbSi_2$	10.7
$CoSi$	12	$ZrSi$	18.5, 17.7		
$CoSi_2$	8.2	$ZrSi_2$	12.9, 11.9	Ta_5Si_3	9.5
				$TaSi_2$	8.7, 9.3
Ni_2Si	11.2, 10.5	$HfSi$			
$NiSi$	10.3	$HfSi_2$		Cr_3Si	7.5
				Cr_5Si_3	8
Pd_2Si	6.9			$CrSi$	7.5
$PdSi$	6.9			$CrSi_2$	7.7
				Mo_3Si	5.6
Pt_2Si	6.9			Mo_5Si_3	8.5
$PtSi$	7.9			$MoSi_2$	8.7, 10.5
$RhSi$	8.1			W_5Si_3	5
				WSi_2	7.3

*J. M. Poate, K. N. Tu, and J. W. Mayer, editors, *Thin Films: Interdiffusion and Reactions*, Wiley–Interscience, New York (1978).

Hence, it is necessary to verify the microscopic layered morphology by cross-sectional TEM. When the dependence of the layer thickness (x) on time (t) at several isothermal annealings is measured, we can plot x versus t and x versus $t^{1/2}$ to see if the rate of the reaction is controlled by diffusion or by interfacial reaction. For a diffusion-controlled reaction, we take the relation

$$x^2 = 4\tilde{D}t \tag{12.29}$$

where $\tilde{D} = \tilde{D}_0 \exp(-\Delta H/kT)$. We can plot x^2/t versus $1/kT$ and the slope of the plot gives the activation enthalpy which controls the reaction. On the other hand, if the reaction is interfacial reaction controlled, we use the relation,

$$x = Kt \tag{12.30}$$

where $K = K_0 \exp(-\Delta H^1/kT)$, and plot x/t versus $1/kT$ to determine the enthalpy ΔH^1. Note that this method of analyzing the reaction is based on isothermal annealing processes. There is another way of performing the annealing, which is by ramping the temperature up (or down) at a constant heating (or cooling) rate. During the ramping an in situ measurement of resistivity or of heat change (by differential scanning calorimetry) can be used to monitor the reaction.

12.6 Analysis of Kinetics by Temperature Ramp

To analyze the ramping data, we need to know whether the reaction is diffusion controlled or interfacial reaction controlled. The ramping technique alone will not give us this information directly. We need RBS and/or TEM to obtain compositional and morphological information. But the advantages of using the ramping technique are that the temperature range in which the reaction occurs can be found quickly, and fewer experimental runs are needed for analyzing the kinetics. For example, if the reaction has been found to be diffusion controlled and the growth of the intermetallic compound follows a layered mode, we begin with

$$x^2 = \int_0^t 4\tilde{D}\, dt = \int_0^T 4\tilde{D} \frac{dt}{dT}\, dT \tag{12.31}$$

where x is the thickness of the compound formed, $\tilde{D}$ is the interdiffusion coefficient in the compound, t is the reaction time, and dT/dt is the constant ramping rate. If we assume that $\tilde{D} = \tilde{D}_0 \exp(-\Delta H/kT)$ where ΔH is the activation enthalpy (which is temperature independent) of interdiffusion, and $\tilde{D}_0$ is the pre-exponential factor and kT has the usual meaning, we obtain the following equation by integration,

$$x^2 = 4\tilde{D}_0 \frac{dt}{dT} \frac{kT^2}{\Delta H} \exp\left(-\frac{\Delta H}{kT}\right) \tag{12.32}$$

In the integration, at $t = 0$ the temperature is typically low where no reaction can occur, so we can take $T = 0$ K. Equation (12.32) shows that if we fix x, we can determine ΔH by plotting ln $(1/T^2)(dT/dt)$ versus $1/kT$. Since a fixed x is equivalent to a fixed change of resistivity, for example, in our measurement, we can choose the midpoint of change of resistivity as shown in Fig. 12.20. The midpoint can be determined rather accurately, and in turn we can determine the temperature T_x corresponding to the chosen x by reading it on the horizontal axis in Fig. 12.20. In the experiment, we can easily vary dT/dt by two orders of magnitude, but T_x typically varies only by 10%, so that it is sufficient to plot ln (dT/dt) versus $1/kT$ for determining ΔH.

Knowing ΔH, the only unknown in Eq. (12.32) is $\tilde{D}_0$, and we can calculate $\tilde{D}_0$ provided that we determine the compound thickness which corresponds to the midpoint change of resistivity. This can be done by RBS or TEM observation of a sample annealed to the midpoint. On the other hand, since we know the original thickness of the thin films in the bilayer sample, we can calculate the total thickness of the compound formed if we know the unit cell or the density of the compound. The temperature in the ramping curve where the reaction is completed can be determined from the resistivity curve at the point where the change starts to flatten out. So $\tilde{D}_0$ can be calculated by knowing x, T_x, dT/dt, and ΔH. Combining the pre-exponential factor and the activation enthalpy, we have determined the interdiffusion coefficient.

For an interfacial-reaction-controlled reaction, we begin with

$$x = \int_0^t K\, dt = \int_0^T K \frac{dt}{dT}\, dT \tag{12.33}$$

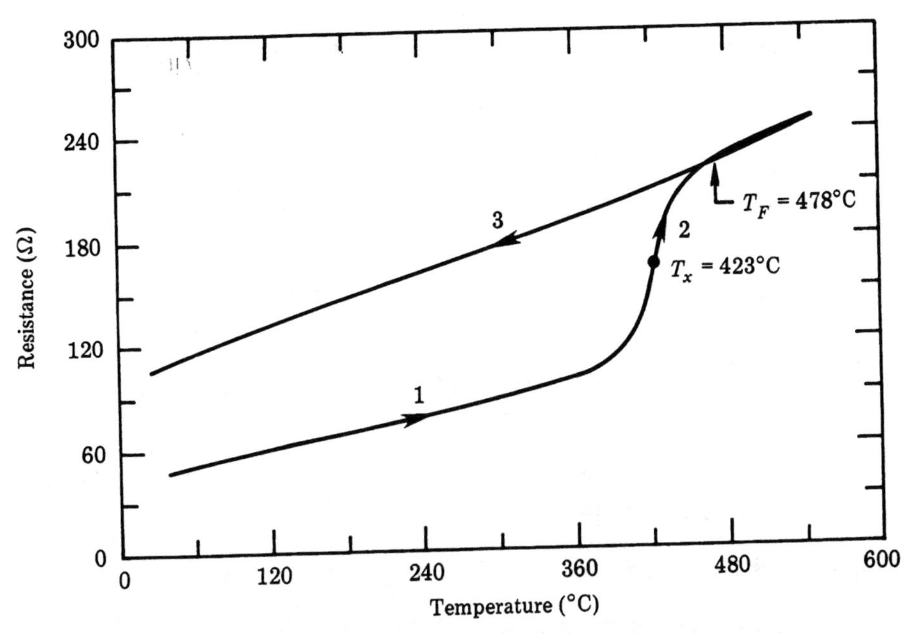

FIGURE 12.20 In situ resistivity change during ramping at a constant rate of 3°C/min, for a thin film sample of Al/Hf. The increase in resistance in region 1 is due to the temperature coefficient of resistivity. The increase in region 2 is due to Al_3Hf formation. The points T_x and T_F are taken to be the midpoint and endpoint of the reaction, respectively (K. Rodbell et. al., 1991).

where $K = K_0 \exp(-\Delta H^1/kT)$ is the interfacial reaction constant. By integration, we obtain the following equation,

$$x = K_0 \frac{dt}{dT} \frac{kT^2}{\Delta H^1} \exp\left(-\frac{\Delta H^1}{kT}\right) \tag{12.34}$$

Again it shows that if we fix x, we can determine ΔH^1 by plotting ln $(1/T^2)(dT/dt)$ versus $1/kT$. The procedure is the same as in the diffusion-controlled reaction. The pre-exponential factor K_0 is calculated from Eq. (12.34) by knowing x, T_x, dT/dt, and ΔH^1.

Comparing Eq. (12.32) to Eq. (12.34), we see that for determining the activation enthalpy of a reaction, whether it is controlled by diffusion or by interfacial reaction, we can just plot dT/dt versus $1/kT$ at a fixed amount of reaction.

In applying the ramping technique, it is implicitly assumed that the mode of the reaction is unchanged and that the product phase is the same in the temperature range scanned. At a high ramping rate the shorter time of annealing must be compensated by ramping to a higher temperature, so the risk of a change in the mode of reaction and in the product phase is much greater. The ramping rate must therefore be kept low to avoid this risk.

References

1. B. Blanpain, J. W. Mayer, J. C. Liu, and K. N. Tu, *J. Appl. Phys.* 68, 3259 (1990).

2. S. U. Campisano, G. Foti, E. Rimini, S. S. Lau, and J. W. Mayer, *Phil. Mag.* 31, 903 (1975).

3. L. J. Chen and K. N. Tu, *Materials Science Reports* 6, 53 (1991).

4. S. H. Chen, L. R. Zheng, C. B. Carter, and J. W. Mayer, *J. Appl. Phys*, S7, 258 (1985a); S. H. Chen, Z. Elgat, J. C. Barbour, L. R. Zheng, J. W. Mayer, and C. B. Carter, *Ultramicroscopy* 18, 297 (1985b).

5. E. G. Colgan, "A Review of Thin Film Aluminide Formation," *Materials Science Reports* 5, 1–44 (1990).

6. H. Foell, P. S. Ho, and K. N. Tu, *Philos. Mag. A* 45, 32 (1982).

7. U. Gösele and K. N. Tu, *J. Appl. Phys*, 53, 3252 (1982).

8. E. L. Hall, N. Lewis, B. D. Hunt, and L. J. Schowalter, *Norelco Reporter* 33 (No. 1), 1 (1986).

9. L. S. Hung and J. W. Mayer, *Thin Solid Films* 109, 85 (1983).

10. E. Ma, C. V. Thompson, L. A. Clevenger, and K. N. Tu, *Appl. Phys. Lett*, 57, 1262 (1990).

11. G. Ottaviani, *J. Vac. Sci. Technol.*, 116, 1112 (1979).

12. E. Philofsky, *Solid State Electronics* 13, 1391 (1970).

13. J. M. Poate, K. N. Tu, and J. W. Mayer, editors, "*Thin Films: Interdiffusion and Reactions,*" Wiley–Interscience, New York (1978).

14. K. P. Rodbell, K. N. Tu, W. A. Lanford and X. S. Guo, *Phys. Rev.* B34, 1422 (1991).

15. R. D. Thompson and K. N. Tu, *Thin Solid Films* 53, 4372 (1982).

16. K. N. Tu, G. Ottaviani, U. Gösele, and H. Foell, *J. Appl. Phys.* 54, 758 (1983).

17. K. N. Tu, "Metal–Silicon Reaction," Chapter 7 in *Advances in Electronic Materials*, edited by B. W. Wessels and G. Y. Chin, American Society for Metals, Metal Park OH (1986).

18. R. T. Tung and J. L. Batstone, *Appl. Phys. Lett.*, 52, 648 (1988); 52, 1611 (1988).

19. L. R. Zheng, L. S. Hung, J. W. Mayer, G. Majni, and G. Ottaviani, *Appl. Phys. Lett.* 41, 646 (1982).

20. L. R. Zheng, PhD thesis, Cornell University (1985).

Problems

12.1 A phase $A_\beta B$ ($\beta = 3$) grows between the pure A phase and the $A_\gamma B$ ($\gamma = 1.5$) phase in a lateral diffusion couple where A is the dominant moving species. Given that $\tilde{D}_\beta = 1.27 \times 10^{-11}$ cm^2/sec, $\Delta C_\beta^{eq} = 2$ atomic percent, $K_\beta^{eff} = 2.79 \times 10^{-7}$ cm/sec, and Ω = atomic volume = 18×10^{-24} cm^3/atom,

(a) Find the changeover thickness of the $A_\beta B$ phase.
(b) Find the $A_\beta B$ growth rate at the changeover thickness.
(c) What is the flux of A through the $A_\beta B$ phase?

12.2 The growth of the single phase Al_3Pd_2, δ phase, between Al and Pd follows the diffusion regime with $x_\delta^2 = K_\delta t$ where $K_\delta = 3.3 \times 10^{-12}$ cm^2/sec at 250°C. The concentration drop $\Delta C_\delta^{eq} = 3.6$ atomic percent, and $G_\delta = 4.17$ where $\Omega = 10 \times 10^{-24}$ cm^3. Calculate the interdiffusion coefficient $\tilde{D}$.

12.3 A diffusion couple of pure A and pure B annealed at 400°C is known to make two phases between the two pure components. The γ phase (A_3B_2) starts to grow first with a growth completely diffusion controlled. The β phase (A_3B_1) starts to grow after the γ phase has reached a certain thickness. The growth of the β phase is completely interface controlled. The table provided gives data obtained from previous anneals at 400°C.

	ΔC_{eq} at %	$\tilde{D}_A$ cm^2/sec	K^{eff} cm/sec
γ phase	.8	5.0×10^{11}	
β phase	3.6		8.2×10^{13}

(a) At what thickness of the γ phase would be the β phase be expected to start growing?
(b) At what thickness of the γ phase would the γ phase growth stop?
(c) Would the γ phase be expected to shrink at some point over the duration of the experiment?

12.4 In the text, we study the case of the growth of two layered compounds where one layer exhibits diffusion-controlled growth and the other layer shows interface-controlled growth. Describe what happens in the two cases when

(a) Both layers exhibit diffusion-controlled growth.
(b) Both layers exhibit interface-controlled growth.

12.5 Given the two sets of concentration profiles for this problem,

(a) Which set of concentration profiles, (a) or (b), is diffusion controlled and why?

(b) What is the growth constant of the diffusion-controlled phase growth in units of ($\mu m^2/min$)?

(c) What is the interface reaction constant of the interface-reaction-controlled phase growth?

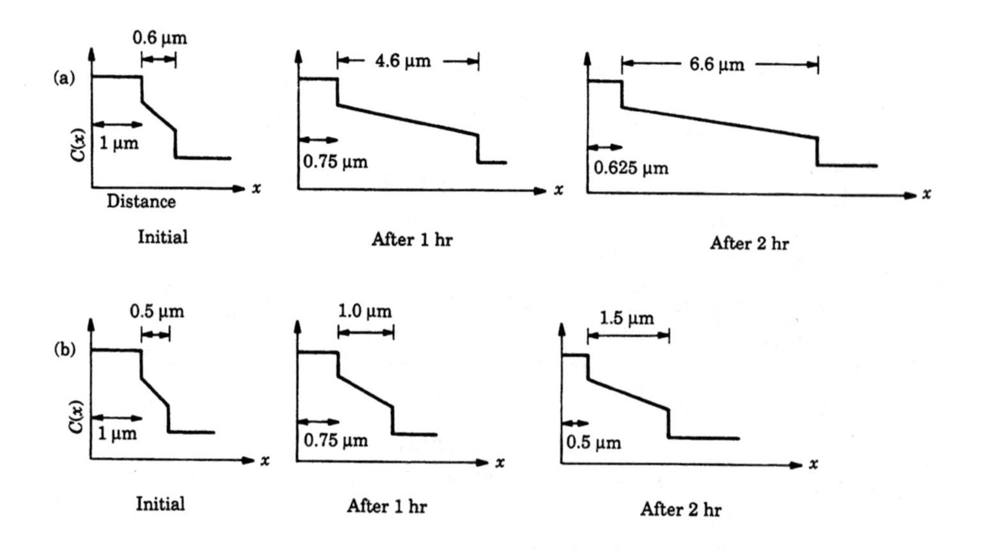

CHAPTER 13

Grain Boundary Diffusion

13.0 Introduction

Most of the metallic thin films used in microelectronic devices are polycrystalline rather than monocrystalline. Grain boundary diffusion is of concern. It has caused two very well-known failure modes in Si devices: electromigration and Al-penetration through diffusion barriers. In electromigration, voids are formed at the triple point of grain boundaries and extend out along grain boundaries. In penetration, Si precipitates decorate the Al grain boundaries.

In general, atomic diffusion along grain boundaries is faster than in the bulk of the grains. This assumes that atoms have a lower activation energy of motion in the boundary layer and that vacancy formation is easier because of the excess volume in the boundary. The effect of grain boundary diffusion has been demonstrated by several experiments. First, by comparing tracer diffusivities of Ag* in single-crystal Ag and in polycrystalline Ag, it has been found that the diffusivity is faster in the polycrystalline Ag at temperatures below 750°C. Second, radioactive tracers deposited on a bulk bicrystal show deeper penetration along the grain boundary in autoradiography images. Third, we compare thin film reactions in two sets of Pb/Ag/Au samples: one was epitaxially grown on rock salt and the other was deposited on fused quartz to grow polycrystalline grains. The latter showed Pb_2Au compound formation at 200°C, but not the former. In the former case the Ag layer was a single-crystal layer grown on Au/NaCl, while the Ag and Au layers on the fused quartz were polycrystalline.

Grain boundaries are rapid paths of atomic diffusion. Many studies have been devoted to the understanding of the subject. However, if a comparison is made to lattice diffusion as discussed in Chapter 3, we find that at the present time our understanding of grain boundary diffusion is only phenomenological. We have yet to form an atomistic picture. This is because the atomic structure and atomic positions of an arbitrary grain boundary are not known. Without this knowledge the atomic jump frequency and jump distance are ill-defined, and we have only a crude and macroscopic analysis of grain boundary diffusion, or a continuum approach.

There have been concerted efforts undertaken to determine the grain boundary structure. For low angle tilt-type and twist-type grain boundaries, dislocation models have been presented and are quite successful. Hence, the problem of diffusion in

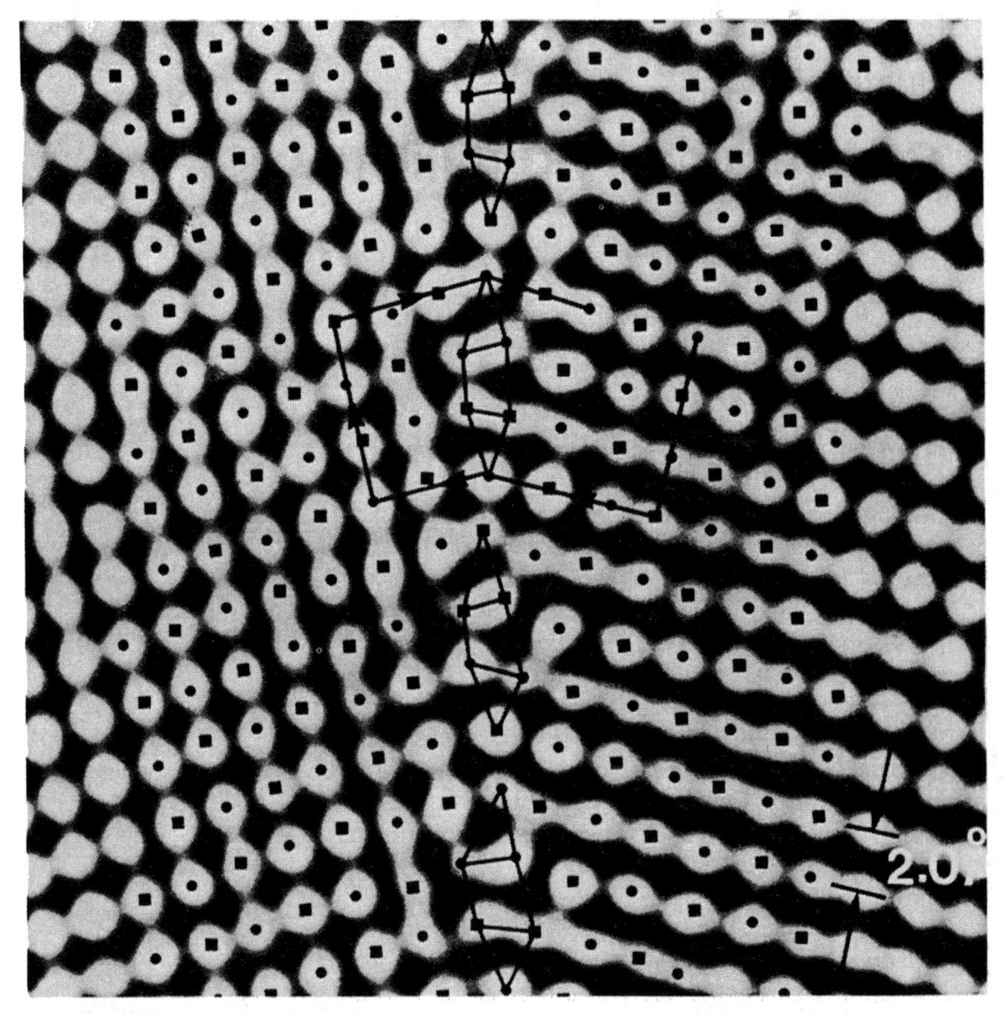

FIGURE 13.1 High-resolution transmission electron microscopic image of a (100) tilt-type large-angle grain boundary in a Au thin film. The dots and squares indicate atomic positions in the two alternating stacking layers. The nearest interplanar distance is 0.202 nm. (Courtesy of W. Krakow, IBM Research Division)

low-angle grain boundaries can be reduced to diffusion in an individual dislocation (i.e., "pipe diffusion") provided that the dislocation cores in the boundary are far apart. In a later section, we shall describe the diffusion measurement in small-angle tilt-type grain boundaries. For large-angle grain boundaries, there have been systematic efforts at developing the "coincidence site lattice" model for the energy and structure of the boundaries. At the same time, high-resolution transmission electron microscopy has been used to observe grain boundary images, and periodic clusters of atoms in the grain boundaries have been found. For example, Fig. 13.1 shows a lattice image of a large-angle (100) tilt-type grain boundary in a Au thin film. Periodicity has also been detected by x-ray diffraction in the large-angle grain boundaries which have a high density of coincidence lattice sites. Still, the direct link to atomic processes in grain boundary diffusion is missing. For example, the basic concept in lattice diffusion is point defects; whether or not we can define a point defect in a grain boundary is unclear. When a grain boundary contains ledges and steps, they can serve as sources and sinks of point defects for grain boundary diffusion.

13.1 Comparison of Grain Boundary and Bulk Diffusion

In order to present an overview of the relative magnitude of grain boundary diffusion to lattice diffusion, the diffusion data of tracer Ag* in Ag are examined. For self-diffusion in Ag, the diffusivities are

$$D_l = 0.67 \times e^{-\frac{1.95\text{eV}}{kT}} \quad \text{(lattice diffusion)} \tag{13.1}$$

$$D_b = 2.6 \times 10^{-2} e^{-\frac{0.8\text{eV}}{kT}} \quad \text{(grain boundary diffusion)} \tag{13.2}$$

At 200°C

$$D_l \simeq 10^{-21}\ \text{cm}^2/\text{sec}$$

$$D_b \simeq 10^{-10}\ \text{cm}^2/\text{sec}$$

If the annealing time is 10^5 sec (28 hours or about one day), we estimate the diffusion distances:

$$x_l^2 \simeq D_l t \simeq 10^{-16}\ \text{cm}^2, \qquad x_l \simeq 1\ \text{Å}$$

$$x_b^2 \simeq D_b t \simeq 10^{-5}\ \text{cm}^2, \qquad x_b \simeq 30\ \text{microns}$$

The ratio of grain boundary to lattice penetration is

$$\frac{x_b}{x_l} \simeq 3 \times 10^5$$

We see a very large difference between D_l and D_b for Ag at 200°C. There is negligible lattice diffusion in Ag at this temperature, yet grain boundary (GB) diffusion is substantial. In Table 13.1, we list the values of D_l and D_b in Ag at three temperatures for comparision.

TABLE 13.1 Comparison of D_l and D_b in Ag

	200°C	400°C	800°C
D_l cm²/sec	10^{-21}	10^{-15}	10^{-10}
D_b cm²/sec	10^{-10}	10^{-8}	10^{-6}
D_l/D_b	10^{-11}	10^{-7}	10^{-4}

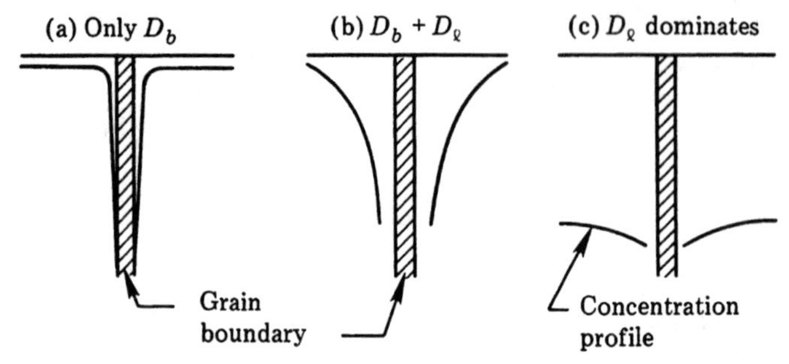

FIGURE 13.2 Schematic penetration profiles of concentration in three cases of combining grain boundary and lattice diffusion: (a) grain boundary diffusion is dominant, (b) both are comparable, and (c) lattice diffusion is dominant.

We see that D_l increases much more rapidly with temperature than D_b, and the same is true for their ratio. We can classify the ratio values into three regions as shown in Fig. 13.2. In the first region, Fig. 13.2a, grain boundary diffusion dominates and the penetration occurs primarily along the grain boundary. This region occurs typically at temperatures of 150–300°C. At intermediate temperatures, there is a noticeable penetration into the lattice of the grains adjacent to the grain boundary, as indicated by Fig. 13.2b. In the last region, although grain boundary diffusion is still faster than lattice diffusion, the effect is negligible because of the drain of atoms from the grain boundary into grains on both sides of the boundary. Therefore, the penetration along the boundary is slower and shallower than if there were no drain into the adjacent grains. This situation, depicted by Fig. 13.2c, occurs at about 800°C for Ag.

The ratio of mass transport through the grain boundary and the grain is

$$R = \frac{J_l A_l}{J_b A_b} = \frac{D_l \pi r^2}{D_b 2\pi r \delta} \tag{13.3}$$

where r is the radius of a grain and δ is grain boundary width. Assuming $r = 50$ nm (grain size of 100 nm in a film of 100 nm thickness), and $\delta = 0.5$ nm, we have at 200°C

$$R = \frac{10^{-21} \times 5 \times 10^{-6}}{10^{-10} \times 2 \times 5 \times 10^{-8}} \simeq 10^{-9}$$

The mass transport in the grain boundary is much larger than that in the grains.

We have a similar situation in Al films, if we take the activation energy of lattice diffusion and of grain boundary diffusion in Al to be 1.3 eV and 0.7 eV, respectively, and the temperature of diffusion is 100°C. The latter is roughly the operating temperature of a Si device. It is easy to see that electromigration in Al films is dominated by grain boundary diffusion, because D_b is much faster and the grain boundary flux is significant.

To illustrate that grain boundary flux is indeed significant and can lead to actual failure in a conducting line or through a diffusion barrier, we consider the case of

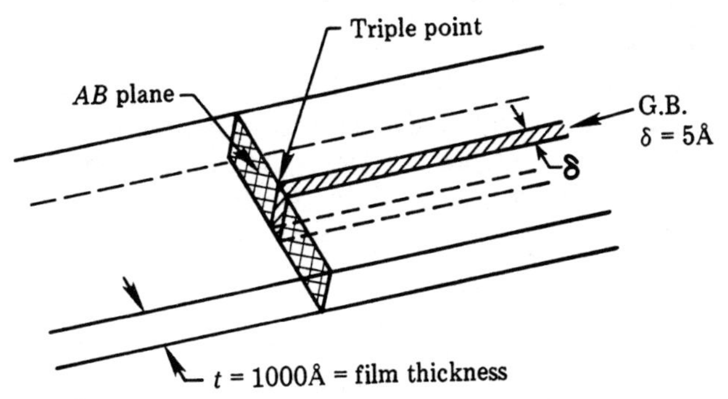

FIGURE 13.3 Schematic diagram of a triple point at the intersection of two grain boundaries. There is a divergence of diffusional flux at the triple point.

mass transport of Al along a grain boundary starting from a triple point as shown in Fig. 13.3.

There is a divergence of the flux at the AB plane. The mass transport M_b along the GB of a cross-sectional area A_b is given by

$$M_b = J_b A_b t \tag{13.4}$$

where $A_b = 0.5 \text{ nm} \times 100 \text{ nm} = 5 \times 10^{-13} \text{ cm}^2$, t is the diffusion time, and

$$J_b = -D_b \frac{\partial C_b}{\partial x} \tag{13.5}$$

If we assume it is an Al line, the grain boundary diffusivity is then, at 127°C

$$D_b \simeq 0.1 \times e^{-0.7\text{eV}/kT} \simeq 10^{-10} \text{ cm}^2/\text{sec}$$

We take

$$\frac{\partial C_b}{\partial x} = \frac{\Delta C_b}{\Delta x} = \frac{0.01 \times 10^{23} \text{ atoms/cm}^3}{10^{-4} \text{ cm}} = 10^{25} \text{ atoms/cm}^4$$

where we assume a one-percent (0.01) change in concentration (ΔC_b) along the grain boundary length of 1 micron, 10^{-4} cm. We approximated the Al concentration to be $10^{23}/\text{cm}^3$ instead of $6 \times 10^{22}/\text{cm}^3$. We take the diffusion time t to be 100 days $= 10^7$ sec, which is about the failure time of Al lines due to electromigration during circuit operation. Therefore

$$\begin{aligned} M_b &= J_b A_b t \\ &= \left(-D_b \frac{\partial C_b}{\partial x}\right) A_b t \\ &= \left(-10^{-10} \frac{\text{cm}^2}{\text{sec}}\right)\left(-10^{25} \frac{\text{atoms}}{\text{cm}^4}\right)(5 \times 10^{-13} \text{ cm}^2)(10^7 \text{ sec}) \\ &= 5 \times 10^9 \text{ atoms} \end{aligned}$$

We have transported 5×10^9 atoms. If all these atoms come from the grain behind the grain boundary, we can deplete a significant amount of Al from the line and create an opening or void. The line has a cross section of $A = 1$ micron $\times$ 100 nm $= 10^{-9}$ cm^2 and there are 10^{15} atoms per cm^2; this means there are only $10^{15} \times 10^{-9} = 10^6$ atoms on the cross section. We have transported about 5×10^3 atomic layers, which is about 1 micron in length and is about the same as the line width. This is a big gap, indicating an opening in the line.

This same calculation can be applied to grain boundary penetration through a diffusion barrier having a grain size of 1 micron. The uncertainty in the above calculation is $\Delta C_b/\Delta x$. If this gradient changes by one to two orders of magnitude, the conclusion is still the same: grain boundary diffusion is significant in thin films at moderate temperatures. Therefore we must consider how to measure D_b in grain boundary diffusion analysis.

13.2 Fisher's Analysis of Grain Boundary Diffusion

We consider the two-dimensional analysis of diffusion taking into account both diffusion along (y-direction) the grain boundary ($D = D_b$) and the drain into the adjacent grains (x-direction) by lattice diffusion ($D = D_l$) (Fig. 13.4). Apply the continuity equation

$$\frac{\partial C}{\partial t} = -(\nabla \cdot \mathbf{J}) = -\left(\frac{\partial J_x}{\partial x} + \frac{\partial J_y}{\partial y}\right) \tag{13.6}$$

Inside the grains,

$$\frac{\partial C}{\partial t} = D_l\left(\frac{\partial^2 C}{\partial x^2} + \frac{\partial^2 C}{\partial y^2}\right) \tag{13.7}$$

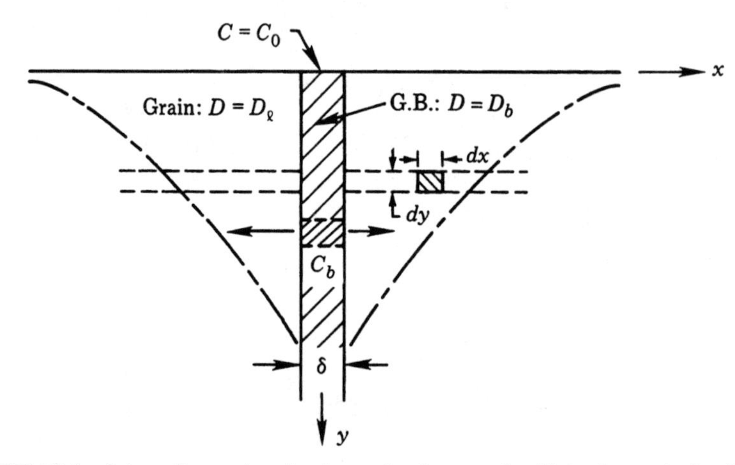

FIGURE 13.4 A two-dimensional schematic diagram for Fisher's analysis of grain boundary diffusion.

For the grain boundary slab, we first consider the continuity equation in a small area $\delta\, dy$ and let

$$\Delta J_x = J_{x_1} - J_{x_2} = 2\left(-D_l \frac{\partial C}{\partial x}\bigg|_{x=\delta/2}\right) \tag{13.8}$$

Therefore in the slab with $\Delta x = \delta$,

$$\frac{\partial C_b}{\partial t} = D_b \frac{\partial^2 C_b}{\partial y^2} - \frac{\Delta J_x}{\Delta x} = D_b \frac{\partial^2 C_b}{\partial y^2} + \frac{2D_l}{\delta} \frac{\partial C}{\partial x}\bigg|_{x=\delta/2} \tag{13.9}$$

For Fisher's analysis, $C = C(x,y,t)$

Initial condition

$$C(x,0,0) = C_0$$

$$C(\pm\delta/2,y,0) = 0$$

Boundary condition

$$C(x,0,t) = C_0$$

$$C(\infty,\infty,t) = 0$$

(1) Assume steady state in the slab

$$0 = D_b \frac{\partial^2 C_b}{\partial y^2} + 2\frac{D_l}{\delta} \frac{\partial C}{\partial x}\bigg|_{x=\delta/2} \tag{13.10}$$

(2) Assume $\dfrac{\partial C}{\partial y} = 0$ in the grains (i.e., diffusion only in the x-direction)

$$\frac{\partial C}{\partial t} = D_l \frac{\partial^2 C}{\partial x^2} \tag{13.11}$$

The physical picture of these assumptions is shown in Fig. 13.5 where we consider the diffusion along a stack of layers parallel to the free surface. We then sum up the diffusion along each of the layers with no communication between layers except at the grain boundary.

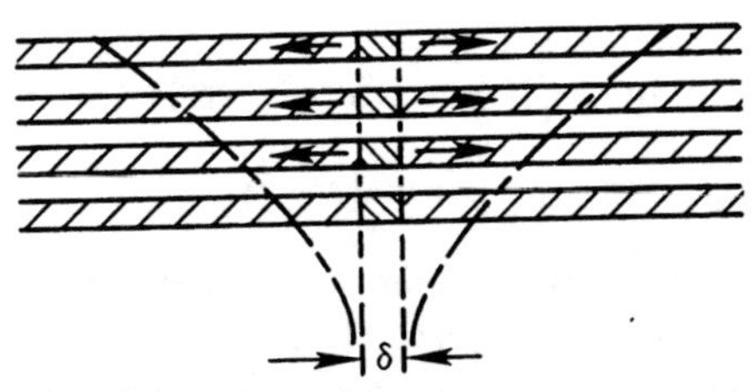

FIGURE 13.5 A two-dimensional schematic diagram of a stack of layers parallel to the free surface. In Fisher's analysis, no communication between layers is assumed except at the grain boundary.

The solution of Eq. (13.11) is

$$C(x,y,t) = C_b(y)\left\{1 - \operatorname{erf}\left(\frac{x - \delta/2}{2\sqrt{D_l t}}\right)\right\} \tag{13.12}$$

The partial derivative of Eq. (13.12) at $x = \delta/2$ gives

$$\left.\frac{\partial C(x,y,t)}{\partial x}\right|_{x=\delta/2} = \frac{C_b(y)}{(\pi D_l t)^{1/2}} \tag{13.13}$$

Substitute Eq. (13.13) into Eq. (13.10) to yield

$$D_b \frac{\partial^2 C_b}{\partial y^2} + \frac{2D_l C_b}{\delta(\pi D_l t)^{1/2}} = 0 \tag{13.14}$$

which has a solution,

$$C_b(y) = C_0 \exp\left\{\frac{-\sqrt{2}y}{(\pi D_l t)^{1/4}\left(\dfrac{D_b\delta}{D_l}\right)^{1/2}}\right\} \tag{13.15}$$

that can be substituted for $C_b(y)$ in Eq. (13.12)

$$C(x,y,t) = C_0 \exp\left\{\frac{-\sqrt{2}y}{(\pi D_l t)^{1/4}\left(\dfrac{D_b\delta}{D_l}\right)^{1/2}}\right\}\left\{1 - \operatorname{erf}\left(\frac{x - \dfrac{\delta}{2}}{2(D_l t)^{1/2}}\right)\right\} \tag{13.16}$$

Let

$$\eta = \frac{y}{(D_l t)^{1/2}} \qquad \xi = \frac{x - \dfrac{\delta}{2}}{2(D_l t)^{1/2}} \qquad \beta = \frac{D_b\delta}{2D_l(D_l t)^{1/2}}$$

Then

$$\eta\beta^{-1/2} = \frac{\sqrt{2}y}{(D_l t)^{1/4}\left(\dfrac{D_b\delta}{D_l}\right)^{1/2}} = y\,\frac{(4D_l/t)^{1/4}}{(D_b\delta)^{1/2}} \tag{13.17}$$

Note that η, ξ, β are all dimensionless. Then Fisher's solution becomes:

$$C(x,y,t) = C_0 \exp[-\pi^{-1/4}\eta\beta^{-1/2}][1 - \operatorname{erf} \xi] \tag{13.18}$$

This solution gives insight about the time dependence of the grain boundary penetration depth and about the method of measuring D_b by sectioning radiotracer diffusion profiles. To understand *penetration depth*, we can evaluate the solution at the edge of the grain boundary. At $x = \delta/2$ we have erf (0) = 0 and

$$\frac{C}{C_0} = \exp[-\pi^{-1/4}\eta\beta^{-1/2}]$$

where β can be obtained if C/C_0 and η (which depends on D_l only) are given. Let C/C_0 = constant, then $\eta\beta^{-1/2}$ = constant and from Eq. (13.17)

$$y \propto (D_b\delta)^{1/2} t^{1/4} \tag{13.19}$$

The penetration depth is proportional to $t^{1/4}$ rather than to $t^{1/2}$. It is slower due to the drain by the side diffusion into the grains adjacent to the boundary.

The average concentration $\overline{C}$ along the layer is measured by the difference in count rate of radioactive species between sequential layer removal steps (which each remove a thickness Δy). The *sectioning method* gives the number of tracer atoms/cm^2 in a layer Δy thick.

$$\begin{aligned} \overline{C}(y,t)\Delta y &= \Delta y \int_{-\infty}^{\infty} C(x,y,t)\, dx \\ &= \Delta y\, C_0 \exp[-\pi^{-1/4}\eta\beta^{-1/2}] \int_{-\infty}^{\infty} \left[1 - \operatorname{erf}\left(\frac{x - \delta/2}{2\sqrt{D_l t}}\right)\right] dx \end{aligned} \tag{13.20}$$

The integral is a constant. Therefore:

$$\ln \overline{C} = -\pi^{-1/4}(\eta\beta^{-1/2}) + \text{constant} \tag{13.21}$$

Consider differentiating Eq. (13.17)

$$\begin{aligned} \frac{\partial \eta\beta^{-1/2}}{\partial y} &= \frac{(4D_l/t)^{1/4}}{(D_b\delta)^{1/2}} \\ D_b\delta &= \frac{(4D_l/t)^{1/2}}{\left(\dfrac{\partial\eta\beta^{-1/2}}{\partial y}\right)^2} = \frac{(4D_l/t)^{1/2}}{\left[\dfrac{(\partial\eta\beta^{-1/2}/\partial \ln \overline{C})}{(\partial y/\partial \ln \overline{C})}\right]^2} \end{aligned} \tag{13.22}$$

Rearrange

$$D_b\delta = \left(\frac{\partial \ln \overline{C}}{\partial y}\right)^{-2} \left(\frac{4D_l}{t}\right)^{1/2} \left(\frac{\partial \ln \overline{C}}{\partial \eta\beta^{-1/2}}\right)^2$$

Note that the last term equals $\pi^{-1/2}$ from Eq. (13.21). Hence

$$D_b\delta = \left(\frac{\partial \ln \overline{C}}{\partial y}\right)^{-2} \left(\frac{4D_l}{t}\right)^{1/2} \pi^{-1/2} \tag{13.23}$$

Equation (13.23) is the key result of Fisher's solution for sectioning. It shows that by measuring $\ln \overline{C}$ as a function of y, and by knowing D_l, we can determine D_b.

A calculation of grain boundary diffusivity using Fisher's analysis can be made with radiotracer data of Ag in polycrystalline Ag as shown in Fig. 13.6. If we take the data for 5 days ($t = 4.3 \times 10^5$ sec) at 479°C (725 K), we obtain graphically from the slope of the line

$$\frac{\partial \ln \overline{C}}{\partial y} = \frac{1}{2.5 \times 10^{-3}\ \text{cm}}$$

and using Eq. (13.1) for the lattice diffusion of Ag,

$$D_l = 0.67 e^{-30.1} = 5.7 \times 10^{-14}\ \text{cm}^2/\text{sec}$$

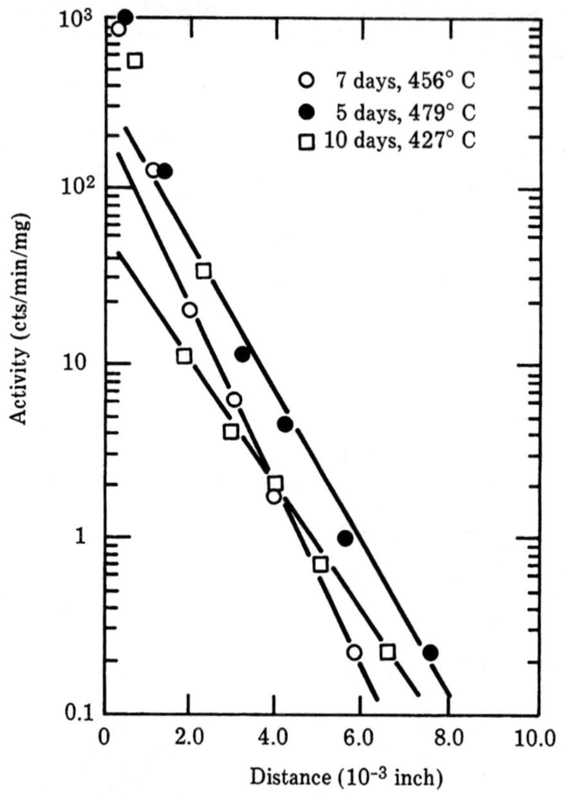

FIGURE 13.6 Radioactivity data of Ag in polycrystalline Ag. The grain boundary diffusivity is obtained by using Fisher's analysis.

We insert these values in Eq. (13.23) and obtain

$$D_b\delta = (2.5 \times 10^{-3})^2 \times \left(\frac{4 \times 5.7 \times 10^{-14}}{4.3 \times 10^5}\right)^{1/2} \times \frac{1}{\pi^{1/2}}$$

$$= 25.6 \times 10^{-16} \text{ cm}^3/\text{sec}$$

If the grain boundary width is 0.5 nm, then

$$D_b = 5 \times 10^{-8} \text{ cm}^2/\text{sec}$$

which is close to the value 8×10^{-8} cm^2/sec obtained from Eq. 13.2 at 725 K.

13.3 Whipple's Analysis of Grain Boundary Diffusion

The grain boundary (GB) has a width δ and grain boundary diffusivity D_b. Outside the GB, we have D_l. The coordinates of the analysis are shown in Fig. 13.7. The initial conditions are

$$C(x,0,0) = C_0$$

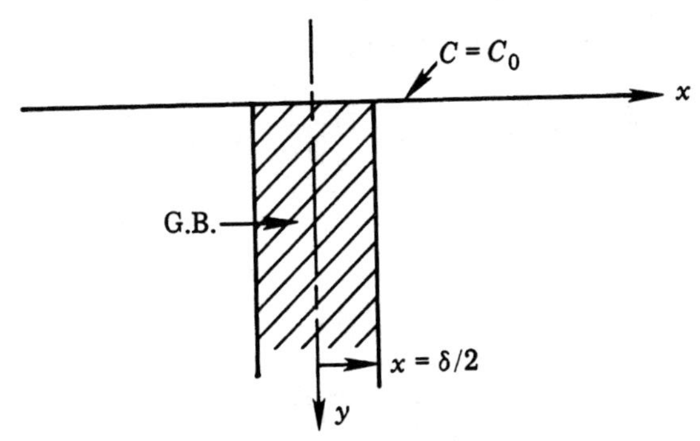

FIGURE 13.7 Coordinates used in Whipple's analysis of grain boundary diffusion.

$$C(x,y,0) = 0$$

The boundary conditions are

$$C(x,0,t) = C_0$$

$$C(\infty,\infty,t) = 0$$

In the grains

$$\frac{\partial C}{\partial t} = D_l \left(\frac{\partial^2 C}{\partial x^2} + \frac{\partial^2 C}{\partial y^2} \right) \tag{13.24}$$

Inside the grain boundary

$$\frac{\partial C_b}{\partial t} = D_b \left(\frac{\partial^2 C_b}{\partial x^2} + \frac{\partial^2 C_b}{\partial y^2} \right) \tag{13.25}$$

It is assumed that at the position $x = \pm \delta/2$, where the grain meets the grain boundary slab,

$$C = C_b \tag{13.26}$$

$$D_b \frac{\partial C_b}{\partial x} = D_l \frac{\partial C}{\partial x}, \tag{13.27}$$

That is, the concentration and flux are continuous.

Whipple considered that within the grain boundary slab, C_b is an even function of x. We note that the solution of the problem of thin film diffusion into a rod, as given in Equation 11.2, is an even function:

$$C = \frac{bC_0}{2(\pi Dt)^{1/2}} \exp\left(-\frac{x^2}{4Dt} \right)$$

It is then assumed that the concentration within the boundary is an even function,

$$C_b = C_b^0 + \frac{x^2}{2} C_b^1(y,t) \tag{13.28}$$

where C_b^0 and C_b^1 are coefficients of the even function.

Substituting Eq. (13.28) into Eq. (13.25)

$$\frac{\partial C_b^0}{\partial t} = D_b \left(\frac{\partial^2 C_b^0}{\partial y^2} + C_b^1 \right) \tag{13.29}$$

At $x = \delta/2$, if δ is very small, we have from Eqs. (13.26), (13.27), and (13.28) that

$$C = C_b^0$$

$$D_l \frac{\partial C}{\partial x} = D_b \frac{\delta}{2} C_b^1$$

Substituting C_b^0 and C_b^1 into Eq. (13.29) we obtain

$$\frac{\partial C}{\partial t} = D_b \left(\frac{\partial^2 C}{\partial y^2} + \frac{2}{\delta} \frac{D_l}{D_b} \frac{\partial C}{\partial x} \right) \tag{13.30}$$

(This is Fisher's Eq. (13.10) if $\partial C/\partial t = 0$.) Now, if we substitute $\partial^2 C/\partial y^2$ from Eq. (13.24) into Eq. (13.30)

$$\begin{aligned} \frac{\partial C}{\partial t} &= D_b \left(\frac{1}{D_l} \frac{\partial C}{\partial t} - \frac{\partial^2 C}{\partial x^2} + \frac{2}{\delta} \frac{D_l}{D_b} \frac{\partial C}{\partial x} \right) \\ &= \frac{D_b}{D_l} \frac{\partial C}{\partial t} - D_b \frac{\partial^2 C}{\partial x^2} + \frac{2 D_l}{\delta} \frac{\partial C}{\partial x} \end{aligned}$$

Rearranging terms, we obtain

$$\left(\frac{D_b}{D_l} - 1 \right) \frac{\partial C}{\partial t} = D_b \frac{\partial^2 C}{\partial x^2} - \frac{2 D_l}{\delta} \frac{\partial C}{\partial x} \tag{13.31}$$

We have a pair of simultaneous Eqs. (13.24) and (13.31), for C. This is the starting point of Whipple's analysis.

Whipple's Analysis (Summary)

Initial condition $\quad C(x,0,0) = C_0$, $\quad C(x,y,0) = 0$

Boundary condition $\quad C(x,0,t) = C_0$, $\quad C(\infty,\infty,t) = 0$

Eq. (13.24) $\Rightarrow \dfrac{\partial C}{\partial t} = D_l \left(\dfrac{\partial^2 C}{\partial x^2} + \dfrac{\partial^2 C}{\partial y^2} \right)$ where $\dfrac{\partial^2 C}{\partial y^2} = 0$ in Fisher's analysis.

Eq. (13.31) $\Rightarrow \left(\dfrac{D_b}{D_l} - 1 \right) \dfrac{\partial C}{\partial t} = D_b \dfrac{\partial^2 C}{\partial x^2} - \dfrac{2 D_l}{\delta} \dfrac{\partial C}{\partial x}$ where $\dfrac{\partial C}{\partial t} = 0$ in Fisher's analysis.

Equation (13.31) can be regarded as a boundary condition of C at $x = \delta/2$.

Solution

$$C(x,y,t) = C_0\left(1 - \operatorname{erf}\frac{\eta}{2}\right) + \frac{C_0\eta}{2\sqrt{\pi}}\int_1^{\Delta(\infty)} \frac{d\sigma}{\sigma^{3/2}} \exp\left(\frac{-\eta^2}{4\sigma}\right) \operatorname{erfc}\left[\frac{1}{2}\sqrt{\frac{\Delta - 1}{\Delta - \sigma}}\left(\xi + \frac{\sigma - 1}{\beta^1}\right)\right] \tag{13.32}$$

where σ is the variable of integration and

$$\eta = \frac{y}{(D_l t)^{1/2}} \qquad \xi = \frac{x - \delta/2}{(D_l t)^{1/2}} \qquad \Delta = \frac{D_b}{D_l}$$

$$\beta^1 = \left(\frac{D_b}{D_l} - 1\right)\frac{\delta/2}{(D_l t)^{1/2}} \sim \beta \qquad \text{when } \frac{D_b}{D_l} >> 1$$

The last result follows from Eq. (13.17). Note that

$$\beta = \frac{D_b\delta}{D_l S} \simeq \frac{\text{flux through grain boundary}}{\text{flux through grains}}$$

where $S = 2(D_l t)^{1/2}$.

For application to the sectioning method, we recall:

$$\eta\beta^{-1/2} = y\,\frac{\left(\frac{4D_l}{t}\right)^{1/4}}{(D_b\delta)^{1/2}} \tag{13.17}$$

$$(\eta\beta^{-1/2})^m = y^m\,\frac{\left(\frac{4D_l}{t}\right)^{m/4}}{(D_b\delta)^{m/2}}$$

$$\frac{\partial(\eta\beta^{-1/2})^m}{\partial y^m} = \frac{\left(\frac{4D_l}{t}\right)^{m/4}}{(D_b\delta)^{m/2}}$$

$$\frac{\dfrac{\partial(\eta\beta^{-1/2})^m}{\partial \ln \overline{C}}}{\dfrac{\partial y^m}{\partial \ln \overline{C}}} = \frac{\left(\frac{4D_l}{t}\right)^{m/4}}{(D_b\delta)^{m/2}} \tag{13.33}$$

$$D_b\delta = \left(\frac{\partial \ln \overline{C}}{\partial y^m}\right)^{-2/m}\left(\frac{4D_l}{t}\right)^{1/2}\left(\frac{\partial \ln \overline{C}}{\partial(\eta\beta^{-1/2})^m}\right)^{2/m} \tag{13.34}$$

We note from Eq. (13.33) that if

$$\frac{\partial \ln \overline{C}}{\partial y^m} = \text{constant}$$

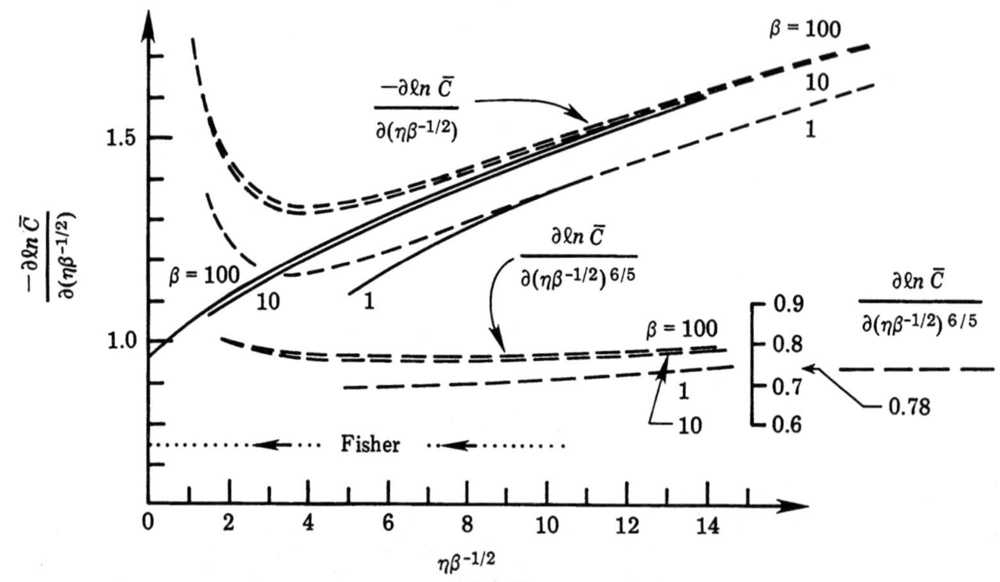

FIGURE 13.8 A plot of the relation of $(\partial \ln \overline{C})/[\partial(\nu\beta^{-1/2})^m]$ = constant. The m and the constant are determined to be 6/5 and 0.78, respectively.

then

$$\frac{\partial \ln \overline{C}}{\partial(\eta\beta^{-1/2})^m} = \text{constant}$$

Therefore by evaluating the condition

$$\frac{\partial \ln \overline{C}}{\partial(\eta\beta^{-1/2})^m} = \text{constant}$$

m is found to be 6/5. In fact, as shown in Fig. 13.8,

$$\frac{\partial \ln \overline{C}}{\partial(\eta\beta^{-1/2})^{6/5}} = 0.78$$

So finally, we have

$$D_b\delta = \left(\frac{\partial \ln \overline{C}}{\partial y^{6/5}}\right)^{-5/3} \left(\frac{4D_l}{t}\right)^{1/2} (0.78)^{5/3} \tag{13.35}$$

We note that Eq. (13.35) is the key result of Whipple's solution. It has a form similar to that of Fisher's solution, Eq. (13.23). It shows that by plotting $\ln \overline{C}$ versus $y^{6/5}$ and by knowing D_l, we obtain $D_b\delta$. The slopes of $\ln \overline{C}$ versus $\eta\beta^{-1/2}$ are shown in Fig. 13.8.

13.4 Diffusion in Small-Angle Grain Boundaries

The effect of grain boundary structure on diffusion has been clearly shown in small-angle grain boundaries. Small-angle grain boundaries of the tilt type can be repre-

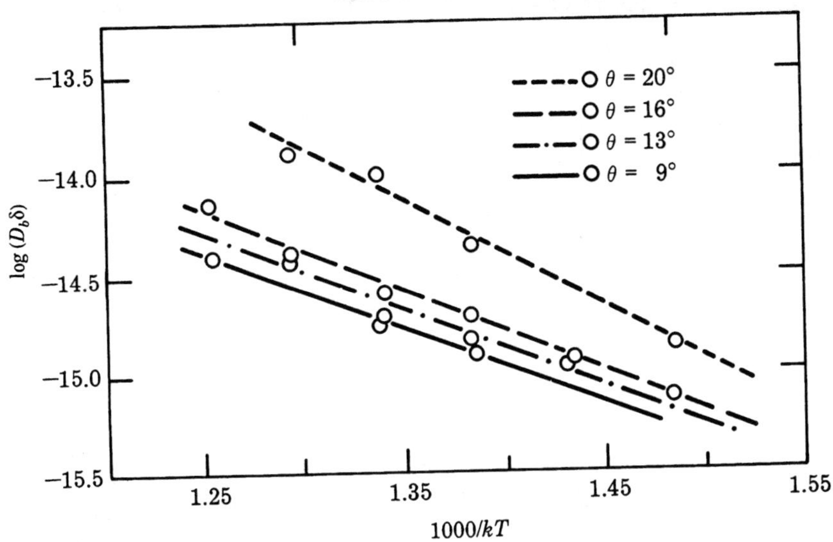

FIGURE 13.9 A plot of ln ($D_b\delta$) versus $1/kT$ for radiotracer diffusion of Ag in various (100) tilt-type grain boundaries in Ag (after Turnbull and Hoffman, 1954).

sented by a parallel array of edge dislocations. When these dislocations are far apart, the diffusion along each of them can be regarded as the diffusion in a pipe, "pipe diffusion." It is highly anisotropic, since diffusion along the pipe and normal to the pipe are very different, and the latter is much slower. Also the activation energy of pipe diffusion is invariant as long as the dislocations are not too close to each other. This is shown in Fig. 13.9; from $\theta = 9°$ to $16°$ of tilt angle, the activation energy is the same, except that at a higher angle, the diffusivity is faster because D_0 is greater.

If we assume that the distance between dislocation cores in a tilt type grain boundary is given by

$$d = \frac{a}{2 \sin \frac{\theta}{2}} \tag{13.36}$$

where a is the lattice constant and θ is tilt angle, then we have

$$D_b\delta = \frac{D_p h^2}{d} = 2D_p h^2 \frac{\sin \frac{\theta}{2}}{a} \tag{13.37}$$

where D_p and h^2 are the diffusivity and the effective cross-sectional area of a dislocation pipe, respectively, and

$$D_p = A \exp\left(-\frac{Q}{kT}\right), \tag{13.38}$$

Q does not depend on θ. When the diffusion is conducted normal to the pipe, the measured diffusivity is much lower but varies with θ (see Fig. 13.10).

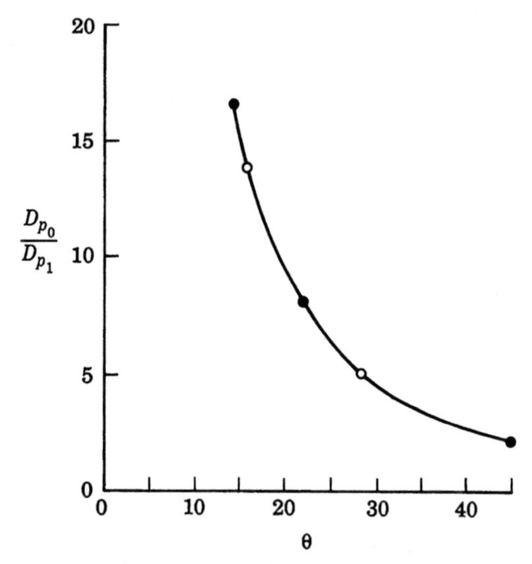

FIGURE 13.10 The dependence of anisotropy of grain boundary diffusions on the tilt angle in (100) tilt boundaries in Ag (after Hoffman, 1958).

13.5 Diffusion-Induced Grain Boundary Motion

We have discussed diffusion in a stationary grain boundary, that is, a grain boundary which does not move at all while atomic diffusion occurs in it. In a fine-grained thin film, the grain boundaries tend to migrate because of curvature and because of a reduction of grain boundary energy. Therefore, we encounter diffusion in a moving grain boundary. We will show later that diffusion in a moving grain boundary is a kinetic process of low activation energy of phase transformation in polycrystalline solids. To consider this kind of diffusion, there are two key issues: (1) the diffusion equation in a moving grain boundary and (2) the driving force which moves the grain boundary. We first discuss the diffusion equation using Fig. 13.11 in which the grain boundary is moving to the right with a constant velocity v. Following the

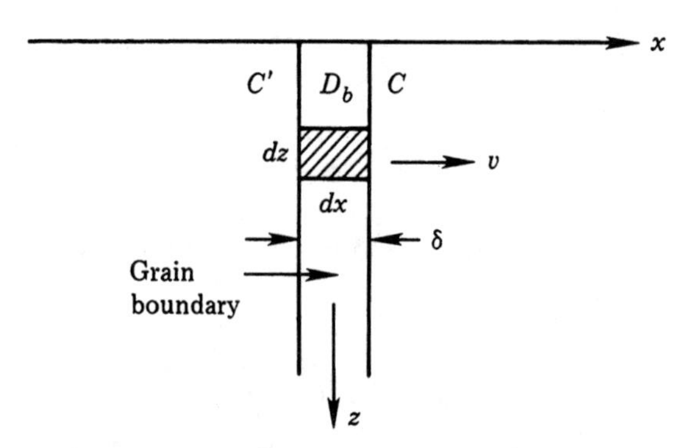

FIGURE 13.11 The coordinates for a grain boundary in motion induced by diffusion along the grain boundary.

derivation of the continuity equation as given in Chapter 3, we consider the fluxes flowing in and out of a tiny square $dx\,dz$ in the grain boundary and we ignore the y-dimension because we assume that this is a two-dimensional problem. To simplify the problem, we further assume that we can ignore lattice diffusion for the moment. This assumption is acceptable for cases where the diffusion takes place at temperatures below half the melting point of the solid.

In the z-direction as shown in Fig. 13.11, we have

$$J_z = -D_b \frac{\partial C_b}{\partial z} \tag{13.39}$$

where D_b and C_b are the diffusivity and concentration in the grain boundary. In the x-direction, we have

$$\Delta J_x = vC - vC' \tag{13.40}$$
$$\Delta x = \delta$$

where C and C' are the lattice concentrations before and behind the GB. Then the divergence in the square $dx\,dz$ is given by

$$\nabla \cdot \mathbf{J} = \frac{\partial J_x}{\partial x} + \frac{\partial J_z}{\partial z} = 0 \tag{13.41}$$

$$\frac{v(C - C')}{\delta} + \frac{\partial}{\partial z}\left(-D_b \frac{\partial C_b}{\partial z}\right) = 0$$
$$\delta D_b \frac{\partial^2 C_b}{\partial z^2} - v(C - C') = 0 \tag{13.42}$$

To solve this steady state diffusion equation, we assume that

$$\frac{C}{C_b} = k \tag{13.43}$$

and the ratio k is called the segregation coefficient. We have then

$$\frac{\partial^2 C}{\partial z^2} - \frac{vk}{D_b\delta}(C - C') = 0 \tag{13.44}$$

The solution of this equation is in the form of a simple exponential function if we let $p^2 = vk/D_b\delta$. The solution is readily available if the boundary conditions are given.

Next we shall consider a simple case of application of the equation in a thin film, as shown in Fig. 13.12. The grain boundary is moving with a velocity v in the film of thickness Z. If we assume that lattice diffusion is negligible, the grain boundary diffusivity can be estimated to be

$$D_b \simeq \frac{Z^2 v}{\delta} \tag{13.45}$$

where δ/v is roughly the time available for the grain boundary diffusion and Z is the distance of diffusion.

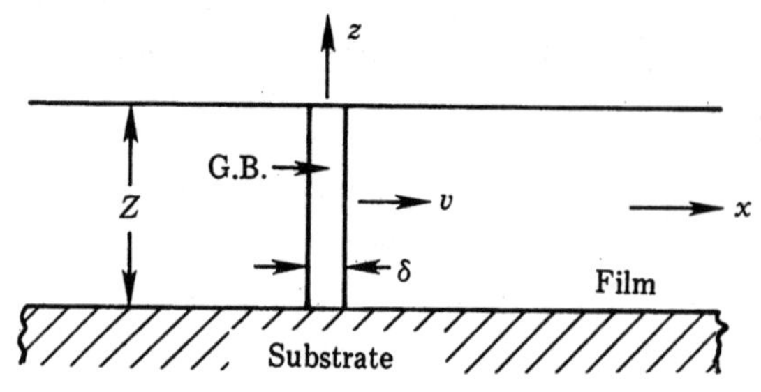

FIGURE 13.12 A schematic diagram of a grain boundary in motion in a thin film.

The above kinetic process has been applied to the explanation of "DIGM" (diffusion-induced grain boundary motion phenomena in metals), of dopant diffusion-induced grain growth in polycrystalline Si films, and of oxygen diffusion-induced phase boundary migration in copper oxide films. An example of DIGM is the interdiffusion between a polycrystalline Cu film and a polycrystalline Au film around 160°C. The temperature is so low that lattice diffusion is negligible in a reasonable period of time. Yet interdiffusion occurs by diffusion of Cu along the moving grain boundaries in Au, leading to the formation of Cu–Au solid solution in regions swept by the moving grain boundaries.

To discuss the driving force of grain boundary motion, we note that a moving grain boundary leads to grain growth when the change across the grain boundary is only the crystallographic orientations of the two grains on either side of the boundary. In such grain boundary motion, no long-range diffusion along the boundary is required; atoms need only move across the grain boundary of width δ and reorient themselves to the crystallographic axes of the growing grain. On the other hand, in DIGM a long-range grain boundary diffusion is required. The major difference between conventional grain growth and the DIGM mode of transition is the driving force. In conventional grain growth (see Section 15.1), the driving force behind the moving grain boundary comes from the reduction of grain boundary energy (or area) and the grain boundary always moves against its curvature. In the case of DIGM, the driving force is chemical in nature; in other words, it is due to the free energy change of a phase transition. Therefore, the grain boundary motion can go with curvature, or it can move a straight grain boundary. This is because the chemical driving force is of the order of 1 eV which is much greater than the driving force of curvature change in grain growth. The latter is of the order of 0.01 eV.

References

1. R. W. Balluffi and J. M. Blakely in *Low Temperature Diffusion and Applications to Thin Films* (p. 363), edited by A. Gangulee, P. S. Ho, and K. N. Tu, Elsevier Sequoia, Lausanne, (1975).

2. J. W. Cahn, J. D. Pan, and R. W. Balluffi, *Script. Met.* 13, 503 (1979).

3. F. J. A. den Broeder, *Acta Met.* 20, 319 (1972).

4. J. C. Fisher, *J. Appl. Phys.* 22, 74 (1951).

5. D. Gupta, D. R. Campbell and P. S. Ho, Chapter 7 in "*Thin Films: Interdiffusion and Reactions*, edited by J. M. Poate, K. N. Tu, and J. W Mayer, Wiley–Interscience, New York (1978).

6. R. H. Hoffman, *Acta Met.* 4, 96 (1956).

7. A. D. LeClaire, *Brit. J. Appl. Phys*, 14, 351 (1963).

8. J. Li, S. Q. Wang, J. W. Mayer, and K. N. Tu, *Phys. Rev.* B39, 12369 (1989).

9. T. Suzuoka, *Trans. Jap. Inst. Met.*, 2, 25 (1961).

10. K. N. Tu, *J. Appl. Phys.*, 48, 3400 (1977).

11. K. N. Tu, J. Tersoff, T. C. Chou, C. Y. Wong, *Sol. State Commun.* 66, 93 (1988).

12. D. Turnbull and R. H. Hoffman, *Acta Met.* 2, 419 (1954).

13. R. T. P. Whipple, *Phil Mag.* 45, 1225 (1954).

Problems

13.1 Diffusion Induced Grain-boundary Migration (DIGM) occurs in the diffusion of Ni into a Cu thin film. Grain boundary velocities are recorded as a function of temperature for a single sample: at 350°C the velocity is 3.2×10^{-11} m/s, while at 900°C it is 4.0×10^{-9} m/s. Estimate the activation energy for the grain boundary diffusion of Ni in Cu.

13.2 At T = 400°C and t = 10 min, diffusion length = 150 nm.
(a) Determine D (assume D_0 = 1.0) and the activation energy.
(b) Decide whether the self-diffusion mechanism is due to lattice diffusion or to grain boundary diffusion. Indicate the reason for your choice.

13.3 From the data for Al in the text, determine whether the shortest diffusion time to go from A to B (in the sketch provided) is by lattice diffusion (0.1 μm) or by grain boundary diffusion (200 μm) at 200°C and at 500°C.

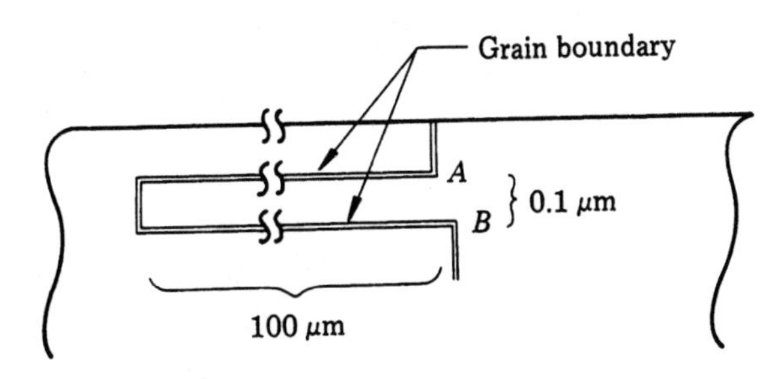

PROBLEM 13.3

13.4 Determine the grain boundary diffusion coefficient D_b (assume grain boundary width $\delta = 0.5$ nm) from the radiotracer data in the figure for polycrystalline material. Single-crystal material, diffused at the same temperature (425°C) and for the same time (10 days), had a diffusion length $\lambda_D = 2 \times 10^{-5}$ cm (0.2 micron).

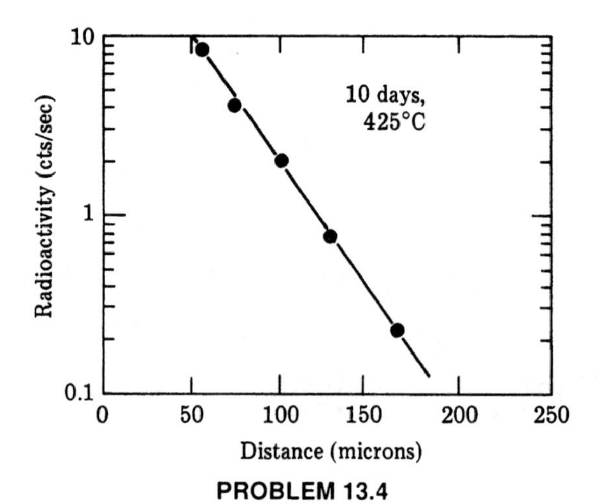

PROBLEM 13.4

13.5 A 50 nm thick polycrystalline Al sample has square grain size of 100 nm, grain boundary thickness of about 1 nm, and grain boundary diffusion coefficient $D_b = 10^{-10}$ cm^2/sec. It was found using an isotope tracer that the solute penetrates 1.78 nm into adjacent grains in about 0.1 second.

(a) Using Fisher's analysis, estimate the amount of time it takes to penetrate the entire grain.

(b) At $t = 0.1$ second, the concentration profile for solute from the grain boundary located at $x = 0$ was given by $\overline{C} = \exp(9 \times 10^6 \text{ cm}^{-2}\, x)$ for $x > 0$. Using Fisher's analysis, find the lattice diffusion coefficient D_l.

13.6 Derive Eq. (13.29).

13.7 Solve Eq. (13.44) with the boundary conditions that $dC/dz = 0$ at $z = 0$; and $C = C_e$ at $z = Z$.

CHAPTER 14

Electromigration in Metals

14.0 Introduction

An ordinary household extension cord conducts electricity without transport of atoms in the cord. The free electron model of conductivity of metals assumes that the conduction electrons are free to move in the metal, unconstrained by the lattice of ions except in scattering interactions. The scattering of electrons by the ions is the cause of electrical resistance. The scattering does not cause large displacements of the ions when the current density is low; the scattering from phonon vibrations generates Joule heating.

At a high current density (above 10^4 amp/cm^2), the transport of current can displace the ions and influence the transport of mass. The mass transport by the electric field and charge carriers is called electromigration. It occurs in interconnecting lines in microelectronic devices where the current density is high. For example, when a 5.0 μm wide Al line of 0.2 μm thick is subjected to a current of 1 m amp, the current density is 10^5 amp/cm^2, which can cause mass transport in the line near ambient temperature, and lead to void and extrusion formations. It is a unique and serious mode of reliability failure in thin film circuits. As device miniaturization requires lines to have smaller and smaller cross-sections, the current density tends to go up, and so does the probability of circuit failure induced by electromigration.

The phenomenon of electromigration can be illustrated by the response of short conducting lines under a high current density as shown in Fig. 14.1a. This is a scanning electron micrograph of the morphology of a short Al line transformed under electromigration. At one end of the line, a large void of missing matter is seen, but at the other end, extrusion is seen. Both void formation and extrusion occur in the same experiment. To discuss the observation, a schematic diagram of the side view of such a sample is shown in Fig. 14.1b.

The short Al line was deposited on a long Mo line which was deposited on a SiO_2 substrate. The applied electrical current in the Mo line takes a detour to go along the Al line because the latter is a path of lower resistance. When the current density is high and the temperature is moderate, ~200°C, electromigration occurs and leads to extrusion at the anode and voids at the cathode. The direction of mass transport is the same as that of the electron flow.

In this Chapter, we discuss the effective charge number in the driving force of

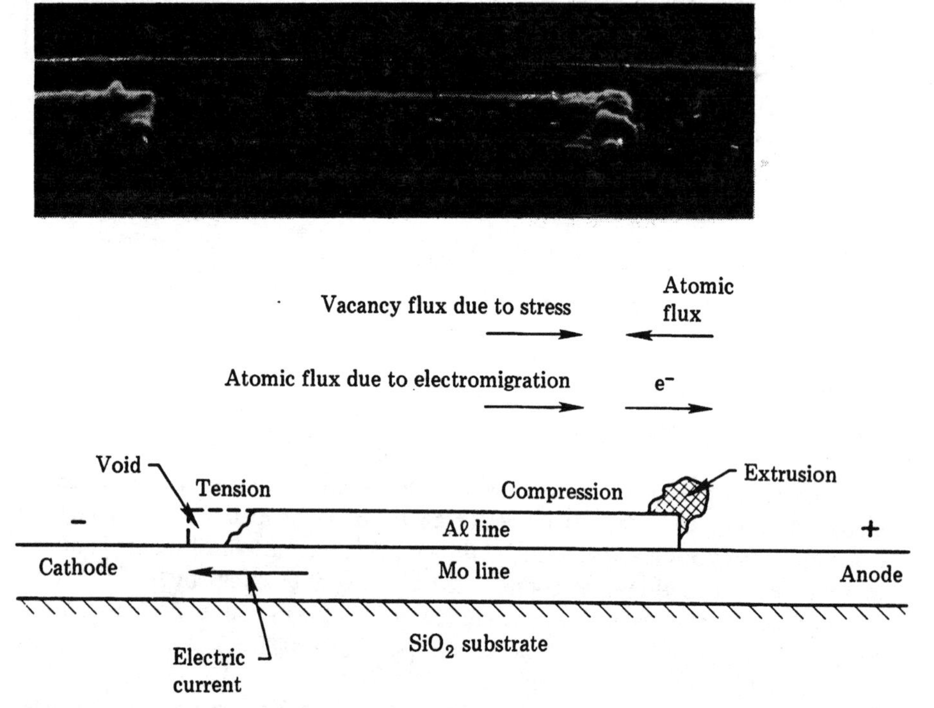

FIGURE 14.1 (a) Scanning electron micrograph of a short Al line undergoes morphological change due to electromigration where depletion occurs at the cathode, and extrusion occurs at the anode (courtesy of P.S. Ho, IBM Research Division). (b) A sketch of the cross-sectional view of morphological changes due to electromigration in the short Al line deposited on a long Mo line.

electromigration, irreversible processes of electromigration under stress, the methods of measurement of electromigration, and the practical aspects of electromigration in fine lines.

14.1 The Driving Force of Electromigration

Electromigration is the result of a combination of thermal and electrical effects on mass motion. If the conducting line is kept at a very low temperature (e.g., liquid nitrogen temperature), electromigration would not occur at a current density of 10^5 amp/cm^2. The contribution of the thermal effect can be recognized by the facts that electromigration in a bulk metal occurs at about three-quarters of its melting point (in absolute temperature) and that electromigration in a metallic polycrystalline thin film occurs at about one-half of its melting point (in absolute temperature). At these temperatures, there is a large number of atoms undergoing random walk processes in the lattice of the bulk and in the grain boundaries of the thin film, respectively, and it is these atoms which participate in electromigration under the applied field.

In Chapter 3, diffusion in solids is shown to be defect-mediated. Electromigration is also defect-mediated. Without an electrical field, the thermal effect alone at constant temperature leads to no net mass transport in a pure metal. With an electrical field, there is a force on those atoms undergoing random walk to displace them in a given direction, hence a net mass transport occurs.

To consider the mass transport by electromigration, we recall in Section 3.3 that atomic diffusion flux in a solid can be represented by

$$J = -D\frac{\partial C}{\partial x} + \sum_i CM_iF_i$$

where the first term comes from the chemical potential gradient and the second term is the sum of various applied forces. For electromigration in a pure metal, the chemical gradient term is removed. The driving force of the net atomic flux comes from the applied electrical field and the force has two parts: the first is the direct action of the electrostatic field on the diffusing atoms and the second is the momentum exchange of the moving charge carriers with the diffusing atoms. For simplicity, an effective charge number Z^* has been introduced in the following manner,

$$F_{\text{em}} = Z^*e\mathscr{E} = (Z^*_{\text{el}} + Z^*_{\text{wd}})e\mathscr{E} \tag{14.1}$$

where e is the charge of an electron and $\mathscr{E}$ is the electrical field. Z^*_{el} can be regarded as the nominal valence of the diffusing ion in the metal when the dynamic screening effect is ignored; it is responsible for the field effect and is often called the *direct force*. Z^*_{wd} is the charge number representing the momentum exchange effect and is commonly called the *electron wind force*. Z^*_{wd} has generally been found to be of the order of ten for a good conductor, so the momentum exchange effect is much greater than the electrostatic field effect for electromigration in metals.

To appreciate the electron wind force, we depict in Fig. 14.2a the configuration of a shaded Al atom and a neighboring vacancy in a fcc structure before they exchange positions along a $\langle 110 \rangle$ direction. They have four nearest neighbors in common, including the two shown by the broken curves, one on top and one on the bottom of the close-packed atomic plane. When the shaded atom is diffusing halfway towards the vacancy as shown in Fig. 14.2b, it is at the activated state, sitting at a saddle point while displacing the four nearest-neighbor atoms. Since the saddle point is not part of the lattice periodicity, the atom at the saddle position will make a greater contribution to the resistance to electrical current than a normal atom. In other words, it experiences a greater electron scattering and hence a greater electron wind force. The diffusion of the atoms is enhanced in the direction of the electron flow.

Several quantum mechanical attempts have been made to estimate the electron wind force, yet none has been widely accepted. The difficulty lies in carrying out the proper and complete treatment of the scattering of electrons by the diffusing atom and its surrounding atoms as shown in Fig. 14.2b. We will not review these attempts; the readers are referred to the work by Bosvieux and Friedel (1962), Sorbello (1973, 1985), and Landauer and Woo (1974). In the following, we discuss the

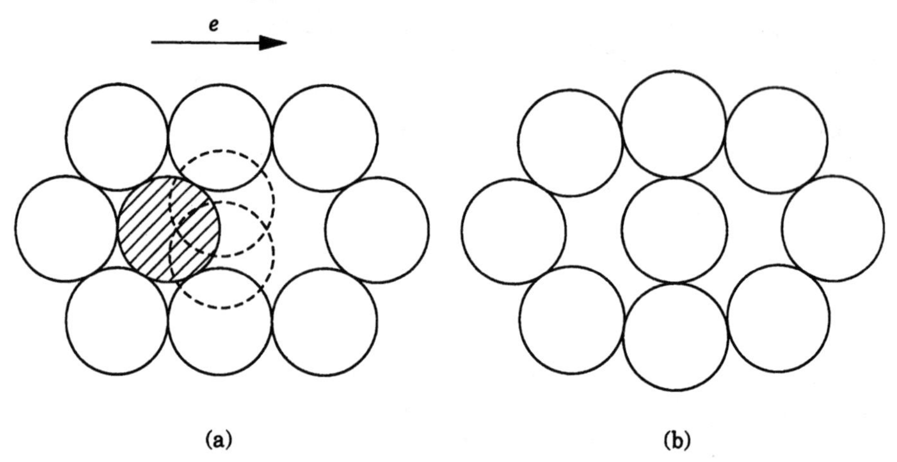

FIGURE 14.2 A sketch of the diffusion of the shaded Al atom to a neighboring vacancy. The pair have four nearest neighbors in common, including the two drawn in broken curves (a) before diffusion and (b) halfway during diffusion.

ballistic approach to the scattering process given by Fiks (1959) and Huntington and Grone (1961).

The idea by Fiks (1959) assumes that the diffusing atom has a scattering cross section σ_d and that the conduction electrons have an average velocity v, with n the concentration of conduction electrons per unit volume. The number of collisions per unit time between the diffusing atom and the electrons is $nv\sigma_d$. For each collision, the moving atom acquires a momentum $F_x^1\tau_d = e\mathscr{E}_x\tau_d$ from one electron during one relaxation time τ_d, where the subscript x indicates that the electrical field and the force are acting in the x-direction. The total change of momentum per unit time (i.e., the wind force) is the product of the number of collisions per unit time and the momentum exchange per collision,

$$F_x = -(nv\sigma_d)(e\mathscr{E}_x\tau_d) = -e\mathscr{E}_x n\lambda_d\sigma_d \tag{14.2}$$

where $\lambda_d = v\tau_d$ is the mean free path of scattering. Let

$$F_x = eZ^*_{\text{wd}}\mathscr{E}_x \tag{14.3}$$

then

$$Z^*_{\text{wd}} = -n\lambda_d\sigma_d \tag{14.4}$$

This is Fiks' equation of the effective charge number. It shows that the electron wind force is proportional to the scattering cross section of the diffusing atom.

The semiclassical model by Huntington and Grone (1961) postulates a transition probability per unit time from one free-electron state to another free-electron state due to the scattering by the diffusing atoms. A formal expression was given for the momentum transfer per unit time (i.e., the wind force) by treating electrons as free particles. Subsequently, it was modified by taking into account electron pseudo-momentum of the electronic states (i.e., the momentum of the electron plus that of the lattice). The force is calculated by summing over the initial and final states of the scattered electrons, or rather by integrating the transition probability over the jumping path of the diffusing atom (i.e., the path to go from the configuration in Fig. 14.2a to that in Fig. 14.2b). The force acting on the diffusing atom is not constant along the jumping path, but the calculation is simplified by assuming that the electron reciprocal relaxation time arising from collision with the moving atoms is constant. The final result shows that the effective charge number can be given in terms of specific resistivities.

$$Z^*_{\text{wd}} = -Z \frac{\dfrac{\rho_d}{C_d}}{\dfrac{\rho}{C}} \frac{m_0}{m^*} \tag{14.5}$$

where $\rho = m_0/ne^2\tau$ and $\rho_d = m^*/ne^2\tau_d$ are the resistivity of the lattice atoms and the diffusing atoms, respectively, m_0 and m^* are the free electron mass and effective electron mass, respectively, and τ and τ_d are relaxation times. We recall that

$$\rho = \frac{1}{ne\mu_e} \tag{14.6}$$

where μ_e is electron mobility and $\mu_e = e\tau/m_0$.

In an fcc lattice, there are 12 equivalent jump paths along <110> directions. For a given current direction, the average specific resistivity of a diffusing atom must be corrected by a factor of one-half. Rewriting Eq. (14.1), we have

$$Z^* = -Z\left[\frac{1}{2}\frac{\dfrac{\rho_d}{C_d}}{\dfrac{\rho}{C}}\frac{m_0}{m^*} - 1\right] \tag{14.7}$$

where Z^*_{el} has been taken as Z, the nominal valence of the metal atom. This is the equation of Huntington and Grone for the effective charge number.

14.2 Calculation of the Effective Charge Number

It is seen from Eq. (14.5) that the magnitude of Z^*_{wd} is dominated by the ratio of the specific resistivities. Since it is hard to calculate the ratio properly, we shall make an estimate of the ratio by assuming that the electrical resistance of a scatterer is proportional to its cross section of scattering. In the case of A1, the scattering cross section $\sigma \cong \langle x^2 \rangle$ where $\langle x \rangle$ is the atomic vibration amplitude of a normal atom. This can be estimated from the Einstein model of atomic vibration (see Section 3.9) in which the energy of each mode is

$$\frac{1}{2} m\omega^2 \langle x^2 \rangle = \frac{1}{2} kT \tag{14.8}$$

where the product $m\omega^2$ is the force constant, and m and ω are atomic mass and angular vibrational frequency, respectively. Let $h\nu = k\theta_E$, where $\omega = 2\pi\nu$ and θ_E is the Einstein temperature; then

$$\langle x^2 \rangle = \frac{\hbar^2 T}{km\theta_E^2} \tag{14.9}$$

For Al, $m = 27$ gm/mole, $\theta_E(\simeq \theta_D) = 428$ K, and we have $\langle x^2 \rangle = 0.6 \times 10^{-4}$ nm^2 at $T = 600$ K.

To obtain the cross section of scattering of the diffusing atom σ_d or $\langle x_d^2 \rangle$, we assume that it has acquired an energy close to the motion energy of diffusion, ΔH_m (see Section 3.7), which is independent of temperature,

$$\frac{1}{2} m\omega^2 \langle x_d^2 \rangle = \Delta H_m \tag{14.10}$$

Then the ratio of Eq. (14.10) to Eq. (14.8) gives

$$\frac{\langle x_d^2 \rangle}{\langle x^2 \rangle} = \frac{2\Delta H_m}{kT} \tag{14.11}$$

The ratio varies inversely with temperature. This dependence can be understood from Eq. (14.9), which bears on the well-known fact that the resistivity of a normal metal varies linearly with temperature above θ_E. If we substitute the specific resistivity ratio in Eq. (14.7) by the cross section ratio in Eq. (14.11), we obtain

$$Z^* = -Z\left[\frac{\Delta H_m}{kT}\frac{m_0}{m^*} - 1\right] \tag{14.12}$$

We can calculate Z^* with Eq. (14.12) and compare it to the measured value. For Au, by taking Z and m_0/m^* to be unity, the measured Z^* in the temperature range of 800 to 1000°C is shown by the solid curve in Fig. 14.3. For comparison, the calculated Z^* for Au using Eq. (14.12) is shown by the broken curve in Fig. 14.3 taking $\Delta H_m = 0.83$ eV for Au self-diffusion from Table 3.1.

Table 14.1 lists the measured Z^* values of Au, Ag, Cu, Al, and Pb taken from the review article by Huntington (1974), and the calculated Z^* values using Eq. (14.12) and the value of ΔH_m from Table 3.1. The agreements are reasonable; the simple model works well for the nearly-free electron metals as evidenced from the basic assumption of scattering used in obtaining Eq. (14.7).

14.3 Stress Effect on Electromigration (Irreversible Processes)

When a stressed thin film line undergoes electromigration, it is a case of interacting transport processes. The stress can affect a long-range mass transport by creep as discussed in Section 4.6, so there are two forces, one due to the stress and the other due to the electric field, which act simultaneously on mass transport. Since the rates of creep and electromigration are comparable and their driving forces are directional,

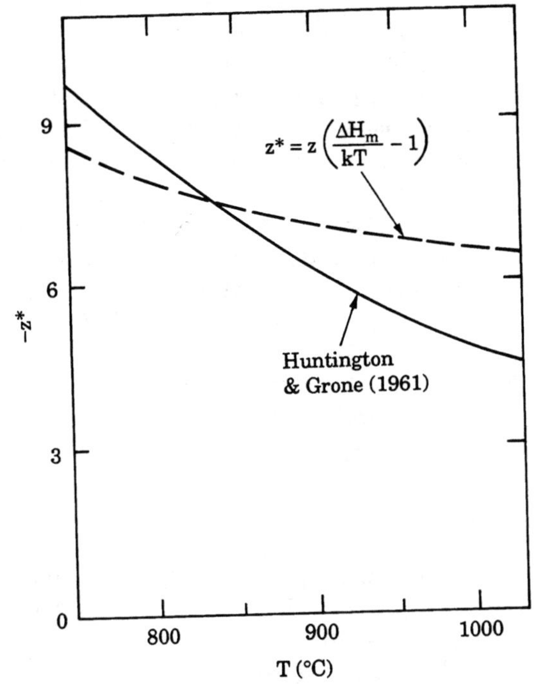

FIGURE 14.3 Plot of Z^* versus T for Au.

they interfere with each other. The interference can be analyzed by the thermodynamics of irreversible processes (e.g., Prigogine, 1961).

The stress in a metallic thin film line comes from dielectric confinement in microelectronic devices. The multilayered Al lines in a large-scale integrated device are embedded in SiO_2 for insulation. Because the processing temperatures for the metal and the dielectric are different, thermal stress exists, as discussed in Section 4.5.

TABLE 14.1 Comparison of the Measured and Calculated Values of Z^*

Metal	[a]Measured Z^*	Temp (°C)	ΔH_m (eV)[b]	[c]Calculated Z^*
Monovalent				
Au	−9.5 to −7.5	850 to 1000	0.83	−7.6 to −6.6
Ag	−8.3 ± 1.8	795 to 900	0.66	−6.2 to −5.5
Cu	−4.8 ± 1.5	870 to 1005	0.71	−6.3 to −5.4
Trivalent				
Al	−30 to −12	480 to 640	0.62	−25.6 to −20.6
Quadrivalent				
Pb	−47	250	0.54	−44

[a]Data of measured Z^* taken from Huntington (1974), where the correlation factor is ignored.

[b]Data of ΔH_m taken from Table 3.1.

[c]Data of calculated Z^* are obtained by using Eq. (14.12).

Stress effect on electromigration has been recognized in the short stripe experiment shown in Fig. 14.4 (Blech, 1976). It was found that no electromigration occurs below a certain length due to the back stress generated in the stripe. Electromigration pushes atoms to the anode and builds up a compressive stress there. Extrusion releases the stress, yet to sustain the extrusion, the compressive stress must be maintained by electromigration. According to the Nabarro–Herring model of defect formation under dilatational stresses, there are more vacancies in the anode end. Hence a gradient of vacancy concentration exists in the stripe. The direction of the gradient is such that it acts against the atomic flux of electromigration, as indicated at the top of Fig. 14.1. When the stripe is very short, the vacancy gradient can be large enough to prevent electromigration from taking place (Blech and Herring, 1976), so that there exists a critical length below which no electromigration damage occurs. The longer the stripe, the lesser the back stress, and in turn the faster the electromigration and the greater the amount of missing volume as shown in Fig. 14.4. It is clear that an applied tensile stress can enhance electromigration at the anode end and that an applied compressive stress can retard electromigration there. The reverse is true at the cathode end.

When there are multiple forces acting on mass transport in a pure metal, the linear phenomenological relations between the fluxes J_i and the forces X_j on the basis of the thermodynamics of irreversible processes are

$$J_i = \sum_j L_{ij} X_j \tag{14.13}$$

where L_{ij} are the phenomenological coefficients. In pairing the forces of stress and electric field, the fluxes of atoms and electrons can be given as

$$\begin{aligned} J_1 &= L_{11}X_1 + L_{12}X_2 \\ J_2 &= L_{21}X_1 + L_{22}X_2 \end{aligned} \tag{14.14}$$

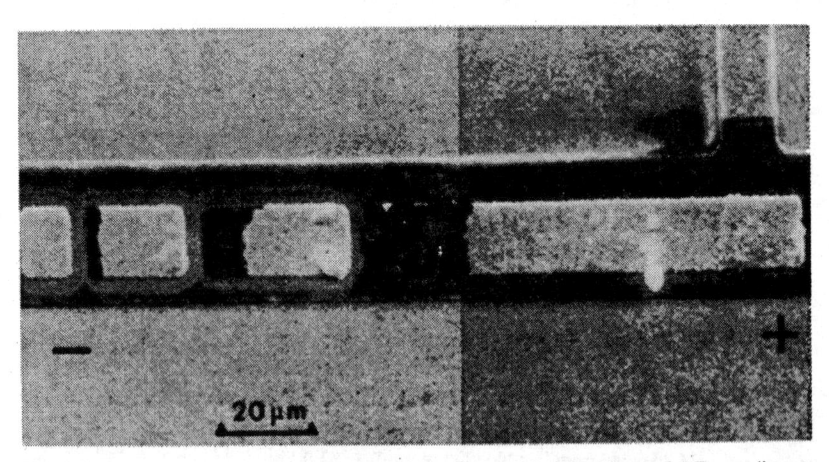

FIGURE 14.4 Drift of four Al stripes with lengths of 10, 20, 30, and 85 μm (heat treated at 350°C, 20 hrs.) after passage of 3.7×10^5 amp/cm^2 for 15 hrs (courtesy of I. Blech, J. Appl. Phys. 47, 1203, (1976)).

where J_1 = atomic flux in units of atoms/cm^2-sec (J_m); J_2 = electron flux in units of coulomb/cm^2-sec (J_e); $X_1 = -\nabla\mu = -d\sigma\Omega/dx$ is the driving force due to stress, σ, in a pure metal, and $\mu = \sigma\Omega$ is the chemical potential; and $X_2 = -\nabla\phi = \mathscr{E}$ is the driving force due to the electric field $\mathscr{E}$, and ϕ is the electrical potential. We rewrite the flux equations as

$$J_m = -C\frac{D}{kT}\frac{d\sigma\Omega}{dx} + C\frac{D}{kT}Z^*e\mathscr{E} \tag{14.15a}$$

$$J_e = -L_{21}\frac{d\sigma\Omega}{dx} + n\mu_e e\mathscr{E} \tag{14.15b}$$

The two terms in J_m have been derived in Section 4.4 and Eq. (14.1), respectively; the first is the creep term and the second is the electromigration term. The last term in J_e is electrical current in a normal metal without stress, and μ_e is electron mobility. The coefficient L_{21} will be discussed later.

In Eq. (14.15a), let $J_m = 0$, and we have

$$\Delta x = \frac{\Delta\sigma\Omega}{Z^*e\mathscr{E}} \tag{14.16}$$

This is the critical length below which no electromigration damage occurs in the short stripe experiment

Experimentally, short stripes of Al of length 10, 20, 30 and 85 μm were annealed at 350°C with a passage of current density of $J_e = 3.7 \times 10^5$ amp/cm^2 (Blech, 1976), and it was found that, except for the shortest sample, the rest showed electromigration damage after 15 hours, indicating that 10 μm $< \Delta x <$ 20 μm (see Fig. 14.4). To calculate Δx with Eq. (14.16), we take the stress at the elastic limit, $\Delta\sigma = 1.2 \times 10^9$ dyne/cm^2, and $\Omega = 16 \times 10^{-24}$ cm^3 for the atomic volume of Al atoms. The resistivity of Al is $\rho = 4.2 \times 10^{-6}$ ohm-cm at 350°C and the electric field is $\mathscr{E} = J_e\rho = 1.54$ volt/cm. Substituting these values into Eq. (14.16), we have

$$\Delta x = \frac{1.2 \times 10^9 \text{ dyne/cm}^2 \times 16 \times 10^{-24} \text{ cm}^3}{Z^* \times 1.6 \times 10^{-19} \text{ coulomb} \times 1.54 \text{ volt/cm}}$$

$$= \frac{78\ \mu\text{m}}{Z^*}$$

If we take $Z^* = 26$ for bulk Al from Table 14.1, we have $\Delta x = 3$ μm which is in the right order of magnitude yet shorter than the experimental value of about 10 μm. Since the Al thin film stripes are polycrystalline, grain boundary diffusion played a dominant role. Z^* for atoms diffusing in grain boundaries might be different from that in lattice, and $Z^* \cong 4$ to 8 would give the right critical length.

The temperature dependence of the critical length is obtained by substituting Z^* from Eq. (14.12) and $\mathscr{E}_x = J_e\rho$ into Eq. (14.16),

$$\Delta x = \frac{\Delta\sigma\Omega}{eJ_e\rho}\frac{-1}{Z\left(\frac{\Delta H_m}{kT} - 1\right)} \tag{14.17}$$

It shows that Δx is insensitive to temperature above the Debye temperature for metals whose electrical resistivity increases linearly with temperature, provided that Z^* of the stripe obeys Eq. (14.12) and that $\Delta H_m >> kT$.

In Eq. (14.15b), let $J_e = 0$, we obtain

$$L_{21} = \frac{1}{\rho\Omega}\left(\frac{d\phi}{d\sigma}\right)_{J_e=0} \tag{14.18}$$

where $(d\phi/d\sigma)_{J_e=0}$ is defined as the electrical potential difference per unit stress difference with zero current—as in a piezoelectric solid under pressure, for instance. For a short stripe of metal on an insulating substrate, we can regard $(d\phi/d\sigma)_{J_e=0}$ as the deformation potential. Its expression is obtained by using Onsager's reciprocity relation, $L_{12} = L_{21}$,

$$\left(\frac{d\phi}{d\sigma}\right)_{J_e=0} = \frac{Z^*D\rho e}{kT} \tag{14.19}$$

in units of volt-meter^2/newton or cm^3/coulomb.

14.4 Measurement of Electromigration

The equation which governs the measurement of electromigration in pure metals is the flux equation

$$J_{\text{em}} = C\,\langle v\rangle = CMF = C\frac{D}{kT}Z^*e\mathscr{E} \tag{14.20}$$

At constant T and $\mathscr{E}$, the flux J_{em} can be measured by the amount of mass transport. The drift velocity can be measured by marker motion or by displacement of radioactive tracers. The product DZ^* is determined and then D and Z^* are separated by knowing D independently.

It is possible to measure DZ^* and D simultaneously in a single experiment; this is achieved by using the tracer technique as shown in Fig. 14.5. A long bar of Au contains a thin middle layer of radioactive tracer of Au annealed at a high temperature for a given time with a current density. The concentration profiles of the tracer before and after electromigration are depicted in Fig. 14.5 by the curves A and B, respectively. The distance between the centroids of A and B divided by the annealing time gives the drift velocity, from which the product DZ^* can be determined. The spreading of curve B with respect to A determines D; the tracer diffusion is a random walk. The difficulty of carrying out such an experiment is to maintain a uniform temperature over the entire sample, for otherwise a correction due to the effect of temperature gradient is required.

Using the short stripe experiment shown in Fig. 14.1, the rate of mass transport can be measured from the time dependence of the missing volume at the cathode end of the stripe. In turn the flux or the drift velocity can be measured. However, to determine Z^*, the diffusivity must be known independently.

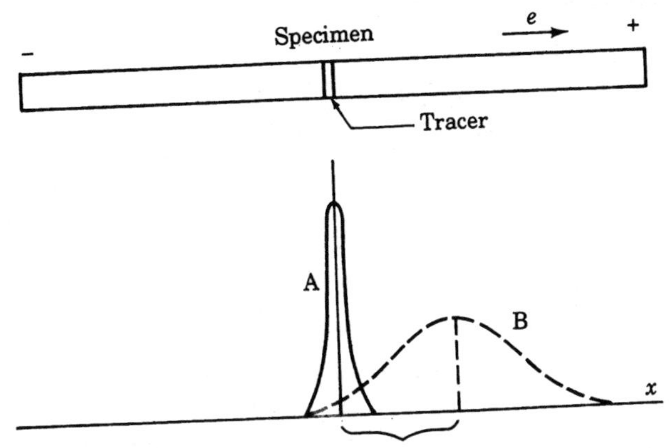

FIGURE 14.5 The isothermal isotope method. The curves A and B show the concentration profiles of the radioactive isotope before and after isothermal annealing at a given current density.

Another technique is to use Eq. (14.16). By rewriting the equation, we have

$$Z^* = \frac{\Omega}{e\mathscr{E}} \frac{\Delta\sigma}{\Delta x} \tag{14.21}$$

We calculate Z^* by measuring $\Delta\sigma$ and Δx. The critical length can be determined by performing electromigration in the short stripe experiment in Fig. 4.4 for a sufficiently long time until the mass transport stops. The stress can be measured by x-ray diffraction of the lattice parameters at the two ends of the stripe during electromigration. The advantage of this technique is that it measures Z^* alone rather than the product of DZ^*. Also, it is possible to study the change of Z^* as a function of microstructure of the thin film stripe and temperature.

14.5 Electromigration in Metallic Fine Lines

In microelectronic devices, the interconnecting Al lines are polycrystalline films. Electromigration can occur at relatively low temperatures (~100°C), mediated by grain boundary diffusion. For Al metallization on Si devices, a temperature of 100°C is easily reached by Joule heating under a high current density. The temperature is sufficient for grain boundary diffusion to take place in Al. The grain boundary electromigration flux can be expressed as

$$J_b^{em} = C_b \frac{D_b \delta}{kTd} Z_b^* e\mathscr{E} \tag{14.22}$$

where the subscript b means a quantity of grain boundary, and d and δ are grain diameter and effective grain boundary width, respectively. The evidence for electromigration along grain boundaries are the measured low activation energy of mass transport and the existence of triple points of grain boundary junctions serving as flux divergences for void and extrusion formations.

To consider void growth, note that the atomic flux generated by electromigration must be balanced by a vacancy flux plus any sink for vacancies. We express

$$J_m + J_v + s = 0 \tag{14.23}$$

where $s = \nabla \cdot \mathbf{J}_v \neq 0$ is a sink (e.g., a growing void). The continuity equation for vacancies can be written as

$$\frac{\partial C_v}{\partial t} = -\nabla \cdot \mathbf{J}_v + \frac{C_v - C_v^0}{\tau} \tag{14.24}$$

where $C_v - C_v^0$ expresses the local supersaturation of vacancy concentration above thermal equilibrium; τ is the average lifetime of a vacancy, which depends on the effectiveness of the sink in annihilating vacancies. In a one-dimensional case,

$$J_v = -D_v \frac{\partial C_v}{\partial x} + C_v \frac{D_v}{kT} Z^* e\mathscr{E} \tag{14.25}$$

where D_v is the diffusivity of vacancies and the second term is the electromigration contribution. We note that Eq. (14.24) is similar to that used to describe the recombination of minority carriers injected into a semiconductor. Under a steady state condition, we have from Eq. (14.24),

$$\nabla \cdot \mathbf{J}_v = \frac{C_v - C_v^0}{\tau} \tag{14.26}$$

The solution to Eq. (14.26) in the one-dimensional case is given as a problem (see problem 14.12).

The growth of a localized void into an opening in a metallic fine line is the most serious failure mode of electromigration in microelectronic devices. Equally, we can consider an extrusion as a sink of the flux of atoms. A local stress concentration point in the line would be a favorable site for nucleating a sink of vacancies or of atoms, depending on the sign of the stress and on its interference with electromigration. The kinetics of void formation and hillock growth are given in Chapter 15.

The prevention of electromigration has been one of the most challenging problems in the microelectronic industry. The electromigration flux can be reduced by lowering Z^* and D, and the flux divergence at grain boundary triple points can be eliminated by using single-crystal thin films. The solutions which have been adopted successfully in actual devices are (1) adding a few atomic percent of Cu into Al and (2) constructing a layered thin film line such as $Al(Cu)/Al_3Hf/Al(Cu)$. The beneficial effect of adding Cu to Al is to slow down the grain boundary diffusion of Al. The formation of θ-phase (Al_2Cu) in grain boundaries is believed to provide sources for Cu replenishment when Cu is depleted from the grain boundaries by electromigration, hence the effect of Cu solution can be kept in the line for a longer time. The advantage of the sandwich structure is to have the intermetallic barrier to prevent voids from growing across the line to form an opening. Other remedies are to use bamboo-type grains to remove divergences from fine lines, to lower the operating temperature, or to use other conducting metals which have a higher activation energy of diffusion. For details about these solutions, readers are referred to review articles by d'Heurle and Rosenberg (1973), d'Heurle and Ho (1978), and Ho and Kwok (1989).

References

1. I. Ames, F. M. d'Heurle, and R. Horstman, *IBM J. Res. Develop*, 4, 461 (1970).
2. M. J. Attardo and R. Rosenberg, *J. Appl. Phys.* 41, 2381 (1970).
3. I. A. Blech, *J. Appl. Phys.* 47, 1203 (1976).
4. I. A. Blech and C. Herring, *Appl. Phys. Lett* 29, 131 (1976).
5. C. Bosvieux and J. Friedel, *J. Phys. Chem. Solids* 23, 123 (1962).
6. S. R. de Groot and P. Mazur, *Non-Equilibrium Thermodynamics*, Dover, New York (1984).
7. F. M. d'Heurle and P. S. Ho in *Thin Films: Interdiffusion and Reactions* (p. 243), edited by J. M. Poate, K. N. Tu, and J. W. Mayer, Wiley–Interscience, New York (1978).
8. F. M. d'Heurle and R. Rosenberg in *Physics of Thin Films* (Vol. 7, p. 257), Academic Press, New York (1973).
9. V. B. Fiks, *Sov. Phys. Solid State* (English trans.) 1, 14 (1959).
10. P. S. Ho and J. K. Howard, *J. Appl. Phys.* 45, 3229 (1974).
11. P. S. Ho and T. Kwok in *Rep. Prog. Phys.* 52, 301 (1989).
12. J. K. Howard, J. F. White, and P. S. Ho, *J. Appl. Phys.* 49, 4083 (1978).
13. H. B. Huntington in *Diffusion in Solids: Recent Developments* (p. 303), edited by A. S. Nowick and J. J. Burton, Academic Press, New York (1974).
14. H. B. Huntington and A. R. Grone, *J. Phys. Chem. Solids* 20, 76 (1961).
15. R. Landauer and J. W. F. Woo, *Phys. Rev. B* 10, 1266 (1974).
16. I. Prigogine, *Introduction to Thermodynamics of Irreversible Processes*, Wiley–Interscience, New York (1961).
17. R. S. Sorbello, *J. Phys. Chem. Solids* 34, 937 (1973).
18. R. S. Sorbello, *Phys. Rev. B* 31, 798 (1985).

Problems

14.1 What is the current density in an extension cord used at home for a 100-watt table lamp? Assume the Cu wire in the cord has a diameter of 0.1 mm and is 10 ft long and the applied voltage is 110 volt.

14.2 Calculate the cross section of scattering of an Au atom at 800°C based on the Einstein model of atomic vibration.

14.3 In an fcc metal, show that the displacement of an atom from its equilibrium position to the activated position (Fig. 14.2b) is $\sqrt{2}a/4$ where a is the lattice parameter. Take $\langle x_d^2 \rangle = (\sqrt{2}a/4)^2$ and show that $\frac{1}{2}m\omega^2 a^2/8$ is a good approximation of ΔH_m.

14.4 Plot Z^* versus T for Al from 350°C to 650°C by using Eq. (14.12).

14.5 Given drift displacement x as a function of time t, calculate $\langle v \rangle$, and also calculate Z^* by knowing D.

14.6 Calculate the electrical force $e\mathscr{E}$ at a current density of 10^5 amp/cm^2 and the chemical potential $\sigma\Omega$ at the elastic limit for Au, and calculate the critical length for Au using Eq. (14.16).

14.7 Two Al short stripes with length of 20 μm and 30 μm undergo electromigration. Calculate the stress at their anode ends when they carry a current density of 10^5 amp/cm^2.

14.8 Show that the critical length ratio of Al to Au

$$\frac{\Delta x_{\text{Al}}}{\Delta x_{\text{Au}}} \cong \frac{Y_{\text{Al}} Z^*_{\text{Au}}}{Y_{\text{Au}} Z^*_{\text{Al}}} < 1$$

where Y is Young's modulus. Explain why the ratio is less than unity.

14.9 Use the Onsager reciprocity relation L_{21} and L_{12} in Eq. (14.16) to show that

$$\frac{d\phi}{d\sigma} = \frac{\mu_m Z^*_\Omega}{\mu_e Z}$$

where μ_m is atomic mobility.

14.10 In Eq. (14.16), assume that the stress is zero, and calculate the ratio of J_m/J_e.

14.11 In conducting electromigration in a Au wire, because of uneven heat conduction, there is a temperature gradient in the wire. Take the forces as $X_1 = -\nabla T$ and $X_2 = -\nabla\phi$, and write down the pair of equations similar to those in Eq. (14.15) and explain the phenomenological coefficients in the equations.

14.12 Substituting Eq. (14.25) into Eq. (14.26), find the general solution for the one-dimensional case.

CHAPTER 15

Morphological Changes in Thin Films

15.0 Introduction

The external form of a thin film can be influenced by internal changes in structure and/or composition. The internal atomic rearrangement is often induced by an applied force as in electromigration. Voids, extrusions, and precipitates are common morphological changes of a thin film. These present serious yield and reliability problems in microelectronic devices. They can cause electrical failures such as an opening in interconnecting lines, shorting of a junction, or breakdown of an interface.

From the point of view of device applications, a homogeneous morphological change such as the formation of many tiny voids in a film is not as serious as an inhomogeneous change (the growth of a single large void). It takes only one opening to cause a circuit to fail, so the localized growth of a large void (or a hillock) at a stress concentration point, for example, is critical. However, whether the morphological change is homogeneous or not depends not only on the driving force, but also on how the thin film microstructure responds. Hence, the control of a thin film microstructure in order to minimize the localized deformation is an important technical issue; in this respect uniform grain size, composition, and stress distribution are desirable.

Owing to miniaturization, the critical dimensions of metallization—such as the contact area, line width, and via diameter—in very large-scale integrated devices are now approaching the size of a grain in thin film microstructure. Since the grain size is typically limited by the film thickness, the distribution of grains in the microstructure tends to be nonuniform when the critical dimensions of metallization are about the same as film thickness.

Because of the large surface-to-volume ratio, a thin film and a fine line possess the intrinsic cause of morphological instability. As shown in Chapter 5, minimizing the surface energy tends to ball up a thin film, especially if the surface energy of the thin film is greater than that of the substrate. Furthermore, minimizing the internal grain boundary energy can lead to morphological changes by grain growth and grooving. Due to the simple two-dimensional nature of grain growth in a polycrystalline thin film, it has been a subject of intensive study by computer simulation. Grain growth relates to the evolution and control of a microstructure.

In this chapter, we study grain growth in thin films, hillock and void formation, pitting in Si by Al-penetration, and finally the corrosion of an AgPd alloy electrode.

15.1 Grain Growth

In a microstructure, grain growth accompanies grain shrinkage. Certain grains grow at the expense of their neighboring grains. In grain growth, a grain boundary migrates against its curvature. This is shown in Fig. 15.1. The overall driving force of grain growth is the reduction of the total grain boundary area or the total grain boundary energy. Locally, the driving force of migration is the reduction of curvature.

In a narrow-base Si bipolar device, a heavily doped polycrystalline Si film has been used as an emitter contact to form a shallow emitter and at the same time to achieve a narrow base in Si. From the grain growth viewpoint, the device has a fine-grained poly-Si film making contact to an extremely large grain of single-crystal Si; grain growth can occur. Indeed, the substrate Si has often been found to grow into the poly-Si. It becomes an issue of controlling the grain growth, and in turn the dimension and performance of the device.

In Chapter 3, the expression for velocity of a diffusing atom is

$$v = MF \tag{3.4}$$

where $M = D/kT$ is the mobility of the atom and F is the driving force. Equation (3.4) is a very general kinetic expression of motion, and it also applies to motion of grain boundaries. In grain boundary motion, which is influenced by its curvature, the pressure acting on the grain boundary per unit area, assuming a spherical shape, is

$$p = \frac{-\dfrac{dE}{dr}}{A} = \frac{-\dfrac{d}{dr}(4\pi r^2 \gamma)}{4\pi r^2} = -\frac{2\gamma}{r} \tag{5.1}$$

where γ and r are the grain boundary energy per unit area and radius, respectively. The last two equations give

$$v = M\frac{2\gamma}{r} \tag{15.1}$$

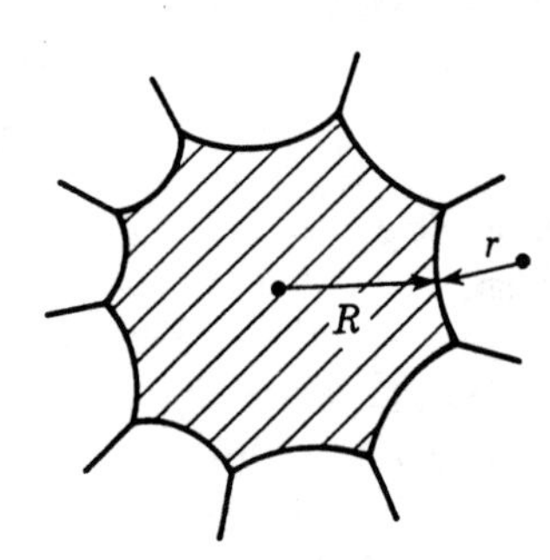

FIGURE 15.1 Schematic drawing of a growing grain of average radius R in contact with grains of radius r.

where M is defined as the mobility of a unit area of grain boundary and the negative sign is removed because the motion is against the curvature. By taking $v = dr/dt$, substituting it into Eq. (15.1), and integrating

$$r = At^{1/2} \tag{15.2}$$

which describes the time-dependent behavior of a grain radius. For the average grain size R in a microstructure, it is found experimentally that

$$R = Bt^n \tag{15.3}$$

where B is a temperature-dependent parameter and contains M and γ, and n is about 0.5 to 0.4. The deviation from 0.5 is expected if grain growth is delayed due to local equilibrium at triple points of grain boundary junctions.

To examine the magnitude of the driving force due to curvature, the work done per atom by grain growth is taken as

$$p\Delta V = \frac{2\gamma}{r}\Omega \tag{15.4}$$

where Ω is the atomic volume $\cong 2.0 \times 10^{-23}$ cm^3 for Si. If we take γ to be 500 ergs/cm^2 or 0.5 Joule/m^2 and r to be 100 nm,

$$p\Delta V = \frac{10^3 \text{ ergs/cm}^2}{10^{-5} \text{ cm}} \times \frac{2.0 \times 10^{-23} \text{ cm}^3}{1.6 \times 10^{-12} \text{ ergs/eV}}$$

$$= 1.2 \times 10^{-3} \text{ eV/atom}$$

The value is about 2 to 3 orders of magnitude smaller than the chemical energy per atom, which is typically 0.1 to 1 eV/atom. In the amorphous-to-crystalline phase transformation of $CoSi_2$ around 150°C, presented in Chapter 10, the driving force to move the amorphous–crystalline interface is therefore 2 to 3 orders of magnitude greater than that to move the grain boundaries between the crystalline $CoSi_2$ grains. While the mobility of the interface and grain boundaries are comparable, the latter do not move because of a very weak driving force.

In general, grain growth will not proceed forever. It stops when the microstructure reaches a metastable state where the boundaries at a triple point make 120° angles (i.e., the forces and torques at the triple point are at local equilibrium). In pure metallic thin films, this tends to happen when the grain size is about several times the film thickness.

Very often some exceptionally large grains are found in films although most of the grains have reached the stable size. These exceptionally large grains are due to *secondary grain growth* under a different driving force; it is due to the reduction in surface energy of those grains having a low-index plane such as (111) as their surface. In this case, the driving force can be expressed as

$$F = \frac{\Delta\gamma}{h} + \frac{2\gamma}{r} \tag{15.5}$$

where $\Delta\gamma$ is the difference in surface energy between the secondary grains and the neighboring grains and h is the film thickness. The growth rate of the secondary grains is obtained by substituting Eq. (15.5) into Eq. (3.4). The driving force is inversely proportional to film thickness, mainfesting the surface-to-volume ratio effect. In a highly textured thin film, the driving force of secondary grain growth is removed between neighboring grains having the same surface-normal axis. In this special case, the grain growth of a large grain may still take place by coalescence in which the smaller grain merges with the large one by a small-angle rigid rotation. This process is limited to neighboring grains with a small-angle tilt-type grain boundary.

Grain growth has been studied by computer simulation, especially in two-dimensional cases. The study begins with a given microstructure and inputs a set of kinetic parameters to allow the microstructure to evolve with time. Then the statistics of the microstructure, such as the number of sides of grains and the diameter of grains, can be collected and their time-dependence analyzed. For example, a continuous random nucleation rate and an isotropic constant growth rate can be programmed to generate a microstructure as shown in Fig. 15.2, which serves as an initial state for microstructure evolution.

If it is assumed that (a) the nucleation is random and continuous, (b) the growth is constant and isotropic, and (c) the grain boundaries do not migrate, all the grain boundaries formed are hyperbolic. Strictly, the boundaries are hyperboloids, but since the grain diameter is much greater than the film thickness the two-dimensional approximation is good. Figure 15.3 represents the successive positions of the growth fronts by two sets of concentric circles centered at A and B. The family of broken lines, including the straight one, which connect the points of intersection of these circles, are hyperbolas and represent the grain boundaries formed when grains nucleated at the points A and B impinge. All the grain boundaries shown in Fig. 15.2 are hyperbolas. Experimentally, this is also true for the grain boundaries of $CoSi_2$ shown in Fig. 10.7. Since the crystallization temperature of amorphous $CoSi_2$ at about 150°C is much lower than the melting point of 1325°C for crystalline $CoSi_2$, the grain boundaries do not move during crystallization.

The bold lines in Fig. 15.3 depict a large grain and a small grain joined by a hyperbolic grain boundary. If the distance between A and B is $2c$, the time interval

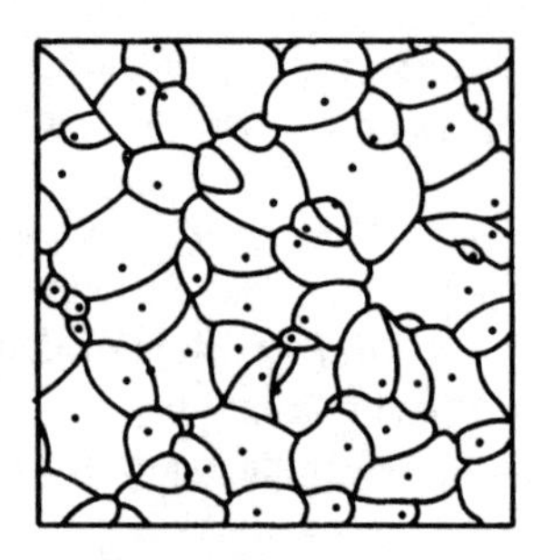

(a) Continuous nucleation (Johnson–Mehl model)

FIGURE 15.2 Computer simulation of polycrystalline microstructures obtained from a mode of continuous nucleation and isotropic growth by the Johnson-Mehl-Avrami model (taken from Frost and Thompson, 1987).

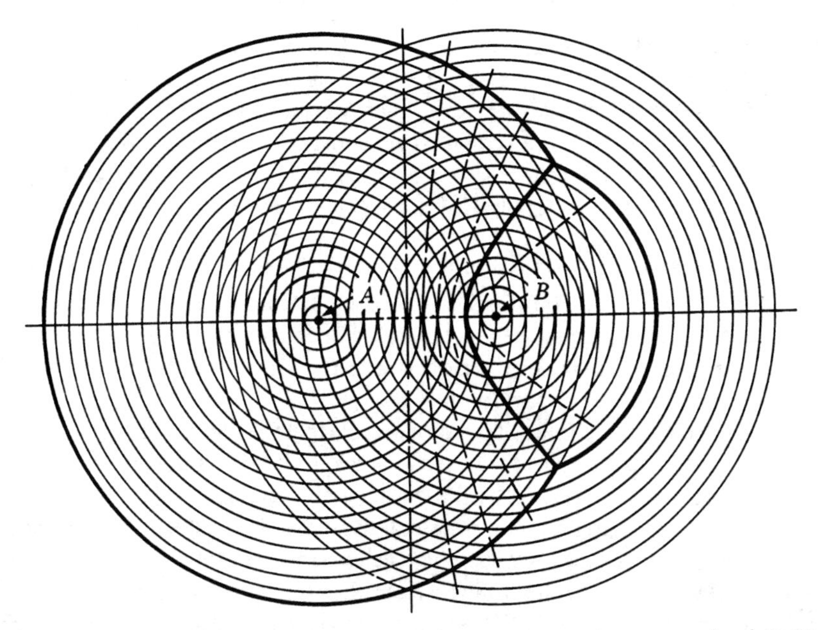

FIGURE 15.3 The family of broken curves, which connect the successive intersecting points of the two sets of concentric circles, are hyperbolas. The bold curve depicts the formation of a hyperbolic grain boundary between a large and a small grain.

between the nucleation events at A and B is Δt, and the growth rate is v, the hyperbola is described by

$$\frac{x^2}{a^2} - \frac{y^2}{b^2} = 1 \tag{15.6}$$

where $a = v\Delta t/2$ and $b = (c^2 - a^2)/2$. The x-axis passes through A and B, and the y-axis passes through their midpoint. All the parameters (i.e., c, v, and Δt) can be measured experimentally. In fact, for a pair of grains, the value of Δt can be calculated by knowing c and v. Thus, nucleation events can be reconstructed in a given microstructure. In the case of three-dimensional growth, the growth fronts are spherical, and the grain boundary can be obtained by a revolution of the hyperbola around the A-B axis.

The hyperbolic grain boundaries themselves are in nonequilibrium states because of the curvature. The curvature of a hyperbolic grain boundary can be obtained from the first and second derivatives of Eq. (15.6). The migration of such a grain boundary is known, provided that the grain boundary energy is measured, which has been done in Section 10.4.

15.2 Hillock Growth

A common mode of releasing compressive stress in metallic thin films is the formation of hillocks. The compressive stress is due to the constraint of the substrate upon warming up. When warming up from liquid nitrogen temperature to room temperature, thin films of Pb and Sn (low-melting-point metals) on quartz show hillocks. In Fig. 15.4, scanning electron microscopic pictures of hillocks in a Pb alloy film confined in SiO_2 are shown. A related phenomenon is the formation of whiskers on a Sn film deposited on top of a Cu film, as shown in Fig. 15.5. The growth of Sn whiskers on Sn-plated Cu capacitors was found to cause shorting problems.

The growth of hillocks is similar to a two-dimensional precipitation. The hillocks are solid mounds and can be regarded as precipitates (i.e., a flux of material is transported to accumulate at the hillocks). In order to analyze the hillock formation from this viewpoint, we consider the driving force of the flux, the nucleation, and the mechanism of mass transport for growth.

Concerning the driving force, it is a prerequisite that the film is under compressive stress—specifically, a biaxial compressive stress. Following the concept of diffusional creep, if there are certain tensile or stress-free regions in the film, we have stress gradient as a driving force. It is natural to assume that hillocks are the regions of tensile stress, or just stress-free regions. The Nabarro–Herring creep model shows that atomic flux will go to the hillocks from their compressive surroundings. We simplify the picture and assume the hillocks are regularly spaced as shown in Fig. 15.6, and we consider the nucleation and growth of one hillock of radius a in a circular area of radius b.

However, we must ask the question why a compressed film chooses the hillock mode to release its stress? Why not relax uniformly instead of using a seemingly localized relaxation process? If we take the viewpoint that it is a kinetic issue, we

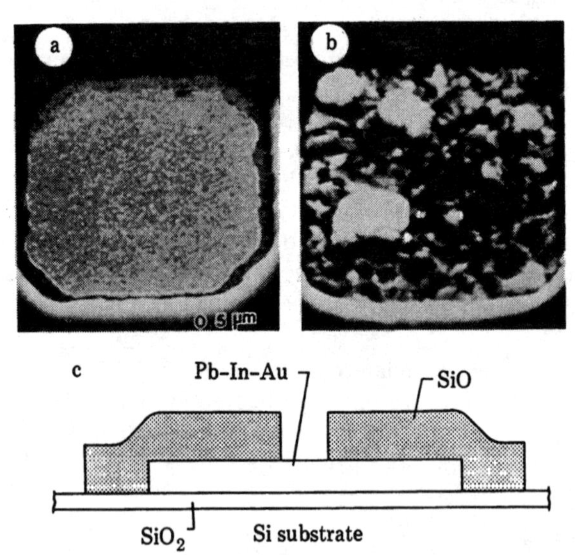

FIGURE 15.4 Scanning electron micrograph of hillocks in a Pb(12 atomic percent)–In (4 atomic percent)–Au film confined in quartz except at the open window. (a) As deposited. (b) After 100 times of thermal cycle between room temperature and liquid nitrogen temperature. (c) Schematic diagram of the cross-sectional view of the sample. (Courtesy of M. Murakami, IBM Research Division)

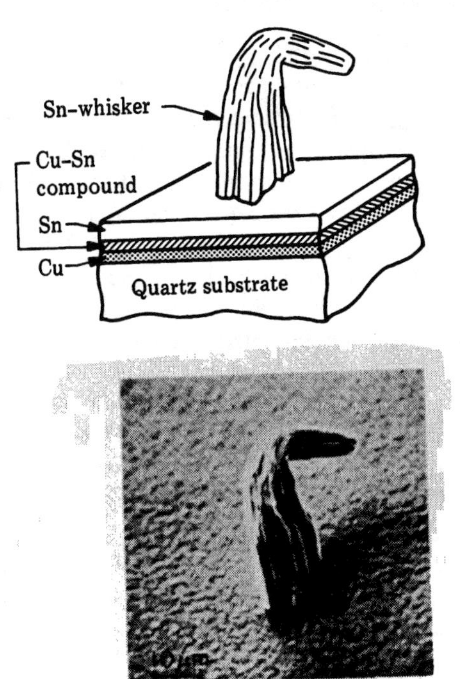

FIGURE 15.5 Growth of whiskers from a Sn film deposited on top of a Cu film. (Courtesy of J. W. Mayer, J. M. Poate and K. N. Tu, *Science*, 190, 228, 1975)

argue that the process of nucleation and growth of hillocks may be the fastest mode of releasing the stress with an oxidized surface.

The nucleation of a hillock is the formation of an unstressed region in the compressively stressed film. We recall that in Chapter 10 the classical nucleation theory gives an expression for a critical nucleus

$$N_{\text{crit}} = \left(\frac{2b}{3}\frac{\gamma_{AC}}{\Delta H_{AC}}\right)^3 \tag{10.36}$$

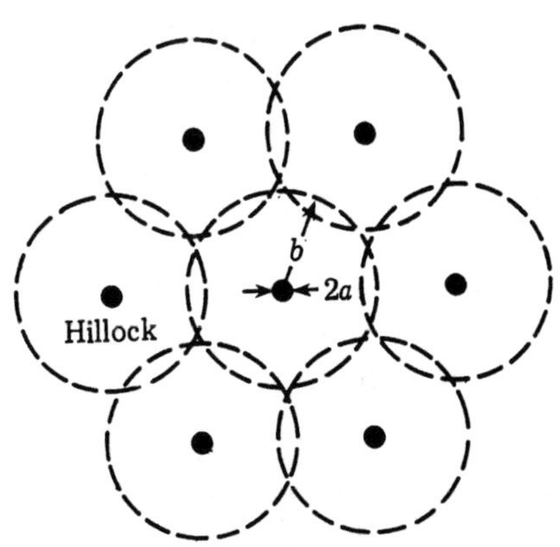

FIGURE 15.6 A sketch of regularly spaced hillocks on a film surface.

where b is a geometrical constant, and γ_{AC} and ΔH_{AC} are interfacial energy per atom and heat of crystallization per atom, specifically for the amorphous-to-crystalline transformation. Physically, the lower limit of N_{crit} cannot be less than one atom, and the higher limit cannot be very large—not so much as a few thousand atoms. Nucleation is a thermal fluctuation phenomenon, and it cannot require a large number of atoms to come together by fluctuation. However, N_{crit} can be very large by having a small value of ΔH_{AC} or a large ratio of γ_{AC} to ΔH_{AC} in Eq. (10.36). We encounter such a case here, in considering the nucleation of an unstressed region in a stressed region, because we have shown in Section 4.3 that the elastic energy is much smaller than chemical energy. If we take the interfacial energy to be 0.1 eV per atom and the strain energy to be about 10^{-4} eV per atom, the value of N_{crit} becomes unreasonably large. Therefore from the viewpoint of nucleation, we must assume that the nuclei are centers of high stress concentration and that the interface involved is coherent. Most likely a much lower activation energy process is involved in nucleating the localized stress-free regions for hillock growth. The relaxation process in these localized regions has been assumed to be the slip of dislocations. The barrier to nucleation of the stress-free region, and how large is the activation energy of the barrier, are unclear. What is known is that there is an incubation period of hillock formation, and the incubation period has been taken to be nucleation of the stress-free regions.

For the growth of hillocks in Pb and Sn films, mass transport by lattice diffusion has been ruled out because (1) it is not a high temperature phenomenon, (2) the distance of diffusion is much greater than the average grain size in the film, and (3) the activation energy of mass transport was measured to be much less than that for lattice diffusion. Therefore, surface diffusion, grain boundary diffusion, dislocation climb, and diffusion along the film–substrate interface have been considered as mechanisms of hillock growth. In the case of Sn hillocks, the deposition of 30 nm of Ni on SiO_2 before the Sn deposition greatly reduced the number of hillocks. Hence, diffusion along the film–substrate interface was identified as the kinetic path of mass transport.

Another question in hillock growth is whether the arriving atoms add themselves to the top of the hillock or to the bottom. By observing that the morphology of the top of the hillock does not change with growth (a similar observation was made for whiskers), it was concluded that the arriving atoms attach themselves to the bottom of the hillock. The back pressure from the substrate at the base of the hillock pushes the hillock out, resulting in its growth out of the film surface. The sides of the hillock may resist the vertical slide motion in the beginning; this could account for part of the incubation time needed to begin growth.

To model the growth, we assume that the hillock is a cylindrical mound with a base of radius a. The stress field which supplies atomic flux for the growth has a radius b as shown in Fig. 15.6. To evaluate

$$F = -\frac{\partial \mu}{\partial r}$$

in the stress field, (μ is the product of the stress and the atomic volume) we consider only the early stage of growth (i.e., we use a steady-state situation for simplicity).

It is clear that over a long period of time, the stress will be released, so it is actually a non-steady-state problem, but the mathematics for that becomes complicated and involves Bessel functions (Chaudhari, 1974). For the purpose of gaining a simple physical picture here, we solve for a cylindrical coordinate, two-dimensional, steady-state continuity equation of the potential $\sigma\Omega$ (see Equation 5.14 for a comparison),

$$\nabla^2\sigma = \frac{\partial^2\sigma}{\partial r^2} + \frac{1}{r}\frac{\partial\sigma}{\partial r} = 0 \tag{15.7}$$

where σ is the stress and Ω is the atomic volume. We note that a density-like function such as σ can be solved by the continuity equation (see Secton 4.5). The solution has the form of $\sigma = A'' \ln r + B''$. Using the boundary conditions at t (time) $= 0$, taken to be the end of the incubation time,

$$\sigma = \sigma_0 \qquad \text{where } r = b$$

$$\sigma = 0 \qquad \text{where } r = a$$

we have

$$A'' = \frac{\sigma_0}{\ln(b/a)} = \sigma_0 A' \tag{15.8}$$

and

$$B'' = -\frac{\sigma_0 \ln a}{\ln(b/a)} \tag{15.9}$$

So

$$\sigma = A'\sigma_0 \ln r + B'' \tag{15.10}$$

Then,

$$F = -\frac{\partial\mu}{\partial r} = -\frac{\partial\sigma\Omega}{\partial r} = -A'\,\frac{\sigma_0\Omega}{r} \tag{15.11}$$

The atomic flux at $r = a$, using Eq. (3.5), is

$$J = \frac{CD}{kT}F = \frac{CD}{kT}\left(-A'\,\frac{\sigma_0\Omega}{a}\right) = -A'\,\frac{\sigma_0 D}{kTa} \tag{15.12}$$

The amount of material accumulated at the base of the hillock in a period t equals the volume of the hillock, or $JAt\Omega = h\pi a^2$, where A is the peripheral area of the base at the interface. $A = 2\pi a\delta_i$ where δ_i is the effective thickness of the interface (~3 to 5 Å). The growth rate of the hillock,

$$\frac{dh}{dt} = \frac{JA\Omega}{\pi a^2} = \frac{2\pi a\delta_i\Omega}{\pi a^2}(-A')\,\frac{\sigma_0 D_i}{kTa} = -\frac{A'2\sigma_0\Omega\delta_i D_i}{kTa^2} \tag{15.13}$$

where D_i is the diffusivity along the film–substrate interface. In Eq. (15.13), the compressive stress σ_0 is negative in sign. Also, since we assume a steady-state process, the hillock growth rate is constant. Obviously this is not true after a longer time of growth. The final height of the hillock h_f can be evaluated using the law of conservation of mass,

$$\pi a^2 h_f = \pi b^2 h' \varepsilon$$
$$h_f = \frac{b^2 h' \sigma_0 (2 - 3\nu)}{a^2 Y} \tag{15.14}$$

where h' is the thickness of the film and ε is the strain of the film, $\sigma_0(2 - 3\nu)/Y$, ν is Poisson's ratio and Y is Young's modulus.

15.3 Void Formation in Fine Lines

In the Nabarro–Herring creep model, the tensile region has a higher chemical potential than an unstressed region. Hence, a film generates excess vacancies under tension. These vacancies can condense into dislocation loops or voids where they can nucleate and can compete effectively for vacancies with other sinks such as the existing dislocations and free surfaces. The concern about void growth in a narrow conducting line is that a void can extend itself across the line and cause an open failure.

The tension in an Al line on a Si device arises from cooling. To reduce the contact resistance of Al to Si or to achieve a solder-joint between the line and the packaging module, the device has to be heated to around 400°C. Upon cooling, the Al line is under tension due to the constraint of the substrate. Then, in device operation, the device may experience a temperature near 100°C, where grain boundary diffusion in Al becomes significant. The line creeps and void formation can occur.

However, we might ask why the line relaxes its tensile stress by void formation rather than by transporting vacancies to free surfaces or to dislocations. In these other ways, the integrity of the line could be preserved upon stress relaxation. But the line usually is coated by a passivation layer such as sputtered quartz to prevent corrosion or to achieve a multilayered construction, so that the surfaces and edges of the line are no longer good sinks for vacancies. And although dislocation climb still provides good sinks for vacancies, the dislocation density may not be high enough.

Figure 15.7a shows a schematic diagram of a section of a line of width $2W$ and thickness h, containing a grain boundary and a cylindrical void of radius r at the center of the grain boundary. The line is under a tensile stress σ. The growth of the void occurs because it serves as a sink for vacancies. In Chapter 5, the potential energy of an atom on a spherical void surface is shown to increase due to curvature by the amount

$$p\, dV = \frac{2\gamma}{r}\Omega$$

where γ is the surface energy per unit area of the void. On a cylindrical void surface, the increase is $\gamma\Omega/r$. Consequently, the vacancy concentration C_{v1} near the void

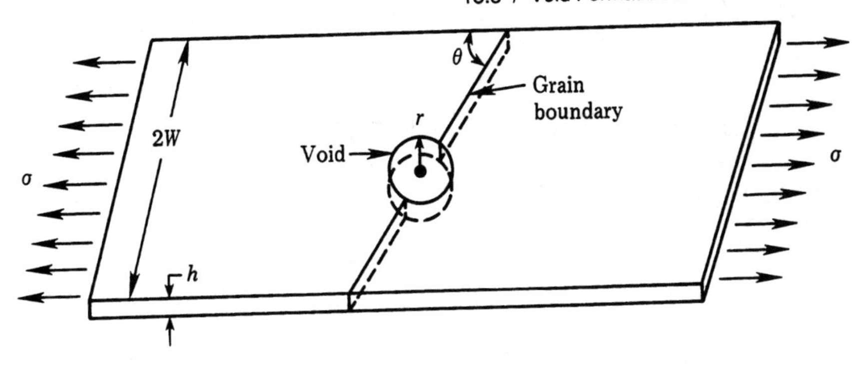

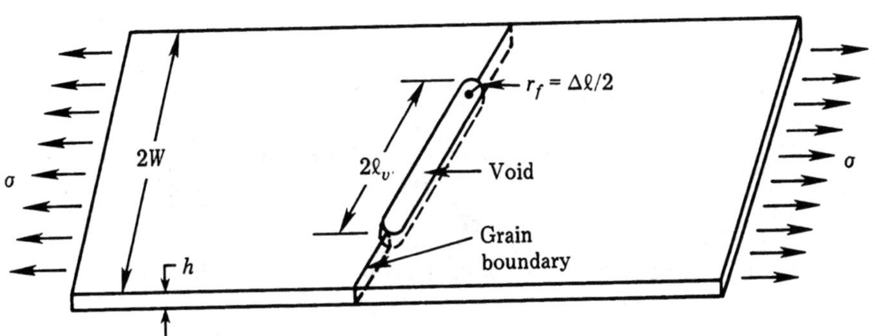

FIGURE 15.7 (a) A schematic diagram of a section of a fine line containing a grain boundary and a cylindrical void, and (b) the same line containing a grain boundary and a sausage void.

surface is increased due to the fact that the formation energy is reduced to $\Delta G_f - \gamma\Omega/r$, where ΔG_f is the formation energy of a vacancy in the unstressed plane surface regions. We have

$$C_{v1} = C \exp \left\{ - \frac{\left(\Delta G_f - \frac{\gamma}{r} \Omega \right)}{kT} \right\} \tag{15.15}$$

The effect of the tensile stress is ignored near the void since it is assumed that the void formation has relaxed the stress, and besides there is no normal stress on the free surface. In the region normal to the grain boundary, the vacancy concentration C_{v2} is increased due to the tensile stress σ,

$$C_{v2} = C \exp \left\{ - \frac{(\Delta G_f - \sigma\Omega \cos \theta)}{kT} \right\} \tag{15.16}$$

It is clear that we must have

$$\sigma\Omega \cos \theta > \frac{\gamma}{r} \Omega$$

in order for the void to grow. The difference in vacancy concentration is

$$C_{v2} - C_{v1} = \Delta C_v = C_v \left[\left(\exp \frac{\sigma\Omega \cos\theta}{kT} \right) - \exp\left(\frac{\gamma}{r} \frac{\Omega}{kT} \right) \right]$$

where $C_v = C \exp(-\Delta G_f/kT)$. If both $\sigma\Omega \cos\theta$ and $\gamma\Omega/r$ are much smaller than kT, we have

$$\Delta C_v = C_v \left(\frac{\sigma\Omega \cos\theta}{kT} - \frac{\gamma\Omega}{rkT} \right) \tag{15.17}$$

The flux of vacancies going to the void is

$$J_v = -D_v \frac{\Delta C_v}{\Delta x} = -D_v C_v \frac{\left(\frac{\sigma\Omega \cos\theta}{kT} - \frac{\gamma\Omega}{rkT} \right)}{W - r} \tag{15.18}$$

If the vacancy flux is converted to an atomic flux,

$$J = D \frac{C\Omega \left(\sigma \cos\theta - \frac{\gamma}{r} \right)}{(W - r)kT} = \frac{D \left(\sigma \cos\theta - \frac{\gamma}{r} \right)}{kT(W - r)} \tag{15.19}$$

Furthermore, if the mass transport (or vacancy transport) takes place along the grain boundary, we must correct the flux by $D_b\delta/W$ (see Chapter 13),

$$J_{\text{gb}} = \frac{\left(\sigma \cos\theta - \frac{\gamma}{r} \right) D_b\delta}{(W - r)WkT} \tag{15.20}$$

where δ and D_b are the effective boundary width and diffusivity, respectively. The volume of matter transported by J_{gb} along the grain boundary is

$$V = J_{\text{gb}} A t \Omega$$

where $A = \delta h$ is the cross-sectional area of the grain boundary. Now if we assume that the void grows in a circular shape and if $r_{\text{max}} = W$, the line is open. We can evaluate the time to failure t_f from

$$\int_0^W 2\pi r h \, dr = \int_0^{t_f} 2 J_{\text{gb}} A \Omega \, dt$$

substituting Eq. (15.20) into the last equation to yield

$$\pi W^2 h = \int_0^{t_f} \frac{2\left(\sigma \cos\theta - \frac{\gamma}{r} \right) D_b \delta^2 h \Omega}{(W - r)WkT} \, dt \tag{15.21}$$

Since $r = r(t)$, it is not straightforward to obtain the integration on the right-hand side. Nevertheless, Eq. (15.21) shows a time dependence of the failure on line width to the power of 3 to 4 (i.e., between W^3 and W^4).

However, in assuming a circular shape of the void which is maintained all the way to failure, we have ignored the fact that it is the tensile stress which drives the void growth. The total volume that can be transported is limited by the tensile strain; that is,

$$V_{\text{total}} = 2Whl \frac{\Delta l}{l}$$

where l is the effective length of the line that contributes to the growth of the void, and $\Delta l / l$ is the strain. The maximum radius of the void is then obtained by

$$\pi r_{\max}^2 h = 2Wh\Delta l$$

$$r_{\max} = \left(\frac{2W\Delta l}{\pi} \right)^{\frac{1}{2}} \tag{15.22}$$

For the line to open, we require that $r_{\max} = W$.

Thus Δl must be of the same magnitude as W. The void needs vacancies from an effective length of line about $500W$ if the strain is taken to be 0.2%. The length required is much greater than the grain boundary separation or grain size in the line. This seems unlikely! On the other hand, the basic assumption that the void grows in a circular shape is due to minimizing the total surface energy. However, the void growth is to release the stress. For this reason, we assume that when the radius of a circular void reaches the size of,

$$r_f = \frac{\Delta l}{2}$$

it will lengthen itself along the grain boundary and grow in the shape of a sausage, as shown in Fig. 15.7b. In this case, the total volume required to be transported away for the opening is

$$V_f = 2W(2r_f)h(1/\cos \theta)$$

so the time to failure is given by

$$\int_0^W 4r_f h(1/\cos \theta)\, dW = \int_0^{t_f} 2J_{gb} A\Omega\, dt$$

so

$$4r_f h W(1/\cos \theta) = \int_0^{t_f} \frac{2\left(\sigma \cos \theta - \frac{\gamma}{r_f}\right) D_b \delta^2 h\Omega}{(W - l_v)WkT}\, dt \tag{15.23}$$

where l_v is half the length of the void and it is a function of time. The main difference between Eq. (15.23) and Eq. (15.21) is that now the time dependence of the failure on the line width is to the power of W^3 or slightly less. It agrees better with experimental observation than the time dependence calculated in Eq. (15.21).

In the above analysis of void growth in a fine line, it is assumed that the void has already nucleated in the grain boundary and that the grains are as big as or bigger than the line width. For Al lines that have experienced annealings up to 400°C, the grains can be micron size. Nucleation of a void in the grain boundary is assumed to occur at a high local stress concentration around a grain boundary precipitate. Under tensile stress, two grains try to slide against each other across a grain boundary which makes a tapered angle θ to the stress. The precipitate becomes an obstacle and stress concentration arises. The local deformation can cause a crack or void formation. Again, we use a localized model to explain relaxation of homogeneous stress in a line. We must ask what would happen if a void cannot be nucleated. The line can no doubt relax its stress in a homogeneous manner by the bulk Nabarro–Herring creep. It will be a process involving lattice diffusion. However, to facilitate a low-temperature process by grain boundary diffusion, a local stress relief center such as a void is required.

15.4 Pit Formation in Si Substrate by Al Penetration

Al contacts to single-crystal Si are known to form pits in the Si upon thermal annealing. Without annealing the contact has a high contact resistance due to the presence of native oxide on the Si surface. Annealing around 400 to 450°C for 15 to 30 minutes removes the oxide and lowers the contact resistance, but also forms pits in the Si. Although Al forms no intermetallic compounds with Si, it dissolves about 0.1% of the Si at the annealing temperature. The dissolution in general does not occur uniformly over the entire contact area; it starts at weak spots (or pin holes) in the native oxide and results in pit formation. The pits are revealed by scanning electron microscopy of the Si surface after etching the Al away as shown in Fig. 15.8a, or by cross-sectional transmission electron microscopy of the contact as shown in Fig. 15.8b and c.

When the depth of a pit is about the depth of the p–n junction below the Si surface, the pit shorts out the junction since it is filled with Al. For this reason, in devices having shallow junctions, Al contacts have been replaced by silicide contacts. The interesting question here is why Al contacts form pits but silicide contacts do not.

Pit formation on a Si surface by chemical etching has been a technique used to identify dislocations in the Si. Etching starts from the dislocation core when a dislocation ends at the surface, and proceeds at the slowest rate on the $\langle 111 \rangle$ surfaces; hence the pit is bound by $\langle 111 \rangle$ surfaces and penetrates into the Si. The anisotropic rate indicates that the etching, like the dissolution of Si into Al, is an interfacial-reaction-controlled process (Chapter 11). When the etching process is combined with a localized starting point, it leads to pit formation. This is clear because when a dislocation-free Si wafer is etched, no pits are seen.

The growth of diffusion-controlled silicides such as Pt_2Si and PtSi on Si forms no pits. Cross-sectional transmission electron microscopy showed flat interfaces between the Si and silicides; although not atomically flat, they do not contain gross defects such as pits. The Si native oxide of 1.5 to 3 nm thick has no effect on the uniform reaction between the Si and the transition metals. The oxide can be removed

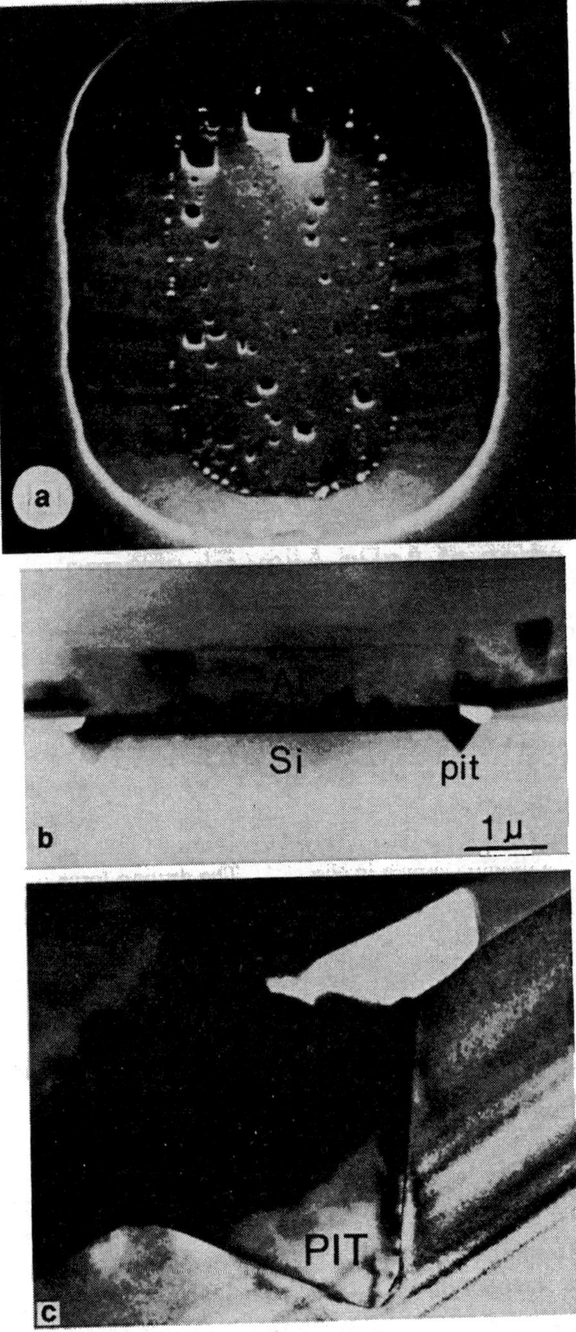

FIGURE 15.8 (a) Scanning electron microscopic image of pit formation in the Si after the Al contact is etched away (from T. Reith, IBM). (b) Cross-sectional transmission electron microscopic image of the pit formed in Si at the edges of the silicide contact. The dark layer between the Al and Si is the image of the silicide. (c) An enlarged image of the pit (from F. LeGoues, IBM).

by interposing a thin Ti film between the Si and the metal. Furthermore, the silicide–Si interfacial energy does not seem to play a role in anisotropic growth.

Solid phase epitaxy occurs in Al/Si contacts during the cooling stage of annealing. In cooling, the supersaturated Si in Al precipitates epitaxially on the substrate where the surface native oxide has been removed. Because Al is a p-type dopant, the expitaxial layer forms a p–n junction on an n-type substrate (Reith and Schick, 1974). Solid phase epitaxy also occurs in the reaction between polycrystalline Si and Al (or Au) films, and results in a mutual penetration. While the small and randomly oriented Si grains are dissolved, the dissolved Si atoms regrow on the $\langle 111 \rangle$ and $\langle 111 \rangle$ oriented grains and penetrate into the Al or Au (Allen et al., 1990).

The use of Al metallization on silicide resurrects the Al-penetration problem. This is because Al can decompose silicide to reach the Si substrate. A diffusion barrier is used between Al and silicide to prevent their reaction. Typically, a slightly oxidized Cr or a TiN layer is used (Nicolet, 1978).

15.5 Corrosion of AgPd Alloy Electrodes

Corrosion in thin film circuitry is a serious reliability issue because there is little material available for corrosion in a thin film line. Furthermore, it only takes localized corrosion to cause an open circuit. For these reasons, thin film circuits are generally coated with a protective layer and sealed hermetically from the ambient. Like the ambient, the unrinsed chemicals left behind from wet chemical processes can be the source of corrosive elements, such as sulfur and chlorine.

The subject of device corrosion is too vast to be covered in detail here. We shall briefly discuss a corrosion problem which has occurred in the AgPd alloy electrodes used in the device module shown in Fig. 4.4a. The driving force of the corrosion is discussed.

The electrodes are made of an alloy of 80 atomic percent Ag and 20 atomic percent Pd; Ag is chosen for electrical conductivity and Pd is added to prevent sulfur corrosion. These electrodes are connected by Pb–10 atomic percent Sn solder joints to the Si chip. It was found that near the solder joints, corrosion occurred at ambient temperatures, forming Ag_2S. Figure 15.9 shows Ag_2S on an electrode surface. The localized corrosion reaction near the solder joint is

$$2Ag + S \rightarrow Ag_2S \tag{15.24}$$

Since Pd has been added to the electrode to form an alloy with Ag, and since the unsoldered part of the electrode does not corrode, we must ask where is the free Ag coming from? The free Ag formation and in turn the corrosion are related to the solder–electrode reaction. In the reaction the molten solder depletes Pd from the AgPd alloy to form a compound Pb_2Pd, leaving behind free Ag which corrodes. This has been confirmed by thin film reactions between AgPd alloy and Pb; they produce Pb_2Pd and Ag. Figure 15.10 shows x-ray diffraction spectra of the thin film samples before and after annealings at 200°C. After the annealings, reflections of Pb_2Pd, Ag, and the remaining AgPd alloy are observed.

FIGURE 15.9 A scanning electron micrograph of Ag_2S on the surface of an AgPd electrode. (Courtesy of N. Koopman, IBM Corp.)

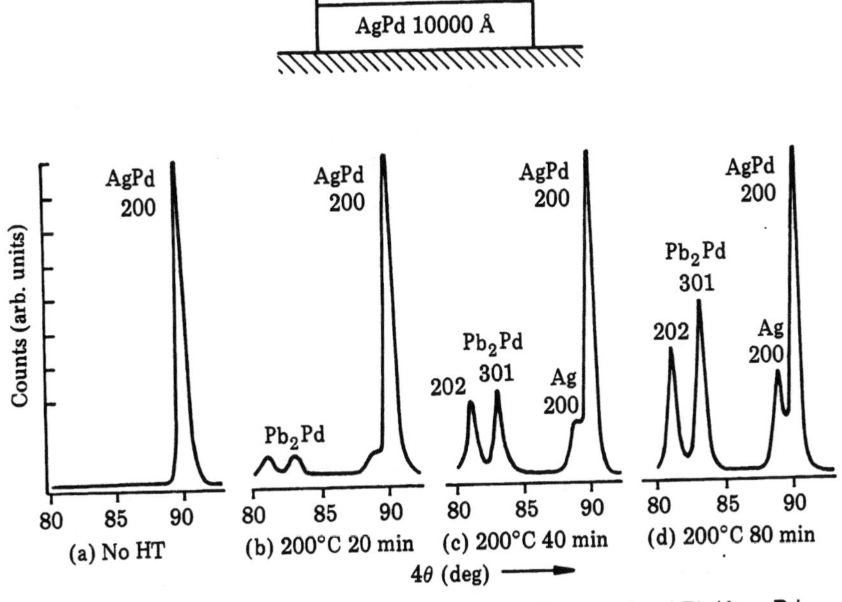

FIGURE 15.10 Sequential x-ray diffraction of a thin film sample of $Pb/Ag_{.80}Pd_{.20}$ deposited on a fused quartz substrate (a) before anneal, and (b) to (d) after annealing at 200°C for 20, 40, and 80 minutes. The formation of Ag and Pb_2Pd is detected in the annealing. The 200 reflection of the AgPd alloy is normalized to the same height and the reflections of Pb are not shown.

To analyze the depletion reaction, we show in Fig. 15.11 the schematic Gibbs free energy curve of AgPd alloys as a function of composition. To remove Pd from the composition $Ag_{.80}Pd_{.20}$, even by a small amount, leads to free energy increase. This is indicated by the arrow in the figure, which is the tangent of the curve at the $Ag_{.80}Pd_{.20}$ composition. It means that there exists a resistance force to the depletion reaction. The reaction can overcome the resistance by gaining energy from forming the compound Pb_2Pd. However, the resistance, like the slope of the tangent, increases with increasing amount of depletion, so the Pd cannot be completely depleted. The net free energy change in the reaction can be calculated by considering

$$Ag_{.80}Pd_{.20} + 2\delta Pb \rightarrow Ag_{.80}Pd_{.20-\delta} + \delta Pb_2Pd \tag{15.25}$$

where δ represents an infinitesimal amount of concentration. By rearrangement, we have

$$\frac{Ag_{.80}Pd_{.20} - Ag_{.80}Pd_{.20-\delta}}{\delta} \rightarrow Pb_2Pd - 2Pb \tag{15.26}$$

Since the alloy has a continuously varing composition, in the limit of $\delta \rightarrow 0$ (i.e., compositional change of an atom) the term on the left-hand side becomes the differential of the alloy with respect to composition, $dAg_{.80}Pd_{.20}/dC$, where C is concentration of Pd in the alloy. In terms of free energy change, the differential means the partial molar free energy (chemical potential per mole) of Pd in the AgPd alloy. Then, the reaction will proceed if

$$\Delta\overline{G}^{Pd}_{AgPd} > G_{Pb_2Pd} - 2G_{Pb} \tag{15.27}$$

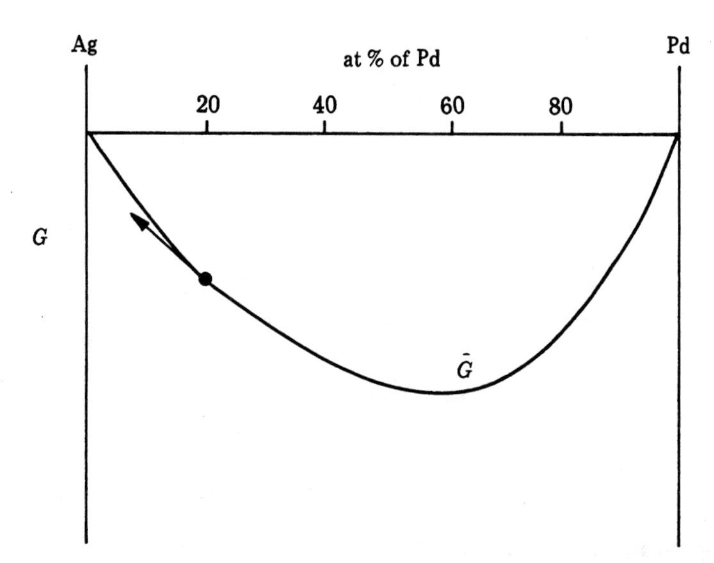

FIGURE 15.11 Gibbs free energy curve of AgPd alloys as a function of composition. The arrow indicates an increase in free energy when Pd is depleted from the alloy $Ag_{.80}Pd_{.20}$.

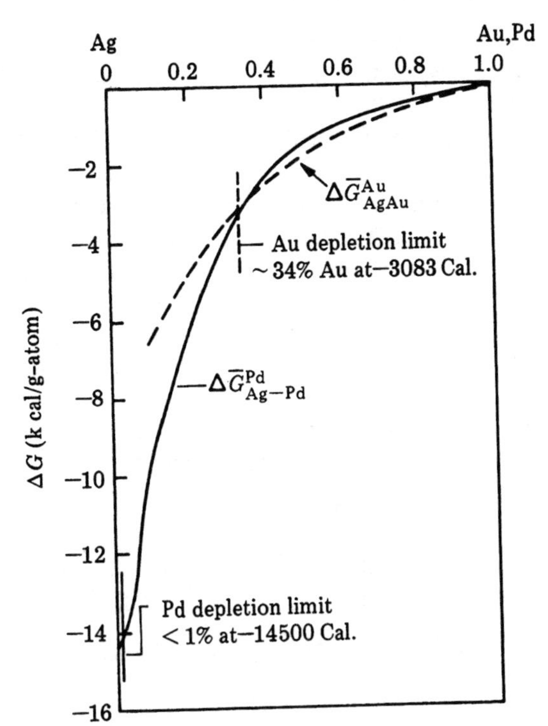

FIGURE 15.12 Partial molar free energy curves of AgPd and AgAu are shown by the solid and broken curves, respectively. The vertical solid and broken lines indicate the depletion limits of Pd and Au by the reaction with Pb.

where $\Delta\overline{G}^{Pd}_{AgPd}$ is the partial molar free energy of Pd in AgPd alloys and G_{Pb_2Pd} and G_{Pb} are the Gibbs free energy per mole of Pb_2Pd and Pb, respectively.

Assuming a reaction temperature of 623 K (at which the Pb−10 atomic percent Sn solder is molten), the partial molar free energy of Pd in AgPd alloy covering the entire solid solution range is plotted by the solid curve in Fig. 15.12 using the existing data measured at 1200 K by Hultgren et. al. (1973). The difference $G_{Pb_2Pd} - 2G_{Pb}$ is calculated to be about − 14.5 Kcal/mole and it is indicated by the vertical solid line which crosses the solid curve of $\Delta\overline{G}^{Pd}_{AgPd}$ at about 1 atomic percent Pd. It shows that most AgPd alloys are thermodynamically unstable in contact with Pb, so that Pb_2Pd forms. Provided that the reaction kinetics are fast enough, the Pd in the alloy can be depleted, resulting in the formation of a free Ag phase which contains no more than 1 atomic percent of Pd and thus can be corroded by sulfur.

To prevent free Ag formation, we compare AgPd alloys to AgAu alloys. If we follow the same calculation, we plot the partial molar free energy of Au in AgAu alloys by the broken curve in Fig. 15.12, and the difference $G_{Pb_2Au} - 2G_{Pb}$ by the vertical broken line. This shows that Pb will not be able to deplete Au from an AgAu alloy below a concentration of 34 atomic percent Au. In this respect, the AgAu alloy is much more stable. The main difference between the two reactions is that the Pb_2Au

compound has a formation energy of about −3 Kcal/mole, which is much less than that of the Pb_2Pd. Hence, by adding some Au to the Ag electrode, its resistance to corrosion is enhanced.

We have discussed the thermodynamic driving force of the depletion reaction. From the point of view of diffusion kinetics, the depletion reaction at 623 K is not surprising. At that temperature, the lattice diffusion of Pd in the AgPd alloy is fast enough for dealloying (creating free Ag). What is unexpected is depletion of Pd in thin film samples at 200°C, as shown in Fig. 15.12. The reaction rate can no longer be explained by the lattice diffusion. Hence the depletion in the thin films must have taken place via a kinetic process with a much lower activation energy than that of lattice diffusion. Naturally, grain boundary diffusion—in particular, diffusion along moving grain boundaries (DIGM, see Section 13.5)—has been considered as a mechanism for the low-temperature dealloying reaction. Low-temperature thin film phase changes are often dominated by grain boundary or interface diffusion because of the large area-to-volume ratio in the polycrystalline microstructure.

References

1. L. H. Allen, J. W. Mayer, K. N. Tu, and L. C. Feldman, *Phys. Rev. B* 41, 8213 (1990).
2. P. Chaudhari, *J. Appl. Phys.* 45, 4339 (1974).
3. H. J. Frost and C. V. Thompson, *Acta Met.* 35, 529 (1987).
4. R. Hultgren, P. D. Desai, D. T. Hawkins, M. Gleiser, K. K. Kelley, and D. D. Wagman, *Selected Values of the Therhmodynamic Properties of the Alloys*, American Society for Metals, Metals Park OH (1973).
5. S. K. Lahiri, *J. Appl. Phys.* 41, 3172 (1970).
6. C. Y. Li, R. D. Black, and W. R. LaFontaine, *Appl. Phys. Lett.* 53, 31 (1988).
7. M. Murakami, *Acta Met.* 26, 175 (1978).
8. M. A. Nicolet, *Thin Solid Films* 52, 415 (1978).
9. T. M. Reith and J. D. Schick, *Appl. Phys. Lett.* 25, 524 (1974).
10. C. V. Thompson, "Grain Growth in Thin Films," *Ann. Rev. Mater. Sci.* 20, 245 (1990).
11. K. N. Tu and D. A. Chance, *J. Appl. Phys.* 46, 3229 (1975).

Problems

15.1 Given the maximum radius of a void is 0.8 μm, film thickness = 0.5 μm, line width = 1 μm, and strain is 0.2%, calculate the effective length of the line that contributes to the growth of the void assuming the void grows in a circular shape only. How large is the effective width compared to the film width? Now assume the void grows in a "sausage" shape and that the volume required to be transported away is one-third the volume required for the circular void. Calculate the radius at which the void will begin to lengthen $\theta = 10°$.

15.2 Polycrystalline aluminum lines with an inherent processing stress (σ_p) had time to 1% failure of t_{f400} when held at 400 K. How would this time change if the temperature were reduced to 350 K? (Assume $\Delta\sigma$ due to the change in temperature is $<< \sigma_p$).

15.3 The growth of Sn hillocks in a thin Sn film (0.5 μm thick) undergoing a 500°C anneal is observed. The growth rate was found to be 10Å/hr. and the average radius and final height of the hillocks were found to be 2 μm and 4μm, respectively. Using the values given in the table of properties for Sn, calculate the average stress in the film and the effective radius of the stress field which gives rise to the hillocks. Assume $D_i = 10^{-9}$ cm^2/s, $\delta = 5.0$ Å, $A' = 1.75 \times 10^{-4}$.

Properties for Sn

At. Wt.	Density	ν	Y
118.69	5.80 g/cm^2	0.5	~1.0 × 10^{11} N/m^2

15.4 What is the strain in a 1 μm thick aluminum film having hillocks 2 μm high with average radius of .5 μm and an average of 40,000 hillocks per cm^2?

15.5 Using Eq. (15.17), find an expression for the time to failure t_f for (a) l_v constant and (b) $l_v = C_V t$.

15.6 In the text void growth was analyzed using the Nabarro–Herring creep model. In that analysis it was assumed that the effect of tensile stress near a void is negligible. Let's relax that assumption. From linear elasticity theory, the stress field in a plate with a hole when subjected to an in-plane stress σ_∞ is given by Timoshenko and Goodier in *Theory of Elasticity* (McGraw–Hill, New York, 1970, 3rd ed., p. 91) as

$$\sigma_r = \frac{\sigma_\infty}{2}\left[\left(1 - \frac{a^2}{r^2}\right) + \left(1 + \frac{3a^4}{r^4} - \frac{4a^2}{r^2}\right)\cos 2\theta\right]$$

$$\sigma_\theta = \frac{\sigma_\infty}{2}\left[\left(1 + \frac{a^2}{r^2}\right) - \left(1 + \frac{3a^4}{r^4}\right)\cos 2\theta\right]$$

$$\tau_{r\theta} = \frac{-\sigma_\infty}{2}\left[1 - \frac{3a^4}{r^4} + \frac{2a^2}{r^2}\right]\sin 2\theta$$

where σ_r, σ_θ are normal stresses, $\tau_{r\theta}$ is shear stress, and a is the radius of the assumed circular void. Assume that a fine line of Al on Si with a circular void has a similar stress distribution as this plate, and show that at the void surface

($r = a$) the tensile stress is not zero. Identify where the chemical potential (stress) would be maximum on the void surface. How would you expect the void to grow? Assume that there are a sufficiently large number of grain boundaries intersecting the void such that the orientation of grain boundaries does not provide a preferred direction. Note that $\theta = 0°$ and $90°$ are aligned with and perpendicular to the applied stress σ_∞.

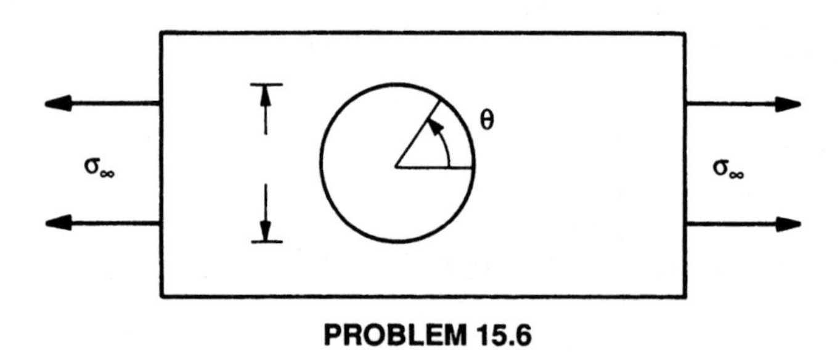

PROBLEM 15.6

APPENDIX A

Maxwell's Velocity Distribution Function

We consider an assembly of gas particles at a low pressure. We assume that the average interaction potential energy between particles is negligible, yet we allow them to interact by collision to maintain thermal equilibrium. Hence we deal only with the kinetic energy of the particles. To describe the kinetic energy, we consider each particle to have the same mass m but different velocity v. The particle velocity has the components v_x, v_y, v_z in the x, y, z directions so that the momenta of the particles are $p_x = mv_x$, $p_y = mv_y$, $p_z = mv_z$, respectively.

At thermal equilibrium, and on the basis of Boltzmann's law of distribution of noninteracting particles, the probability of finding the x-component of velocity of a particle to lie between v_x and $v_x + dv_x$ is

$$P(v_x)\, dv_x = B \exp\left(-\frac{mv_x^2}{2kT}\right) dv_x \tag{A.1}$$

where the constant B is determined by the integral that

$$\int_{-\infty}^{\infty} P(v_x)\, dv_x = 1 \tag{A.2}$$

and we obtain $B = (m/2\pi kT)^{1/2}$. Hence

$$P(v_x) = \sqrt{\frac{m}{2\pi kT}} \exp\left(-\frac{mv_x^2}{2kT}\right) \tag{A.3}$$

Fig. A.1 shows the plot of $P(v_x)$ against v_x; the area under curve equals unity according to Eq. (A.2). Following the same argument, the probability of finding a particle in thermal equilibrium having its velocity components in between v_x and $v_x + dv_x$; v_y and $v_y + dv_y$; v_z and $v_z + dv_z$ is

$$\begin{aligned} P(v_x, v_y, v_z)\, dv_x\, dv_y\, dv_z &= P(v_x)\, dv_x\, P(v_y)\, dv_y\, P(v_z)\, dv_z \\ &= \left(\frac{m}{2\pi kT}\right)^{\frac{3}{2}} \exp\left[\frac{-m(v_x^2 + v_y^2 + v_z^2)}{2kT}\right] dv_x\, dv_y\, dv_z \\ &= \left(\frac{m}{2\pi kT}\right)^{\frac{3}{2}} \exp\left(\frac{-mv^2}{2kT}\right) dv_x\, dv_y\, dv_z \end{aligned} \tag{A.4}$$

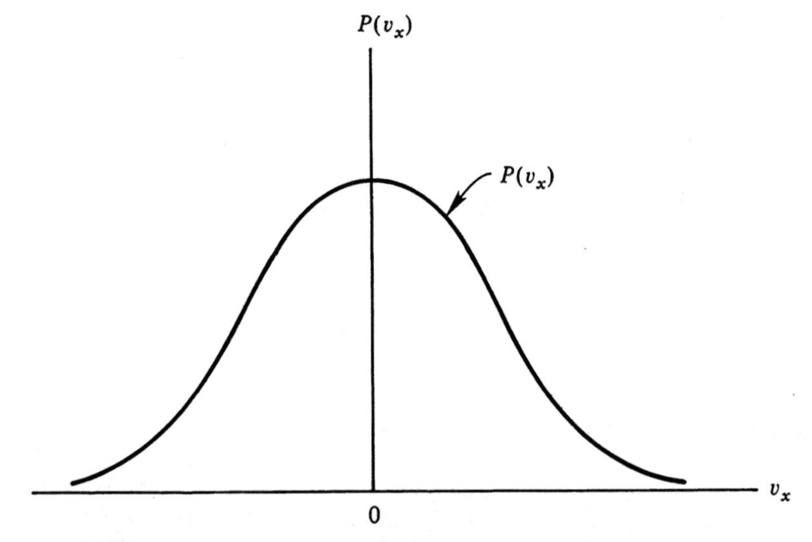

FIGURE A.1 Probability distribution function $P(v_x)$ plotted against v_x.

where $v = (v_x^2 + v_y^2 + v_z^2)^{1/2}$ is the speed of the particle. Now let us perform a transformation of coordinates from (v_x, v_y, v_z) to (v, θ, ϕ) as shown in Fig. A.2, and let us consider an infinitesimal volume. We can write

$$
\begin{aligned}
dv_x\, dv_y\, dv_z &= (dv)(vd\theta)(v \sin\theta\, d\phi) \\
&= v^2 \sin\theta\, dv\, d\theta\, d\phi
\end{aligned}
\tag{A.5}
$$

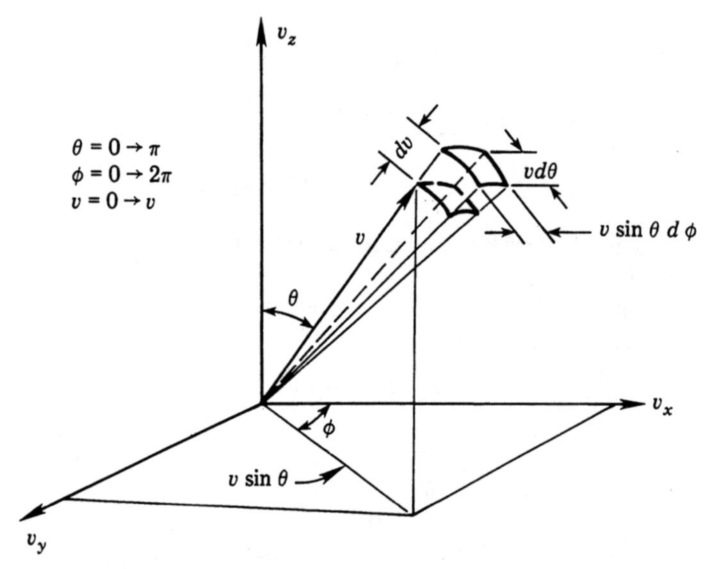

FIGURE A.2 The volume element (dv) $(v\, d\theta)$ $(v \sin\theta\, d\phi)$ in the coordinates of v_x, v_y, v_z.

Substituting Eq. (A.5) into Eq. (A.4), we obtain

$$P(v, \theta, \phi)\, dv\, d\theta\, d\phi = \left(\frac{m}{2\pi kT}\right)^{\frac{3}{2}} v^2 \exp\left(\frac{-mv^2}{2kT}\right) \sin\theta\, dv\, d\theta\, d\phi \quad \text{(A.6)}$$

For a large number of particles, we can assume that the distribution of v has spherical symmetry so that the variable v is independent of the other two coordinates, θ and ϕ. Noting that

$$\int_0^{2\pi} d\phi \int_0^{\pi} \sin\theta\, d\theta = 4\pi \quad \text{(A.7)}$$

we obtain the Maxwell's distribution function of velocity,

$$P(v)\, dv = 4\pi \left(\frac{m}{2\pi kT}\right)^{\frac{3}{2}} v^2 \exp\left(\frac{-mv^2}{2kT}\right) dv \quad \text{(A.8)}$$

Figure A.3 shows a plot of the distribution function at two temperatures. Again the area under each curve equals unity. At the higher temperature, the plot shows that the distribution moves to higher velocities.

Maxwell's distribution function allows us to calculate the "mean" velocity and "mean" kinetic energy of the gas particles. The mean velocity $\bar{v}$ is defined by

$$\begin{aligned} \bar{v} &= \int_0^{\infty} vP(v)\, dv \\ &= \int_0^{\infty} 4\pi \left(\frac{m}{2\pi kT}\right)^{\frac{3}{2}} v^3 \exp\left(\frac{-mv^2}{2kT}\right) dv \quad \text{(A.9)} \\ &= \sqrt{\frac{8kT}{\pi m}} \end{aligned}$$

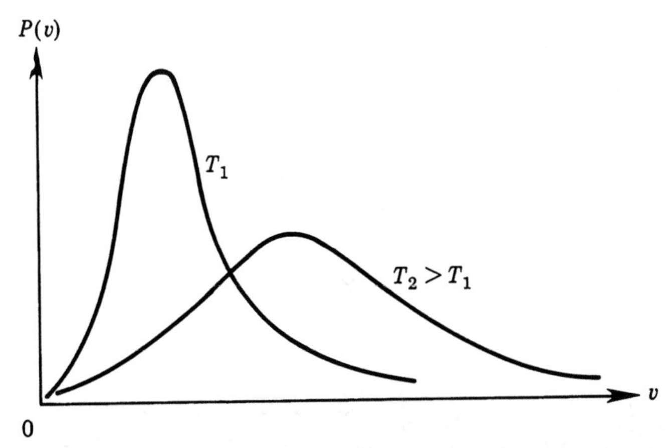

FIGURE A.3 The schematic Maxwell distribution function at two temperatures.

The integration is carried out by substituting

$$x^2 = \frac{mv^2}{2kT}$$

and by using the integral of

$$\int_0^\infty x^3 \exp(-x^2)\, dx = \frac{1}{2} \tag{A.10}$$

Now we return to the flux equation of residual gas, Eq. (1.4), discussed in Chapter 1. If we assume the normal of the substrate is in the x-direction as shown in Fig. A.4, a particle with velocity v_a having component v_x directed towards the substrate will hit the substrate, but a particle with a velocity in the opposite direction will not. Then, the flux hitting the substrate is

$$J = nv_a = n \int_0^\infty v_x P(v_x)\, dv_x$$

where $P(v_x)$ is given by Eq. (A.3). Hence,

$$\begin{aligned} J &= n \int_0^\infty \sqrt{\frac{m}{2\pi kT}}\, v_x \exp\left(\frac{-mv_x^2}{2kT}\right) dv_x \\ &= n\sqrt{\frac{kT}{2\pi m}} = \frac{1}{4} n\bar{v} \end{aligned} \tag{A.11}$$

The factor of $\frac{1}{4}$ comes in when we substitute $\bar{v}$ from Eq. (A.9).

Furthermore, if we assume that the substrate surface is elastic (i.e., a particle with a momentum mv_x hitting the surface will bounce back elastically) there is no loss of energy and the momentum change is $2mv_x$. Since the number of particles hitting the surface of area A in a unit of time is JA, the force on this area is the rate of change of momentum,

$$F = 2mv_x JA \tag{A.12}$$

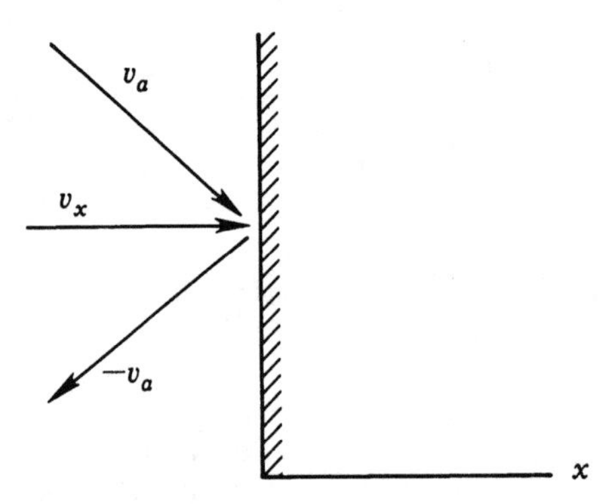

FIGURE A.4 Impingement of a particle at a velocity v_a with a component of v_x normal to the wall.

So the pressure p is given by

$$p = \frac{F}{A} = 2mv_xJ$$
$$= 2mv_xnv_a = 2mnv_x^2P(v_x)\,dv_x$$

If we now integrate over all v_x, we have the total pressure given by

$$p = 2mn\left(\frac{m}{2\pi kT}\right)^{1/2}\int_0^\infty v_x^2 \exp\left(\frac{-mv_x^2}{2kT}\right) dv_x$$

$$= 2mn\left(\frac{m}{2\pi kT}\right)^{1/2}\left(\frac{2kT}{m}\right)^{\frac{3}{2}}\frac{1}{4}(\pi)^{1/2} \qquad \text{(A.13)}$$

$$= nkT = \frac{N_A}{V}kT$$

which is the ideal gas law, given in Eq. (1.1).

Regarding the "mean" kinetic energy of the gas particles, we leave it as an exercise to show that the "mean square velocity" is

$$\overline{v^2} = \int_0^\infty v^2P(v)\,dv = \frac{3kT}{m} \qquad \text{(A.14)}$$

The "root mean square velocity,"

$$v_{\text{rms}} = (\overline{v^2})^{1/2}$$

is close, but not equal to, the mean velocity $\overline{v}$ as given by Eq. (A.9). The mean kinetic energy is equal to $\frac{1}{2}m\overline{v^2}$, not $\frac{1}{2}m(\overline{v})^2$.

APPENDIX B

Thermodynamic Functions

We begin by reviewing some general thermodynamic relations of closed systems having a fixed number of particles. The four energy functions, internal energy E, enthalpy H, Helmholtz free energy F, and Gibbs function G are related to one another and to the four thermodynamic variables, pressure p, volume V, entropy S, and temperature T by

$$\begin{aligned} H &\equiv E + pV \\ F &\equiv E - TS \\ G &\equiv E - TS + pV \equiv F + pV \equiv H - TS \end{aligned} \tag{B.1}$$

The first law of thermodynamics relates changes in heat, dQ, and work done to the system, dW, to the change in internal energy dE

$$dE = dQ + dW \tag{B.2}$$

If the change in heat is reversible, the second law of thermodynamics defines the increase in entropy to be

$$dS = \frac{dQ}{T} \tag{B.3}$$

If only mechanical work is done to the system, we have $dW = -p\,dV$ where the work is positive when the volume of the system decreases. Then

$$dE = T\,dS - p\,dV \tag{B.4}$$

By using Eq. (B.4), the differentials of H, F, and G can be obtained from Eq. (B.1) as

$$\begin{aligned} dH &= T\,dS - V\,dp \\ dF &= -S\,dT - p\,dV \\ dG &= -S\,dT + V\,dp \end{aligned} \tag{B.5}$$

The four variables appear in Eq. (B.4) and Eq. (B.5) in pairs: p and V, T and S. Experimentally, it is easier to carry out a process at constant pressure or constant temperature, but harder at constant volume or constant entropy. A process occurring at constant entropy is called an adiabatic process, meaning that heat is isolated. We note that while it is hard to measure dS, we can measure $dQ = T\,dS$ at constant volume or at constant pressure; these are heat capacity measurements. From Eq. (B.4), we define the heat capacity at constant volume,

$$c_v = \left.\frac{\partial E}{\partial T}\right|_v \tag{B.6}$$

and from the first equation in Eq. (B.5), we define the heat capacity at constant pressure,

$$c_p = \left.\frac{\partial H}{\partial T}\right|_p \tag{B.7}$$

The value of c_p is easier to measure for solids than c_v.

For a chemical reaction which occurs at atmospheric pressure (a constant pressure process) to form an intermetallic compound, we have

$$dH = T\,dS|_p = dQ \tag{B.8}$$

The enthalpy of formation is often called the ''heat'' of formation. When a chemical process occurs at constant temperature, we have

$$\begin{aligned} dF &= -p\,dV = dW \\ &= dE - dQ \end{aligned} \tag{B.9}$$

The work done in an isothermal reversible change is equal to the energy change $(dE - dQ)$, which is called the free energy.

The Gibbs function has T and p as the two independent variables. They are easy to control, and do not depend on the size of the system. Hence in considering equilibrium reactions at constant temperature and pressure, we use the Gibbs function.

The Gibbs function is often called the Gibbs free energy. When we consider a chemical process of liquid or solid phases which occurs at one atmosphere pressure, the pV term is negligibly small, so the Gibbs free energy and Helmholtz free energy are practically the same. To calculate the pV term, we recall that 1 atm corresponds to a pressure of 1.013×10^6 dyne/cm^2 and to an energy density of 1.013×10^6 erg/cm^3. If we assume that there are approximately $3 \times 10^{22}\ \Omega$ atoms in 1 cm^3 where Ω is the atomic volume, we have

$$\begin{aligned} p\Omega &= \frac{1.013 \times 10^6}{3 \times 10^{22} \times 1.6 \times 10^{-12}}\ \text{eV/atom} \\ &= 0.21 \times 10^{-4}\ \text{eV/atom} \end{aligned}$$

This value is much smaller than a typical binding energy of about 0.1 to 1 eV/atom in liquid or solid.

For surfaces there may be another contribution to the work, namely

$$dW_S = \gamma\,dA \tag{B.10}$$

This is the increase in the work against surface forces, by increasing the area of the surface.

We consider a system with material M_1 on top, M_2 on the bottom, and interfacial area A (Fig. B.1).

Component 1:

$$dE_1 = T\,dS_1 - p_1\,dV_1 \tag{B.11}$$

Component 2:

$$dE_2 = T\,dS_2 - p_2\,dV_2 \tag{B.12}$$

And for the energy for the interfacial surface:

$$dE_S = T\,dS_S + \gamma\,dA \tag{B.13}$$

The total energy:

$$\begin{aligned} dE &= dE_1 + dE_2 + dE_S \\ &= T\,d(S_1 + S_2 + S_S) - p_1\,dV_1 - p_2\,dV_2 + \gamma\,dA \\ &= T\,d(S_T) - p_1\,dV_T + (p_1 - p_2)\,dV_2 + \gamma\,dA \end{aligned} \tag{B.14}$$

If the entropy is constant and the total volume dV_T is a constant, then

$$(p_1 - p_2)\,dV_2 + \gamma\,dA = 0$$

and

$$(p_2 - p_1) = \gamma\,dA/dV_2 \tag{B.15}$$

The Gibbs free energy change dG is given by

$$dG = -S\,dT + V\,dp \tag{B.16}$$

and if a surface contribution is added,

$$dG = -S\,dT + V\,dp + \gamma\,dA \tag{B.17}$$

Consider the free energy for each of the separate components:

$$\begin{aligned} dG_1 &= -S_1\,dT + V_1\,dp \\ dG_2 &= -S_2\,dT + V_2\,dp \end{aligned} \tag{B.18}$$

Then the free energy associated with the surface is:

$$\begin{aligned} dG_S &= d(G - G_1 - G_2) \\ &= -S_S\,dT + \gamma\,dA \end{aligned} \tag{B.19}$$

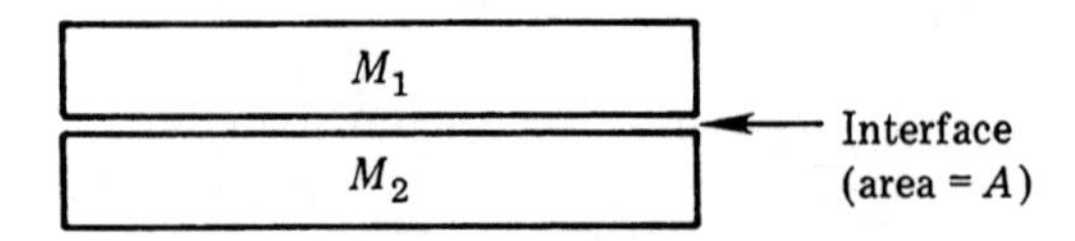

FIGURE B-1. The interface between M_1 and M_2.

At a constant temperature, dT equals zero,

$$dG_S = \gamma\, dA + A\, d\gamma$$

Use, once again,

$$dG_S = \gamma\, dA - S_S\, dT$$

so

$$S_S = -A\, d\gamma/dT \tag{B.20}$$

and the entropy is always positive so that $d\gamma/dT$ is always < 0.

The energy of the surface is

$$E_S = \gamma A - T\frac{d\gamma}{dT}A \tag{B.21}$$

which relates the thermodynamic energy of the surface to γ and a temperature dependence. Tables of experimental values of $d\gamma/dT$ are listed in Table 2.4.

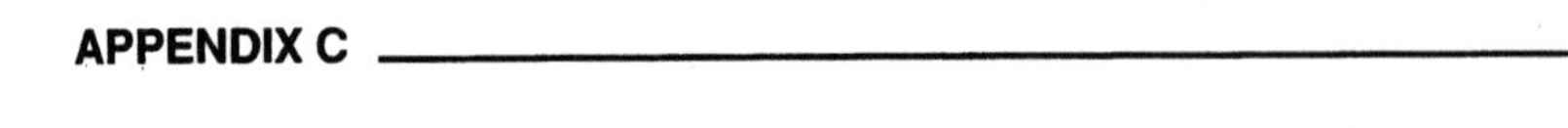

APPENDIX C

Defect Concentration in Solids

We introduced the subject of diffusion in the beginning of Chapter 3 by stating that diffusion in crystalline solids is mediated by defects in the solid, especially the point defects of the vacancy and interstitial. Indeed, the point defect mechanism of atomic diffusion in metals is well developed. A key question is what is the defect concentration in the solid, since the quantity will no doubt control the flux of diffusion. Point defect concentration is a thermodynamic equilibrium quantity, unlike that of complex defects such as dislocations, twins, and grain boundaries. These complex defects may serve as sources and sinks of point defects but their presence and their concentration in a sample depend on the sample history of thermal and mechanical treatments. However, a well-annealed sample will contain the equilibrium or fixed concentration of vacancies. Since a thermal equilibrium quantity can be calculated statistically, we give a simple example in the following.

We begin by considering the entropy of a lattice of N lattice sites in which N_v sites are vacant (vacancies) and $N - N_v$ sites are taken by atoms. The configurational entropy of the lattice is given by

$$S = k \ln \Gamma \tag{C.1}$$

where k is Boltzmann's constant and Γ is the number of possible states of the lattice or the number of ways to arrange the vacancies and the atoms in the lattice.

$$\begin{aligned}\Gamma &= \frac{N(N-1)......(N-N_v+1)}{N_v!} \\ &= \frac{N!}{(N-N_v)!N_v!}\end{aligned} \tag{C.2}$$

Using Stirling's approximation, $\ln x! = x \ln x$ when x is large,

$$S = k[N \ln N - (N - N_v) \ln (N - N_v) - N_v \ln N_v] \tag{C.3}$$

The Gibbs function of the lattice having N_v vacancies at temperature T is

$$\Delta G = N_v \Delta H_f - T\Delta S \tag{C.4}$$

where ΔH_f is the enthalpy of formation of a vacancy in the lattice and ΔS is the entropy increase from the ground state. If we take the ground state to be unique, its entropy is zero (the 3rd law of thermodynamics), and furthermore if we ignore the vibrational entropy contribution, we have

$$\Delta S = S - 0 = k \ln \Gamma$$

At thermal equilibrium at temperature T, the chemical potential of a vacancy or of an atom in the lattice is the same, so we have

$$\frac{\partial \Delta G}{\partial N_v} = 0$$

$$\Delta H_f - T\frac{\partial \Delta S}{\partial N_v} = 0$$

Since

$$\frac{\partial \Delta S}{\partial N_v} = k \ln \frac{N - N_v}{N_v}$$

we obtain

$$\frac{N_v}{N - N_v} = \exp\left(-\frac{\Delta H_f}{kT}\right) \tag{C.5}$$

Since $N >> N_v$, this equation shows that the equilibrium vacancy concentration obeys the Boltzmann distribution.

In Chapter 3, we showed that in Al, the value of $\Delta H_f = 0.76$ eV. Therefore near the melting point of 660°C, we have

$$\frac{N_v}{N} \cong 10^{-4}$$

which is the experimentally measured vacancy concentration. Since the concentration decreases with temperature, we can quench a sample from near its melting point to a lower temperature to produce a supersaturation of vacancies in the sample. The rate of reestablishing equilibrium can be measured by resistivity change in the sample and it enables us to determine the activation energy of motion of vacancies.

Terrace Size Distribution in Si MBE

In Chapter 6 we discussed the step periodicity (Section 6.5) in Si MBE. In this appendix we show that the standard deviation of the terrace size distribution reduces upon deposition of a small amount of material, under the assumption that the material grows via step growth. In this simplified argument we further assume that growth is entirely one way (i.e., attachment of all atoms occurs at the up-step only). This is a critical assumption which may not always apply, although the successful growth of step-mediated structures illustrates that the assumption is applicable in some cases.

Consider a terrace size distribution, representing an irregular terrace size array. The average terrace width $\bar{l}$ is given by

$$\bar{l} = \frac{1}{N} \sum_{i=1}^{N} l_i \tag{D.1}$$

where we will exclude the first and last step for convenience. In general the number of elements N is so large that there is no difficulty with this assumption.

The standard deviation SD of the terrace size distribution is given by

$$SD = \frac{1}{N} \sum_{i=1}^{N} (l_i - \bar{l})^2 \tag{D.2}$$

and is a measure of the uniformity of terrace sizes. If SD approaches zero we have an infinitely sharp terrace size distribution and a perfectly periodic step spacing. Note that SD can also be written as

$$SD = \frac{1}{N} \sum_{i=1}^{N} (l_i^2 - 2l_i\bar{l} + \bar{l}^2) = \left(\frac{1}{N} \sum_{i=1}^{N} l_i^2\right) - \bar{l}^2 \tag{D.3}$$

since

$$\frac{1}{N} \sum_{i=1}^{N} 2l_i\bar{l} = \frac{2\bar{l}}{N} \sum_{i=1}^{N} l_i = 2\bar{l}^2$$

Assume that we add a fraction of a monolayer of material, Q, which diffuses to the nearest step riser (Fig. D.1). Then the terrace l_i acquires a new size l_i' given by

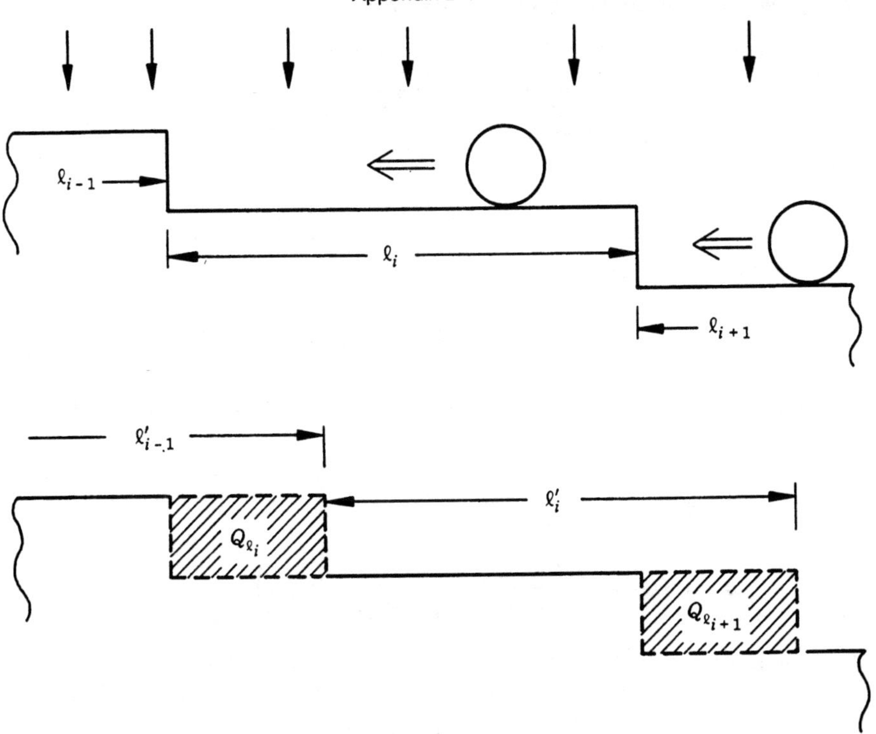

FIGURE D.1 Representation of the surface cross section during step-mediated growth, where a fraction Q of a monolayer of material leads to an increase in terrace length.

$$l'_i = l_i - Ql_i + Ql_{i+1}$$

or

$$l'_i = (1 - Q)l_i + Ql_{i+1} \tag{D.5}$$

The original terrace becomes smaller by a fraction proportional to the original terrace size and larger by a fraction proportional to the adjacent terrace size. The new average terrrace width $\bar{l}'$ becomes

$$\bar{l}' = \frac{1}{N} \sum_{i=1}^{N} [(1 - Q)l_i + Ql_{i+1}] = \bar{l} \tag{D.5}$$

where we have taken $l_1 = l_{N+1}$ for convenience. This result shows that the average terrace width of the new distribution is the same as that of the original distribution. This is intuitively sensible. The average terrace width is determined by the miscut angle and the lattice constant, $a/\tan\theta$, where θ is the miscut angle. Adding epitaxial material can never change this geometric relation, so the average terrace width remains the same. This is equivalent to saying that homoepitaxy neither increases nor decreases the number of steps per length.

Now consider the standard deviation of the new terrace size distribution,

$$SD' = \frac{1}{N}\sum_{i=1}^{N}(l'_i - \bar{l}')^2 = \left(\frac{1}{N}\sum_{i=1}^{N} l'^2_i\right) - \bar{l}'^2 \tag{D.6}$$

$$SD' = \frac{1}{N}\sum_{i=1}^{N}[l_i(1 - Q) + l_{i+1}Q]^2 - \bar{l}'^2 \tag{D.7}$$

Expanding the squared term,

$$SD' = \frac{1}{N}\sum_{i=1}^{N}[l_i^2 - 2Ql_i^2 + Q^2l_i^2 + Q^2l_{i+1}^2 + 2l_il_{i+1}(Q - Q^2)] - \bar{l}'^2 \tag{D.8}$$

The sum of the first term and last term are simply the standard deviation SD of the original distribution ($\bar{l}' = \bar{l}$) from Eq. (D.3), so

$$SD' = SD - \frac{2}{N}\sum_{i=1}^{N} l_i^2(Q - Q^2) + \frac{2}{N}\sum_{i=1}^{N} l_il_{i+1}(Q - Q^2)$$

$$SD' = SD - \frac{2(Q - Q^2)}{N}\left(\sum_{i=1}^{N}(l_i^2 - l_il_{i+1})\right) \tag{D.9}$$

Our goal is to show that $SD' < SD$, which follows if the quantity following the negative sign is positive. The factor $2(Q - Q^2)/N$ is certainly positive since $0 < Q < 1$. For the second factor consider the quantity:

$$\sum_{i=1}^{N}(l_i - l_{i+1})^2 = 2\sum_{i=1}^{N}(l_i^2 - l_il_{i+1}) \tag{D.10}$$

since

$$\sum_{i=1}^{N} l_i^2 = \sum_{i=1}^{N} l_{i+1}^2 \qquad l_1 = l_{N+1}$$

Since the left-hand side of Eq. (D.10) is positive, the right-hand side is also positive, thus

$$SD' < SD \tag{D.11}$$

This is the main result of our proof; the standard deviation decreases upon deposition in (preferential) step-mediated growth. It is clear that one can continue to deposit material and each time the resulting standard deviation will decrease and approach zero. This corresponds to approaching the perfect periodic step distribution.

The step distribution can never become perfect due to at least two effects. First is *statistics of the growth process*. The number of deposited atoms M in any one interaction is assumed to be proportional to the terrace size. Since this is a "Poisson-like" process, there will be an uncertainty in this number of the order of $\sqrt{M}$, giving rise to imperfection in growth. The second reason the step distribution cannot become perfect is because of *thermal fluctuations*. A step is not perfectly sharp but has a roughness due to thermal fluctuations. This imperfect step, originating from thermal processes, also yields an imperfect template resulting in a lack of perfect periodicity.

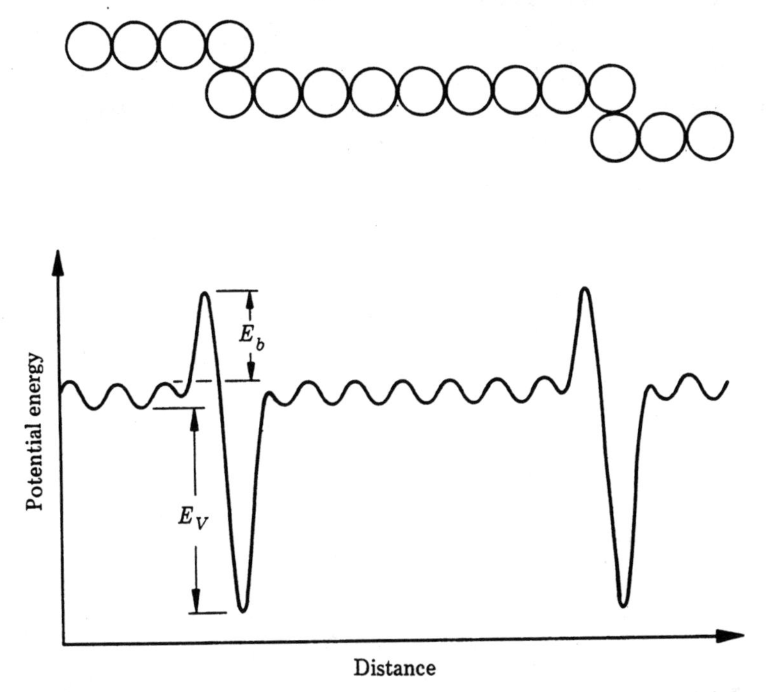

FIGURE D.2 Schematic of a stepped terrace (side view) and the potential energy of an adsorbed atom.

A more complete derivation (Gossmann et al., 1990) shows that after S interactions of deposited material Q, the standard deviation SD^S in the terrace width relative to the initial standard deviation is,

$$SD^S/SD = \frac{1}{(2\pi\Theta)^{1/4}} \tag{D.12}$$

where $\Theta = SQ$. Note that the approach to periodicity is slow because the driving force for reorganization depends on the adjacent terrace size differences which decrease as growth proceeds.

One more word about terrace size distributions and the approach to periodicity: the derivation given assumes that atoms only stick to an up-step and do not stick at a down-step. This assumption is investigated more thoroughly in the work of Gossmann and others (1990); its ultimate justification must be related to the atomic structure at a step.

The atomic parameters which determine this situation can be viewed as a potential energy diagram (Fig. D.2). E_b is the energy barrier, which may or may not be present, and E_V is the potential well associated with the enhanced binding energy. Some limiting cases are immediately apparent: if $E_b = 0$ and $E_V \rightarrow \infty$ then it is no more probable for an atom to stick at an up-step than at a down-step, and periodicity is not achieved. If $E_b \rightarrow \infty$ and $E_V \rightarrow \infty$ then atoms stick only to an up-step. It is important to recognize that if $E_b = 0$, there is no preferential sticking and the terrace size distribution remains the same upon deposition. There is no decrease of the standard deviation in this case. If sticking is only partially preferential there is an approach to periodicity, but the derivation is considerably more complicated than that given here.

References

H.-J. Gossmann, F. W. Sinden and L. C. Feldman, *J. Appl. Phys. 67,* 745 (1990).

Elastic Constants, Tables and Conversions

A complete discussion of elasticity theory is given in the references at the end of this appendix. The purpose of this section is to give a brief review of the formulae and relationships among the commonly used elastic constants and provide tables of elastic constants used in the design of electronic materials.

I. Formulae and Definitions

a. Stiffness and Compliance Tensors

For the electronic materials of interest the stress and strain can be related through a 6 × 6 matrix as shown below.

$$\left\|\begin{matrix} c_{11} & c_{12} & c_{13} & c_{14} & c_{15} & c_{16} \\ c_{21} & c_{22} & c_{23} & c_{24} & c_{25} & c_{26} \\ c_{31} & c_{32} & c_{33} & c_{34} & c_{35} & c_{36} \\ c_{41} & c_{42} & c_{43} & c_{44} & c_{45} & c_{46} \\ c_{51} & c_{52} & c_{53} & c_{54} & c_{55} & c_{56} \\ c_{61} & c_{62} & c_{63} & c_{64} & c_{65} & c_{66} \end{matrix}\right\| \qquad \left\|\begin{matrix} s_{11} & s_{12} & s_{13} & s_{14} & s_{15} & s_{16} \\ s_{21} & s_{22} & s_{23} & s_{24} & s_{25} & s_{26} \\ s_{31} & s_{32} & s_{33} & s_{34} & s_{35} & s_{36} \\ s_{41} & s_{42} & s_{43} & s_{44} & s_{45} & s_{46} \\ s_{51} & s_{52} & s_{53} & s_{54} & s_{55} & s_{56} \\ s_{61} & s_{62} & s_{63} & s_{64} & s_{65} & s_{66} \end{matrix}\right\|$$

In short-hand these tensors relate the stress and the strain via Hooke's Law:

$$\overleftrightarrow{\varepsilon} = \|s\| \, \overleftrightarrow{\sigma}$$

$$\overleftrightarrow{\sigma} = \|c\| \, \overleftrightarrow{\varepsilon}$$

where $\overleftrightarrow{\varepsilon}$ and $\overleftrightarrow{\sigma}$ are the strain and stress tensors, respectively. The s_{ij} are referred to as the elastic compliance and the c_{ij} referred to as the elastic stiffness constant.

b. Cubic Materials—*c*'s and *s*'s

Most of the semiconductors and metals of interest are cubic materials, i.e., the single crystal form is a cubic crystal. Cubic materials are characterized by three independent values: c_{11}, c_{12}, c_{44}. For cubic crystals: $c_{ij} = c_{ji}$ and

$$c_{11} = c_{22} = c_{33};\ c_{12} = c_{13} = c_{23};\ c_{44} = c_{55} = c_{66};\ c_{45} = c_{46} = c_{56} = 0$$

$$\begin{aligned} s_{11} &= (s_{11} + s_{12})/(s_{11} - s_{12})(s_{11} + 2s_{12}) \\ c_{12} &= -s_{12}/(s_{11} - s_{12})(s_{11} + 2s_{12}) \\ c_{44} &= 1/s_{44} \end{aligned}$$

$$\begin{aligned} s_{11} &= (c_{11} + c_{12})/(c_{11} - c_{12})(c_{11} + 2c_{12}) \\ s_{12} &= -c_{12}/(c_{11} - c_{12})(c_{11} + 2c_{12}) \\ s_{44} &= 1/c_{44} \end{aligned}$$

$$\begin{aligned} 1/(s_{11} + 2s_{12}) &= c_{11} + 2c_{12} \\ 1/(s_{11} - s_{12}) &= c_{11} - c_{12} \end{aligned}$$

c. Isotropic Materials

Polycrystalline materials and amorphours materials are usually described as "isotropic." They are characterized by two independent parameters: c_{11} and c_{12} or s_{11} and s_{12}.

The relations between the c's and the s's are the same as for the cubic material with the additional relation:

$$c_{44} = (c_{11} - c_{12})/2$$

$$s_{44} = 2(s_{11} - s_{12})$$

d. Common Elastic Constants (Isotropic Materials)

Young's Modulus (Y):

$$Y = (c_{11} - c_{12})(c_{11} + 2c_{12})/(c_{11} + c_{12})$$
$$Y = 1/s_{11}$$

Bulk Modulus (K)($-V \cdot dp/dV$)

$$K = (c_{11} + 2c_{12})/3$$
$$K = 1/3(s_{11} + 2s_{12})$$

Poisson's Ratio (ν)

$$\nu = c_{12}/(c_{11} + c_{12})$$
$$\nu = -s_{12}\, Y = -s_{12}/s_{11}$$

Linear Compressibility, (β)($l \cdot dp/dl$)$^{-1}$

$$\beta = 1/(c_{11} + 2c_{12})$$
$$\beta = s_{11} + 2s_{12}$$

Shear Modulus (μ)

$$\mu = c_{44} = (c_{11} - c_{12})/2$$
$$\mu = 1/s_{44} = 1/2(s_{11} - s_{12})$$

$$\mu = Y/[2(1 + \nu)]$$

e. c's in terms of Y, ν, μ (Isotropic Materials)

$$c_{11} = Y(1 - \nu)/(1 + \nu)(1 - 2\nu)$$

$$c_{12} = Y\nu/(1 + \nu)(1 - 2\nu)$$

$$c_{44} = Y/2(1 + \nu) = \mu$$

f. Lamé Constants (Isotropic Materials)

A commonly used notation involving the stiffness constants is the Lamé constants λ, μ for isotropic materials.

$$c_{12} = \lambda, \quad c_{44} = \mu, \quad c_{11} = 2\mu + \lambda.$$

g. Young's Modulus—Single Cubic Crystals

In single crystals Young's modulus is not isotropic but depends on the crystallographic direction of the applied stress. For a cubic system:

$$1/Y_{l_1l_2l_3} = s_{11} - 2(s_{11} - s_{12} - s_{44}/2)(l_1^2l_2^2 + l_2^2l_3^2 + l_3^2l_1^2)$$

where the l_i represent the direction cosines referred to the $\langle 100\rangle$ axes. For a [100] orientation $(l_1^2l_2^2 + l_1^2l_3^2 + l_2^2l_3^2) = 0$ and $Y = 1/s_{11}$, the same value as for an isotropic material.

Direction	$l_1^2l_2^2 + l_2^2l_3^2 + l_3^2l_1^2$	$Y_{l_1l_2l_3}$
$\langle 100\rangle$	0	$1/s_{11}$
$\langle 111\rangle$	1/3	$3/(s_{11} + 2s_{12} + s_{44})$
$\langle 110\rangle$	1/4	$4/(2s_{11} + 2s_{12} + s_{44})$

h. Elastic Strain Energy

The energy of a thin film, thickness h, with strain ε parallel to the film plane is given by Cahn (1962) as

$$E_\varepsilon = B\,\varepsilon^2 h.$$

For common electronic materials growth is usually along one of the principal crystalline directions: $\langle 100\rangle$, $\langle 111\rangle$, $\langle 110\rangle$. In this case B is given exactly by:

$$B = \frac{1}{2}(c_{11} + 2c_{12}) \cdot \left[3 - \frac{c_{11} + 2c_{12}}{c_{11} + 2(2c_{44} - c_{11} + c_{12})(l_1^2l_2^2 + l_2^2l_3^2 + l_3^2l_1^2)}\right]$$

where l_1, l_2, l_3 are the direction cosines that relate the direction normal to the interface or cube axes.

	$l_1^2l_2^2 + l_2^2l_3^2 + l_3^2l_1^2$	B
⟨100⟩	0	$\dfrac{(c_{11} + 2c_{12})(c_{11} - c_{12})}{c_{11}}$
⟨111⟩	1/3	$\dfrac{6(c_{11} + 2c_{12})c_{44}}{c_{11} + 2c_{12} + 4c_{44}}$
⟨110⟩	1/4	$\left(\dfrac{c_{11} + 2c_{12}}{2}\right)\left(\dfrac{c_{11} - c_{12} + 6c_{44}}{c_{11} + c_{12} + 2c_{44}}\right)$

Note that there is a difference between $Y(l_1, l_2, l_3)$ and $B(l_1, l_2, l_3)$. The former is the correct cubic crystal value, the latter is the appropriate strain factor for a film under biaxial stress. B is occasionally referred to as "Young Modulus under biaxial stress." In an "isotropic approximation" $B(100) = 2\mu\left(\dfrac{1 + \nu}{1 - \nu}\right)$, often used in strain energy calculations.

II. Tables of Elastic Constants

a. Group IV Semiconductors (units of 10^{11} dynes/cm^2 at 300°K)

	c_{11}	c_{12}	c_{44}
Diamond	107.6	12.5	57.6
Si	16.56	6.39	7.90
Ge	12.88	4.83	6.71

From Gray (1972)

	c_{11}	c_{12}	c_{44}
Diamond	107.64	15.2	57.4
Si	16.577	6.393	7.962
Ge	12.40	4.13	6.83

From Böer, (1990)

b. Group III-V Semiconductors (10^{11} dynes/cm^2 at 300°K)

	c_{11}	c_{12}	c_{44}
AlAs	12.02	5.70	5.89
AlSb	8.77	4.34	4.976
GaP	14.050	6.203	7.033
GaAs	11.90	5.38	5.95
GaSb	8.834	4.023	4.322
InP	10.11	5.61	4.56
InAs	8.329	4.526	3.959
InSb	6.669	3.645	3.020

From Böer (1990)

	c_{11}	c_{12}	c_{44}
$Al_xGa_{1-x}As$	$11.88 + 0.14x$	$5.38 + 0.32x$	$5.94 - 0.05x$

From Adachi (1985)

c. Common Metals (10^{11} dynes/cm^2 at 300°K)

	c_{11}	c_{12}	c_{44}
Al	10.82	6.13	2.85
Ag	12.40	9.34	4.61
Au	18.6	15.7	4.20
Cr	35.0	6.78	10.08
Cu	16.84	12.14	7.54
Ni	24.65	14.73	12.47
Mo	46	17.6	11.0
W	50.1	19.8	15.14

From Huntington (1958)

d. Common Insulators (10^{11} dynes/cm^2)

	Y	μ	ν
Fused Quartz[a]	7.26	3.10	0.17
Vitreous Silica[b]	7.29	3.13	0.17
Silicon Nitride[c]	3.00	—	0.22

[a]Huntington (1958).
[b]Bansal and Doremus (1986).
[c]Battelle (1976).

III. Useful Combinations for the Common Semiconductors (10^{11} dynes/cm^2)

	$Y\langle 100\rangle$	$Y\langle 111\rangle$	$Y\langle 011\rangle$
GaAs	8.53	14.12	12.13
GaP	10.34	16.69	14.47
Si	13.02	18.75	16.89
Ge	10.37	15.51	13.80

From Brantley, (1973).

IV. Conversion Factors

To convert from dynes/cm^2

to:	multiply by:
Atmospheres	9.87×10^{-7}
Bars	1×10^{-6}
Pounds/sq in	1.45×10^{-5}
Newtons/m^2	0.1
Pascal (= 1 Newton/m^2)	0.1
mm of Hg	7.5×10^{-4}

$$1 \text{ Dyne/cm}^2 = 1 \text{ erg/cm}^3 = 6.24 \times 10^{11} \text{ eV/cm}^3$$

$$1 \text{ N/m}^2 = 1 \text{ J/m}^3 = 6.24 \times 10^{18} \text{ eV/m}^3$$

References

1. S. Adachi, *J. Appl. Phys. 58*, R1 (1985).
2. N. P. Bansal and R. H. Doremus, *Handbook of Glass Properties*, Academic Press, Orlando, 1986.
3. Battelle-Columbus Laboratories, Engineering Property Data on Selected Ceramics, Vol. 1, Nitrides, Metals and Ceramics Information Center, Battelle's Columbus Laboratories (1976). Internal Report-M.C.I.C.-HB-07 Vol I.
4. K. W. Böer, *Survey of Semiconductor Physics*, Von Nostrand Reinhold (1990).
5. W. A. Brantley, *J. App. Phys. 44*, 534 (1973).
6. J. W. Cahn, *Acta. Met. Vol. 10*, 179 (1962).
7. D. E. Gray, Coord. Ed., in *American Institute of Physics Handbook*, McGraw-Hill Book Company (1972).
8. H. B. Huntington in *Solid State Physics* Vol. 7, ed. by F. Seitz and D. Turnbull, Academic Press, New York (1958) p. 213.
9. C. Kittel, *Introduction to Solid State Physics* 2nd Ed., John Wiley & Sons, New York (1953).
10. J. F. Nye, *Physical Properties of Crystals*, Oxford, Clarendon Press, London (1957).
11. G. Simmons and H. Wing, *Single Crystal Elastic Constants and Calculated Aggregate Properties*, M.I.T. Press, Cambridge (1971).

Answers to Selected Problems

CHAPTER 1

1.1 **a)** $a(\text{Ge}) = 5.65 \times 10^{-8}$ cm, $a(\text{Al}) = 4.05 \times 10^{-8}$ cm
b) $N_s(\text{Ge}) = 6.24 \times 10^{14}/\text{cm}^2$, $N_s(\text{Al}) = 1.22 \times 10^{15}/\text{cm}^2$
c) $h(\text{Ge}) = 1.41 \times 10^{-8}$, h(Al) $= 2.02 \times 10^{-8}$
d) Ge: 22.6×10^{-24} cm^3, Al: 16.6×10^{-24} cm^3

1.2 **a)** $J = 1.02 \times 10^{14}$ atoms/m^{2}sec.
b) $n = 4.45 \times 10^{11}$ atoms/m^3
c) $v = 9.98 \times 10^2$ m/sec

1.3 $N_s = 2.3/a^2$

1.4 $\theta(\text{Si}) = 0.156°$

1.5 **a)** $a = 3.52 \times 10^{-8}$ cm, $h = 1.78 \times 10^{-8}$ cm
b) $N_s = \sqrt{2}/a^2$

1.6 **a)** $d(100) = a$, $d(110) = a/\sqrt{2}$, $d(111) = a\sqrt{3}$
b) (211)

1.7 –

1.8 **a)** $E(\text{Si}) = 1.34$ keV, $E(\text{Cu}) = 0.394$ keV
b) $E(\text{Si}) = 1.59$ keV, $E(\text{Cu}) = 6.93$ keV
c) $E(\text{Si}) = 1.74$ keV, $E(\text{Cu}) = 8.05$ keV

1.9 **a)** –
b) –
c) –

CHAPTER 2

2.1 **a)** $\gamma(\text{Ni}) = 3491$ ergs/cm^2
b) –
c) $h = 5.35$ cm

2.2 –

2.3 **a)** $\Delta E_s = 1$ eV/atom
b) $\varepsilon_b = 0.17$ eV/atom

2.4 23.52 erg/cm^2

2.5 Ratio $= 0.68$

2.6 **a)** –
b) –

2.7 **a)** 142 erg/cm^2
b) –
c) –

2.8 –

2.9 Surface tension = 0.68 eV/atom, heat of sublimation = 1.36 eV/atom

2.10 14.8 cm

2.11 –

CHAPTER 3

3.1 a) $D(\text{Al}) = 8.37 \times 10^{-13}$ cm²/sec, $D(\text{Cu}) = 6.64 \times 10^{-19}$ cm²/sec
b) $D(\text{Al}) = 5.7 \times 10^{-9}$ cm²/sec, $D(\text{Cu}) = 3.2 \times 10^{-9}$ cm²/sec
c) –

3.2 a) $D = 2.08 \times 10^{-12}$ cm²/sec
b) $C = 2.06 \times 10^{19}$ atoms/cm³
c) $J = 2.9 \times 10^{11}$ atoms/cm² sec

3.3 a) Al, 1.5×10^{-5}; Ge, 4.74×10^{-14}
b) –
c) $D(\text{Al})/D(\text{Ge}) = 1.45 \times 10^5$

3.4 0.674

3.5 Einstein: Cu = 1.22×10^{13}/sec, Ag = 0.87×10^{13}/sec
Debye: Cu = 0.71×10^{13}/sec, Ag = 0.47×10^{13}/sec

3.6 a) $D = 3.04 \times 10^{-11}$ cm²/sec, $t = 0.82$ sec
b) $F = 9.2 \times 10^3$ eV/cm, $v = 3.04 \times 10^{-6}$ cm/sec

3.7 $t \simeq 6 \times 10^5$ years

3.8 a) $D = 1.5 \times 10^{-9}$ cm²/sec
b) $D_0 = 0.123$ cm²/sec

3.9 –

3.10 36 at.%

CHAPTER 4

4.1 a) $F_{max} = 1.11 \times 10^{-4}$ dyne/atom
b) $Y = 5.48 \times 10^{12}$ dyne/cm²
c) $E_{el} = 3.37 \times 10^{10}$ dyne/cm²

4.2 –

4.3 a) strain = 2.2×10^{-3}
b) –

4.4 a) $C_v(\text{Al}) = 2.79 \times 10^{18}/\text{cm}^3$, $C_v(\text{Cu}) = 1.66 \times 10^{16}/\text{cm}^3$
b) Al : 1.31; Cu : 1.24

4.5 $N = 6.6 \times 10^6$ atoms, $V = 0.78 \times 10^{-18}$ cm³

4.6 –

4.7 breaks

4.8 Stress (dual) = 2×10^9 dyne/cm², stress (nitride) = 1.6×10^9 dyne/cm²

4.9 a) –
b) stress = 3.84×10^6 N/m²

CHAPTER 5

5.1 –

5.2 71.2 ergs

5.3 –

5.4 3.66 eV

5.5 –

5.6 $T = 527°C$

5.7 $t = 75$ sec

5.8 –

5.9 1.88×10^{12} eV

5.10 $p = 0.12$ atm

CHAPTER 6

6.1 $N_s = 6.78 \times 10^{14}/cm^2$, $\Theta = 0.078°$, $\Theta = 0.78°$

6.2 $D_s = 1.32 \times 10^{-4}$ cm^2/sec, $\tau_0 = 1.46 \times 10^{-8}$ sec

6.3 –

6.4 –

6.5 –

6.6 $T = 1140°C$

6.7 $T = 560K$, $p = 1.02 \times 10^{-6}$ Pascal

6.8 $R = 1.3 \times 10^{-8}$ cm/sec

6.9 $L_0 = 2.27 \times 10^{-6}$ cm

6.10 $W = 2.56$ eV

6.11 **a)** $t = 4.61$ hours
b) $v = 1.7 \times 10^{-5}$ cm/sec

6.12 $T = 730K$

6.13 **a)** 0.428 μm/hr
b) 1.428 μm/hr
c) 1.4/sec
d) $= 2.38 \times 10^{-5}$ Pascal

CHAPTER 7

7.1 –

7.2 $X_c = 98 \times 10^{-8}$ cm

7.3 –

7.4 **a)** $x = 0.45$
b) $b = 0.384$ nm, $S = 20.2$ nm, $E_d = 33.4$ eV/nm

7.5 **a)** 0.035
b) 0.017 rad

7.6 **a)** 3.9%
b) $\Delta\Theta = 1.06°$

7.7 **a)** 0.6%
b) 0.37%

7.8 **a)** $S = 15.57$ nm
b) $S = 5.45$ nm

7.9 **a)** 0.0188
b) 29.3 nm

7.10 0.4

7.11 Si: AlP, GaP, ZnS; GaAs: ZnSe, AlAs, Ge

7.12 **a)** $x = 0.14$
b) –
c) 36 layers

7.13 **a)** $x = 0.47$
b) 2.8 nm

7.14 **a)** –
b) –
c) 1.7×10^{-6}

7.15 –

CHAPTER 8

8.1 –

8.2 –

8.3 **a)** –
b) –
c) –

8.4 **a)** Reflectivity = 0.31
b) Reflectivity = 0.60

8.5 $R(n = 1) = 0.131$; $R(n = 10) = 0.99$

8.6 $V_{TOT} = \left(-\dfrac{2eZ_2}{r_0}\right)(1 + (\Delta/r_0)^2)$

8.7 **a)** 1.4×10^{-2} eV
b) 0.28×10^{14}/eV cm^2
c) 4.2×10^{-2} eV

8.8 $E(L = 50) = 1.684$ eV, $E(L = 400) = 1.425$ eV

8.9 **a)** $g_{th} = 158$/cm
b) $J = 3.6 \times 10^3\ A/cm^2$
c) $I = 7.2 \times 10^{-3}\ A$

8.10 $E = -0.36$ eV

CHAPTER 9

9.1 $m_e^*/m_0 = 0.07$, $m_h^*/m_0 = 0.43$

9.2 $2.5 \times 10^{19}/\text{cm}^3$, $7.9 \times 10^{17}/\text{cm}^3$; 0.2 eV, 0.112 eV

9.3 0.24 eV, 0.32 eV

9.4 $E^{-1/2}$, Energy independent

9.5 a) 9.38 eV
b) 9.4×10^6

9.6 a) 0.5 Volt
b) 8.5×10^{-5} cm
c) $8.5 \times 10^{11}/\text{cm}^3$
d) 1.24×10^{-12} F

9.7 a) at 300 K, 1.8×10^{-5} Amp/cm^2
b) –

9.8 –

9.9 a) –
b) –
c) –

CHAPTER 10

10.1 $t = 1.71$ hours, $\Delta H = 0.892$ eV

10.2 $t = 0.974$ hours

10.3 a) 3.06×10^3 nm/min
b) 2.58×10^8 nm/min

10.4 a) –
b) 8.5 and 7.9 hours

10.5 a) Ratio $= 7 \times 10^{-3}$
b) 37 atoms

10.6 transport

10.7 $X(t) = 1 - \exp(-kt^3)$

10.8 1.3 eV

10.9 $\frac{4\pi}{3}(r^*)^3 = \frac{2\Delta H_N^*}{\Delta H_{\alpha\beta}}$

CHAPTER 11

11.1 $D_A = 3.1 \times 10^{-14}$ cm^2/sec, $D_B = 1.1 \times 10^{-14}$ cm^2/sec

11.2 $D_A = 2.44 \times 10^{-10}$ cm^2/sec, $D_B = 8.62 \times 10^{-11}$ cm^2/sec

11.3 a) unchanged
b) twice the original

11.4 a) 12.5 at.%, 37.5 at.%
b) $t = 160$ hours

11.5 **a)** $D = 1.81 \times 10^{-11}$ cm²/sec
b) $D_A = 2.48 \times 10^{-13}$ cm²/sec, $D_B = 0.88 \times 10^{-13}$ cm²/sec

11.6 $D_A/D_B = 2.0$, $\tilde{D} = 1 \times 10^{-14}$ cm²/sec

11.7 3.06×10^{-6} cm

11.8 $\tilde{D} = 4.3 \times 10^{-10}$ cm²/sec, $D_{In} = 7.1 \times 10^{-10}$ cm²/sec,
$D_{Pb} = 3.6 \times 10^{-11}$ cm²/sec

CHAPTER 12

12.1 **a)** 4.55×10^{-5} cm
b) 3×10^{-8} cm/sec
c) 1.6×10^{14} atoms/cm² sec

12.2 $\tilde{D} = 11 \times 10^{-12}$ cm²/sec

12.3 **a)** 8.4×10^{-4} cm
b) 1.69×10^{-3} cm
c) No

12.4 **a)** –
b) –

12.5 **a)** A
b) –
c) 0.25 μm/hour

CHAPTER 13

13.1 0.59 eV

13.2 **a)** $D = 3.75 \times 10^{-13}$ cm²/sec, $E_A = 1.5$ eV
b) –

13.3 grain boundary

13.4 $D_b = 2.5 \times 10^{-9}$ cm²/sec

13.5 **a)** –
b) $D = 5.2 \times 10^{-12}$ cm²/sec

CHAPTER 14

14.1 $J = 1.2 \times 10^2$ amp/cm²

14.2 1×10^{-4} (nm)²

14.3 9.45 eV/atom

14.4 –

14.5 $Z^* = 21.5$

14.6 length = 39 μm

14.7 stress = 2×10^9 and 3×10^9 dyne/cm²

14.8 –

14.9 –

14.10 $J_m/J_e = \mu_m Z^*/\mu_e Z$

14.11 –

14.12 –

14.13 –

CHAPTER 15

15.1 length = 5×10^{-2} cm, radius = 1.6×10^{-6} cm

15.2 $t(350) = 160\, t(400)$

15.3 radius = 5.6×10^{-3} cm

15.4 strain = 8×10^{-4}

15.5 –

15.6 max stress at 90, 270; void would elongate

Physical Constants, Conversions, and Useful Combinations

Physical Constants

Avogadro constant	$N_A = 6.022 \times 10^{23}$ particles/mole
Boltzmann constant	$k = 8.617 \times 10^{-5}$ eV/K $= 1.38 \times 10^{-23}$ J/K
Elementary charge	$e = 1.602 \times 10^{-19}$ coulomb
Planck constant	$h = 4.136 \times 10^{-15}$ eV · s
	$= 6.626 \times 10^{-34}$ joule · s
Speed of light	$c = 2.998 \times 10^{10}$ cm/s
Permittivity (free space)	$\epsilon_0 = 8.85 \times 10^{-14}$ farad/cm
Electron mass	$m = 9.1095 \times 10^{-31}$ kg
Coulomb constant	$k_c = 8.988 \times 10^9$ newton-m²/(coulomb)²
Atomic mass unit	$u = 1.6606 \times 10^{-27}$ kg
Acceleration of gravity	$g = 980$ dyn/gm

Useful Combinations

Thermal energy (300 K)	$kT = 0.0258$ eV $\simeq$ (1/40) eV
Photon energy	$E = 1.24$ eV at $\lambda = 1$ μm
Permittivity (Si)	$\epsilon = \epsilon_r\epsilon_0 = 1.05 \times 10^{-12}$ farad/cm

Conversions

1 nm $= 10^{-9}$ m $= 10$ Å $= 10^{-7}$ cm
1 eV $= 1.602 \times 10^{-19}$ joule $= 1.602 \times 10^{-12}$ erg
1 eV/particle = 23.06 kcal/mol
1 newton $= 0.102\ kg_{force}$ = 1 coulomb · volt/meter
10^6 newton/m² = 146 psi = 10^7 dyn/cm²
1 μm $= 10^{-4}$ cm
0.001 inch = 1 mil = 25.4 μm
1 bar $= 10^6$ dyn/cm² $= 10^5$ N/m²
1 weber/m² $= 10^4$ gauss = 1 tesla
1 pascal = 1 N/m² $= 7.5 \times 10^{-3}$ torr
1 erg $= 10^{-7}$ joule = 1 dyn-cm
1 joule = 1 newton · meter = 1 watt · second
1 calorie = 4.184 joules

Index

T